tHe science of EVOLUtiON

William D. Stansfield

California Polytechnic State University

MACMILLAN PUBLISHING CO., INC.
new york

COLLIER MACMILLAN PUBLISHERS
london

To my wife, Janis

Macmillan Publishing Co., Inc.
866 Third Avenue, New York, New York 10022

Collier Macmillan Canada, Ltd.

Library of Congress Cataloging in Publication Data

Stansfield, William D   (date)
   The science of evolution.

  Bibliography: p.
  Includes index.
  1. Evolution. I. Title.
QH366.2.S7   575   76-7471
ISBN 0-02-415750-3

Printing: 1 2 3 4 5 6 7 8   Year: 7 8 9 0 1 2 3

Preface

A study of organic evolution is essential for the biologist because only through an understanding of the changes (and the processes which have brought about these changes) in living systems can such disciplines as microbiology, botany, and zoology be integrated into a unified life science. For the nonscientist also, the study of evolution is important for an understanding of the origin of variations and of the development of races, species, and higher taxonomic groups in the biological world. Modern man is naturally inquisitive about his own origin, and the science of evolution is beginning to provide some insights into this subject.

This textbook is written for the undergraduate college or university course in evolution and assumes as minimal preparation at least one introductory course in college biology. The mathematics presented herein can easily be comprehended with the elements of algebra. Some institutions prefer to offer evolution as a freshman level course because of its role in unifying the diverse disciplines of biology. Other institutions prefer the student to become somewhat mature in the biological sciences before attempting a grand synthesis through a study of evolution. In either case, the design of this textbook should prove to be very helpful to the student in an initial study of the subject.

The design of this text is quite different from that of other books on evolution in several respects. Probably the most distinctive features of this text are its attempt to

(1) condense major principles into a form that allows the student to review rapidly the broad sweep of evolution in minimal reading time.

(2) clearly delineate between the presentation of basic concepts and their supporting examples.

(3) present a balance between evidence and processes, as well as between classical theories and mathematical models.

(4) provide thought-provoking questions at the end of each chapter.

(5) prepare the student with the background of facts and theories required for scholarly discussions of evolution.

The text is divided into four major sections. Part I is devoted to a review of some general concepts required as a basis for a detailed study of evolution. This review naturally includes the intellectual legacy left to us by those who have grappled with problems of changes in populations. It also includes a survey of the events that may have been instrumental in the synthesis of "life" on our planet. However, most of the chapters in this section attempt to answer the question, "What is the evidence that has led biologists to the acceptance of the concept of evolution?"

Part II provides the student with the essentials of heredity in an attempt to answer the question, "What are the origins of genetic diversity in natural populations upon which the evolutionary processes depend?"

Part III establishes a static baseline in the theoretical equilibrium population. Then, one by one, relaxation of the restrictions implicit in such a population is allowed, and their effects in causing genetic changes are defined mathematically. This section attempts to answer the question, "What are the mechanisms that cause evolution to occur?"

Part IV investigates special aspects of the evolutionary process that generate such divergence between populations that they become candidates for different taxonomic status. Some possible origins of man are presented in the concluding chapter. The basic question this section attempts to answer is, "What are the origins of species and higher taxonomic groups?"

This text is most effectively used in a class situation under the guidance of a skilled instructor who can lead discussions of the questions raised in the text or from the students themselves as a by-product of their readings. Much can be learned, however, through an independent study of this text without the benefit of interactions with other scholars. A word of caution: mastery of this subject usually requires that it be thoroughly studied (not merely read). New terminology appears in italics; mastery of terminology is important for efficient communication in discussions of evolution. Exemplary information is set off from the main body of the text so that the reader can clearly differentiate facts, principles, and theory from the examples. A set of objective questions (true-false, multiple choice, and fill-in) and their answers are available for each chapter as a teacher's supplement to this text. The instructor can aid the student in self-evaluation by providing access to the questions in this supplement.

A diligent study of this text may encourage the student to continue scholarship in this field of biology beyond the level presented herein. The brief but comprehensive exposure to the major principles presented in this book should allow the student to read biological journals or advanced treatises concerning evolution with a high level of understanding. The student should be at ease in discussions concerning evolution because of the breadth of subject matter encountered in the study of this text. Finally, by giving serious thought to the discussion questions at the end of each chapter and by prior exposure to the objective questions in the teacher's supplement, the student should be well prepared for classroom examinations.

W. D. S.

Contents

part IV
THE GENERATION OF TAXA
421

REFERENCES
580

INDEX
589

part

I

THE FRAMEWORK OF EVOLUTION

INTRODUCTION

It was sweltering that July 20, 1925 on the lawn of the courthouse in Dayton, Tennessee at the climax of the famous "monkey trial." On the stand, as an authority on the Bible and witness for the prosecution, sat William Jennings Bryan (lawyer, lecturer, journalist, editor, Secretary of State, and thrice-defeated candidate for President). Before him stood the attorney for the defense, Clarence Darrow (master jury-pleader and noted agnostic; Figure 1.1). The defendent was a young science teacher in the Dayton High School named John Thomas Scopes (Figure 1.2) who was indicted for teaching evolution. It was really not Scopes that was on trial, but rather the legality of the Tennessee law prohibiting the teaching of evolution in the public schools.

Evolution is a scientific theory proposing that higher forms of life have descended from lower forms by a gradual process of modification through natural mechanisms. This was essentially the thesis of a book entitled *"The Origin of Species"* (1859) by Charles Darwin. In Britain, Darwin's thesis immediately aroused violent opposition from many quarters, especially from the clergy (e.g., Samuel Wilberforce, the Anglican Bishop of Oxford). The impact of Darwin's theory in the United States was delayed by the Civil War. But thereafter, American religious leaders became increasingly disturbed about the concept of evolution. Following the passage of legislation to outlaw liquor after World War I, a religious group

Figure 1.1. Clarence Darrow (left) and William Jennings Bryan at Scopes' trial in July 1925. (Wide World Photos)

called the Fundamentalists began actively lobbying for similar legislation to stifle the "heretical theory of evolution." During the early 1920s several bills prohibiting the teaching of evolution were narrowly defeated in several states. An antievolution bill, introduced by John Washington Butler in the Tennessee legislature, became law in March 1925. Almost immediately, the American Civil Liberties Union prepared to contest the constitutionality of the Butler Act. John Scopes was coaxed by some interested townspeople to be the "guinea pig" in a test case. He agreed to this because of his belief in the theory of evolution, and the fact that he was only 24 years old, unmarried, and could afford to risk losing his job. Furthermore, there were doubts that the law would be enforced or that the penalties, if he lost, would be severe; also the ACLU was picking up the tab.

The trial began on July 10, 1925 with Judge John T. Raulston on the bench. This event attracted nationwide attention as the "monkey trial," because one of the major objections of

Figure 1.2. John Scopes and his father in July 1925. (Wide World Photos)

the Fundamentalists to the theory of evolution was their notion (not Darwin's) that humans descended from monkeys. Several scientists and Biblical scholars had been assembled by the defense to show that evolution was a scientific fact and that it did not necessarily conflict with a liberal interpretation of the Bible. Judge Raulston, however, ruled that the testimony of such experts was irrelevant to the case. Darrow lost his temper, insulted the Judge, and was promptly cited for contempt of court. The next day Darrow apologized to the court and the trial resumed. Raulston had the court moved to the lawn because of the heat and from fear that the courthouse floor could not bear the weight of spectators and reporters. The trial was not going well for the defense until Darrow decided to fight fire with fire. This was how Bryan, a member of the prosecution team, came to be asked by the defense to take the witness stand as an authority on the Bible. Bryan accepted the challenge. Under a blistering sun, Darrow grilled Bryan for an hour and a half with such questions as, whether he believed that Joshua had lengthened a day by making the sun stand still; . . . that prior to the construction of the tower of Babel, everyone spoke a common language; . . . that the earth was created in seven days of twenty-four hours each. Bryan conceded that the days of creation might have been longer than 24 hours. Under intensive questioning, Bryan confessed ignorance of several subjects on which he had earlier claimed to be an expert. When Darrow inquired of the witness how the Serpent had walked prior to the time that God commanded that it should crawl on its belly and suggested that perhaps it jumped about on its tail, the audience broke into raucous laughter. Bryan then lost his temper, Darrow also became abusive, and a near riot occurred among the spectators; as a result Judge Raulston was forced to adjourn the court. On the following morning, Raulston ruled that Bryan's testimony be stricken from the record.

Darrow broadly hinted in his closing address that he wanted a verdict of guilty so that the case could be appealed to the Supreme Court of Tennessee. Scopes was found guilty and fined $100. In January 1927, the Supreme Court seized on a technicality (the manner in which the fine had been levied) and remanded the case to a lower court with an admonition that the District Attorney not continue the prosecution of "this bizarre case." The D.A. complied, leaving the law intact and Scopes unpunished. Meanwhile Scopes had left teaching to become a graduate student at the University of Chicago; following that he became a geologist with the Gulf Oil Company. Five days after the conclusion of the trial, Bryan died in his sleep. It has been suggested that the strain of the trial had taxed his heart, but it was well known that he was a voracious eater despite a diabetic condition.

Following the initial triumph of the antievolutionists in the Scopes trial, several states initiated bills similar to the Butler Act. However, only one of the twelve bills voted on during 1926 and 1927 passed (this was the one in Mississippi). Despite strong pressure from Fundamentalists, politicians (largely for fear of ridicule) tended to avoid becoming embroiled in legislation that could lead to a similar fiasco. In April 1967, another science teacher in Jacksboro, Tennessee was fired for discussing the theory of evolution with his students. (He was also accused of insubordination, unprofessional conduct, and failing to meet his teaching responsibilities.) Later that same month, the Tennessee State Senate voted to amend the law that led to the Scopes trial forty-two years earlier. According to the amendment, teachers could refer to evolution but only as a theory and must maintain that God's creation is a fact consistent with the "Good Book."

A similar kind of movement also appeared in California during 1969. California is one of the few states that prints its own textbooks for all primary and secondary grades through high school. In November 1969, the California State Board of Education unanimously approved a set of guidelines directing that the creation story should be taught as an alternative to the theory of evolution in science classes. Henceforth, science textbooks in California should also present both concepts. The California Supreme Court decided against this policy in 1972, but compromised by ordering that textbooks should refrain from "scientismic bias" in fostering the notion that evolution is the only possible explanation.

In January 1969, three Roman Catholic nuns (the principal and two teachers) were dismissed from their posts in a Staten Island parochial school for teaching "evolution vs. creation." In January 1970, a bill to repeal Mississippi's ban on teaching evolution was defeated 70 to 42 in the state House of Representatives after an emotion-charged debate. It then had the dubious distinction of being the only state in the nation with laws still discriminating against the theory of evolution.

In August 1975, a U.S. District Court and the Tennessee Supreme Court declared unconstitutional a 1973 law requiring biology textbooks to provide equal space to Biblical and scientific theories. This decision apparently was based on the premise that giving equal space to religious theories in public school texts would violate First Amendment guarantees of separation of church and state. Furthermore, it was beyond the court's comprehension how all the theories or beliefs of every religious sect (from the worshipers of Apollo to the followers of Zoroaster) could be included in any textbook of reasonable size. These objections are all valid, but it appears that the crucial issue (from the standpoint of the scientist) has still been missed; *viz.* the distinction between science and other branches of philosophy (including religion).

Scopes was interviewed in November 1969 shortly after the ruling of the United States Supreme Court on the unconstitutionality of the Rotenberry Act, which forbade teaching the evolution theory in public schools and colleges of Arkansas. Then retired at the age of 68, Scopes told reporters "This country must not rest with the overturning of the 'monkey law,'" warning that "our personal freedom and liberties are never laid to rest because if individuals are not constantly aware of the dangers and pitfalls in the path of maintaining liberties, they become complacent and they'll lose every one of them." He also expressed his dissatisfaction with the educational system, saying "I think that education is for understanding of human emotions and human society, which has not been stressed (in schools) too much. You can learn (in universities) to be the best electrical engineer and still maintain all your prejudices."

Prejudices commonly arise from misunderstandings. There is no room for bigotry in science (although there is ample room for it to occur in other intellectual spheres). Therefore, a major purpose of this text is to delineate clearly the role of science in modern societies apart from other aspects of philosophy. If it is successful in this regard, we can avoid entanglement in useless debates involving religious beliefs and scientific theories. Furthermore, in order to enhance comprehension of Darwin's thesis, the reader should be acquainted with the basic principles of the science of ecology. With the realization that not all readers have backgrounds in philosophy and ecology, this introductory chapter attempts to present enough fundamentals to enable further discussions of evolutionary theory to be more fruitful. Those who are well versed in these subjects may wish to skip the remainder of this chapter.

Science and Philosophy

The 1969 guidelines on science teaching adopted by the California State Board of Education advocated that the words "or designed for" be added after each phrase "is (or are) adapted to" in descriptions of biological attributes in science textbooks. Scientific organizations across the nation, as well as eminent individual scientists (including some Nobel Prize winners) vigorously lobbied to block the passage of such laws. Their objection was uniformly one against mixing science with other forms of philosophy (including religion). The general public is largely unaware of the distinguishing features of science, and until the public becomes educated in this regard, we will never have an end to "monkey wars."

Philosophy encompasses all human thought directed toward a critical evaluation of both sensory input (experience) and extrasensory constructs of the mind (extrasensory perceptions). The term derives from the Greek *philosophia,* which means "love of wisdom." *Science* is a specialized discipline within the realm of philosophy that is bound to a procedural design known as *scientific methodology.* The purpose of science is not to find "facts" or discover "truth," but rather to formulate and use theories in order to solve problems and ultimately to organize, unify, and explain all the material phenomena of the universe. Scientists attempt to avoid the use of "fact," "proof," and "truth," because these words could easily be interpreted to connote absolutes. Nothing in science is deemed absolute. Science deals only with theories or *relative* "truth,"—a temporary correctness so far as can be ascertained by the rational mind at the present time. Science attempts to answer questions such as "Of what structural units are organisms composed?" and "How does an ocean fish function to prevent loss of water?" It is not within the realm of science to find answers to questions such as "What is the purpose of human existence?" and "Why does the universe exist?" These questions are, however, the proper objects of philosophical discourse.

The intellectual approaches used to various degrees by philosophers of all ages include observation, reason, faith, and intuition. Some philosophers favor empirical methodology. *Empiricism* holds that all knowledge comes only from experience. Scientists also use empiricism by relying on sensory observations and experimentation. They use the *inductive method* of reasoning, i.e., the gathering of specific information and postulation of general principles consistent with those specific observations. Scientists realize that their principles (theories) may need to be revised if and when new observations of a phenomenon are found inconsistent with the old general principle under which it would logically fall. Other philosophers, using the *deductive method,* begin with *axioms* (self-evident general principles accepted without proof) and use reason to derive specific conclusions based on those axioms. Obviously, the conclusions of deductive reasoning can be no sounder than the axioms that serve as a starting point. Thus, deductive reasoning proceeds from the general phenomenon to the specific, whereas inductive reasoning proceeds from the individual case to the universal. Both instances require a degree of faith (reliance) in the ability of the methodology to provide answers to our questions. Some philosophers claim that certain things can be known directly without reasoning or sensual verification. This is *intuition,* an inexplicable type of personal revelation forbidden to the scientist as an explanatory method.

All scientific methodology has, at its core, the observation of phenomena in the natural world. The initial steps in the science of biology were taken when direct observation of organisms was substituted for philosophical speculations based on intuition and/or deductive reasoning. Describing and classifying organisms is a pioneering step in any biological discipline. Scientists always attempt to be as objective as possible in their observations, although it is too much to expect human beings to be entirely free of unsuspected biases. Furthermore, to have scientific validity the observations must be capable of verification by others using the same observational techniques, i.e., they must stand the test of repeatability. This is why it is important that all steps in a research project be carefully documented in publications for consumption by the scientific community. Many biologists are still involved in this most basic task of making descriptive observations, especially for taxonomic groups with very large numbers of species such as protozoans (single-celled animals), nematodes (roundworms), and insects.

Experimental science, as distinguished from purely descriptive science, also begins with observations. These experiences provoke questions that define a problem area. Questions amenable to scientific investigation must be both relevant and testable. For example, the question "Does God exist?" is scientifically irrelevant because it does not pertain to the material world and because it is not subject to an experimental test in which He is present in one situation and absent in another. The next step is the formulation of a hypothesis, i.e., an educated guess as to a possible answer to the problem. The hypothesis may be stated as an either-or alternative—for example, "The oxygen liberated by photosynthesizing plants is derived either from water or from carbon dioxide." An experiment is then designed, with appropriate controls, to provide the evidence by which the validity of the hypothesis can be evaluated. The results of the experiment are then analyzed in an attempt to reach a decision. It is often impossible to answer scientific questions with a simple yes-or-no conclusion. Much experimental work in biology is quantitative and requires statistical treatment by which it may be possible to conclude, with a certain degree of confidence, that the hypothesis under test is consistent with the data (or unacceptable, as the case may be). If the original hypothesis is invalidated, new hypotheses may be generated and subjected to redesigned experimentation and analysis in a search for an acceptable explanation. Once an answer to a specific problem is available, it may be possible to postulate the ramifications of the verified hypothesis to a broader base of application. Thus emerges a scientific theory that serves as a catalyst for further specific investigations to test its general validity. A good theory has predictive value. A few scientific theories have such a high degree of predictive value that they are called *natural laws*. This is why science has proved to be so useful in both its pure and applied aspects. Most scientific theories, however, are ephemeral. Exceptions will likely be found that invalidate a theory in one or more of its tenets. These can then stimulate a new round of research leading either to a more comprehensive theory or perhaps to a more restrictive (i.e., more precisely defined) theory. Nothing is ever completely finished in science; the search for better theories is endless.

The interpretation of a scientific experiment should not be extended beyond the limits of the available data. In the building of theories, however, scientists propose general principles by extrapolation beyond available data. When former theories have been shown to be inadequate, scientists should be prepared to relinquish the old and embrace the new in their never-ending search for better solutions. It is unscientific, therefore, to claim to have "proof of the truth" when all that scientific methodology can provide is evidence in support of a theory.

> For example, the assertion that populations of organisms can change in their genetic composition from one generation to another (i.e., evolve) is undisputed, even by the creationists. To say without qualification that "all present life has evolved from more primitive forms" is unscientific because such a statement is an absolute. A scientifically acceptable restatement is that "scientists have found a great deal of evidence from many sources which they have interpreted to be consistent with the theory that all present life has evolved from more primitive forms."

As scientists, humans are bound not only to scientific methodology, but also to a set of premises accepted on faith. This probably comes as a shock to many people who thought that they knew what science was all about. Scientists believe that every phenomenon results from a discoverable cause. They assume that the behavior of the universe is not capricious, but describable in terms of constant laws, such that when two sets of conditions are the same, the same consequences may be expected. Scientists believe that the forces now operating in the world are those that have always operated (*uniformitarianism*), and that the universe is the result of their continuous operation. They think of the world and the phenomena in it as consisting of sets of relationships rather than absolutes. They do not regard generalizations as final, but are willing to modify them if they are contradicted by new evidence. Scientists must rely on material and mechanical explanations of phenomena rather than on nonmaterial or supernatural factors. They often think in terms of continua, distrust sharp boundary lines, and expect to find related classes of natural phenomena grading imperceptibly into one another. They expect nature to be dynamic rather than static and to show variation and change. They expect that in any situation involving competition among units of varying potentialities, those that work best under existing circumstances will tend to survive and be perpetuated. Scientists prefer simple and widely applicable explanations of phenomena. They attempt to reduce their view of the world to as simple terms as possible. They attempt to incorporate all phenomena into a single consistent natural scheme, but recognize that contradictory generalizations may be necessary to describe different aspects of certain things as they appear to us (e.g., the corpuscular vs. the wave theories of light). These are the boundaries and guidelines within which science is conducted. Scientists must attempt to explain phenomena *mechanistically,* i.e., in terms of the laws of chemistry and physics. They should not resort to *vitalism,* the assumption that material phenomena are controlled by supernatural powers. They also should not resort to *teleology,* the notion of purpose in natural events. Rather, their thinking must be *causalistic,* that events occur only as previous events permit them to occur. In scientific reasoning, scientists are careful to avoid *anthropocentric* explanations, which attribute human-like qualities to other species.

> The statement "Ground squirrels store nuts in their burrows in order that they will have something to eat during the winter when food is scarce" is both anthropocentric and teleological because it ascribes a human-like mind to the animal, allowing it to think and plan ahead for a winter it may not yet have experienced and to take appropriate action for a specific purpose in the future. A scientifically acceptable restatement is that "ground squirrels appear to be stimulated by a shortened length of daylight to store nuts in their burrows; this food helps sustain the animals through the winter."

Scientific explanations of natural events are inherently without absolute truth, without innate value, and devoid of purpose (nonteleological). This is the framework within which science operates. It is really because of these restrictions that science has been so productive in problem solving.

Now we return to the philosophers. They too seek answers to questions, but their questions may range far beyond the realm of science. *Philosophy* is concerned with the rules and methods of correct thinking (*logic*); with the origins, nature, and limitations of knowledge (*epistemology*); with the understanding of reality other than through sense perceptions (*metaphysics*); and with the study of values (*axiology*). Philosophy is involved in science, religion, government, education, personal assessment, and indeed in every aspect of human endeavor.

The scope and methodologies of philosophy are so different from those of science that arguments derived along philosophical lines of reasoning often cannot be accepted by the scientist; likewise, conclusions derived through scientific methodology are not always acceptable to the philosopher (depending on the basic premises of his or her particular beliefs). *Creationists* are vitalistic philosophers who believe that all "kinds" of life were created *de novo* by a supernatural power (God) and that these life forms have not changed substantially since their advent. They acknowledge that fossils are the remains of prehistoric life forms, but do not accept the theory that some of them were ancestral to quite different forms living today. The creationist view may well be absolute truth, but this is not within the realm of science to determine. The evolutionist is a scientist who works within the confines of scientific methodology, trying to explain mechanistically how populations of organisms change today and then, by extrapolation of these processes back through time, to explain the origin of species and other taxonomic categories in a coherent theory of evolution. Both the creationist and the evolutionist have faith in their respective methodologies to provide answers to questions. However, the creationist can easily explain any phenomenon by simply saying "God did it." This approach, though it may be perfectly correct in an absolute sense, does not foster further inquiry and is therefore intellectually emasculated. The scientist has no ready answer for most questions and must labor for solutions, and must continually monitor theories for their congruity with objective data.

A law that would require inclusion of the creationist view in science textbooks might only serve to confuse the reader, unless a lengthy explanation of the kind we have just been through is presented at the outset. Such a book might more properly be termed a philosophy text rather than a science text.

Whereas science is "godless," the people we call scientists need not be! When attempting to answer questions outside the realm of science, the scientist must become a philosopher. Therefore, it is philosophically permissible for a scientist to believe in God or not, as he or she finds the rationale to make the choice. Those who claim that a person cannot believe in God and practice science obviously do not understand the complementary nature of these two major realms of human thought. Perhaps many scientists actually believe that a supernatural "Force/Being" is maintaining the natural laws of the universe according to His purposes, but vitalism and teleology should be avoided in their scientific explanations.

When we come to the origin(s) of life, both creationists and evolutionists are forced into the role of speculators. Laboratory experiments conducted with presumed primitive earth atmospheric conditions (methane, ammonia, hydrogen, water) and various energy sources (electrical discharge, ultraviolet radiation, high energy radiation, and heat) have yielded small amounts of amino acids (the building blocks of proteins) and nucleic acid precursors (the building blocks of genetic information and components of the protein synthesizing machinery of contemporary cells). Certain conditions have seen the formation of microspheres, which are chemically complex entities surrounded by a double-layered membrane suggestive of the gross structure of certain cellular components. Creationists have looked forward to the day when science may actually create a "living" thing from simple chemicals. They claim, and rightly so, that even if such a man-made life form could be created, this would not prove that natural life forms were developed by a similar chemical evolutionary process. The scientist understands this and plods on testing theories.

In some instances, the evidence for evolution is meager and/or equivocal. Creationists focus attention on any tendency to acceptance of such evidence *carte blanche*. Perhaps the greatest contribution creationists are currently making to science is their recognition of "creeping dogmatism" in the science of evolution. Through their efforts, it is likely that science textbooks in California will have to retreat from such dogmatic statements as "Life began in the primordial sea at least three billion years ago." An acceptable revision of this concept might be "Most scientists have interpreted from the fossil record that life began in the primordial sea at estimates exceeding three billion years ago." This is as it should be. Absolutes have no place in science. The scientist should carefully avoid dogmatic statements, couching all conclusions in relativistic terms. When the scientist fails to do this, other members of the scientific community must be ready to correct such errors. If evolutionists do not keep their own house in order, the creationists stand ready to attack their veracity.

Ecology

Population biology is broadly divided into two disciplines: ecology and evolution. *Ecology* is the study of how organisms interact with one another and with their physical environment. To understand how processes such as natural selection function to cause genetic changes in populations (evolution), one must first have a working knowledge of fundamental ecological principles.

Natural Groupings

Species. The "species" concept will be rigorously dissected in Chapter 17. For now, a species can be considered as an assemblage of freely interbreeding organisms exhibiting characteristics which are distinctive from other assemblages.

> For example, the coyote and the wolf are both dog "kind" but are easily recognized as distinctive species. The tiger and the lion are both cat "kind," but again have distinctive features by which individuals can easily be classified into one or the other species. Because creationists do not rigorously define the term "species," it could relate to the coyote in one discussion and in another to dog "kind" in general. Scientific discussions cannot be fruitful when such fuzzy thinking is employed.

Population. A biological population is a group of individuals belonging to the same species and occupying a common geographical area. The members of a given species may be distributed in one or more populations, either geographically contiguous or disjunct from one another.

Community. A community consists of all the biological populations of the different species that live within a common geographical area. All the plants in a community comprise its *flora;* all the animals constitute its *fauna.*

Ecosystem. All the biotic (living) and abiotic (physical) components of a defined geographical area constitute an *ecosystem.* At least three terms relate to the province in which an organism is to be found. The *habitat* is the set of ecological conditions that can be tolerated by members of the species as evidenced by their natural occurrence in these environments. The *niche* is the life style of the organism, i.e., the manner in which it obtains food, escapes from being eaten, builds its home, and in other ways exploits its environment. A simplistic way of distinguishing between a niche and a habitat is to consider a niche an organism's ecological profession and a habitat its ecological address. The *geographical range* of a species encompasses all the regions of the earth in which it is naturally found. Each individual within an animal population usually has a *home range* or area within which it searches for food and/or other means of sustenance. Many animals have within the home range a *territory* or area that they actively defend and protect from intrusion by other members of the same species.

Biome. A relatively large, easily recognized terrestrial life zone consisting of communities located within a rather distinctive climatic belt constitutes a *biome.* Because plants are so widespread and conspicuous, a biome is usually classified by its most distinctive (numerous and prominent) vegetation.

> For example, the *grassland biome* of North America consists of plants such as buffalo grass, bluestem, and gamma grass, covering large areas of the western prairies. The *deciduous forest biome,* which characterizes much of the eastern United States, consists of trees such as maples, elms, beeches, and oaks. Spruce, firs, cedars, and pines are found within the *coniferous forest biome* of the cool, moist Pacific northwestern United States. Shrubs with thickened evergreen leaves such as chamise and manzanita constitute the *chaparral biome* of the central California coast. Hardy, water-conserving plants like cactus are typical of the xeric (dry), hot *desert biome* in southwestern United States and Mexico. A treeless area of sparse vegetation found in cold northern latitudes near the Arctic circle characterizes the *tundra biome* (mosses and lichens). A *tropical rain forest biome* covers much of lowland South America in regions of high humidity and warm temperatures.

One biome usually intergrades with an adjacent biome over a broad transitional zone called an *ecotone.* An *edge effect* is a tendency for both higher density populations and more complex communities (greater numbers of species) to develop in an ecotone because of its more variable ecological conditions. It is sometimes possible to pass through several ecotones as one travels up a mountain, because the various climatic conditions that determine biomes on a latitudinal scale are also produced on an altitudinal scale. Thus, it might be possible to be in a grassland biome at the foot of a mountain and as one ascends the mountain to pass through a broad-leaved deciduous forest, followed by a coniferous forest, and near the top an *alpine zone* similar to the tundra (Figure 1.3).

Marine Life Zones. Though not designated as biomes, various life zones also exist in the marine environment (Figure 1.4). Many of these same life zones apply to *limnology* (study of fresh water ecological situations such as lakes and rivers) as well as to *oceanography* (salt water ecology).

> Water filling the ocean basins constitutes the *pelagic* habitat; the sea floor is the *benthic* habitat. The *neritic zone* of the pelagic habitat consists of the relatively shallow waters over the continental shelf. This is the richest biological zone in the marine environment. Some organisms are specialized to live in the splash zone above the high-tide mark (*supratidal*); others are adapted to live in the *intertidal* area between high and low tides; still others cannot survive even brief exposure to air and live in the *subtidal* region below the low-tide mark. The *littoral zone* is a loosely defined area near the shore that, in practice, could apply to any or all divisions of the neritic zone. The open sea beyond the continental shelf is the *oceanic zone* of the pelagic habitat. The depth to which sunlight penetrates and supports photosynthesis determines the *euphotic region*. This is a relatively shallow region because, under average conditions, light penetrates only 250 feet; however, it can reach 600 feet under ideal conditions. Below this is darkness in the *aphotic zone*. The *bathyal zone* over the continental slope extends from the edge of the continental shelf to the deep oceanic planes. Water more than approximately 6,000 feet deep constitutes the *abyssal zone*. Relatively few forms of life can exist under the adverse conditions of temperature and pressure found in the abyssal zone.

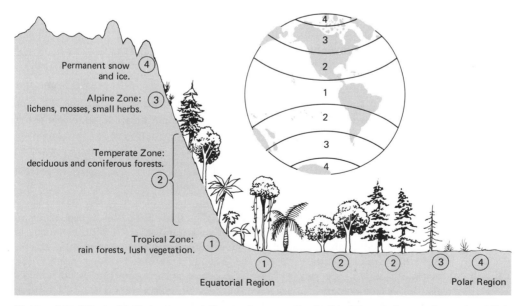

Figure 1.3. Effect of latitude on vegetation is shown on the horizontal scale from polar region (4) to equatorial region (1). The same pattern of vegetation may be observed by ascending a mountain in the equatorial region (vertical scale).

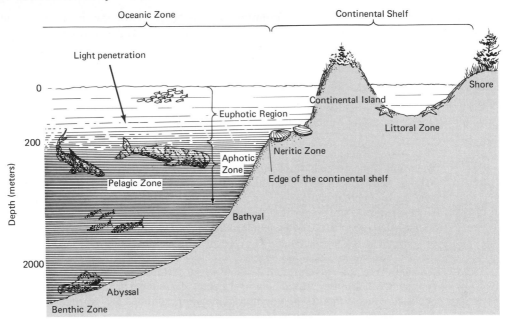

Figure 1.4. Marine life zones.

Aquatic forms of life can be classified by style of locomotion. Organisms that cannot swim actively, or do so very poorly, are called *plankton;* active swimmers such as fish are in the category *nekton;* species that live on or in the bottom are termed *benthos*. Organisms that inhabit quiet waters such as lakes and ponds are said to be *lenitic;* those that live in rapidly moving waters (waves, currents, streams) are described as *lotic*. Obviously, most lotic species have evolved adaptations for maintaining their position in the water (e.g., devices for clinging to the substrate, a tendency to swim against the current, etc.).

It is easy to see how various forms of aquatic life exhibit *vertical stratification,* i.e., creatures with different adaptations tend to live at different depths in the water. Some species live mainly at or near the surface where light is more plentiful, temperatures are usually warmer, and hydrostatic pressures are reduced. Some species are adapted to live at interme- diate depths, either eating dead or living plants or animals that drift down from layers above, or preying on other species living at the same depth. A few species can live at very great depths in, on, or near the bottom substrate. Vertical stratification also exists in terrestrial ecosystems (Figure 1.5).

For example, we can find organisms that make their livelihood by digging underground (earthworms, moles); others live mainly on the ground surface (cattle, elephant); some species live mainly above ground at varying heights in bushes or trees (sloths, monkeys); still others can exploit the air above the ground (birds, insects). Most plants are *sessile* (without motion) because they are fixed by roots to their substrate. Some animals are also sessile (sea anemones, barnacles), but most animals exhibit some kind of motility. Some are best adapted to swimming, or to digging underground (*fossorial*), or to running (*cursorial*), or to jumping (*saltatory*), or to swinging from branch to branch (*brachiation*), or to gliding (*volant*), or to active flight (*aerial*). These various forms of

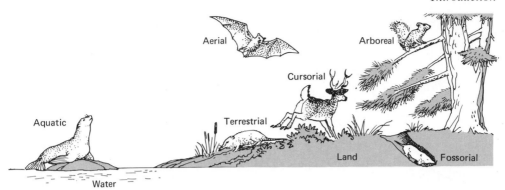

Figure 1.5. Vertical stratification and limb adaptations for locomotion and occupation of particular niches.

animal movements are useful in gathering food, finding mates, escaping predation, etc. Motility and various behavioral adaptations make it possible for two species to avoid complete competition with each other. For example, although a frog and a bird may eat the same kind of insect, most frogs feed from the ground or in the water, whereas birds can feed on insects in trees or in flight.

Owls and hawks are both carnivores, but avoid direct competition because hawks hunt during the day and owls by night. No two species can exist indefinitely in the same community if they are pursuing identical ways of life. Ecologists would say they are in the same niche. This is the so-called *competitive exclusions principle.* The species that has any advantage that allows it to better exploit its environment will eventually drive the other species to extinction unless one of the species evolves a new characteristic that allows it to ameliorate a completely competitive situation. This basic fact is summarized in *Gause's Principle:* only one species can occupy an ecological niche in a community at a given time.

Biosphere. All of the ecosystems of the earth collectively form the biosphere.

Population Characteristics

Biotic Potential. Consider a very simple unbalanced ecosystem consisting of a tube of nutrient broth richly supplemented with amino acids, vitamins, nucleic acid precursors, etc. into which we introduce a single bacterium. This single cell reproduces by fission, yielding two cells within thirty minutes under ideal conditions. Each cell divides again and we have four cells; the third division produces eight cells, the fourth yields 16 cells, and so forth. The generalized formula for the size of the population under these conditions is 2^g, where g is the number of elapsed generations. The population increases geometrically for a time until the nutrients in the broth become scarce and the accumulation of metabolic waste products has a toxic effect on the population. Growth slows down, the generation time increases, and the population reaches its numerical peak. Soon after it peaks, a "population crash" occurs in which all the individuals die. The conditions in the tube have become incompatible with life. The foregoing events are anticipated in any artificial, unbalanced, closed system (Figure 1.6).

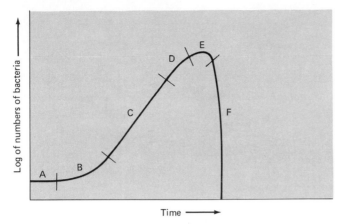

Figure 1.6. An unbalanced ecosystem (e.g., a culture tube of bacteria). A. lag phase; B. positive acceleration growth phase; C. log (logarithmic) or exponential phase; D. negative acceleration growth phase; E. stationary or equilibrium phase; F. death or decline phase.

   Most natural ecosystems are balanced and open to renewable resources because of the cyclical nature of most physical phenomena. Visualize a small group of individuals establishing a population in an unoccupied area. With plenty of food, shelter, and space available, the population begins to expand exponentially until one or more of these necessities become limiting. The population tends to stabilize when the *carrying capacity* of its habitat is reached. A sigmoid (S-shaped) growth pattern is typical of such natural populations (Figure 1.7). A balance tends to be struck between the use of resources and the renewal of same in most ecosystems. Thus, populations tend to remain stable (after the carrying capacity of the ecosystem is reached) until the balancing factors are disturbed. A change in any of the biotic

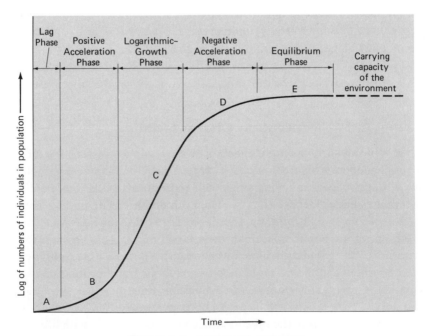

Figure 1.7. A balanced ecosystem.

or abiotic factors of the ecosystem can destroy such an ecological balance. For example, if the climate becomes colder, less plant growth occurs, and fewer animals can be supported. Or perhaps another species of animal comes into the area to compete for the same kind of food and/or other necessities.

The term *homeostasis* refers to the ability of an individual or group of individuals to resist change. Ecologists generally concede (admittedly without adequate scientific evidence) that large, complex ecological groups tend to be relatively stable entities, because the integrated relationships of their subcomponents weave an ecological fabric difficult to disrupt. Thus, homeostasis is thought to be more easily maintained at the higher levels of the community and biome than at the lower levels of the population or individual. For example, a climatic change may cause a decrease in the number of individuals within one species of a community. This decrease may be balanced by an increase in another species better adapted to the changed conditions. Thus, the total biomass of the community tends to remain stable. Relatively simple biomes, such as the desert or tundra, are probably more precariously balanced because of their relatively small numbers of interdependent species. Disturbance in any species is likely to have a relatively large impact on the entire biome. On the other hand, a large, complex biome such as a tropical rain forest is not so easily disrupted because large numbers of species interactions create numerous pathways by which stabilizing adjustments can be made.

Birth Rate and Death Rate. Birth rate or *natality* is often defined as the number of new individuals produced (born or hatched) per year per 1,000 individuals of all ages in the parental population. Death rate or *mortality* may be defined as the number of individuals that die per year per 1,000 individuals of all ages in the existing population.

Sex Ratio. The sex ratio, defined as the numbers of males relative to females in a population, is approximately unity for many animal populations (e.g., 100 males/100 females = 1). During the reproductive period, however, the number of males actively engaged in the procreation of offspring may be considerably smaller than the number of females. This causes the sex ratio among reproducing members of the population to become less than unity.

> For example, a male within the deer family will often fight other bulls seemingly for maintenance of conjugal rights to its harem of females (does). Under these conditions, the few males that become herd sires make a disproportionate contribution to the next generation compared to the females. The evolutionary ramifications of this situation will be discussed in a later chapter.

Age Distribution. The number of individuals in each age group can be indicative of the reproductive vigor of a population.

> For instance, if a population consists of a disproportionately large number of older individuals, the population size is likely to be dwindling and may be on the verge of extinction. A disproportionately large number of young individuals may indicate an expanding population. This latter evaluation would need to be tempered with a knowledge of average mortality by age groups. It is known that in most populations a

proportionately greater death loss is expected in younger age groups than at sexual maturity.

Density. The total number of individuals in a population constitutes its degree of *aggregation*. The area occupied by a species is seldom uniformly inhabited. Rather, the species tends to be fragmented into clusters of localized groups (populations) called *demes*. The *density* of a terrestrial population is usually given in terms of the number of individuals within a given land area, or of an aquatic population in terms of the number of individuals within a given volume of water. As population density increases, competition for the raw materials of existence is expected to increase. Disease, parasitism, and abnormal behavior usually increase with population density. Excessive crowding can cause experimental rat populations to stop breeding, and the population may expire. On the other hand, some bird populations need a certain minimal density in order to stimulate normal breeding behavior and to maintain population size. These facts are summarized in *Alle's principle:* the size and density of a population that produces optimum population growth and survival varies with the species and ecological conditions; therefore, lack of sufficient aggregation, as well as overcrowding, may act as a limiting factor.

Limiting Factors

Any condition or set of conditions that prevents a population from reaching its biotic potential is classified as *environmental resistance* or a *limiting factor*. In a bacterial culture, the depletion of nutrients and the accumulation of poisonous waste products are factors that ultimately limit the size of the population.

Biotic Factors.   Plants have photosynthetic ability,* i.e., the capacity to take simple minerals and water from the soil and carbon dioxide from the air and, in the presence of sunlight, convert these substances into the myriad complex molecules of the living plant. Animals ultimately depend on plants for their energy. *Herbivores* are animals that eat plants. *Carnivores* are animals that eat other animals. *Predators* are carnivores that actively seek and capture other animals (prey). *Scavengers* are carnivores that eat the remains of dead animals (carrion). *Omnivores* are animals that have a mixed diet of both plants and animals. When plants and animals die, they are subject to decomposition by bacteria and fungi. These breakdown products, together with the excreta from animals, can enrich the soil and make it possible for another cycle of plants to grow.

Energy does not cycle on earth, but rather flows from the thermonuclear fusion reaction of our sun. Sunlight drives the photosynthetic reaction of plants. Energy flows to animals via *food chains* (Figure 1.8), which always begin with the *producer* organisms (plants). The herbivores are *primary consumers* that feed directly on the producers and therefore have a single step in their food chain. The lion is a *secondary consumer* that preys directly on primary consumers such as zebras and wildebeests. Other secondary consumers may have food chains of varying lengths.

---

*Some bacteria are also photosynthetic, containing either green, red, or purple pigments that function as chlorophyll does in higher plants.

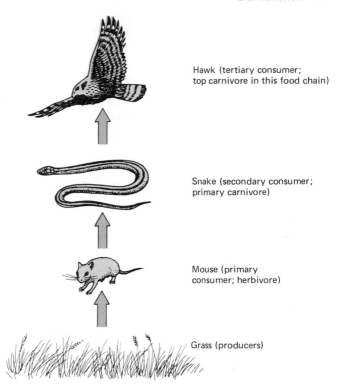

Hawk (tertiary consumer; top carnivore in this food chain)

Snake (secondary consumer; primary carnivore)

Mouse (primary consumer; herbivore)

Grass (producers)

Figure 1.8. A simple terrestrial food chain.

Consider what happens when a person consumes tuna fish (Figure 1.9). Tuna eat smaller fish such as minnows; minnows eat tiny floating animals (*zooplankton*) that are often the immature (larval) stages of marine invertebrates; the zooplankton feed on tiny floating plants (*phytoplankton*), mainly diatoms. At each stage of the food chain, only approximately 10 per cent of the energy is used; 90 per cent is eventually lost primarily as heat. Therefore, when a person gains 0.1 pound as a result of eating tuna, that pound of tuna was produced at the expense of consuming about 10 pounds of minnows, which in turn probably required 100 pounds of zooplankton, which likely ate 1000 pounds of phytoplankton. Zooplankton and minnows are also involved in food chains of numerous

Figure 1.9. An energy or food pyramid. All numbers represent pounds.

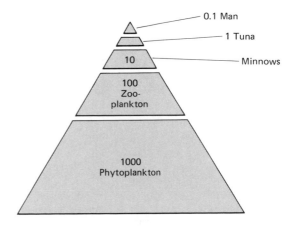

0.1 Man

1 Tuna

Minnows

10

100
Zoo-
plankton

1000
Phytoplankton

other large fishes and marine mammals such as the whale. Most individual food chains are therefore linked together in what is called a *food web* for the community (Figure 1.10).

A relatively close association of organisms is called a *symbiotic* relationship (*symbiosis*, "living together"). If this association is absolutely essential to both parties, it is called *mutualism*. The association of certain flagellated protozoans in the guts of termites is a mutualism. The protozoans cannot live outside their host, and the termite cannot digest wood without these flagellates. If one species benefits from the association, but the second is unaffected, it is called a *commensalism*. A relatively small fish called the remora attaches to sharks and gets a free ride as well as access to scraps leftover from shark kills; the sharks apparently are neutral in this association. When both species benefit from, but neither depend on this association, it is called *protocooperation*. The association of algae and fungi, called a lichen, is an example of protocooperation. The algae photosynthesize and provide the fungi with nourishment; the fungal mat absorbs moisture and offers physical support for the growth of algae. *Parasitism* exists when one species benefits from the association but the other species is harmed. *Ectoparasites*, such as ticks, mites, and leeches, live on the surface of their hosts; *endoparasites*, such as tapeworms, flukes, and the malarial protozoan, live inside their hosts. When the host dies, the parasite may also perish. It would therefore seem advantageous for a parasite to restrict the amount of damage inflicted on its host. As the host may evolve more elaborate defenses against parasites (as well as other pathogenic agents), parasite populations are under pressure to evolve new characteristics in order to continue to exploit the host, or perhaps to expand its host range (i.e., the number of species in which it can survive).

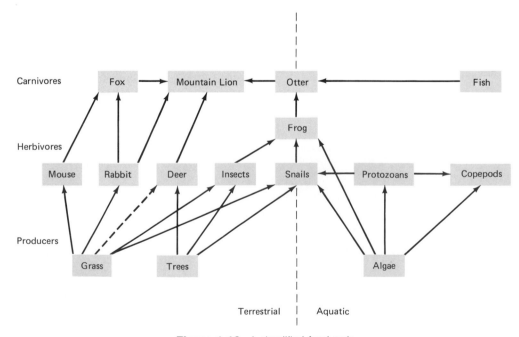

Figure 1.10. A simplified food web.

Physical (Abiotic) Factors. Our sun is the ultimate source of all energy used in the biosphere. Only light energy in the wavelengths near 475 nm (blue-green) and 650 nm (orange-red) are useful in driving photosynthesis. Therefore, the quality as well as the quantity of light can act as limiting factors to the growth of plants. Approximately 0.03 per cent of the atmosphere is carbon dioxide. It is a product of respiration from plants and animals. This molecule provides carbon from which all organic compounds of the plant are derived. Oxygen is required by animals as the ultimate electron acceptor in the process of cellular respiration. At high elevations on mountains (as well as in stagnant pools of water) oxygen can become a limiting factor for both terrestrial and aquatic life. Most animals live within a modest temperature range (5–45°C). However, some insect larvae dehydrated to 8 per cent of their normal water content survive at -134°C. Some blue-green algae live in hot water springs at 70°C. In aquatic environments, the concentration of dissolved salts (*salinity*) can be a limiting factor. Saltwater fish can seldom survive in fresh water, and *vice versa*. Water can be considered a nutrient for plants and animals. Desert plants such as the cactus are well adapted to drought conditions. Likewise, the desert kangaroo rat can live entirely on water liberated by its cellular metabolism of nutrients (metabolic water).

Atmospheric moisture in the form of humidity is important at varying optima for the growth of certain plants and animals. Hydrostatic pressure may be a limiting factor to the survival of certain kinds of fish at great ocean depths. The size of soil particles can vary from clays to sands. *Edaphic* factors include the chemical compositions and particle sizes of different soils. These attributes determine important properties for plant growth such as acidity, fertility, water-holding capacity, and friability.

The character of the ocean bottom is also important for the growth of specific benthic (bottom) life in both its immature and adult forms. Currents can also be important in determining where certain species can live. Strong water currents might prohibit many fresh water fish from inhabiting streams or marine fish from inhabiting near shore (littoral) habitats. Strong water currents can stir the bottom and keep so much particulate matter suspended that the amount of sunlight becomes ineffective in photosynthesis except at very shallow depths.

Air currents can sometimes move soil particles rapidly, as in shifting sand dunes, creating conditions hostile to many forms of life. Wingless insects are sometimes commonly found on oceanic windswept islands, because such forms tend to stay on the ground and are less likely than their winged relatives to be blown out to sea and lost from the population. Even fire can be a limiting environmental factor. The cones of the lodgepole pine tree require great heat to open and shed seed. Special factors, both living and dead, can be essential to the survival of certain species. For example, some species of birds will nest only in hollow logs (or holes in trees); many upland game birds avoid regions without grass or brush (cover) in which to hide.

Cyclical Processes

Nature is characterized by cyclical events on earth. The *water cycle* (Figure 1.11) is a familiar example. Water evaporates from the surface of oceans, lakes, etc. As moisture-laden air rises and is carried over land by wind currents, it cools and condenses the water from a gas into a liquid and falls as rain or snow. Much of the water returns via streams or rivers to

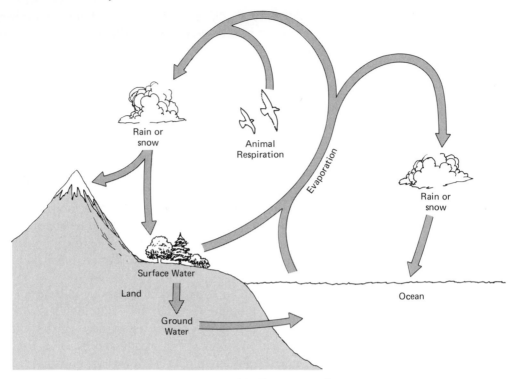

Figure 1.11. The water cycle.

larger bodies of water such as lakes and oceans, from whence the cycle repeats. In the *carbon cycle* (Figure 1.12), we begin with carbon dioxide in the atmosphere. This is the source of all organic chemicals made by photosynthesis in plants. Some plants are eaten and digested by animals. Both plants and animals produce carbon dioxide as a result of metabolic respiration. When plants and animals die, they are decomposed by microorganisms and the carbon is again returned to the atmosphere as a waste product of metabolism, carbon dioxide. Bacteria also play an important role in the *nitrogen cycle* (Figure 1.13). Atmospheric nitrogen can be converted to nitrites (NO_2) or nitrates (NO_3) by certain nitrogen-fixing bacteria and algae. The nitrogenous wastes of animals as well as the nitrogenous components in the remains of both plants and animals can be converted to ammonium compounds through decomposition by putrifactive bacteria. The ammonium compounds are then transformed by nitrite and nitrate bacteria into forms that can be used by most plants. Nitrogen is returned to the atmosphere by the action of denitrifying bacteria on nitrates, nitrites, and ammonia.

Populations tend to cycle over somewhat regular intervals varying with the species. There is a 17-year cycle associated with the growth and development of certain cicadas. This insect lives in the ground in the immature (nymph) stage for 17 years, before it finally molts into an adult for a short period during which it mates, lays eggs, and dies. Many plants have annual cycles of growth and reproduction. Short-lived organisms may have seasonal cycles. Predator-prey cycles are common.

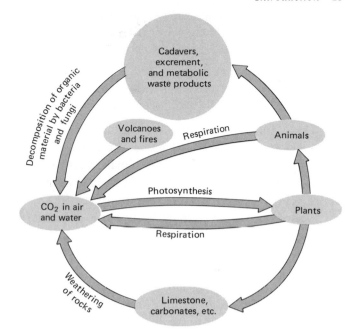

Figure 1.12. The carbon cycle.

Figure 1.13. The nitrogen cycle.

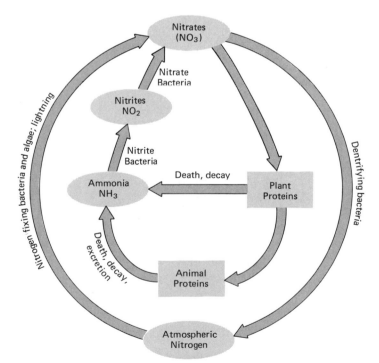

For example, the Canadian lynx (a member of the cat family) preys on the snowshoe hare. When hares are plentiful, the lynx finds good hunting and would be expected to raise larger litters than in lean years. A lag of several years occurs before the number of lynx becomes relatively large. During this interval, the hare population has begun to decline because more lynx are preying on them. The size of lynx litters born and/or weaned begins to decline as does survival of the adult cats. As lynx numbers begin to decline, the hares again rejuvenate, and the 9- to 10-year cycle repeats (Figure 1.14).

Occasionally a very large population is produced, as exemplified by the unpredictable plagues of locusts in Africa. These events are noteworthy, because they are exceptions to the rule that most populations tend to remain relatively stable over long periods of time. This is true even for populations that exhibit cyclical changes in number, because over many years their average size is relatively unchanged.

Various animals exhibit cyclical periods of activity characteristic of their species. Certain ground squirrels enter a torpid state of reduced metabolic activity during the cold winter months. This inactive period is called *hybernation*. A few amphibians and fish can likewise become torpid during the summer months when ponds tend to dry up. This period of quiescence is termed *aestivation*. Two physiological classes of vertebrate animals are easily recognized: *poikilothermic* animals (fish, amphibians, reptiles) cannot regulate their body temperatures and tend to assume the temperatures of their environments: *homoiothermic* animals (birds, mammals) maintain characteristic normal body temperatures despite great variations in the temperatures of their environments. Poikilotherms thus tend to be more active during warmer weather, whereas homeotherms can often exploit their environment year-round. Perhaps this is a major reason why birds and mammals are dominant forms of terrestrial life today, while the amphibians and reptiles, which flourished in geological ages past before the advent of homeotherms, are now relatively inconspicuous life forms.

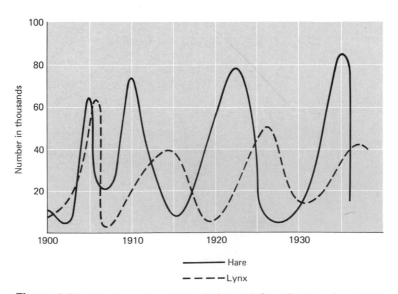

Figure 1.14. Numbers of snowshoe hares and Canadian lynx from 1900 through 1935 according to the number of pelts received by the Hudson's Bay Company.

Some plants require a minimal number of daylight hours each day (*photoperiod*) in order to flower. Certain animals also need specific photoperiods to stimulate their reproductive systems. Poultrymen know that chickens lay more eggs if lights are kept on in the hen house after dark, giving them a longer photoperiod. Many animals exhibit *circadian rhythms* of approximately 24-hour periodicity. *Diurnal* organisms have their greatest activity during the day; *nocturnal* organisms are most active at night. These natural circadian rhythms have been referred to as "biological clocks." These hereditarily determined periods of activity allow different species to exploit the same environment at different times and thereby avoid direct competition.

Summary

This brief discussion has focused on the central fact of any ecological situation, *viz.* every part is intricately linked with every other part. The term *neutralism* has been coined to denote the absence of symbiotic associations between two species in a community. Although there might appear to be no immediate relationship between the earthworm and the lion in a community, we could perhaps ascribe some portion of the productivity of the land to the tilling activities and fertilization of the soil accomplished by earthworms; from the increase in plant tissue thereby attained, perhaps more pounds of wildebeest will be available as food for lions. In like manner, it has been proposed that an ecological link exists between the number of house cats and the strength of the British Navy. The tongue-in-cheek logic is as follows: sailors need beef; beef is raised on clover; clover is cross-pollinated by bumblebees; these insects nest in the ground; their larvae are eaten by field mice; and mice are preyed upon by cats. Hence, the more numerous cats become, the more beef can be raised to support the Navy. Therefore, in its ultimate aspects, a term such as "neutralism" is misleading. Every community is an extremely complex set of interactions; to disturb one part ultimately has ramifications throughout the whole. This important aspect of ecology must be borne in mind during a study of evolution. A change in one species is likely to have subsidiary effects on one or more of the other species in that same community.

Essay Questions

1. *Have we really learned anything from the Scopes trial about the adjudication of what should and what should not be taught in science courses or in science textbooks?*

2. *What are the basic premises and purposes of science, and in what respects do they differ from those of other philosophical disciplines?*

3. *Discuss the nature of scientific inquiry and how it differs from that of other branches of philosophy.*

4. *How can a person be theistic (believe in a Supreme Being) and still believe in evolution and conduct scientific investigations pertaining thereto?*

5. *What do you consider to be the most important concept for the study of evolution to be gleaned from your knowledge in the discipline of ecology?*

HISTORY OF THE EVOLUTION CONCEPT

Pre-Darwinian Contributions

Long before the Greeks became prominent in the sphere of philosophy, some men in the ancient Chinese and Egyptian empires had speculated on the possible origins of biological diversity. However, these early ideas (including those of the Greeks) were very crude and lacking in scientific rigor.

One of the most prescient Greek philosophers was Empedocles (504–433 B.C.), the originator of the humoral theory of disease. Although most of his biological concepts were primitive, in at least one of them he approached very closely the modern view: (1) higher forms of life gradually evolve, (2) plants evolved before animals, and (3) better adapted forms tend to replace the less adapted ones.

Aristotle (384–322 B.C.) is supposed to have written 146 books on virtually everything known at that time. Ironically, not a single original word that Aristotle wrote exists today. One of his most useful contributions to biology was his "ladder of nature," a taxonomic scheme whereby nature is viewed as a continuum from inanimate matter through plants to lowly animals to higher animals ending with mankind. He asserted that this progression in complexity was under the guidance of divine intelligence; hence, his concept of biology was both vitalistic and teleological. Aristotle's "ladder" was a device for classification rather than a theory to explain the origins of organic diversity. This hierarchical system might have suggested an evolutionary history behind all living things were it not that constancy of "form" (or as we would now say, his belief in the immutability of species) constituted such a fundamental part of his philosophy. He also subscribed to the prevailing view of the origin of biological "form" through spontaneous generation. This idea was obviously still popular in the seventeenth century, because Van Helmont prescribed how mice could be generated from a mixture of filthy rags and fermenting grain.

The subject of evolution was freely discussed by philosophers and theologians until the early part of the Christian era. For example, St. Augustine (354–430 A.D.) favored an allegorical interpretation of the Biblical account of beginnings and proposed evolution as an alternative to special creation. During the intellectual stagnation that characterized the Middle Ages, inquiry and original thinking along these lines virtually ceased. The Rennaissance saw the rebirth of creativity, but very little of it was applied to biological problems.

Perhaps the first to propose a general theory of evolution was Pierre Louis Moreau de Maupertuis (1698–1759). Many of his ideas were far ahead of their time. He studied the inheritance of a trait (polydactyly) through four generations of a human pedigree and came to the conclusion that hereditary material was particulate and transmitted through both the maternal and paternal sides of the family. His theory of embryology and gamete formation is essentially the same as the pangenesis theory evoked by Charles Darwin in the nineteenth century. Maupertuis thought that the hereditary particles could be changed by the environment (acquired characteristics) and that they were sometimes distributed irregularly. He seems to have had some appreciation of the role natural selection played in evolution and of isolation as an important principle in speciation. Maupertuis' ideas, however, had little effect on contemporary or later naturalists, perhaps because of the ridicule and defamation of his character at the hands of Voltaire.

Rapid strides were made in the seventeenth and eighteenth centuries by systematists in classifying biological diversity. Most prominent among the taxonomists was Swedish scientist Carl von Linné (Carolus Linnaeus; 1707–1778). So useful was his system that the tenth edition (1758) of his *Systema Naturae* is still the basis for zoological nomenclature, and his *Species Plantarum* (1753) is still the basis for modern plant taxonomy. Unfortunately, Linnaeus was strongly biased by his conviction that species were immutable, fixed, and constant entities. In accordance with a literal interpretation of the Bible, each species had been specially and perfectly created. According to this view, there would be no need for evolution to occur and transitional forms should not be found. Guided by this philosophy, Linnaeus and other early taxonomists were engrossed in pigeonholing each species into a rigid, inflexible classification scheme. Since taxonomy is the starting point of all biological research (one must know the organism with which one is working), the Linnaean philosophy became an impediment to the rebirth of evolutionary concepts.

The history of evolutionary theory has been markedly influenced by developments in the science of *geology* (the study of rocks and the forces involved in changing the shape of the earth's crust). A Scottish physician, James Hutton (1726–1797), noted that volcanic activity brings magma up from the earth's molten interior and upon solidification it forms new igneous rocks. Many oceanic islands, such as the Hawaiian Islands, were formed by this process. He also noted that rocks were being eroded by the forces of wind, ice (including glaciers), water (e.g., rain and surf), and the activities of plants and animals. Eroded particles may be transported by water and deposited in layers in the ocean and elsewhere. These layers become compressed into sedimentary rocks. It is in this class of rock that most fossils are found. It seemed reasonable to Hutton that the contemporary forces that were altering the face of the earth before his own eyes were very likely the same forces that have molded and reshaped the planet throughout its history. His idea of gradual geological changes wrought by natural processes later became known as *uniformitarianism*. This principle was presented in his book *Theory of the Earth* (1785), but the concept was neglected by some and attacked as anti-Scriptural by others. People were generally appalled by thoughts of immense periods of geological time and the indeterminancy of the future of the world, which the Huttonian theory conjured in their minds.

An English clergyman, Thomas Robert Malthus (1766–1834; Figure 2.1), published in 1798 *An Essay on the Principle of Population*. This essay was to be very influential in directing Darwin's thoughts toward his theory of natural selection. Malthus saw that human populations have the capacity to increase at a much greater rate than food resources. He proposed three major factors holding human population growth in check: war, pestilence, and famine. He could be considered an early ecologist, who pointed out the biotic potential in human populations and some of the limiting factors that prevent its realization.

The most outspoken critic of uniformitarianism during the early nineteenth century was the Frenchman Georges Cuvier (1769–1832; Figure 2.2). He has been called the "father of modern comparative zoology," because he introduced a new way of looking at the interrelationships of organismal structure and function. He has also been called the "father of paleontology," because of his expertise with fossils. It was said that he could reconstruct an

Figure 2.1. The Rev. Professor Thomas Robert Malthus (1766–1834). (From *The Travel Diaries of Thomas Robert Malthus,* edited by Patricia James, The Syndics of the Cambridge University Press, 1966)

Figure 2.2. Georges Cuvier (1769–1832).

entire creature from a single bone. However, Cuvier was violently opposed to the idea of evolution. He saw species as immutable and fixed since their creation, and he refused to recognize the existence of intermediate forms. In attempting to explain the fossil record, he proposed that a series of sudden and violent geological and climatic changes had periodically exterminated great numbers of species. This theory has come to be known as *catastrophism*. So great was Cuvier's status as a scientist throughout Europe that few dared to disagree with him.

One countryman who did dare to disagree with Cuvier's catastrophic theory was Jean Baptiste Lamarck (1744–1829; Figure 2.3). Lamarck believed that species were not static

Figure 2.3. Jean Baptiste, Pierre Antoine de Monet, Chevalier de Lamarck (1744–1829). [From *Genetics,* **29**: Frontispiece (1944)]

groups, but were derived from pre-existing species. He systematically placed organisms in a series from the simplest to the most complex. In 1802 he published the first truly mechanistic theory of evolution called the *theory of "use and disuse"* or the *"inheritance of acquired characteristics."* According to this theory,

1. The ecological conditions in which an organism lives causes it to have certain "needs."
2. These needs are met by modification of old organs or by production of new rudimentary organs.
3. By continual use of these rudimentary structures, they increase in size and functional capacity and by disuse they may degenerate and become lost.
4. Character changes wrought by the environment during the life of an individual become hereditary and thus can be transmitted to the next generation.

Lamarck would say that cave fish are often blind because their ancestors had lived for several generations in dark conditions and in each generation the eyes had progressively degenerated through lack of use. The long neck of the modern giraffe could also be explained by the Lamarckian theory (Figure 2.4). During times when food was scarce, the short-necked ancestors needed to stretch their necks to get at leaves higher up in the trees. This stretching modified their heredity and allowed the production of offspring with slightly longer necks. In each generation, the exercising of the neck in this fashion caused an increment of hereditary change to be added to that contributed by previous generations.

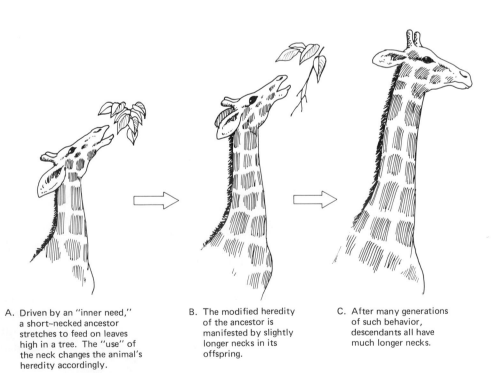

A. Driven by an "inner need," a short–necked ancestor stretches to feed on leaves high in a tree. The "use" of the neck changes the animal's heredity accordingly.

B. The modified heredity of the ancestor is manifested by slightly longer necks in its offspring.

C. After many generations of such behavior, descendants all have much longer necks.

Figure 2.4. Example of the evolutionary concept of *acquired characteristics.*

Lamarck's theory may sound plausible, but it is unsupported by any unequivocal scientific evidence and, indeed, much experimental evidence argues very strongly against it. Nonetheless, it was a scientific breakthrough in that a mechanism had finally been proposed to account for changes in living populations from one generation to the next. Cuvier used his vast influence to belittle and malign Lamarck and thus became responsible for much of the personal hardship Lamarck endured during his life. In later life, Lamarck went blind perhaps partly from his intense microscopic work. He was forced to give up his lectureship in 1825 and to sell his snail collection to obtain money to buy food. After his fourth wife died, his only daughter cared for him until he died. Lamarck was buried in a grave bought on a five-year lease, but its whereabouts is now unknown.

A modern champion of Lamarckism arose in the 1930s in Russia. His name was Trofim D. Lysenko. In one of the most bizarre incidents in the history of biology, this man gained the confidence and backing of the Central Committee of the Communist Party through his association with Premier Stalin. It was decreed that the Lysenko style of biology, genetics, and especially agriculture be adopted throughout the country. Needless to say, the folly of pursuing an erroneous theory led to disastrous effects on Russia's agricultural productivity and thoroughly disrupted its progress in basic biological research for almost two decades. In 1965 Lysenko was demoted from his positions of authority and genetics once again became a respectable science. This case serves as an outstanding example in the argument for keeping science free from political manipulation.

In 1813 three physicians and Fellows of the Royal Society of London (Wells, Prichard, and Lawrence) announced in almost identical terms a theory of natural selection and a repudiation of Lamarckism. Lawrence's book was condemned by the Lord Chancellor who forced him to make a decision between his medical career and his philosophical stance on evolution. He decided in favor of the former and went on to become President of the Royal College of Surgeons.

A Scottish geologist named Charles Lyell (1797–1875) began his academic career as a law student, but soon became interested in geology. His *Principles of Geology,* first published in 1830, was attacked mercilessly, but ran through twelve editions nonetheless. Lyell had long resisted the theory of evolution. He argued that it was only the imperfections of the fossil record that suggested that higher forms of life were absent from the earlier strata. This book offered evidence for a prolonged and natural origin of the earth by processes that are still going on and renounced the notion of worldwide breaks in the geological record. It gave to Charles Darwin the "gift of time" during which his proposed evolutionary process of natural selection could operate to produce the organic diversity now seen in the living world. Lyell had considered ecological concepts such as the struggle for existence, and he even toyed with the idea of natural selection without grasping its creative potential. Darwin once said "I feel as if my book (*Origin of Species*) came half out of Sir Charles Lyell's brain."

In the same year that Darwin set off on his voyage around the world (1831), Scottish botanist Patrick Matthew published a book entitled *On Naval Timber and Arboriculture.* This was a treatise on methods of raising trees for use in the construction of Royal Navy ships. In the appendix of this book, Matthew alludes to a "natural process of selection" and other superficially Darwinian concepts. After reading a report on the *Origin of Species* in *The Gardener's Chronicle* of March 3, 1860, Matthew assumed that his views were similar to those of Darwin and wrote a letter to the magazine to make the public aware of this. When Darwin learned of Matthew's work, he wrote to Alfred Russel Wallace that Matthew had "most clearly anticipated our view of natural selection." Neither Darwin nor Matthew seems to have completely understood each other's theory. Matthew discusses natural selection as an axiom; no theory is proposed and no evidence is marshaled to support it. He was a catastrophist in his geological thinking, but rejected the idea that the world was repopulated by new creations. Apparently he thought that some organisms survived the catastrophies; these remnants evolved very rapidly into new forms and then remained stable for "millions of ages." Therefore, Matthew viewed the natural world as one in which evolutionary change was a relatively rare event. This is far removed from the Darwinian concept of evolution! Furthermore, Matthew's concept of species was that of static Aristotelian classes defined on the basis of similar appearance. Darwin saw species as units of dynamic interaction composed of interbreeding individuals comprising populations. Matthew did not conceive of the role that natural selection could play in the extinction of species. Natural selection to him was the guiding hand of Providence in the production of specific designs; thus, his thinking was both vitalistic and teleological. Though portions of some passages in the appendix of his book may sound like those of Darwin, it is obvious that their views of natural selection and evolution are worlds apart. The amazing thing about Matthew was his ability to unite natural selection and evolution without destroying any of the basic philosophical assumptions that had dominated biological thought since Aristotle. In light of this, we are better able to appreciate the full significance of the Darwinian revolution.

An anonymous publication entitled *Vestiges of the Natural History of Creation* appeared in 1844. Its author, Robert Chambers, was a scientific amateur. His intention apparently was to present a developmental analogy as an explanation for biological diversity, and he was probably motivated by a desire for a comprehensive theodicy. His concept of evolution was one of a divinely planned linear developmental progression from simple to complex according to universal laws. If the laws of gravitational and centrifugal forces are sufficient to account for various phases of concentration from unorganized nebulae to the perfection of the common star, then there must also be laws of development that guide the advances of organisms toward biological perfection. Chambers makes an analogy between the embryological stages of animals and the geological histories, both of which he viewed as a progression from simple to complex. His theory left no room for the notion of common descent of allied species so vital to the Darwinian concept of a branching evolution. Natural selection was a foreign concept to Chambers; all that was needed was an analogical scaling-up of the developmental processes seen in the embryo. According to his theory, offspring may occasionally develop beyond the level of the species to which their parents belonged and thus become, as adults, members of a different and taxonomically higher group.

In 1845 Chambers wrote *Explanations: A Sequal to the Vestiges.* Having read Darwin's new 1845 edition of the *Voyage of the Beagle,* he noted that there are no mammals indigenous to the Galapagos Islands. His explanation for this fact was that there simply had not been enough time for the development of mammals there. He therefore rejected the Darwinian concept of migratory colonization of the islands from the nearest mainland followed by speciation through natural selection in isolated groups. Though Chambers is widely cited as one who entertained the evolutionary concepts later expounded by Darwin, nothing could be further from the truth! Chambers' theories met with widespread disapproval. Charles Darwin may have delayed the publication of his own theories partly from fear that they would be received with the same hostility as those of Chambers.

The Darwin–Wallace Era

Charles Darwin (Figure 2.5) was born in 1809 at Shrewsbury, England. His authoritarian father, Robert Darwin, was a physician impressive in position and stature (he weighed over 300 pounds during most of his adult life). Charles' grandfather, Erasmus Darwin (1731–1802), also a physician, had written a book in 1794 entitled *Zoonomia*. In this volume are to be found many of the essential concepts later popularized by Charles, especially the struggle for existence and the concept of differential reproduction. It has been claimed that every volume written by Charles has a corresponding chapter in *Zoonomia*. There is no evidence that Charles ever acknowledged the profound philosophical legacy left to him by his grandfather.

Figure 2.5. Charles Robert Darwin (1809–1882) at the age of forty. (From *Molecular Evolution and the Origin of Life* by Sidney W. Fox and Klaus Dose. W. H. Freeman and Company. Copyright © 1972)

Charles exhibited, at an early age, an interest in the natural world by becoming an avid collector of various creatures. Above all he enjoyed the sport of hunting. Robert Darwin admonished his son about spending so much time hunting and chasing rats; he was going to bring shame on himself and his family. Charles loved his father, but was somewhat intimidated by his authoritarian manner. To please him, Charles entered medical school, but soon found this not to his liking. For one thing, he could not stand the sight of blood. At his father's suggestion, Charles then began studies for the clergy, although this was not really where his interests lay either. His real love was the natural world. When he heard that a naturalist's position was available on a ship about to sail around the world, he committed the next five years of his life to exploration without compensation, save board and room.

The *H.M.S. Beagle* sailed from England in December 1831 with young Captain Robert Fitz-Roy (26 years old) in command. The voyage of this 242-ton bark was designed to chart waters important to the commerce of the British Empire especially along the coasts of South America (Figure 2.6). Seventy-six people were crowded aboard, but Darwin was the only naturalist. He shared cramped quarters with Capt. Fitz-Roy. When the Beagle arrived in Brazil, Darwin saw a tropical rain forest for the first time. In it he had a sublime feeling of being a part of nature. He made collections, whenever ashore, of plant and animal life and fossils. When aboard ship and the weather permitted, he took samples of marine life. He took copious notes on everything, including geological formations and coral reefs. Darwin suffered chronic seasickness, which severely restricted the amount of work he could perform. In addition, some historians believe he contracted a tropical illness or perhaps Chagas disease, which plagued him the rest of his life.

When the Beagle arrived at the Galapagos archipelago, about 600 miles off the coast of Equador, Darwin made some of his most important observations. He noticed that each island in the archipelago was populated by its own kind of finch, found nowhere else in the world (*endemic*). Some had beaks adapted to eating large seeds, others to eating small seeds; some had parrot-like beaks adapted to feeding on buds and fruits; still others had slender beaks for

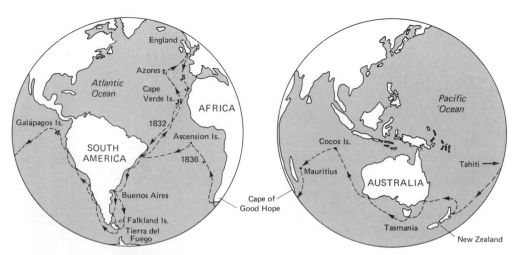

Figure 2.6. The voyage of H.M.S. *Beagle* (1831–1836).

feeding on small insects (Figure 2.7). One of the finches (which Darwin did not observe) had developed the habit of using a thorn as a probe for insect larvae in wood, thereby invading the niche occupied by long-tongued woodpeckers on the mainland. All the Galapagos finches closely resembled certain species of finches on the nearby mainland. To Darwin it appeared reasonable to assume that these closely related species of finches had evolved from an original colonization of the islands by similar members of a mainland population. Over many years, the semi-isolated finch populations on each island evolved slightly different behavioral and structural modifications (mainly involving the beak), which allowed them to exploit different ecological niches.

In the process of natural selection, he later saw a mechanism that could drive organic evolution. Darwin knew that most traits are not uniformly expressed in natural populations and that variations are common. He was not aware, as we are today, that the ultimate source of hereditary variations lies in the process of mutation. However, given the fact of hereditarily determined variation, Darwin reasoned that any slight modification that aided an individual in finding or capturing food, escaping from predators, and in other ways exploiting its environment would favor the survival of that individual. Those who survive the longest tend to leave the most offspring. Such offspring would tend to inherit the capacity to develop these adaptations to a greater degree than was possible in their parents.

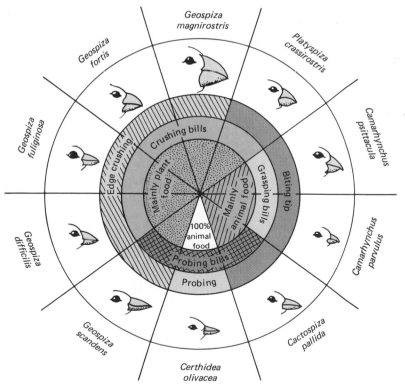

Figure 2.7. Adaptations for various feeding niches exhibited by the bills of ten species of finches (*Geospizinae*) found on a single island of the Galapagos archipelago.

Each generation would be subject to this *selective process* whereby individuals with the more favorable variations (*fitter* individuals) would tend to leave more highly adapted progeny than those with less favorable variations. The population would thereby gradually evolve to become better equipped to exploit its environment. Individuals lacking in *fitness* (i.e., the relative ability of an organism to survive and leave progeny) would be at a *selective disadvantage,* and individuals with these traits would thereby tend to appear less frequently, or perhaps disappear entirely, in subsequent generations. This idea of evolution by natural selection, operating over long periods of geological time, seemed to provide an explanation for the origin of different species.

On his return to England in 1836, Darwin began to collect all the facts that might bear on his theory. He corresponded widely with other naturalists and performed breeding experiments with plants and animals (especially pigeons). From his own experience as well as that of other breeders he gained knowledge of the changes that selection by man (*artificial selection*) can cause in populations of domestic plants and animals. If artificial selection could be responsible for changing the ancestral rock pigeon (Figure 2.8) into such diverse domestic forms as homing pigeons, tumblers, fantails, jacobins, pouters, and more than 100 other different breeds in the very short time that man has been selectively breeding for these extreme forms, is it not reasonable that nature could have accomplished even more remarkable changes through long periods of geological time?

For more than 20 years Darwin amassed data supporting both the concept of evolution and his mechanistic theory of natural selection. During this interval he discussed his ideas with many friends and other naturalists. Then in 1858, Darwin received a letter from Alfred Russel Wallace (Figure 2.9), a professional collector of specimens in the Malay archipelago. While suffering from a tropical fever, Wallace apparently was seized with an idea of evolution so similar to that of Darwin's that it placed Charles in an embarrassing situation. If he published his own theory now, he might be accused of plagiarism. Was all his work for naught? Darwin immediately wrote to his geologist friend, Lyell, "I never saw a more striking coincidence; if Wallace had my manuscript sketch written in 1842, he could not have made a better short abstract." The famous botanist Joseph Hooker (Figure 2.10) and Lyell both knew of Darwin's prescience in this matter. These two men offered to present jointly the papers of Darwin and Wallace at the meeting of the Linnean Society in London on July 1, 1858. In this way, the independence of discovery would be vouched for in a way most acceptable to all concerned.

Thereafter, Darwin began working at top speed to compose a manuscript for publication. This task consisted mainly of greatly reducing his larger work into an abstract of some 350 pages. It was titled *On the Origin of Species by Means of Natural Selection, or the Preservation of Favored Races in the Struggle for Life,* and was published on November 24, 1859. All 1,250 copies of the first printing were sold on the first day. The furor created by this book lasted well into the twentieth century. Even today, Fundamentalists (creationists) are still active in the fight against Darwinism. Darwin himself avoided public confrontations, allowing his "bulldog" Thomas Henry Huxley (1825–1895; Figure 2.11) to defend the thesis of evolution by natural selection in public debate. Richard Owen (1804–1892), director of the Natural History Section of the British Museum, was the most violent opponent of the Darwinian theory in England. The opposition in America was led by Louis Agassiz (1807–1873), the famed

Wild Rock Pigeon

Continuous selection for different extreme characteristics

Pouter   Jacobin   Fantail

Figure 2.8. Evolution of some modern varieties of domestic pigeon through artificial selection. By continuously breeding only from individuals that show the most extreme characteristics desired, humans have been able to ''create'' new forms of pigeons unknown in the ancestral Rock Pigeon population. The creative potential of natural selection is analogous.

Harvard ichthyologist. Agassiz proposed that the whole of Northern Europe had been covered with ice during the recent geological past. He modified the doctrine of catastrophies, however, by postulating that God had performed creative acts on a number of occasions following these catastrophies. According to *progressive creation,* God began His recreation where He left off ''in order to improve His creatures progressively until they reached the final goal, man made in God's image.'' William Thomson (Lord Kelvin), perhaps the most eminent physicist of the nineteenth century, argued strongly that geological time was much shorter than Darwin

Figure 2.9. Alfred Russel Wallace (1823–1913) in 1869. (From *My Life* by A. R. Wallace, Dodd, Mead & Company, 1905)

presumed. Calculations of solar energy production based on conversion rates of conventional fuels disallowed a lengthy history for the earth and the solar system. Thomson, of course, was unaware of the enormous energy locked in the atomic nucleus.

One of the most damaging omissions to a complete theory of evolution was the lack of knowledge at that time of the mechanism of heredity and the origin of variations through genetic mutation. To fill this void, Darwin proposed a "pangenesis theory," according to which all parts of the body generate minute particles called "gemmules," which were carried by the blood (in animals with blood or by similar means in other organisms) to the gonads or reproductive organs (Figure 2.12). The gemmules were somehow concentrated in the gonads and packed into the sex cells (*gametes*). The union of gametes (*fertilization*) produces a single cell (*zygote*) whose gemmules retain memory of the sites from which they were derived and have the capacity to reproduce that same kind of cell, tissue, and organ. Thus, there was a two-way relationship between body cells and sex cells.

The pangenesis theory was a break with the contemporary concept of fluid heredity. Most naturalists were thinking in terms of blood or other body fluids as the source of biologically transmissible materials. Even today there are those who are apprehensive about receiving a blood transfusion from a member of a different race for fear that they will acquire some traits of that race through genetic information carried by the blood. We know now that the hereditary material (DNA) is not transmitted through body fluids but is located in the chromosomes of the cell nucleus. Chromosomes are transmitted as units from one generation to another. Darwin's "gemmules" were presumed to be developmental units or particles and in this sense were closer to reality than were the "hereditary fluids." However, because the gemmules were presumed to derive from all major parts of the body, it was easy to imagine how an acquired characteristic could be inherited. Environmental modification of any part of the body could change the hereditary information carried by the gemmules from that region, and the change could then be transmitted to the next generation. Darwin claimed to be anti-Lamarckian, yet he occasionally slipped into Lamarckian-type thinking due in part to the frailties of the gemmule hypothesis. Thus, Darwin was disconcerted from several sources: (1) the Church branded him an atheist and his theory antireligious; (2) the fact of biological variations was recognized, but completely inexplicable; all biological variation was assumed to be inherited; (3) the gemmule hypothesis of heredity was unsupported by any scientific evidence; and (4) Lamarckian interpretations were sometimes difficult to avoid.

Nonetheless, Darwin continued to investigate nature and wrote books on a wide variety of subjects: *On the Various Contrivances by Which British and Foreign Orchids Are Fertilized by Insects and On the Good Effects of Intercrossing* (1862), *The Movements and Habits of Climbing Plants* (1864), *The Variation of Animals and Plants Under Domestication* (1868), *The Descent of Man and Selection in Relation to Sex* (1871), *The Expression of the Emotions in Man and Animals* (1872), to mention only a few. His productivity is especially remarkable because of the chronic illness that plagued him after his voyage on the Beagle. It is said that he could only manage about four hours of work per day. Charles Darwin died in 1882 at the age of 73,

Figure 2.10. J. D. Hooker (1817–1911) at the age of thirty-two. (From *Pioneer Plant Geography* by W. B. Turrill, Martinus Nijhoff, The Hague, Netherlands, 1953)

but left a legacy deemed to be one of the most magnificent scientific contributions of all time—the theory of evolution by natural selection.

The logic by which Darwin came to his natural selection theory can be summarized in four major steps.

(1) The reproductive potential of any population is far greater than that necessary to reproduce the existing population.

(2) The size of most natural populations nonetheless tends to remain fairly constant over long periods of time because certain factors in the environment are in short supply. Competition for limited resources is inevitable.

(3) Biological variation is a common feature of natural populations for a large number of structural (anatomical and morphological), physiological, and behavioral traits.

(4) Any hereditary variation that gives an organism even a slight advantage in the competition for the means of sustenance would tend to be perpetuated. Those with a selective advantage tend to survive and reproduce their kind in greater proportion than the less-adapted individuals in the population. The accumulation of advantageous traits in this manner may eventually produce new species (Figure 2.13).

It should be clear from the previous discussion that Darwin was not the originator of the concept of organic evolution. This thought had been entertained before the rise of the Greek empire. Neither can Darwin lay claim to initiating a concept of natural selection; he was preceded in this by Lawrence, Matthew, and others. Charles Darwin's contributions were mainly twofold: (1) he succeeded in collecting a great amount of data in support of the theory of evolution, and (2) he showed that most of it was consistent with the concept of natural selection as the prime mover.

Figure 2.11. Thomas Henry Huxley (1825–1895) at the age of thirty-two. (From *Scientist Extraordinary* by Ciril Bibby, Pergamon Press Ltd., New York, 1972)

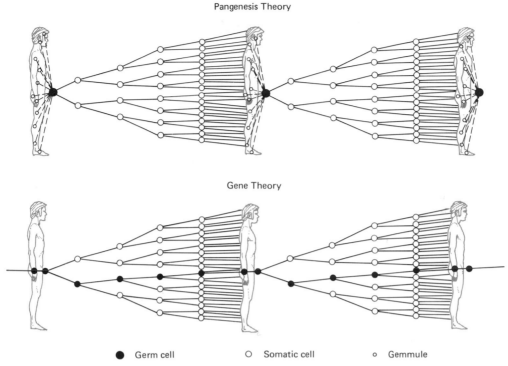

Figure 2.12. Comparison of the "pangenesis theory" (top) with the "gene theory" (bottom). According to the "pangenesis theory," the sex cells are composed of particles (gemmules) derived from various parts of the body, each particle endowed with the capacity to regenerate the whole structure from which it was derived. The "gene theory" proposed that cells of the germ line reproduce other germ line cells and also give rise to all other cells of the body.

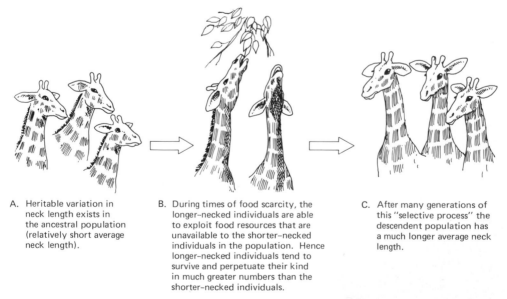

A. Heritable variation in neck length exists in the ancestral population (relatively short average neck length).

B. During times of food scarcity, the longer–necked individuals are able to exploit food resources that are unavailable to the shorter–necked individuals in the population. Hence longer–necked individuals tend to survive and perpetuate their kind in much greater numbers than the shorter–necked individuals.

C. After many generations of this "selective process" the descendent population has a much longer average neck length.

Figure 2.13. Example of the Darwinian concept of evolution by *natural selection*.

The Romantic Period

During the interval from about 1860 to 1902, the life sciences were enthralled by the Darwinian theory. Six years after the initial publication of the *Origin of Species,* an Augustinian monk named Gregor Mendel (1822–1884; Figure 2.14) presented the results of eight year's breeding experiments with the garden pea (*Pisum sativum*) to the Natural History Society of Brünn, Austria (now Brno, Czechoslovakia). From these investigations, Mendel had deduced that

1. Hereditary material is particulate (rather than fluid).
2. Each individual has two particles (factors) for each trait.
3. Parents make equal contributions to their offspring.
4. Only one member of each pair of hereditary factors is carried by a gamete (now called the *principle of allelic segregation*).
5. One factor may produce its effect when in single dose (*dominance*) and the other member of the pair when in double dose (*recessiveness*).
6. Each factor is distributed to a gamete independently of other factor pairs (now called the *principle of independent assortment*).
7. The results of crossing (mating) individuals of known heredity can be statistically predicted on a population basis to occur in ratios such as 3:1, 1:1, and 9:3:3:1.

Mendel had discovered the most basic laws of heredity. Historians can only speculate why the scientific community did not immediately grasp the tremendous implications his principles had for evolution and biology in general. There is no indication that Darwin ever knew of Mendel's experiments. It is doubtful that Darwin would have been any better equipped to realize the importance of Mendel's work than his contemporaries were. Biologists

Figure 2.14. Gregor J. Mendel (1822–1884). Austrian botanist as Prelate, Portrait, *ca.* 1880. (The Bettmann Archive)

of that time were not used to thinking quantitatively (statistically) and, to many people, Mendel's approach undoubtedly appeared to be numerology bordering on mysticism.

Mendel wrote several letters to a leading botanist, Karl Nägeli, seeking support for his theories. Nägeli was unimpressed and wanted Mendel to provide evidence that his "principles" were applicable to organisms other than peas. At Nägeli's suggestion, Mendel began experiments with the hawkweed (*Hieracium*) but failed to demonstrate the same kind of hereditary pattern as in peas. It was not known at that time that hawkweed was *apomictic,* i.e., sets seed without fertilization. Mendel was later promoted to an administrative position in the Church and did no further research. He died in 1884, unaware that his principles of inheritance were to prove of virtually universal application in sexually reproducing organisms.

Another great impediment to the immediate acceptance of Mendel's principles was the lack of a mechanical basis for the regular distribution of his hypothetical "factors." During the years 1875–1900, rapid strides were made in the discipline of *cytology* (the study of cells). The nucleus and chromosomes were discovered, the chromosomes were seen to split prior to cell division, and the processes of mitosis and meiosis were gradually worked out. By the turn of the century, a mechanistic process became available as a basis for the regular distribution postulated for Mendelian factors. The scientific community had apparently forgotten about Mendel's work for thirty-five years.

In 1900, three men, working independently in different countries, had come to essentially the same conclusions as those reached by Mendel. In surveying the older literature, they came across the 1866 paper and gave Mendel credit for prior discovery. All three men were botanists: a Dutchman, Hugo deVries; a German, Carl Correns; and an Austrian, Erich von Tschermak. Soon many biologists used Mendelian methodology and provided evidence for the validity of the Mendelian principles. Thomas Hunt Morgan (Nobel Prize winner, 1933; Figure 2.15) and his colleagues at Columbia University began genetic studies with the fruit fly (*Drosophila melanogaster*) about 1910 and soon had convincing evidence for the *chromosomal*

Figure 2.15. Thomas Hunt Morgan (1866–1945); Nobel Prize laureate 1933. [From *Genetics,* **32**: Frontispiece (1947)]

Figure 2.16. Sir Francis Galton (1822–1911). [From *Genetics*, 2: Frontispiece (1917)]

theory of heredity. The name *"gene"* was coined in 1903 to replace Mendel's "factor". In 1906 Bateson suggested the name *genetics* for the new discipline. The phenomenon of genetic *linkage* emerged from the *Drosophila* studies, consistent with the idea that each chromosome carried many genes constituting a *linkage group*. The number of linkage groups were shown by genetic experiments to be identical with the number of chromosome pairs cytologically visible in the cell. Thus, the hybrid discipline of *cytogenetics* was born.

During the latter part of the nineteenth century, a *biometrical school* of heredity developed under the leadership of Sir Francis Galton (1822–1911; Figure 2.16), a cousin of Charles Darwin. Galton had been particularly concerned with the inheritance of intelligence. We now know that traits such as intelligence and body configurations (including stature, weight, shape of face, etc.), as well as the economically important traits of domestic stock (yield of grain, feed conversion efficiency in livestock, etc.) are *quantitative traits*. These metric characters have a complex inheritance that cannot be deciphered through classical Mendelian methods. When Mendel's paper was rediscovered in 1900, the biometricians had much evidence that could not be reconciled with the way *qualitative* Mendelian traits (traits of kind, rather than degree) were inherited. These two schools battled for almost twenty years. They were finally united by the *multifactorial theory*, i.e., the concept of many genes contributing cumulatively to a metric trait, but the individual effects of each gene being too small to be detected by classical Mendelian techniques.

An alternative to Darwin's pangenesis theory and Lamarckism was developed in the 1890s by August Weismann, a German (1834–1915; Figure 2.17). According to this new *germ plasm* (or *germ line*) *theory*, there were two kinds of cells: *germ plasm* (cells of the reproductive line) and *somatic cells* (all cells of the body exclusive of the germ plasm). Germ plasm is potentially immortal through transmission from one generation to the next; somatic cells are produced from germ plasm, but expire with the death of the individual in whom they reside. Weismann hypothesized that development of the embryo was dependent on a process of sorting out the "determinants" originally in the fertilized egg, until every cell, tissue, and organ came to consist of only the "determinants" peculiar to them. According to this view, all the "parts" of the embryo were already present in the fertilized egg; all that was required was to arrange them in their proper relationships to one another. Only the germ plasm of the

Figure 2.17. August Weismann (1834–1914). [From *Genetics*, 7: Frontispiece (1922)]

gonads contained all the determinants of the soma. Somatic cells could be derived from germ plasm, but not the reverse. The germ line is only derived from other germ lines.

If this mechanism of heredity were true, it would argue against Lamarckism as an explanation for evolution. To show the continuity of the germ plasm and its sequestration from the soma, Weismann cut the tails off mice for over 20 generations. All mice born to these somatically modified parents developed normal tails. It could be argued that if gross modification of body parts by external agents (such as the removal of tails) cannot be shown to

Figure 2.18. Hugo de Vries (1848–1935). [From *Genetics*, 4: Frontispiece (1919)]

A

B

C

Figure 2.19. (A) Ralph Erskine Cleland (born 1892) shown with his favorite research material, the evening primrose (*Oenothera*). [From *Study Guide and Workbook for Genetics* by Irwin H. Herskowitz, McGraw-Hill Book Company, 1960] (B) Albert Francis Blakeslee (1874–1954). [From *Genetics,* **44**: Frontispiece (1959)] (C) Otto Renner (1883–1960). [From *Genetics,* **53**: Frontispiece (1960)]

change the heredity of mice, neither would the more subtle influences of the natural environment. Perhaps Weismann could have proved the same point by merely documenting centuries of experience by husbandmen in docking (cutting off tails) lambs, or in marking livestock by branding or cutting notches in the ears, etc. Despite many generations of such mutilation, domestic stock do not naturally bear these marks. Hebrews since biblical times have practiced the rite of circumcision (removal of the foreskin from the penis). Yet all boys today still are born with a prepuce.

   Hugo deVries (Figure 2.18), one of the rediscoverers of Mendel's paper, had made observations on the evening primrose (*Oenothera*), a weedy plant growing in abandoned fields, which led him to propose a new theory of evolution in his book entitled *The Mutation Theory* (1901). According to Mendelian laws, crosses between two varieties of peas should produce progeny of uniform appearance, but inbreeding of the progeny (such as self-fertilization) should produce various character combinations. DeVries found that crossing various kinds of pure-breeding evening primroses immediately produced a splitting progeny with various character combinations. Each particular combination bred true to type. It appeared to deVries that new species were arising before his very eyes. This certainly was not compatible with the Darwinian scheme wherein many generations of accumulating small modifications were postulated eventually to produce new species. The peculiar behavior of these plants remained a mystery for about 25 years. When the cytogenetics of *Oenothera* was finally deciphered by Cleland, Blakeslee, Renner (Figure 2.19), and others, it was shown that a combination of gross chromosomal aberrations (especially reciprocal translocations and polyploidy), rare recombinations in balanced lethal systems, and obligate self-fertilization were important factors. The knowledge gained from these experiments contributed evidence for the chromosomal theory of heredity. *Oenothera's* behavior in inheritance did not invalidate Mendelian principles nor the theory of natural selection. Rather, it demonstrated that there was more to heredity and evolution than Mendelism or Darwinism.

The Agnostic Period

Following the publication of deVries' *Mutation Theory,* a wave of skepticism and disillusionment swept over many students of evolution. Was Darwin's theory of natural selection, or deVries' mutation theory, or some yet undiscovered principle the propelling force behind evolution? Did the Mendelians or the biometricians or some yet to be formed school have the secret of heredity? No one knew. There were those who claimed that evolution could never be called a science because "there is no way to experiment with over three billion years of life history." Danish botanist Wilhelm Johannsen (1857–1927) reported in 1903 on his artificial selection experiments with beans. It appeared to him that selection could not extend the limits of previously established variability within a genetically pure line. Among those who agreed with Johannsen on this were William Bateson (1861–1926), the most active figure in spreading Mendelism in the early days of genetics, and T. H. Morgan who was later to gain fame through his genetic studies with the fruit fly. Eventually, genetically pure lines were shown to be a rarity in natural populations. Selection from a mixed population of beans, such as a commercial mixture of bean varieties, was shown by Johannsen to be effective in changing average bean size. These data lend credence to the theory of natural selection because most natural populations are now known to be highly heterogeneous in their genetic composition.

 For approximately three decades (*circa* 1903 to *circa* 1935) the study of evolution fell into decline, while rapid strides were being made in the new science of genetics. Early in this period (1908), G. H. Hardy (an English mathematician) and Wilhelm Weinberg (a German physician) concerned themselves with the theoretical results of random mating in large populations with natural selection inoperative. They independently came to essentially the same conclusions, *viz:* the frequency of a given gene is not expected to change from one generation to the next. This principle, now called the *Hardy-Weinberg Law,* defines a static (nonevolving) situation and has become the cornerstone of *population genetics.* The great importance of the Hardy-Weinberg principle for evolutionary theory will be discussed in detail in a later chapter. One of the most important discoveries of this period was the fact that genetic changes (*mutations*) could be induced by X rays at rates much higher than those occurring spontaneously. This was the finding of Herman J. Muller (Nobel Prize 1946), one of the original *Drosophila* team at Columbia under T. H. Morgan. Biometricians of the caliber of Sir Ronald A. Fisher and J. B. S. Haldane in England, Sewall Wright in the United States, and S. S. Chetverikov in Russia turned their attention to evolution. Mathematical models were developed that predicted the effects of selection, migration, inbreeding, and mutation on changing the genetic composition of populations. A solid theoretical foundation for a unified concept of evolution began to emerge.

Neo-Darwinism (The Modern Synthesis)

Theodosius Dobzhansky (Figure 2.20) is generally acknowledged as among the first to review and synthesize observations and experiments involving the evolution of Mendelian populations. In his book *Genetics and the Origin of Species* (1937), he presented the early work on chromosomal studies of *Drosophila* populations and sharpened the biological species concept. This book demonstrated how to close the gap between theory and observation.

About 1940, Richard B. Goldschmidt reintroduced the idea of large genetic changes (systemic mutations) being more important in species formation than the slow accumulation of many small variations postulated by Darwin. Goldschmidt did not gain many supporters for his "hopeful monster" theory, but he did make many valuable contributions in the fields of physiological and developmental genetics. A systematic ornithologist, Ernst Mayr, was a leader in the application of taxonomic knowledge to the problems of evolution. Detailed studies in the plant genus *Crepis* by E. B. Babcock lent much botanical support to the neo-Darwinian theory. Edgar Anderson introduced the concept of introgressive hybridization as an important evolutionary principle. G. L. Stebbins wrote a definitive work entitled

Figure 2.20. Theodosius Dobzhansky (1900–1975).

Variation and Evolution in Plants (1950). James Watson and Francis H. C. Crick discovered the structure of DNA in 1953. The genetic code was cracked in 1961 by Marshall Nirenberg and J. H. Matthaei. Knowledge applicable to the solution of problems in evolution continues to be in a rapid state of development.

Natural selection is still at the heart of neo-Darwinism, but now we understand both the physical and chemical bases of heredity, we know how new biological variations are produced through mutations, and we have numerous mathematical models that predict the genetic changes expected in populations by various evolutionary processes. The study of evolution has unquestionably become a rigorous scientific discipline.

Essay Questions

1. *What is the essential logic behind Darwin's theory of evolution by natural selection?*

2. *Each step in the progress of science draws on the accumulated knowledge and ideas of earlier times. To whom is Charles Darwin indebted for his natural selection theory?*

3. *Besides that of natural selection, explain two other major mechanistic theories of evolution.*

4. *In what respects does neo-Darwinism differ from Darwinism?*

5. *How has evolutionary theory been affected by the theories of geology?*

ORIGIN OF LIFE

Speculations concerning the origin of life have revolved about two basic ideas: (1) life is produced from nonliving components of the environment by natural processes (the theory of *spontaneous generation*), and (2) life is produced by supernatural (vitalistic) powers (the theory of special creation). As explained in Chapter One, the latter theory is outside the realm of science to investigate. Life may have had a singular origin or it may have had multiple origins at approximately the same or at quite different times, in approximately the same location or in quite different locations. All that is necessary, however, for a coherent scientific theory of evolution is evidence that life could have been produced at least once through natural processes.

Aristotle (384–322 B.C.) is generally credited as the one who was most influential during early times in stimulating philosophical discussions of the spontaneous generation theory. It seemed obvious from his casual observations that maggots were naturally produced by meat left out in the open or that muddy puddles of water could bring forth frogs. Aristotelian philosophy remained unchallenged by experimental science until the seventeenth century.

Italian physician Francesco Redi (1626–1697) is commonly acknowledged as the first to marshal experimental evidence against the concept of spontaneous generation. Two pieces of meat were set out; one was covered with fine mesh muslin and the other was left open to the air as a control. After several days, fly larvae (maggots) appeared in the control, but none appeared in the covered meat. Redi concluded that flies are not spontaneously generated by the meat itself, but rather are produced by adult flies that lay their eggs on unprotected meat. Apparently this experiment did not change Redi's belief in the spontaneous origin of other insects.

Grinding lenses was a favorite hobby for Dutch merchant Antony van Leeuwenhoek (1632–1723). His skill was better than that of anyone else at that time. Consequently, he was able to observe in his single lens microscopes what no one else had ever seen—microorganisms (protozoans, bacteria, yeasts, fungi, sperms) that he called "animalcules" and "wee beasties." He found them to be ubiquitous—in pond water, in decaying matter, and in scrapings from his teeth. As the notion of spontaneous generation of insects and other relatively large forms of life visible to the naked eye began to wane, the idea that Leeuwenhoek's tiny creatures might be spontaneously generated became fashionable. During the eighteenth and nineteenth centuries, many attempts were made to test this theory in microorganisms (by Joblet, Needham, and Spallanzani), but the results were inconclusive. Boiled broth sealed in hot containers sometimes failed to produce any signs of life. In these cases the vitalists claimed that sterilization temperatures destroyed the life spirit present in air.

Evidence against the spontaneous origin of microorganisms was provided in the 1860s by a series of elegant experiments designed by French chemist Louis Pasteur (1822–1895). One of his most famous experiments involved drawing out the neck of a flask into a long sine curve (Figure 3.1). The end was left open to the air. Microorganisms on dust particles settled in the bottom of the curved neck and could not reach the boiled broth in the flask; the broth remained clear. When the neck was broken near the top of the flask, dust particles could settle directly into the broth and soon the broth became turbid with the growth of millions of microorganisms. The Pasteur experiments did not disprove the concept of spontaneous generation, although many individuals choose to interpret his results as such. Rather, his work

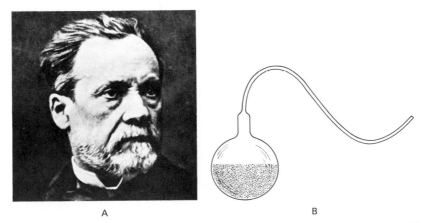

A   B

Figure 3.1. (A) Louis Pasteur (1822–1895). (From *Louis Pasteur—The Man and His Theories,* Hilaire Cuny, Paul S. Eriksson, Inc., 1963). (B) Diagram of one of the flasks used by Pasteur in his experiments on spontaneous generation.

invalidated the incorrect interpretation of spontaneous generation resulting from the "sloppy" experiments of other scientists. He merely demonstrated a failure to find contemporary cells arising from his special broth under certain conditions and over a relatively small observation period. Pasteur's experiments are seen now to lack relevance to the question of the origin of life, because he failed to employ primitive environmental conditions.

Whence, then, did the first cells arise? Some scientists clung to the theory of *panspermia, viz.* that life originated elsewhere in the universe and traveled through space to colonize the earth. This theory, however, still fails to answer the riddle of the origin of life; it merely transfers the problem to some other planet.

The idea of a materialistic origin for living systems was revived in 1936 by Russian biochemist A. I. Oparin (Figure 3.2) in his book *The Origin of Life on Earth.* Oparin called attention to the fact that environmental conditions on earth were quite different when living systems were being formed. The atmosphere at that time very likely contained little, if any, oxygen, but was rich in reducing gases such as ammonia, methane, and hydrogen. Oparin proposed that the carbon (organic) compounds from which contemporary life is built were produced by the energy of sunlight, electrical discharge, and volcanic heat acting on components of the primitive atmosphere. Inexplicably, Oparin's hypothesis waited almost thirty years for experimental confirmation. Finally in 1953, Stanley Miller subjected four gases (hydrogen, methane, ammonia, and water vapor) to the energy of an electric spark and circulated the mixture in a closed system through a boiling water flask (Figure 3.3). He succeeded in producing several organic compounds including amino acids (nitrogen-containing organic building blocks of proteins). No purines or pyrimidines (essential constituents of the building blocks of nucleic acids) or other heterocyclic or aromatic (ring) compounds were produced in the initial experiments, partly because the gas mixture was too rich in hydrogen. This experiment did not adequately represent geological reality, because the primitive earth atmosphere probably contained little free hydrogen. Subsequent experiments by Miller and

Figure 3.2. Aleksandr I. Oparin (born 1894). (From *Molecular Evolution and the Origin of Life* by Sidney W. Fox and Klaus Dose. W. H. Freeman and Company. Copyright © 1972)

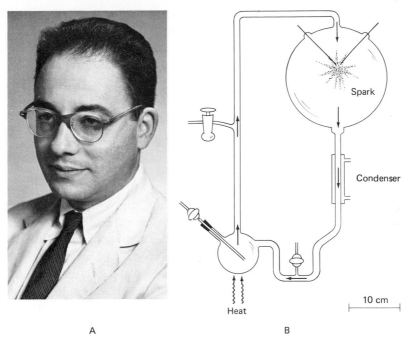

A   B

Figure 3.3. (A) Stanley L. Miller and (B) his apparatus for synthesizing organic molecules under presumptive primitive earth conditions.

many others took this into consideration and succeeded in producing a host of heterocyclic compounds (adanine, uracil, porphyrins, ATP, etc.) as well as numerous aliphatic (straight carbon chain) compounds (formaldehyde, urea, deoxyribose, ribose, etc.). Research along these lines is greatly increasing at present as scientists continue to search for a mechanistic explanation of *protobiogenesis* (origin of life).

Cosmology

Cosmology is the branch of philosophy that considers the nature and governing principles of the universe. *Cosmogony* refers to the origin of the universe and theories applicable thereto. Albert Einstein discovered mathematically that mass and energy are interconvertible. However, science cannot account for the primary origin of matter or energy. Materialists choose to believe that matter and/or energy have always existed. Deists choose to believe that matter and energy (as we know them) were brought into existence *de novo* by a supernatural act of creation. Science simply acknowledges that a material universe exists and that it is endowed with certain invariate natural laws that have been revealed through scientific methodology.

Two major hypotheses are currently popular in the explanation for evolution of matter in the universe: (1) the "big-bang" hypothesis and (2) the "steady-state" hypothesis. According to the "big-bang" hypothesis, the universe originally consisted of a relatively small, dense mass of protons, electrons, and neutrons that became unstable at temperatures in the billions of degrees as it condensed under its own gravitational field and finally exploded. The "big bang" very quickly produced nuclei of the lighter chemical elements. As these elements cooled, some of them gradually condensed to form large diffuse protostars. Large gravitational fields began to form and generate heat in the central regions. Hydrogen atoms began to lose their electrons (ionize), and at temperatures in excess of ten million degrees Centigrade collisions between hydrogen nuclei formed helium nuclei (alpha particles). Each fusion reaction released a great quantity of energy. During most of the life of a star, an equilibrium exists wherein the production of thermonuclear energy is balanced by energy loss through radiation. Toward the end of a star's life, its outer parts expand as it enters the "white dwarf" stage, possibly ending with an explosion as a nova.

One method for estimating the time of the "big bang" employs a principle of light (called the *"redshift"*) akin to a principle of sound (the *Doppler effect*). The Doppler effect is the familiar phenomenon of a lowered pitch in sound as an audible object moves away from us. For example, the high pitch of a train whistle abruptly becomes lower as it passes us. In like manner, the wavelengths of light become lengthened as a star recedes from us. Spectral analysis of the light (and radio waves) from distant galaxies shows a shift toward the red end of the spectrum (longer wave lengths). Higher velocities of recession are theoretically correlated with larger "redshifts." By dividing the average distance between neighboring galaxies by the velocity of their recession, estimates for the age of the universe in the neighborhood of ten billion years have been obtained. The most distant object known in our universe is a quasar (a small galaxy emitting tremendous quantities of radio energy) ten billion light years (or more) away. Accordingly, the universe must be at least ten billion years old; latest estimates place the "big-bang" event at about 15 billion years.

The "steady-state" hypothesis assumes a universe without beginning or end, and one in which the density remains constant. The size of the universe seems to be expanding. The apparent recession of distant galaxies from the observable part of our universe leaves a material void, which some astrophysicists feel compelled to fill with newly created matter (neutrons). While some galaxies are dying, others are being born so that the major features of the universe remain relatively unchanged. This theory is contradicted by modern astrophysical and astronomical data and has relatively few supporters. There is some evidence that the rate of expansion of the universe is slowing down. It has been suggested that at some distant time the universe may cease expanding and enter a contraction phase, eventually arriving at the superdense condition postulated as the initial state in which the "big bang" occurred. This concept of a *pulsating universe* is in keeping with the evolutionary view. However, the data from several lines of investigation (age of stars, mass of galaxies, abundance of chemical elements, and observed rate of expansion of the universe) can be interpreted by a cosmological model of an infinite universe that will expand forever. The weight of the evidence presently appears to strongly support this concept.

Modern particle physics theory (symmetry of elementary particles) requires that each piece of matter in the universe has a corresponding piece of antimatter. In 1932 the first piece of antimatter was discovered. This was the positron (a particle with the same mass as an electron, but with a charge of $+1$ instead of -1). The antiproton (a particle with the same mass as a proton, but with a negative charge rather than a positive charge) was first detected in 1955. When matter and antimatter collide, they annihilate each other with the release of gamma radiation. Newly discovered pulses of gamma radiation and radio waves are hypothesized to be emanating from the boundaries between matter and antimatter in the cosmos. It is difficult to visualize how matter and antimatter could both have resulted from the "big bang," because collisions, inevitable in their initial state of close proximity, should have annihilated both, leaving only a universe of gamma rays. Therefore, another theory has been proposed that allows for antimatter in the universe and also accounts for the "redshift" phenomenon. This theory assumes that a primordial mass of gas was rather uniformly distributed over an immense space. The gas began to condense under its own gravitation into galaxies of compatible matter. A mechanism has been proposed that uses some of the energy of annihilation to separate matter from antimatter, but the details are too involved for our purposes here. The contraction phase continued to force the galaxies toward a central region (perhaps a billion light years in diameter). Because all galaxies do not move at exactly the same speed, it is possible that they could have passed this center at different times. They could have passed through this center about 10 billion years ago and emerged along hyperbolic paths that took them to their present positions. Further discussion of cosmological problems are deferred to advanced treatises on the subject.

Origin of the Earth

Our sun is a relatively young star, approximately five billion years old. The earth is also estimated to be 4.5 to 5 billion years old. It is believed that most, if not all, of the heavier elements found on the sun and elsewhere in our solar system were not generated by the sun, but rather have been acquired from another source. The iron/silicon ratio is so different between the earth and the sun that a solar origin for the earth is seriously questioned. No agreement exists in the scientific community regarding the origin of the earth and other planets in our solar system. The most popular hypotheses today favor a cold initial state of aggregation. Condensation of clouds of gas and dust or of nonvolatile stony matter (perhaps the size of asteroids) eventually formed sufficient mass to generate much heat through gravitational energy. Most contemporary geologists, however, are convinced that the temperature at the surface of the earth was never high enough to cause a worldwide melting (greater than 900°C) of the crust. It is very possible that temperatures were sufficient to volatilize into space lighter elements of the atmosphere such as hydrogen and helium. The retention of relatively dense atmospheres on larger planets such as Saturn and Jupiter and the loss of an atmosphere from smaller planets such as Mercury and our moon are attributed to differential temperatures and gravitational fields.

The average ratio of carbon/silicon in the universe is more than 1000 times greater than that on earth. Presumably a major portion of proto-earth carbon was in the gaseous forms of methane (CH_4), ethane (C_2H_6), formaldehyde ($H_2C{=}O$), carbon monoxide (CO), carbon dioxide (CO_2), hydrogen cyanide (HCN), etc. and was lost from the primitive atmosphere. It is estimated that only 10 per cent of the water of contemporary earth was initially present; 90 per cent has been subsequently supplied by outgassings from the interior (volcanoes, geysers, fumaroles). When the temperature of the earth's crust dropped below the boiling point of water, atmospheric water vapor condensed into torrential rains, which revolatilized upon reaching the hot crust in a vigorous water cycle. Eventually the crust cooled sufficiently to allow liquid water to accumulate in the first oceans. It is likely that the pH of the oceans was slightly alkaline (pH 8) during most of their history. It is highly unlikely that biochemical compounds in the primitive oceans ever reached the consistency of a "thick soup" as so many "popular science" books intimate.

Localized aqueous concentrations of organic matter are conceivable by several mechanisms including regression of the ocean, evaporation, density gradients in the tropical thermocline, adsorption onto minerals, etc. In addition, it is possible for chemical reactions to occur in the molten state, thus avoiding the dilution effects of aqueous solutions. Volcanic activity must have been much more extensive early in the earth's history than at present. After the earth cooled sufficiently to retain gasses in its atmosphere, volcanoes supplied a new set of gasses probably including CO_2, nitrogen (N_2), hydrogen sulfide (H_2S), CH_4, CO, and ammonia (NH_3).

There is now general agreement that the primitive atmosphere was relatively free of oxygen; it was a strongly *reducing atmosphere*. The present level of oxygen in air (approximately 21 per cent at sea level) has been generated by photosynthesis after its evolution in biological systems (Figure 3.4). It was formerly thought that carbon dioxide was produced almost exclusively by the respiration of organisms. Most authorities now believe that both carbon dioxide and carbon monoxide were constituents of the primitive atmosphere and hydrosphere. Carbon monoxide reacts with water at high temperatures (greater than $1200°K$) to produce carbon dioxide and hydrogen. Carbon monoxide also reacts with water to produce formic acid. Because of its chemical reactivity, carbon monoxide very likely participated rapidly in the formation of prebiotic compounds. Ozone in the modern atmosphere prevents

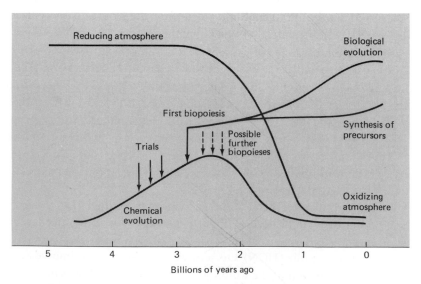

Figure 3.4. Diagram of some important events in the origin of life.

penetration by solar ultraviolet light (wavelengths less than 3200 Å), but in the oxygen-poor primitive atmosphere there was virtually no protection from ultraviolet radiation.

Visible light is not absorbed by simple organic molecules, so it is doubtful that this source of energy was involved in photochemistry of the primitive atmosphere. Visible light may have been important at a later stage when organic compounds were in aqueous systems together with minerals that do absorb light. Although other forms of energy (ionizing radiations, electrical discharges) were undoubtedly available for driving chemical reactions on the primitive earth, infrared radiations (heat) presumably played the major role in producing prebiotic molecules. Only heat can occur in quantities small enough for the synthesis of macromolecules, yet also occur in quanta large enough (greater than 500°C) to produce effects comparable to those of more violent forms of energy. Thus, the quality of energy represented by heat together with its widespread availability on the primitive earth are important attributes for prebiotic syntheses.

Evolution of Organic Substances

Creationists continually refer to the *laws of thermodynamics* in their arguments against a natural origin for living systems. The *First Law of Thermodynamics,* sometimes called the *Law of Conservation of Energy,* states that energy can be transformed from one kind to another, but it can neither be created nor destroyed. Since matter and energy have been shown to be interconvertible, the First Law can be modified to state that neither matter nor energy can be created or destroyed. The *Second Law of Thermodynamics* states that in converting one form of energy to another, some of it is lost as unusable heat. *Entropy* is the thermodynamic quality of randomness or disorder within a system. The Second Law therefore implies that as energy is being transformed throughout the universe, entropy is increasing. These Laws argue strongly for a created universe. The nonrandomness of the newborn universe has apparently been inexorably degrading since that event. Presumably the natural end state will be a heat equilibrium wherein entropy is maximized. However, the Second Law of Thermodynamics applies specifically to closed systems. *Closed systems are uncommon in nature!* Thus, the conclusion that this Law offers proof that biomacromolecules could not be synthesized through abiotic natural processes is scientifically unacceptable. Early experiments in abiotic synthesis of relevant organic molecules met with limited success, until it was realized that hydrogen normally escapes from the earth (an open system). When provision was made to remove hydrogen from the reaction vessel as it was formed, reducing conditions were maintained and a multitude of biologically important molecules appeared.

Amino acids are among the most easily produced micromolecules in the laboratory under presumed primitive earth conditions (gaseous or aqueous). Eighteen of the twenty amino acids found in contemporary proteins (cystine and cysteine excepted) have been thus produced. Furthermore, these amino acids are usually the same types of alpha-amino acids found in organisms. Quantitatively, the most common amino acids in proteins (such as glycine, alanine, aspartic acid, and glutamic acid) are also those most commonly produced in these laboratory experiments. All proteins today consist entirely of amino acids that rotate plane-polarized light to the left (L forms). However, none of the prebiotic compounds synthesized in the laboratory under primitive earth conditions have been reported to have optical activity because of racemization (a mixture of D and L forms). The experiments with the greatest evolutionary plausibility are ones that yield a family of related molecules by virtue of common intermediates in multiple pathways.

Polymerizations

Violent forms of energy (such as high energy atomic radiation, electrical discharge, ultraviolet light, and high temperatures) usually result in the decomposition of proteins rather than in their synthesis. Organic compounds are generally less stable in the presence of water. With this knowledge, many successful polymerization experiments have been conducted under anhydrous conditions using relatively mild heat as a source of energy. A molecule of water is released by the formation of a *peptide bond* involving the linkage of one carboxyl (COOH) group of one amino acid to the amino group (NH_2) of the next. This water is usually bound and not free to act as solvent water of the kind that enhances denaturation. The resulting chain of amino acids is called a *thermal protenoid,* i.e., a macromolecular preparation of mean molecular weight in the thousands containing most of the twenty amino acids found in contemporary proteins. Condensation at lower temperatures (130°C) with polyphosphoric acid has resulted in the inclusion of serine, threonine, cysteine, and cystine in the thermal protenoid, whereas at higher temperatures these amino acids are usually destroyed.

Creationists have repeatedly challenged evolutionists with the admonition that order does not naturally flow from disorder (Second Law of Thermodynamics). Experience in the production of thermal protenoids, however, reveals that their composition is not merely a function of the initial concentration of amino acids, but is decidedly influenced by other chemicals such as phosphoric acid. In addition, there is a tendency for one amino acid to be incorporated into a proteinoid as a function of the kinds and amounts of other amino acids in the reaction mixture. Moreover, the arrangement of amino acids in proteinoids have been shown to be organized (nonrandom), presumably through preferential formation of thermo-dynamically stable configurations. The degree of heterogeneity exhibited by thermal proteinoids arising from a given starting mix of amino acids is very sharply limited, i.e., they are highly uniform in composition. Therefore, it appears that nucleic acids were not required to specify specific amino acid sequences in predecessors of modern proteins (protoproteins). The molecular weights of most proteinoids tends to fall near the lower end of weights for contemporary proteins (4,000–10,000 daltons). Those with the lowest molecular weights are rich in acidic amino acids, whereas those with the highest molecular weights are basic proteinoids. For example, copolylysines have average molecular weights much greater than 10,000. Evidence that peptide linkages (as in proteins) exist in proteinoids is derived from positive biuret tests, infrared spectra, and enzymatic tests. The ninhydrin test applied to hydrolyzed proteinoids is positive, indicating free alpha-amino acids of the kind expected from proteins. Some of these proteinoids have catalytic (rate-enhancing) activities in reactions classified as hydrolysis, decarboxylation, amination, deamination, and peroxidation (oxidoreduction). The specific activity of proteinoids appears to increase approximately in proportion to their molecular weights. Larger molecules are more likely to be retained by cellular membranes. Thus, high molecular weight polymers with catalytic activity would have been advantageous for evolution. An early proteinoid may have been able to catalyze several reactions, albeit with reduced efficiency. Perhaps metabolism could have evolved much more effectively from a relatively nonspecific proteinoid than from a highly specific contemporary enzyme. A thermal proteinoid with some of the hormonal activity of melanocyte-stimulating hormone (MSH) has been produced. Several other properties are shared by proteinoids and contemporary proteins (e.g., range of solubilities, lipid qualities, nutritive qualities, etc.).

It is also possible that proteins could have been formed on the primitive earth without the intervention of amino acids. When methane and ammonia are subjected to ultraviolet light or any high-energy source in liquid water, hydrogen cyanide is produced together with a messy brown compound. Upon degradation, the latter yields amino acids. This model of protein evolution has two virtues: (1) it allows the linkage of amino acids without the assistance of enzymes (which are themselves proteins), and (2) it is possible to visualize the widespread occurrence of protein-like material, some of which may have assisted catalytically in the formation of nucleic acids.

Only limited success has been obtained with thermal polymerization of nucleotides, the monomers ("building blocks") of nucleic acids. To date, oligomers (groups of fewer than ten monomers) of cytidylic acid are the only ones that show the 3′–5′ phosphodiester linkages characteristic of contemporary nucleic acids (DNA, RNA). The ease with which proteinoids can be produced relative to polynucleotides suggests a "proteins-first" hypothesis. The progenitor of contemporary cells (protocell) could have survived with protein alone; it could not have survived with nucleic acids alone. A genetic system involving nucleic acids as molecules of stored information for making specific proteins could have been added to the protocell at a later time.

Protocell Models

In 1956 F. O. Schmitt demonstrated that collagen (the major structural protein of the extracellular fibers of connective tissue) on precipitation from solution spontaneously assembles itself into fibers. This was probably the first experimental evidence that macromolecules have the innate property of self-assembly. More recent experiments have shown that viral components (heads, tails, tail fibers) will assemble themselves spontaneously, under appropriate conditions, into fully infective virus particles.

As early as 1930, a protocell model became available in *coacervate droplets*. These droplets are usually produced by mixing solutions of oppositely charged colloids (e.g., gum arabic and gelatin). The concentration of organic polymer in a coacervate droplet is much greater (sometimes a hundredfold) than that in the aqueous phase. These droplets represent open systems that passively concentrate materials from their environment. If an enzyme is added to the reaction mixture together with monomer substrate, polymers are both formed and retained within the droplet. For example, the inclusion of phosphorylase enzyme in a mixture of histone and gum arabic colloidal suspensions converts glucose-1-phosphate into starch, causing the droplet to increase 50 per cent in weight and 150 per cent in volume. Modern biochemical reactions typically occur in association with membranes or other surfaces within cells. The availability of surfaces (as in droplets) may have provided a selective advantage to protoenzymes in enhancing their catalytic activity. Coacervate droplets tend to disintegrate with time. This lack of stability detracts from their use as a protocell model.

A much better protocell model exists in *proteinoid microspheres* (Figure 3.5). These microscopic spherical structures are easily produced by adding water to thermal proteinoids. Most of them appear similar to coccoid bacteria and may even tend to form chains of varying length similar to the bacteria called streptococci. Microspheres tend to be highly uniform in size, suggesting that innate control mechanisms are involved in the processes of aggregation and disaggregation. Unlike coacervate droplets, proteinoid microspheres are quite stable structures. Microspheres produced from polynucleotides in association with proteinoids exhibit even greater stability (at high pH or in boiling water) than those from proteinoids alone. Perhaps the initial associations between nucleic acids and proteins were selected because of the stabilizing influence the former confers on the latter.

Electron micrographs of proteinoid microspheres reveal a double-layered boundary suggestive of contemporary cell-membrane structure, albeit two to four times as thick. Bacteria are broadly classified into two groups on the basis of ability to take up and retain Gram stain. Both Gram positive and Gram negative microspheres have been produced. As with soap or oil droplets, acidic proteinoid microspheres tend to divide in two by surface tension forces (Figure 3.6). Some microspheres tend to divide in gross appearance similar to

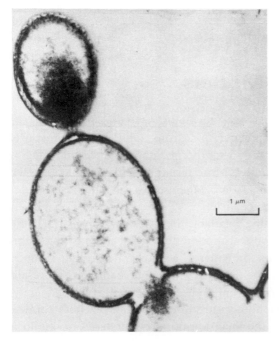

1 μm

Figure 3.5. Electron micrograph of a section of osmium tetroxide-stained proteinoid microsphere after pH has been elevated. Double layer in boundary is prominent. (From *Molecular Evolution and the Origin of Life* by Sidney W. Fox and Klaus Dose. W. H. Freeman and Company. Copyright © 1972)

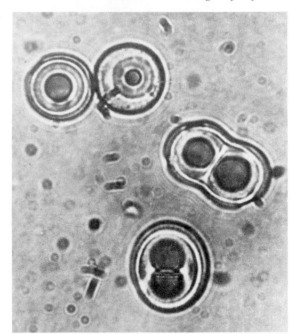

Figure 3.6. A model of incipient binary fission of cells seen in proteinoid microspheres. Note the various stages in the formation of septa. (From *Molecular Evolution and the Origin of Life* by Sidney W. Fox and Klaus Dose. W. H. Freeman and Company. Copyright © 1972)

budding in yeast (Figure 3.7). The buds can be released from the parent spheroids by electrical shock or heat. Furthermore, the first generation of buds can grow by accretion of proteinoids from the environment and proliferate a second generation of buds.

Microspheres have been discovered that exhibit nonrandom motility; this phenomenon can be enhanced by the addition of adenosine triphosphate (ATP). They can be seen to repel each other in some cases and to join together in other instances. The connections appear to be hollow because internal structures (endomicroparticles) have been observed to migrate from one microsphere to another across the junctions. Thus, information could be communicated from one part to another within an aggregate of microspheres. Upon transfer from isotonic solutions to hypertonic solutions, the microspheres shrink; transfer to hypotonic solutions results in swelling. This behavior simulates osmosis without any lipid being in the system. Hydrophobic side chains of amino acids such as phenylalanine and leucine endow proteinoids with some lipid qualities. Therefore, although lipids are a major constituent of contemporary cell membranes, they need not have been present in protocells. Some proteinoid microspheres seem to selectively retain polysaccharides while allowing simple sugars (monosaccharides) to diffuse out freely. The enzyme-like activities of proteinoids are retained by the microspheres.

In summary, it appears that coacervate droplets are poor models for protocells because (1) they are produced from mixtures of contemporary bioproteins, and (2) they are notoriously unstable structures. On the other hand, microspheres appear to be excellent models for protocells because (1) they originate from proteinoids produced under the natural conditions attributable to the primitive earth, and (2) they exhibit many of the structural and functional attributes of contemporary cells.

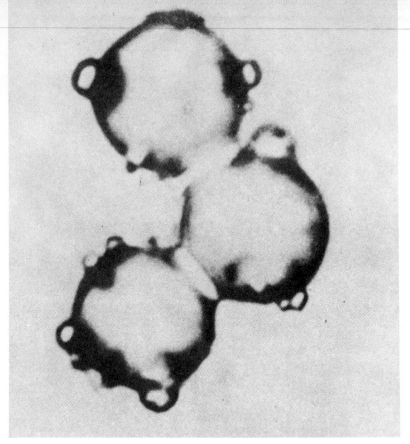

Figure 3.7. Buds on proteinoid microspheres. (From *Molecular Evolution and the Origin of Life* by Sidney W. Fox and Klaus Dose. W. H. Freeman and Company. Copyright © 1972)

Later Embellishments

The monomers of proteins and nucleic acids are known to interact with each other. For example, during protein synthesis in contemporary cells, amino acids become "activated" by ATP, forming amino-acyl adenylates (amino acid plus adenosine plus phosphate). Adenine is a purine base found in nucleic acids (DNA, RNA), as well as in the energy-storage compound adenosine triphosphate (ATP). It is therefore possible that information could have been passed from proteins to nucleic acids or vice versa. Experimental evidence shows that some degree of specificity exists in the shapes of particles (fibrous or globular) produced by various proteinoids and contemporary nucleic acids. Specificity in the composition of microparticles has also been shown to be dependent on the kinds of amino acids in the proteinoid or the kinds of purines or pyrimidines in synthetic nucleic acids. The globular nucleoproteinoid microparticle appears promising as a model for an early ribosome ("protoribosome"). As stated previously, the more stable multifunctional systems probably had a selective advantage early in molecular evolution. Therefore, the protoribosome may have been selectively evolved because of the stabilizing influence nucleic acids confer on proteins and because it had the ability to perform both messenger and transfer functions. A genetic system, specialized for the storage and retrieval of macromolecular information, may have formed relatively late in molecular evolution.

Protocells were *heterotrophic* ("other nourished"), i.e., obtained complex organic nutrients from their surroundings. As the more complex nutrients began to disappear from the environment, those protocells that had metabolic pathways that could use less complex nutrients probably enjoyed a selective advantage. They were able to grow and multiply (reproduce), while protocells without these earlier pathways stopped growing and multiplying. They need not have disappeared in comparable fashion to the death of living cells. Remember, under primitive earth conditions, there was neither oxygen nor "organisms" to cause their degradation. Even today, relatively simple organisms such as yeast can use the energy in sugar under anaerobic (oxygen free) conditions by a process called fermentation.

$$C_6H_{12}O_6 \xrightarrow{\text{enzymes}} 2CO_2 + C_2H_5OH + 28 \text{ Kcal}$$

| *sugar* | *carbon dioxide* | *ethyl alcohol* | *energy* |

A great evolutionary advance was made when the protocell appeared with the capacity to trap energy released by such metabolic processes. Contemporary cells store usable energy in the form of terminal phosphate bonds in ATP. However, the greatest advance in gaining independence from the environment was made with the advent of a metabolic system that could trap the energy from the sun and use it to generate complex molecules from simpler ones. This would be an *autotrophic* (*prototrophic*) system ("self-nourishing") akin to the photosynthetic system of plants.

$$6CO_2 + 6H_2O + \text{light} \xrightarrow[\text{chlorophyll}]{\text{enzymes}} C_6H_{12}O_6 + 6O_2$$

The chlorophyll molecule, which is the green photosynthetic pigment of plants, consists of a complex porphyrin ring compound and a relatively simple phytol chain. Pyrroles, the most likely precursors of porphyrins, have been produced by a variety of experiments with prebiotic compounds (e.g., passing acetylene and ammonia through a heated tube). Porphyrin-like materials have been experimentally condensed, under geologically relevant conditions, from pyrrole. Porphyrin-dependent photosynthesis evolved early in living systems as evidenced by fossil vanadium-porphyrin complexes dated at least two billion years old. Chlorophyll is a magnesium-containing porphyrin. Heme, part of the oxygen transport molecule in animals called hemoglobin and part of the cytochrome pigments involved in the electron transport system of oxidative metabolism, is an iron-containing porphyrin. After sufficient oxygen was released to the atmosphere by the photosynthetic activity of plants, a porphyrin-dependent respiration could have developed. Animals appear in the fossil record 600 million years ago, attesting to the antiquity of this process.

Eventually the mechanisms for synthesis of both proteins and nucleic acids moved internally. A genetic system probably evolved from refinements of the primitive recognition shown by laboratory experiments to exist between proteins and nucleic acids. Compartmentalization and specialization of function within the protocell was very likely favorable to the increasing complexity evolving within living systems. Organelles such as nuclei, mitochondria, and chloroplasts need not have been present in early cells. Indeed, present-day *prokaryotes* (bacteria and blue-green algae) get along very well without these membrane-bound subcellular particles.

Biologists are not uniformly in agreement on the criteria necessary to define life. Among the more important factors are the following:

1. Metabolism—the ability to extract energy from the environment and use it in the synthesis of biological molecules for maintenance, repair, growth, and reproduction.
2. Irritability—the ability to perceive changes in the environment (stimuli) and to make adaptive responses to these stimuli.
3. Homeostasis—the ability to regulate physiological processes for the maintenance of a relatively stable internal milieu.
4. Reproduction—the capacity to transmit macromolecular information from one discrete chemical system to another such that two units of similar or identical chemical composition are derived from a single unit.

At which point in time one chooses to recognize the existence of "life" on earth appears to be arbitrary. The attributes presently ascribed to the living state are viewed as having gradually accumulated by selection continually favoring the most chemically stable, the most environmentally independent, the most metabolically heterogeneous, the most energetically efficient, and the most highly reproductive systems. Each small increment of increase in biological complexity commonly generates a geometric increase in adaptive capabilities through cybernetic mechanisms (synergistic interactions, feedback systems, etc.). Thus, even minor modifications during organic evolution could have greatly enhanced the opportunities for exploitation of the environment.

Conclusion

In 1955 H. Fraenkel-Conrat and R. C. Williams startled the world with the news that a virus had been synthesized in the laboratory. They had fractioned tobacco mosaic virus (TMV) into its protein and RNA components. Neither protein nor RNA alone was capable of infection. On mixing these two components, however, protein coats spontaneously organized in a helical pattern around the spiraling RNA core to reconstitute fully infective TMV particles. Newspaper headlines proclaimed that "life had been created in the test tube." No such event had happened! Their experiment, however, revealed that self-assembly was an innate quality of biomacromolecules. Biologists are not all in agreement about the "living status" of viruses. For one thing, viruses normally reproduce only within living cells, commandeering the metabolic machinery of the cell for this process. Science is a long way from creating life, as we know it today. However, many experiments have been conducted, under geologically relevant conditions, which lend support to the theory of molecular evolution as a scientifically valid explanation for the origin of life. The proposed major steps in this process are shown in Figure 3.8.

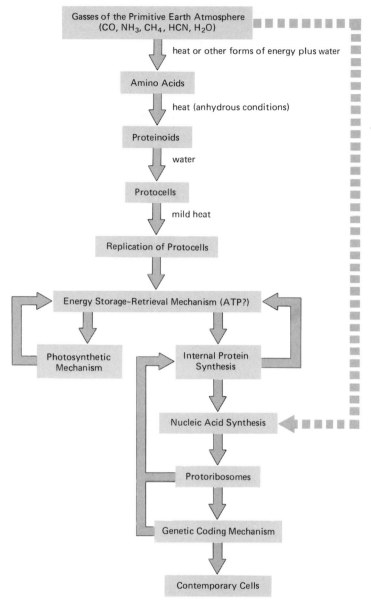

Figure 3.8. Outline of the major steps postulated for the origin of modern cells.

Essay Questions

1. *Outline the major historical events in the development of the theory of spontaneous generation.*

2. *According to most authorities at the present time, what was the condition of the*

   prebiotic earth with respect to lithosphere (crust), hydrosphere (water), and atmosphere (air)?

3. *Beginning with simple gases, explain the steps by which one can produce replicating protocells.*

4. *In what respects are proteinoid microspheres similar to contemporary cells?*

5. *At what point in the evolution of contemporary cells did life begin?*

EVIDENCE OF EVOLUTION— PALEONTOLOGY AND BIOGEOGRAPHY

To have scientific validity, a theory must be supported by observable factual evidence. The status of a scientific theory is directly correlated with its predictive value. Thus, before accepting a theory such as that of evolution, we want to examine its credentials. Let us look at both the quantitative and qualitative aspects of its supporting evidence, as well as at its predictive value. This chapter and the next present some major lines of evidence for organic evolution. Scattered through the rest of the text are examples of agreement between observations and expectations based on modern evolutionary theory.

Fossils

Have the species of organisms presently inhabiting the earth existed since the origin of life, or have entirely different species existed at different points in geological time? The answer to this question is to be found in the science of *paleontology* (the study of fossils). *Fossils* are the remains or traces of former life. Under present conditions, the remains of a dead organism are usually subject to rapid decomposition by bacteria and fungi. The soft parts of animals and plants would have little chance of being preserved as fossils. Only the hard (inorganic) parts of bones, teeth, and shells would be expected to survive in the fossil record. Furthermore, the conditions favoring preservation are far from universal. The cadaver should be covered quickly by dirt or sediment to prevent it from being eaten (and prevent its parts from being scattered) by scavengers and to exclude air, thereby retarding decomposition. The marine environment usually offers a greater opportunity for preservation of organic remains than does the terrestrial environment. Consequently, much of our knowledge concerning past life has been discovered in marine deposits.

The finding of fossil sea shells on the tops of present mountains indicates that at one time the mountains were below sea level, and that geological forces have subsequently thrust the mountains out of the ocean or that the level of the sea has regressed. Comparable conditions favoring preservation of terrestrial organisms would occur near bodies of water such as lakes or rivers. An animal drinking at a muddy water hole could become stuck and die there. Its remains might be covered by silt and some of its parts preserved in the resulting shale (Figure 4.1). Once the organism is trapped in the rock, it must be protected from erosion or fracturing by geological changes in the earth's crust. For example, once a marine deposit is elevated above sea level it is subject to wearing away by wind, ice, and water, and eventually the fossils are degraded, broken beyond recognition, and scattered among new deposits. The fossil record, therefore, does not constitute a random sample of prehistoric life. Rather, it is highly biased in the direction of certain organisms (i.e., those with hard parts) with particular ecological requirements (for marine or fresh water) that happen to live in geological zones relatively undisturbed since their formation.

Fossils can be preserved by several methods. Organic remains can become *petrified,* i.e., turned to stone (Figure 4.2). Dissolved minerals can infiltrate the air spaces of bones and shells and precipitate there without distorting the original shape; this is called *permineralization*. Petrified bones are usually produced by this process. Alternatively, as water dissolves the organic matter, it can be replaced by minerals such as silica, calcium carbonate, and iron pyrites; this process is termed *mineral replacement*. In some petrified woods this substitution has been so delicate that even its microscopic structure is preserved in detail (Figure 4.3).

Volatile organic matter may be removed by *distillation,* leaving a carbon residue revealing the limits of the soft parts. Leaves are commonly preserved in this fashion. Organic matter embedded in rock may be dissolved by percolating ground water, leaving a space called a *mold*. If the object is thin such as a leaf, the mold is called an *imprint* (Figure 4.4). Some imprints show veins and pores in leaves. Others show the fins of extinct fishes, the membranous wings of extinct flying reptiles, and the feathers of extinct primitive birds. If the mold is subsequently filled with other mineral matter (often quartz), it becomes a *natural cast* (Figure 4.5). An artificial cast can be made by filling a natural mold with plaster or plastic. Amber is the dried and hardened resin of ancient conifer trees. Some insects and spiders were trapped in amber as long as 25 million years ago. Their organic matter dried almost to nothing, leaving a natural mold so perfect that in some specimens even the delicate filaments on the antennae are visible. Footprints of gigantic dinosaurs and tiny holes of extinct burrowing worms are also preserved as fossils. Even fossil excrements, called *coprolites,* have been found that sometimes reveal the shape of the intestine (e.g., the spiral valve of sharks) or give clues as to the kinds of food ingested.

(1) An animal drinking at a water hole accidentally slips and drowns.

(2) The carcass is buried by silt and mud; the soft parts usually decay leaving a skeleton trapped in the sediments. Geological processes turn the sediments to shale or sandstone.

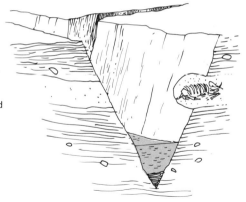

(3) Much later (perhaps millions of years) erosion cuts through the sediments and uncovers the fossil.

Figure 4.1. Three major steps in the formation and discovery of a fossil.

Figure 4.2. Petrified tree trunk section. Minerals such as calcium carbonate, silica, and iron sulfide react with the decomposition products of the wood to form insoluble materials, which help to prevent further decay of the cellular structure. Petrification leaves the wood with some organic matter remaining. Wood with silica infiltration can be cut very thin and prepared on a slide for microscopic structure of cellular details.

Entire organisms, including soft parts, are rarely preserved. At least one woolly rhinoceros with parts of the skin and flesh intact has been preserved in the oil seeps of Poland. The famous tar pits of Rancho LaBrea in southern California have trapped a great many animals and preserved their bones, because most of the bacteria that cause decay cannot live in asphalt (Figure 4.6). Several well-preserved specimens of the woolly mammoth have been found in the frozen earth of Siberia, the meat still edible if thawed quickly. In some instances, entire forests have been rapidly buried by volcanic ash and the upright trunks of the trees later preserved by petrification.

Figure 4.4. Impression-type fossil (*Metasequoia* branch). (Courtesy Carolina Biological Supply Company)

Figure 4.5. Cast-type fossil (fern). (Courtesy Carolina Biological Supply Company)

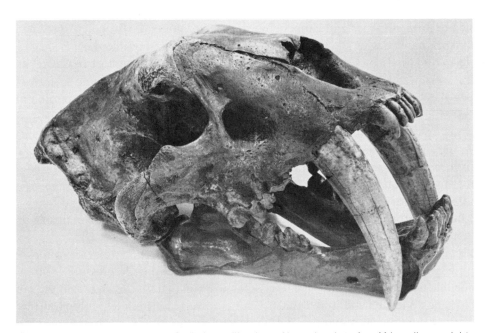

Figure 4.6. Saber-toothed cat, *Smilodon californicus*. About the size of an African lion; weight to about 500 pounds. From La Brea tar pits in Los Angeles County. Saber-tooths lived in Southern California until about 10,000 years ago, probably feeding on such thick-skinned animals as mastodons and ground sloths. The greatly developed canine teeth were used to stab and slash their prey.

Rocks

Three basic kinds of rocks are recognized by geologists: (1) igneous, (2) sedimentary, and (3) metamorphic. *Igneous* rocks are formed by the solidification of molten matter (magma) originating within the earth. These rocks are classified according to texture and mineral composition. The most common of the dark lavas are basalts. Granites are composed largely of the minerals quartz and feldspar. Obsidians are one kind of volcanic glass produced by rapid cooling of silicic magmas. Pumice is frothed glass. Many other kinds are recognized, but the important point for our purpose is the fact that relatively few fossils are found in rocks of igneous origin. As noted previously, volcanic ash sometimes kills and quickly buries plants and animals. Lava flows are known to have covered trees and quickly solidified without consuming all the wood. Alternatively, chunks of solidified lava torn from the crater walls are blown into the air and rain down as "sediments" to bury organisms. The coarser material, when converted into rock, is termed volcanic breccia; the finer material is called tuff. The magnificent fossil forests of Yellowstone National Park were preserved by being buried in tuff and volcanic breccia.

Sedimentary rocks can be formed (1) by deposition of fragments of older rocks transported by wind, water, or ice, or (2) by precipitation of materials in aqueous solution or by settling of shells of aquatic organisms. Conglomerate consists of gravel or pebbles cemented together. Sandstone consists of cemented sand grains. Particles finer than sand can become cemented to form siltstones, mudstones, and shale. Limestones are formed by precipitation of calcium carbonate ($CaCO_3$) from solution or may be composed of microscopic shells. Weakly cemented white limestones are commonly called chalks. Marl is a mixture of $CaCO_3$ and clay. Wind-blown sands are not conducive to fossilization, because oxygen and water can easily penetrate them to enhance decay and dissolution. Nonetheless, some nests of dinosaur eggs found in Mongolia were apparently preserved by drifting sands. The great majority of all fossils are preserved in water-borne sediments. Flooding streams often kill and bury organisms in the shifting channel sands and in mud or clay on the valley floor. Because of the vast extent of oceanic waters, the sea bottom is the richest source of fossils. Marine formations are seldom unfossiliferous. Fossils are usually less common in pure sandstone and most abundant in calcareous shales and limestones. Footprints have been found in red sandstones and red shales, but few other fossils are preserved therein because the red color indicates complete oxidation (of iron), and oxidation destroys organic compounds.

Metamorphic rocks result from the transformation of older, pre-existing rocks by the deforming and recrystallizing forces of high temperature and/or pressure. Marble is the metamorphic derivative of limestone. Glassy appearing quartzites result from the metamorphism of sandstones. Fine-textured shales can be converted into slates with parallel arrangement of their component minerals. The tendency to split parallel to these planes is termed *foliation*. Gneisses are coarse-grained, imperfectly foliated rocks. Schists are visibly crystalline to the unaided eye; their thinly spaced foliation is not plane-parallel. Metamorphic rocks may show blurred traces of fossils (as in many marbles), but no remnants remain if crystalization is complete.

Geological Dating

Archbishop Ussher in the seventeenth century calculated from Biblical lineages that the earth was created in the year 4004 B.C. The concept of a young earth was still in vogue during the nineteenth century. Sir Charles Lyell's (Figure 4.7) *Principles of Geology,* first published in 1830, supported Hutton's principle of uniformitarianism with observations and calculations that argued strongly for a very long geological history. About 1900, most geologists accepted the earth's age as approximately fifty million years; by 1960 the estimate had grown to about five billion years. The oldest meteorite has been dated by radiometric analysis at 4.6 billion years. The oldest rocks on the earth's crust date at about four billion years. To account for these age differences, it has been proposed that some catastrophe destroyed rocks older than four billion years. The old fight between the "catastrophists" and the "uniformitarians" seems to be resolving into a modern geological theory in which elements of both play a part.

The major intervals of geological time, called *eras* (Table 4.1), are separated on the basis of geological *revolutions,* i.e., times of great crustal movements characterized by the formation of mountain ranges, radical changes in sea levels, the widespread extinction of many species, and the sudden appearance of many new species. Geological revolutions radically altered the climate and other aspects of the environment, placing strong selective pressures on many organisms to rapidly evolve new adaptations. Under these conditions, the fossil composition of strata on either side of a major geological event would be expected to be quite different. Indeed, the major breaks in the biological record are relied on for locating breaks between series and systems of rocks. Revolutions probably required millions of years, but might be considered as almost instantaneous catastrophies when viewed over an earth history of five billion years.

Figure 4.7. Sir Charles Lyell (1797–1875) British geologist. (The Bettmann Archive)

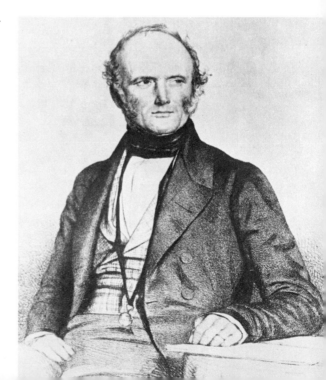

TABLE 4.1. Geological Time Table

Era	Period	Epoch	Approximate Beginning (millions yrs. B.P.)	Approximate Duration (millions yrs.)	Plants	Animals
Cenozoic	Quaternary	Recent	0.01	0.01	Decline of woody plants; rise of herbaceous plants	Modern humans
		Pleistocene	3	3	Extinction of many species	Ice ages; evolution of humans; giant mammals
	Tertiary	Pliocene	10	7	Forests decline; grasslands spread	Early hominids
		Miocene	25	15		Spread of anthropoid apes
		Oligocene	40	15	Monocotyledons become common	First anthropoids; some modern genera of mammals
		Eocene	60	20		Many modern mammalian families appear
		Paleocene	70	10	Rise of monocots; dicots common	Most modern orders of mammals appear
Mesozoic	Cretaceous		135	65	Flowering plants spread; gymnosperms decline	First modern birds; some modern orders of mammals; dinosaurs become extinct at end of period
	Jurassic		180	45	Cycads and conifers common; probable origin of angiosperms	Dinosaurs dominant; first birds and mammals
	Triassic		225	45	Extinction of seed ferns; gymnosperms dominant	Labyrinthodont amphibians; mammal-like reptiles; first dinosaurs

Era	Period	Millions of years ago	Duration	Plants	Animals
Paleozoic	Permian	270	45	Decline of lycopods and horsetails; seed ferns and gymnosperms dominate	Reptiles replace amphibians as dominant land animals
	Carboniferous	350	80	Tropical coal forests; lycopods and horsetails dominant	First reptiles; amphibians dominant; insects common
	Devonian	400	50	First seed plants (gymnosperms); forests	Fish dominant; sharks abundant; first amphibians
	Silurian	440	40	Evidence of primitive land plants	Marine arachnids dominant; arthropods invade land
	Ordovician	500	60	Probable origin of land plants	First vertebrates (jawless fishes); invertebrates dominate the seas
	Cambrian	600	100	Marine algae diversify	All major invertebrate phyla appear suddenly; trilobites, brachiopods dominant
Pre-Cambrian		From origin of earth 4.5–5.0 billion years ago		Primitive aquatic plants—algae, fungi	Marine protozoa; few molluscs, some worms and other soft metazoans

Between revolutions were times of considerable (though less extensive) crustal alterations and widespread retreat of the seas from the continental masses. These events serve to separate geological *periods*. Within these periods, geological *epochs* are separated by times of more localized modifications. Today we know, in various places, that glaciers are creeping slowly across the land. We see old volcanoes erupting and new ones being born both on land and in the sea. We experience periodic earthquakes and flooding. And we observe the age-old processes of erosion gradually changing the face of the earth. Because geological changes usually occur at such a slow rate, from the human perspective, it would be difficult to ascertain at any point in time whether the changes were sufficiently rapid and vast enough to be considered part of a geological "revolution" or whether they were merely the minor modifications that typify the lulls between major divisions of geological time.

Most people find it difficult to visualize the time during which evolutionary processes have been at work. It is probably just as hard to grasp the meaning of a million years of earth history as it is to imagine how far light travels in a year. Perhaps the following analogy will be helpful in placing geological time in its proper perspective. Suppose a motion picture was taken at the rate of one frame per year for the last 757 million years. When the film is run at the normal speed of twenty-four frames per second, twenty-four years of earth history would be shown per second. Now if we let the film run continuously, night and day, for one entire year we would review the events of the last 757 million years. Assume that the show begins at midnight on New Year's Eve and ends at midnight New Year's Eve one year later. For the first three months there are no signs of life on earth. Early in April, single-celled organisms appear; later in the month, multicellular organisms are flourishing. The first vertebrates arrive late in May. There are no land plants until the middle of July. It is late August before vertebrates (amphibians) come out onto the land. Reptiles come on the scene about the middle of September; dinosaurs become the ruling land vertebrates for about the next seventy days (until about mid November). Meanwhile, birds and mammals have evolved. Near the end of November, the geological revolution that raised the Rocky Mountains may have so profoundly altered the climate and other aspects of the environment that the dinosaurs could no longer survive. The mammals become the dominant land vertebrates by December. By Christmas, the Colorado River begins to cut its Grand Canyon. About noon of the last day (December 31), the first signs of humanity are seen. During that afternoon, four ice ages come and go in rapid succession. It is eleven o'clock in the evening when we first see "Old Stone Age" people become prominent. Agriculture begins about 11:45 P.M. The dawn of civilization is only five or six minutes before midnight. The Christian era begins one minute and seventeen seconds before midnight. Twenty seconds before the end, Columbus discovers America. The Declaration of Independence is issued with only seven seconds to go. The length of our own lives could be cut from the film and scarcely be noticed.

Major Biological Events

During the Cambrian Period there suddenly appeared representatives of nearly all the major groups of animals (phyla) now recognized. It was as if a giant curtain had been lifted to reveal a world teeming with life in fantastic diversity. The Cambrian "curtain" has become the touchstone of the creation theory. Darwin was aware of the problem this created for

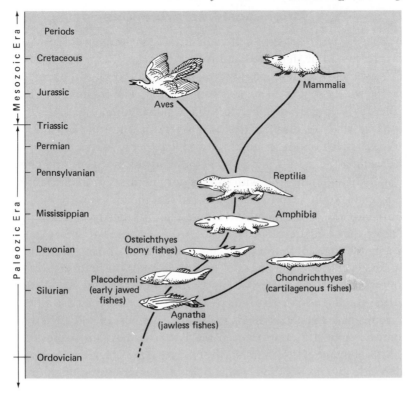

Figure 4.8. A phylogeny of the vertebrates. (After Romer)

evolutionists and it remains a problem today. Evolutionists keep hoping that new discoveries will eventually fill in the missing pieces of the fossil puzzle, but the chances of success may be less than those of finding the proverbial "needle in the haystack." The great revolution that separates the Proterozoic (Pre-Cambrian) from the Paleozoic eras apparently was of such a cataclysmic nature that many of the earlier fossils were destroyed.

Land plants formerly were believed to have first made their appearance in the Ordovician period, but recent finds of spores and pollen in rocks of Cambrian age indicate that land plants were present much earlier. It is more difficult to follow the evolution of plants in the rocks than that of animals, because classification of plants is largely dependent on their reproductive structures (e.g., flowers in angiosperms) and these are not at all well represented in the fossil record.

The evolution of vertebrates is clearly one of the best lineages represented in the fossil record (Figure 4.8). The most primitive vertebrates were the jawless armored fishes called *ostracoderms* of Ordovician age. In the later Silurian period, jawed (gnathostome) fishes called *placoderms* appeared. Amphibians evolved from the gnathostomes in the Devonian period. The first reptiles appeared in the Pennsylvanian period as the next major evolutionary advance from the amphibians. Both mammal-like reptiles and bird-like reptiles are known from Jurassic times.

Geological Interpretations

The *Law of Superposition* is the self-evident fact that the lowest strata of a geological formation was the first to be deposited and is therefore the oldest. The flora and fauna that serve as index fossils allow correlations to be made between rocks of the same age in different outcrops, even in widely separated regions of the world. The fragmentary local geographic records can thereby be pieced together to form a global composite.

Several major kinds of animals live exclusively in a marine environment, including corals, sea urchins, brachiopods, crinoids, and cephalopods. A rock bearing such creatures must have been under the sea during the time these organisms lived. Analogously, finding a rock containing fossils of land plants with stumps or roots intact or finding abundant bones or shells of land animals would lead to the conclusion that the rock was formed above sea level. The sudden appearance of elephant-like animals in North America during the Miocene epoch, long after they had developed in Eurasia, indicates that the two continents were connected by a land bridge (probably across the Bering Strait) during that time. When we find rocks bearing fossils of identical species of marine animals on both sides of the Isthmus of Panama, it follows that a strait had existed at the time these species lived. At the present time, almost all marine specimens on either side of the Isthmus of Panama are different, indicating that the two groups have been evolving in isolation for a considerable period of time. During the several times a strait did exist between the Pacific and Atlantic sides of Panama, land animals could not migrate easily from North to South America. Consequently we find them evolving independently in isolation while marine animals were migrating rather freely through the strait.

If we assume that ancient palms and crocodiles were restricted to tropical or subtropical climates, as they are today, then the presence of these fossils in the Oligocene deposits of North Dakota indicates that the climate there was formerly much warmer than it is now. This assumption may not be valid, because most organisms demonstrate great adaptive flexibility through evolution. It is therefore possible that their physiology could have been modified to function in a more severe climate. Consequently, we have more confidence in estimating past climatic conditions when we use groups of similarly adapted organisms rather than only one species. The oldest fossiliferous strata contain remnants only of relatively inconspicuous, simple organisms. As we move up the geological column through progressively more recent strata, we find the fossils becoming increasingly more complex and diversified. This is probably the most profound evidence for the theory of organic evolution. Furthermore, fossils of species that closely resemble each other commonly occur in adjacent layers of rock (strata). This fact is difficult to explain by the theory of special creation.

> The evolution of "horse types" found in the Cenozoic beds of western United States is an excellent example. *Eohippus,* the "dawn horse" of the Eocene epoch, had four toes on the fore feet and three toes on the hind feet (with vestiges of the first and fifth toes called "splint bones"). By the Oligocene epoch, through evolution, a new species arose called *Mesohippus,* with only three toes of the fore and hind feet touching the ground. Only the fifth digit remained as a splint bone on the fore feet. By the ensuing Miocene epoch, another species called *Merychippus* appeared with three toes fore and aft, but only the central digit touched the ground. The evolutionary lineage produced *Pliohippus* in the Pleiocene epoch and finally the modern genus *Equus* during the Quarternary

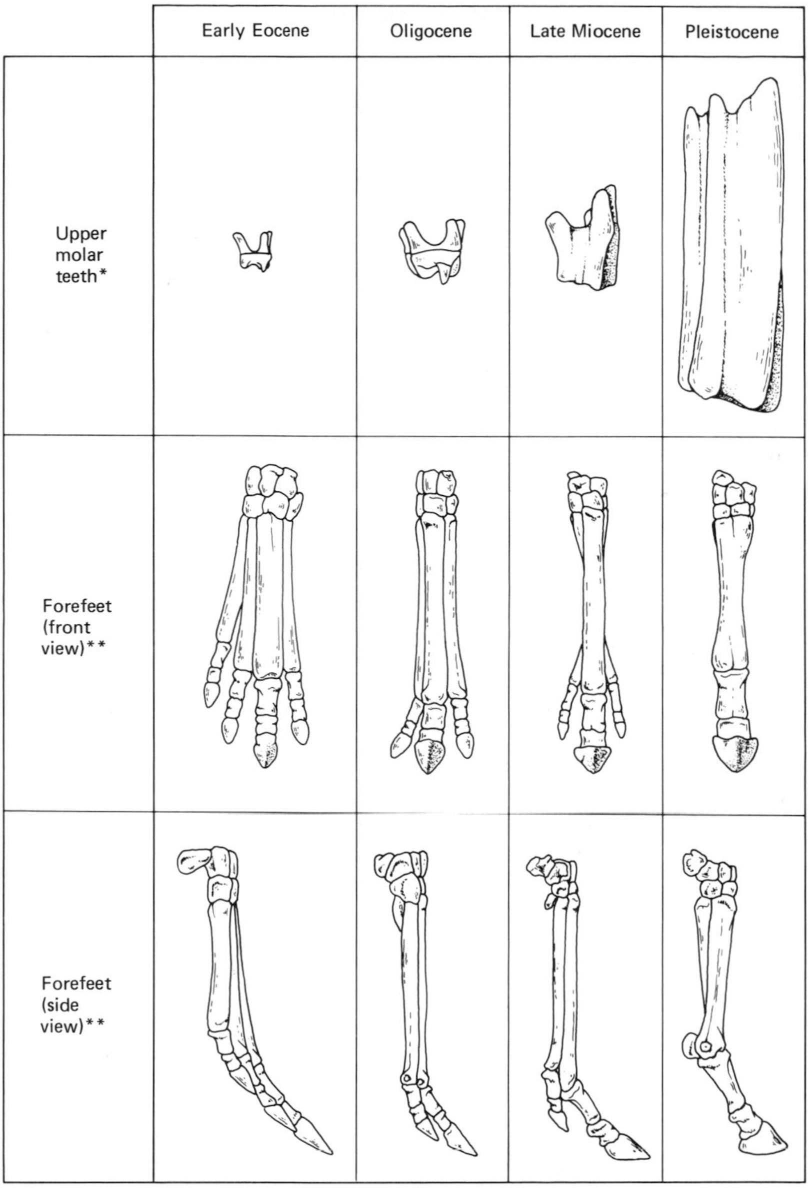

	Early Eocene	Oligocene	Late Miocene	Pleistocene
Upper molar teeth*				
Forefeet (front view)**				
Forefeet (side view)**				

*Drawn to uniform scale   **Not drawn to uniform scale

Figure 4.9. Evolution of the modern horse as inferred from fossils of the Cenozoic era.

period. Both species stood on the third toe, front and rear, but retained splint bones as vestiges of the second and fourth digits. In the evolution from *Eohippus* to *Equus* there was a general trend toward an increase in size. *Eohippus* was about the size of a large dog. There was also a gradual change from simple, short-crowned teeth (compatible with a forest-browsing niche) to the high-crowned teeth with complex enamel-covered ridges typical of the modern horse. These latter teeth were adaptations for shifting from a browsing to a grazing niche (an aid to using the harsher grasses) (Figure 4.9).

Chronometry

Several methods have been devised for estimating the age of the earth and its layers of rocks. These methods rely heavily on the assumption of uniformitarianism, i.e., natural processes have proceeded at relatively constant rates throughout the earth's history. Some estimates of geological age are derived from calculations based on the present average rate of sediment deposition in oceans, deltas, and lakes. Dividing the thickness of a rock layer (*stratum*) by the average rate of deposition for that kind of sediment gives an estimate of the length of time required for its formation. The rate of sedimentation, however, is known to vary widely from one locality to another, so that the use of an average rate would very likely lead to considerable errors of estimate. Probably no more than one percent of geological history can be accurately read in rocks. Furthermore, no single location contains the complete geological record. Gaps in the record can sometimes be pieced together using information from other localities that share strata. Certain fossils appear to be restricted to rocks of a relatively limited geological age span. These are called *index fossils*. Whenever a rock is found bearing such a fossil, its approximate age is automatically established.

> For example, trilobites lived only during the Paleozoic era. Those of the genus *Olenellus* are restricted to the early Cambrian period. Horned dinosaurs of the genus *Triceratops* are found only in rocks of the Cretaceous period (Figure 4.10). Three-toed horses appear only in mid-Cenozoic times.

This method is not foolproof. Occasionally an organism, previously thought to be extinct, is found to be extant. Such "living fossils" obviously cannot function as index fossils except within the broader time span of their known existence.

Geologists now assume that most of the water in the oceans was produced by volcanic outgassings. It has been estimated that seventy volcanoes the size of Mexico's Paricutin producing 0.001 cubic mile of water per year for 4.5 billion years of the earth's history could

A horned dinosaur; an index fossil of the Cretaceous period.

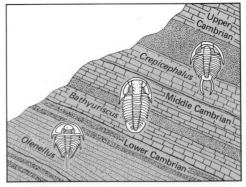

Idealized diagram of Cambrian strata showing index trilobite fossils of the genera *Olenellus* (Lower Cambrian), *Bathyuriscus* (Middle Cambrium), and *Crepicephalus* (Upper Cambrian).

Figure 4.10. Examples of index fossils.

account for the 315 million cubic miles of water in the oceans today. There are now approximately 600 active volcanoes and about 10,000 dormant ones. Six hundred volcanoes comparable to Paricutin could account for the present oceans in approximately 0.5 billion years. Since volcanic activity presumably was much greater during early earth history than at present, creationists argue that the age of the oceans would appear to be considerably less than 0.5 billion years. By this methodology, creationists stand guilty of the "crime" they ascribe to evolutionists, namely uniformitarianism. Perhaps the earth is now experiencing greater volcanic activity than the average; there is no accurate way of knowing. How much water vapor was lost to space during the early warmer stages of geological history is also a big unknown.

Uranium salts presently appear to be accumulating in the oceans at about one hundred times the rate of their loss. It is estimated that 6×10^{10} grams of uranium is added to the oceans annually. Under uniformitarian rules, the total concentration of uranium salts of the oceans (estimated at less than 10^{17} grams) could be accumulated in less than one million years. Again there is no way of knowing if present rates have been operating constantly throughout geological time or if the estimates of rate and total uranium content are accurate.

The atmospheric content of helium-4 (the most abundant isotope of helium) has accumulated from the radioactive decay of uranium and thorium in the earth's crust and oceans, from nuclear reactions caused by cosmic rays, and from the sun. If the present rate of accumulation of helium has been constant throughout four billion years of the earth's history, there should be thirty times as much helium in our atmosphere as is presently there. Little is known about how helium escapes from our atmosphere. Until more light is shed on this problem, arguments of atmospheric age based on these kinds of calculations are highly suspect.

One estimate for the amount of meteorite dust settling to earth places it at 14.3 million tons annually. If this rate has been constant throughout five billion years of geological history, one might expect over fifty feet of meteorite dust to have settled over all the surface of the earth. Some creationists suggest that the failure to find a fifty foot layer of such dust in the geological record argues against a long earth history. It is obvious that as the dust settles on the earth it would become incorporated with terrestrial materials into the geological strata and therefore would not appear as a discrete band. The average meteorite contains about three hundred times more nickel than the average earth rock. Perhaps our entire crustal content of nickel and iron could have a cosmic origin. If one assumes that the meteorite fallout rate was much greater earlier in the earth's history, one would argue for an earth considerably younger than five billion years. No meteorites have been found in the geological column. Creationists submit that evidence of meteorite showers should have appeared by now; their absence should be interpreted as support for the concept of a very young earth (perhaps only 5,000 to 10,000 years old). The rarity with which contemporary meteorites are found, however, makes it much less likely that ancient ones would be found by geologists.

It has been estimated that just four volcanoes spewing lava at the rate observed for Paricutin and continuing for five billion years could almost account for the volume of the continental crusts. The Columbian plateau of northwestern United States (covering 200,000 square miles) was produced by a gigantic lava flow several thousands of feet deep. The Canadian shield and other extensive lava flows indicate that volcanic activity has indeed followed an accelerated tempo in the past. The fact that only a small percentage of crustal rocks are recognizably lavas has been offered in support of the "young earth" concept. However, it is unlikely that the smaller lava deposits would escape extensive erosion, and many of them could have been largely weathered away.

Some geologists find it difficult to understand how the great pressures found in some oil wells could be retained over millions of years. Creationists also use this currently puzzling situation as evidence that oil was formed less than 10,000 years ago.

If humanity is really about 2.5 million years old (as claimed by Dr. Louis Leakey), creationists calculate from conservative population estimates (2.4 children per family, average generation and life span of forty-three years) that the world population would have grown from a single family to 10^{2700} people over one million years. The present world population is about 2×10^9, an infinitesimal part of the 10^{2700}. They conclude that humanity was created only a few thousand years ago. It was pointed out in the first chapter that although populations tend to have large reproductive potentials, the limiting factors of the environment prevent unlimited geometric increase. The size of a population may fluctuate over various lengths of time, but the long-term picture is one of stability. Populations need not continually expand in order to survive over long periods of time. Many more limitations to population growth were undoubtedly imposed on primitive humans than are faced by modern people. For one thing, primitive humans were gatherers and hunters. Scarcity of food was probably severely restrictive of human population growth until relatively recent times (a few thousand years ago), when humans learned how to raise their own crops and domesticate animals (the dawn of agriculture).

All the above methods for dating the age of the earth, its various strata, and its fossils are questionable, because the rates are likely to have fluctuated widely over earth history. A method that appears to have much greater reliability for determining absolute ages of rocks is that of *radiometric dating*. For some inexplicable reason, the nuclei of certain elements become unstable and spontaneously release energy and/or particles.

> For example, when the isotope of uranium with mass number 238 and atomic number 92 (written $_{92}U^{238}$) emits an alpha particle (two protons and two neutrons, equivalent to the nucleus of a helium atom), it becomes a thorium atom ($_{90}Th^{234}$). When radioactive carbon ($_{6}C^{14}$) decays, a neutron changes into a proton, a neutrino (a neutrino has no charge and almost no mass), and a high-speed electron (a beta particle or beta ray). Thus, carbon-14 becomes nitrogen-14 ($_{7}N^{14}$). Gamma radiation (X rays) consists of very high-energy photons (particles of electromagnetic radiation) released by various nuclear reactions. For example, when a proton (hydrogen nucleus) reacts with a nitrogen atom ($_{7}N^{14}$), it forms an oxygen atom ($_{8}O^{15}$) containing more energy than it does in its most stable state. It becomes stable by immediately releasing gamma radiation.

When atoms emit particles of alpha or beta radiation, they transmute (change or decay) into atoms of a different element. Gamma rays are the result of, rather than the cause of, transmutation. Each radioactive isotope (radioisotope) decays at a characteristic rate, measured in terms of its *half-life*. This is the time required for half of its atoms to decay. A radioisotope of carbon (C^{14}) is produced in the atmosphere by the action of cosmic rays on nitrogen (N^{14}). The half-life for C^{14} is about 5730 years. If we start with ten grams of C^{14}, in

5730 years half its atoms would be expected to decay back to stable N^{14}, leaving five grams of C^{14}. After another 5730 years, only one quarter of the original amount of C^{14} would remain (i.e., $2\frac{1}{2}$ grams); at the end of three half-lives (17,190 years) there would be $1\frac{1}{4}$ grams C^{14} and $8\frac{3}{4}$ grams N^{14}, etc. Carbon-14 reacts with oxygen-yielding radioactive carbon dioxide (CO_2) and enters the carbon cycle worldwide. Plants take up carbon dioxide and thereby incorporate both carbon-14 and normal carbon-12 into their tissues in the same proportion as they occur in the atmosphere. Carbon-14 would pass along the various food chains to enter animal tissues. After death, C^{14} would cease to enter the organism through active processes and would begin to decay in the manner previously outlined. The age of a recent fossil (up to approximately 25,000 years old) can be estimated under the assumption that the C^{14} content of the atmosphere has remained constant during this time span. The fossil can be analyzed for the ratio of C^{14} to C^{12} and its age determined from the known rate of radioactive decay.

It now appears that the C^{14} decay rate in living organisms is about 30 per cent less than its production rate in the upper atmosphere. Since the amount of C^{14} is now increasing in the atmosphere, it may be assumed that the quantity of C^{14} was even lower in the past than at present. This condition would lead to abnormally low C^{14}/C^{12} ratios for the older fossils. Such a fossil would be interpreted as being much older than it really is. Various correction factors have been calculated to take into account the disequilibrium between C^{14} production and decay. The correction is small for relatively young fossils, but amounts to about a 25 per cent reduction from an uncorrected age of 10,000 years. Creationists argue that since C^{14} has not yet reached its equilibrium rate, the age of the atmosphere must be less than 20,000 years old. It is possible that a greater concentration of water vapor existed prior to the Biblical flood (presumably about 5,000 years ago). This water vapor may have retarded neutron production by cosmic rays and consequently diminished the yield of C^{14}. This would give early fossils a low C^{14}/C^{12} ratio and therefore the appearance of great antiquity. On the other hand, if a lower concentration of water vapor existed, there may have been a greater amount of C^{14} produced; the increased C^{14}/C^{12} ratio in the fossil would be interpreted as a relatively young age. The water-vapor content of the atmosphere has varied considerably in the past, thereby disturbing any C^{14} equilibrium that may have been attained. Because the C^{14} method is useful only for very recent fossils, it has been mainly used by archaeologists who have found it to yield ages in fairly good agreement with historically dated materials.

Radioisotopes with half-lives in the millions or billions of years are used in dating events in the geological column. For example, half of uranium-238 (U^{238}) decays to its nonradioactive daughter isotope lead-206 (Pb^{206}) in 4.5 billion years. If one gram of U^{238} produces $(1.54 \times 10^{10})^{-1}$ gram of Pb^{206} per year, then x grams of uranium in y years should produce $(xy)/(1.54 \times 10^{10})$ gram of lead. Hence, $y = \dfrac{Pb^{206}}{U^{238}}(1.54 \times 10^{10})$. Suppose that the ratio of lead to uranium in a crystal of uranium-bearing mineral was determined to be 6.5×10^{-3}. The age of the rock in which the crystal was found is estimated to be $y = (6.5 \times 10^{-3})(1.54 \times 10^{10}) \approx 1 \times 10^8$ or approximately 100 million years. More precise calculations would apply correction factors for the gradual decrease in uranium content, for the production of thorium, and for other complicating factors, but the general outline of the methodology is straightforward.

If we assume that (1) a rock contained no Pb^{206} when it was formed, (2) all Pb^{206} now in the rock was produced by radioactive decay of U^{238}, (3) the rate of decay has been constant, (4) there has been no differential leaching by water of either element, and (5) no U^{238} has been transported into the rock from another source, then we might expect our estimate of age to be fairly accurate. Each assumption is a potential variable, the magnitude of which can seldom be ascertained. In cases in which the daughter product is a gas, as in the decay of potassium (K^{40}) to the gas argon (Ar^{40}), it is essential that none of the gas escapes from the rock over long periods of time.

It is obvious that radiometric techniques may not be the absolute dating methods that they are claimed to be. Age estimates on a given geological stratum by different radiometric methods are often quite different (sometimes by hundreds of millions of years). There is no absolutely reliable long-term radiological "clock." The uncertainties inherent in radiometric dating are disturbing to geologists and evolutionists, but their overall interpretation supports the concept of a long history of geological evolution. The flaws in radiometric dating methods are considered by creationists to be sufficient justification for denying their use as evidence against the young earth theory.

In 1972 Jeffrey Bada introduced a new method of absolute dating that does not involve radioactive elements. His method is based on the following principle. Amino acids occur in two optical forms; the L (laevo) isomer rotates the plane of polarized light to the left (counterclockwise); the D (dextro) isomer rotates it to the right (clockwise). A *racemic* substance does not rotate the plane of polarized light, because it contains equal quantities of dextrorotatory and laevorotatory isomers. All living organisms possess only the L-amino acid isomers in their proteins (excepting glycine). After death, the L isomers are slowly converted to D isomers (racemized). By measuring the ratio of D to L forms, the age of a bone can be estimated (provided the environmental temperature is known and it has not varied greatly). Radiocarbon dating is of little use for fossils older than about 40,000 years. *Racemization dating* has been applied to Middle Stone Age artifacts (110,000 years old) as well as to middle Pliocene fossils ($8\frac{1}{2}$ million years old). This method holds promise for dating Cenozoic fossils that are too old for dating by C^{14} and too young for accurate dating by U-Pb or K-Ar methods.

Continental Drift

Francis Bacon (*circa* 1620) was one of the first to notice the striking complementarity between the eastern shoreline of South America and the western boundary of Africa. They would fit together like pieces of a puzzle if they could be moved. Benjamin Franklin speculated in 1782 that ". . . the internal parts (of the earth) might be a fluid more dense, and of a greater specific gravity than any of the solids we are acquainted with; which therefore might swim in or upon that fluid. Thus the surface of the globe would be a shell, capable of being broken and distorted by the violent movements of the fluid on which it rested. . . ." Franklin's idea was very similar to the modern geological theory of *continental drift* or, as it is now called, *plate tectonics*.

Between 1915 and 1929 German meteorologist and theorist Alfred I. Wegener (Figure

Figure 4.11. Alfred Lothar Wegener (1880–1930). Drawing based on a photograph made in the 1920s.

4.11) published four editions of *Die Entstehung der Kontinente und Ozeane* (The Origin of Continents and Oceans) in which he made a strong case for the theory of continental drift. This notion attracted few supporters, because it was inconceivable that continents could move through ocean floors of solid rock. After World War II and the development of sophisticated sonar equipment for mapping the ocean bottoms, it was discovered that a huge submarine mountain range 40,000 miles long winds around the earth analogous to the seam of a baseball. This oceanic ridge has a narrow, deep valley along its center line. Midocean earthquakes seemed to originate near this ridge. Many other earthquakes appeared to cluster along some continental edges and island chains. Since earthquakes result from the slippage between blocks of earth, it was obvious that great geological activity was occurring along the ocean rifts and along the deep slopes. From these observations, the idea of *sea-floor spreading* developed. This concept proposed that molten magma from the earth's interior welled up through the rifts in the ocean floor, solidified, and moved outward to form new crust. The sea floors were spreading, carrying, or pushing the continents ahead of them.

The current theory of plate tectonics proposes that the lithosphere is divided into six major plates and about a dozen minor ones. These huge sections of crust float on the semiplastic upper layer of the earth's mantle. The crust under the continents is granitic, approximately twenty miles thick; under the oceans the crust is basaltic and only about three miles thick. A zone of change in chemical composition, called the *Mohorovičić discontinuity,* separates the more siliceous crust from the more basic material in the mantle (largely olivine, a magnesium iron silicate). The lithosphere extends from the surface to an average depth of about forty miles, including all the crust and the upper layer of mantle. The lithosphere is cool enough to be rigid and can break into plates or spherical caps. The *asthenospheric* layer of the mantle, underlying the lithosphere, is hot enough to be semiplastic, allowing the crustal plates to slide across its surface.

Along the edges of some continents (e.g., west coast of South America) and certain island chains (e.g., those in the northern and eastern Pacific), the old crust would be forced downward in *subduction zones* to disappear in deep trenches and under continents, there to be incorporated into the plastic core of the earth (Figure 4.12). As the Pacific plate moved under the South American continent, it uplifted the Andes Mountains. Earth slippage in subduction zones would account for the high incidence of earthquakes in these areas.

As molten iron-bearing rocks cool or as particles settle to form sedimentary rocks, they become slightly magnetized along the lines of the earth's magnetic field. Rocks of the same age from different continents seemed to be magnetized in different directions. This problem could be solved by assuming that the continents themselves had moved. For example, the magnetic pole recorded in England's rocks could be matched with those of North America by closing up the Atlantic. It is known that the magnetic field has reversed many times during geological history. It is estimated that at least 171 reversals have occurred in the past seventy-six million years. F. J. Vine and D. H. Matthews decided to check the theory of sea floor spreading by gathering magnetic data along both sides of the oceanic rifts. They found alternating zones of rock in which the magnetic field was sharply reversed. For each zone on one side of the rift there was a corresponding zone of identical character on the other side. By measuring the age of the rocks and the distance they were found from the rift, the rate of sea floor spreading could be estimated. The Atlantic Ocean appears to be expanding at the rate of one or two inches per year; the sea floor along the west coast of South America may be moving four times as fast. By extrapolation back through time, the continents must have formed one giant land mass, called *Pangaea* ("all lands"), about 200 million years ago (Figure 4.13). This left one giant ocean called *Panthalassa* ("all seas"). By 135 million years ago Pangaea had split into two major land masses, a northern one called *Laurasia* and a southern one called *Gondwana* (named for a geological region in India). Laurasia consisted of the ancient land mass of North America (called *Laurentia*) plus Europe and Asia. In Gondwana were the ancient land masses of South America, Africa, Australia, Antarctica, and India. A branch of Panthalassa called the *Tethys Sea* separated Laurasia from Gondwana. A rift appeared between South America and Africa about 135 million years ago, and South America began sliding westward. Then Africa separated from Antarctica. India also broke loose and sailed 5,000 miles northward and rammed into Asia about forty million years ago. This collision is thought to be responsible for uplifting the Tibetan Plateau and the Himalayan Mountains. North America became isolated from Europe only about eighty million years ago.

A British team under the leadership of Sir Edward Bullard used a computer to optimize the fit of Africa, North and South Americas, and Eurasia along the 3,000 foot depth of the continental shelves. They found that over most of this boundary the average mismatch was no more than a degree of arc.

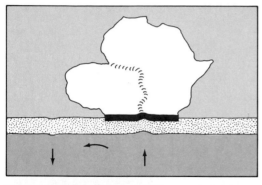

(1) A cross-section through a portion of the ancestral "supercontinent" from which South America and Africa were derived. Arrows indicate convection currents deep within the earth. An ascending current appeared under the supercontinent and a descending current existed to the west of this land mass.

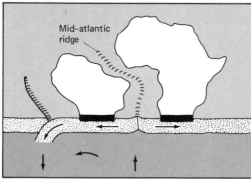

(2) The ascending current splits the supercontinent and creates the South Atlantic Ocean. Continued addition of new crustal material to the sea floor under the mid-Atlantic ridge causes the plates carrying Africa and South America to drift apart. A trench or subduction zone to the west of South America receives the old crustal material brought down by the descending current.

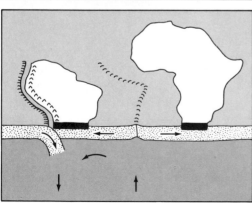

(3) South America has drifted westward to the edge of the trench. Now the subduction zone receives crustal material from the west. The edge of the eastward-moving plate disappearing underneath the west coast of South America was probably responsible for the elevation of the Andes Mountains, and slippage along this frontier still causes many earthquakes in this area today. Because of this process, the sea floor is usually less than about 150 million years old, whereas some continental rocks have been dated at 3.5 billion years old.

Figure 4.12. Diagrams of the geological forces thought to produce splitting of land masses and "continental drift."

Recall that scientific theories gain validity in proportion to their predictive value. One of the most striking predictions of the continental drift theory was verified by P. M. Hurley. A sharp discontinuity was known to exist between a 2,000 million year old geological province in Ghana on the Ivory coast of Africa and the 600 million year old province in Dahomey, Nigeria to the east. This discontinuity continues southwesterly into the ocean near Accra in Ghana. On the assumption that Brazil was formerly joined to Africa, the discontinuity should have entered South America near São Luís. Hurley went to São Luís and discovered that same discontinuity in Brazilian rocks exactly where it had been predicted by the continental drift theory.

(1) About 200 million years ago (near the beginning of the Mesozoic era) geologists believe that there was a singular land mass (Pangaea) which probably looked something like this. The ancestral Pacific Ocean (Panthalassa) and the ancestral Mediterranean Sea (Tethys Sea) are shown. Two modern geographic reference points are shown as arcs (A = Antilles arc in the West Indies; S = Scotia arc in the South Atlantic). The central meridian in all of these reconstructions is 20 degrees east of the Greenwich meridian.

(2) About 180 million years ago (near the end of the Triassic period) a northern land mass (Laurasia) had separated from a southern land mass (Gondwana). Gondwana had begun to fragment along rifts (solid lines), liberating India to move northward. The hatched lines represent the Tethyan trench (a zone of crustal uptake or subduction zone). Vector motions of the continents are indicated by solid arrows. Sea floor spreading is indicated by shaded areas.

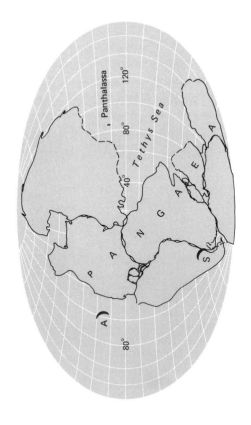

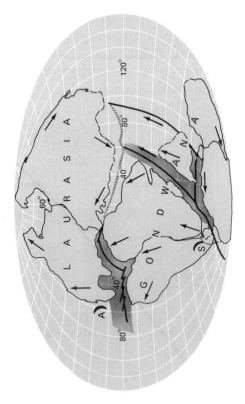

(3) About 135 million years ago (near the end of the Jurassic period), the North Atlantic and Indian Oceans have grown considerably. The eastern end of the Tethys Sea is narrowed by rotation of the Eurasian land mass; dotted lines along continental boundaries indicate questionable limits. A rift between South America and Africa initiates the South Atlantic Ocean.

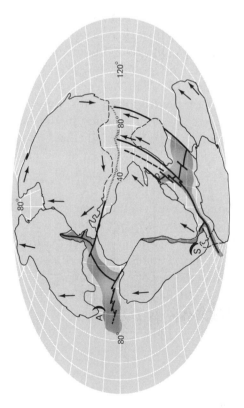

(4) About 65 million years ago (near the end of the Mesozoic era) the South Atlantic had become a major ocean. A new rift had separated Madagascar from eastern Africa. Australia was still connected to Antarctica. An extensive north–south trench (not shown) must have existed to accommodate the westward drift of the plates carrying the North American and South American land masses.

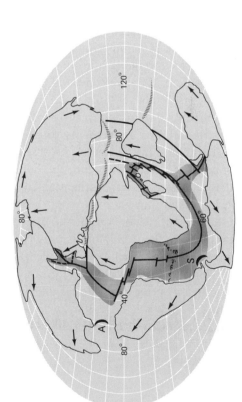

Figure 4.13. Inferred geological history of major modern land masses by "continental drift."

Additional evidence to support this theory has come from the fossil record. For example, the fossils of a sheep-size reptile known as *Lystrosaurus,* which flourished about 200 million years ago, were recently found in the Coal Sack Bluff of the Transantarctic Mountains. They appear identical to those found in Africa, India, and China at that same time. *Lystrosaurus* could not have crossed even narrow spans of open ocean. The conclusion is inescapable that Antarctica was formerly joined to these other land masses.

The overwhelming evidence that has accumulated since the 1960s has converted most "antidrifters" into staunch supporters of the theory. The conceptual revolution the continental drift theory has recently induced among geologists may be considered analogous to the conceptual revolution among biologists induced by Darwin's evolutionary theory during the last century. Both theories suddenly made sense out of what was formerly a jumbled mass of disconnected facts.

About 450 million years ago the region that is now the Sahara Desert was probably at the South Pole. Algeria is estimated to have drifted 8,000 miles since that time. When Africa was at the South Pole, fossil evidence indicates that the equator ran diagonally across North America some 6,000 miles away. Yet 250 million years later these two continents were joined in Pangaea. The inference is that there was at least one earlier Atlantic. This ocean closed between 450 and 350 million years ago. The colliding continents that formed Pangaea uplifted the Appalachian Mountains of eastern North America. When the separate continents of Europe and Asia collided during the formation of Pangaea, the Ural Mountains were similarly uplifted. Relatively little is known about the continents prior to Pangaean times.

Geologists do not yet know what forces drive the continents. Alfred Wegener speculated that the earth's spin has driven the broken pieces of Pangaea toward the equator. Hess and others attribute their movements to slow but relentless convection currents in the hot, semiplastic mantle just under the crust. According to another hypothesis, as the heavy cold leading edge of a plate sinks into the hotter, less-rigid mantle, it pulls the rest of the plate behind it. The process of continental drift is so slow that we are usually unaware of any change in crustal movements except during occasional earthquakes.

If the present rate of continental drift continues for another fifty million years, what will the world be like then? Geologists predict that the narrow coastal strip of California west of the San Andreas fault will have moved far north; Los Angeles will be near the Aleutian Islands; Central America will have disappeared; North America and South America will be separated by an ocean. East Africa will have separated from the rest of Africa, and its present Great Rift Valley will be a long water-filled gulf. The Red Sea will have widened; the Mediterranean Sea will have shrunk. Australia will have moved north beyond what is now Indonesia. The geological world is a dynamic (changing) system; the biological world is also in a state of evolution.

Biogeography

The distributional patterns of organisms in space and time constitute the subject of *biogeography.* Rather than being content with merely the descriptional aspect, modern biogeographers are also concerned with the causal factors of these distributional patterns. In the search for these explanations, biogeographers must use knowledge from many scientific disciplines including geology, geophysics, climatology, paleontology, biosystematics, taxonomy, physiology, ecology, and evolution.

There is a historical aspect fundamental to biogeographical distributions. Each species evolved in a certain locality and later spread from this *center of origin* into other regions with suitable ecological conditions. Some organisms (e.g., plants with airborne seeds), could be rapidly and widely dispersed from these centers of origin; others with limited mobility (e.g., land snails) are more likely to be found only in the vicinity of their origins.

The presence of a persistent water barrier between major land masses tends to limit the dispersal of a species. Terrestrial organisms generally find it difficult to traverse large bodies of water. Similarly, land masses tend to hinder the spread of aquatic organisms. The flora and fauna of North America resemble that of Europe much more than that of other parts of the world, because these continents became isolated from one another only recently (about eighty million years ago).

Climate can have an isolating effect just as profound as that of land masses. Each species has a range of tolerance for ecological conditions. In order to exploit fully a given set of ecological conditions, each population must specialize. A population cannot simultaneously be well adapted to opposite extremes of heat and cold or wet and dry conditions. Therefore, when a species attempts to invade a region that has different ecological conditions and is already occupied by other locally adapted organisms, it must become adapted itself to the changed conditions and do so in successful competition with the natives of the new region. These factors can be formidable barriers to dispersal and penetration into new territories. Such topological, biotic, and climatic factors as mountain ranges, valleys, rivers, forests, and deserts can retard the spread of a species from its center of origin.

Because of continental drift and other dynamic geological processes, land mass barriers to dispersal need not be permanent. Occasionally these barriers can be bridged. Simpson recognizes three types of bridges: (1) corridors, (2) filter bridges, and (3) sweepstakes bridges. *Corridors* are broadly continuous connections of land masses existing for prolonged periods of time. Such bridges apparently existed during much of the Paleozoic and early Mesozoic eras when the single land mass of Pangaea existed. Today, Asia and Europe are broadly connected as a corridor and this accounts for the great similarity of their floras and faunas. Asia was also broadly connected to North America until early in the Tertiary period and the climate was much warmer than at present, allowing extensive exchange of organisms across a north Pacific corridor. During the Tertiary period this corridor was largely submerged, the mountains of western North America were elevated, and the climate of the north became colder. These events caused widespread extinction of temperate and subtropical organisms throughout much of their former ranges, leaving in many cases only remnant populations isolated continents apart.

Filter bridges are narrower connections of a more temporary nature. Climatic conditions are more restrictively uniform on a filter bridge than on a broad corridor; hence it tends to "filter" the organisms that attempt to cross it. The Bering Strait functioned as a filter bridge repeatedly during the Pleistocene epoch when several extensive glaciations occurred. Mammoths, bison, bears, cats, and deer crossed from Asia to North America; horses, camels, and dogs crossed in the other direction. There was also a nearly continuous filter bridge between Asia and Australia during the Mesozoic era. This allowed primitive mammals such as monotremes and marsupials to reach Australia. But this bridge was broken before the evolution of more modern placental mammals. Accordingly, the Australian mammalian fauna is poorly represented by placentals (a few small rodents and bats probably made the journey via a sweepstakes route). The Isthmus of Panama has functioned as a filter bridge between North and South America on at least two occasions. Primitive marsupials crossed from North to South America on this filter bridge during the Cretaceous and early Paleocene. Then the isthmus was submerged and evolution occurred independently on the two continents. A chain of islands emerged during the late Eocene and Oligocene forming a sweepstakes bridge, but disappeared during the late Oligocene. Islands again appeared in the Pliocene, and the isthmus was reestablished in the Pleistocene. From North to South America went mastodons, horses, camels, deer, cats, weasels, racoons, tapirs, bears, and dogs. In the other direction went porcupines, armadillos, capybaras, and ground sloths. Many of the marsupials of South America undoubtedly became extinct in competition with the more advanced placentals. Today, South America is the only place outside of Australia that supports a marsupial fauna (the only marsupial in North America is the opossum).

Sweepstakes migrations do not involve land connections. They are accidental transportations usually involving small animals (particularly arboreal types) that perhaps drifted on logs across a water barrier. Chance, rather than the nature of the organisms themselves, is largely responsible for this type of dispersal. Whereas corridors and filter bridges should allow reciprocal migrations of numerous species, a sweepstakes route usually functions unidirectionally for a lucky few. Colonization of islands is typically via the sweepstakes method. For example, this method explains the unusual fauna of Madagascar (Malagasy Republic), which includes some mice, the fossa (a cat-like carnivore), lemurs and other primitive primates, the tenrec (an insectivore), and a pigmy hippopotamus (now extinct). Many closely related forms on the nearby African mainland could probably live just as well on Madagascar, but have not yet been accidentally transported across the Mozambique Channel. Rabbits introduced by humans to Australia and starlings introduced to the United States from Europe have become serious pests. It is obvious from the many examples of this nature that the exclusion of plants and animals from biogeographical realms to which they are very well adapted is difficult to explain on any basis other than an evolutionary one.

Some striking differences between the distributions of certain floras and faunas can be explained by evolutionary and continental drift theories. Today's floral realms are defined on the distributions of angiosperms (flowering plants). Angiosperms arose during the Jurassic, but did not become dominant until the mid-Cretaceous. It seems that the angiosperms had spread through Gondwanaland before the breakup of that supercontinent during the Cretaceous. The floras of the three major southern temperate regions (Australia and southern parts of Africa and South America) actually show more similarities to one another than they do to the floras neighboring them to the north. Primitive mammals appeared in the Triassic, long before the earliest angiosperms. However, differentiation of most mammalian orders did not occur until after the decline of the dinosaurs in the late Cretaceous. Since the adaptive radiation of the mammals occurred after the break-up of Laurasia and Gondwanaland had commenced, the modern faunal realms (based on the distribution of mammals) are more clearly determined by recent geological and climatic events than that of the floral realms.

Six major biogeographical realms (Figure 4.14) are recognized among the continental

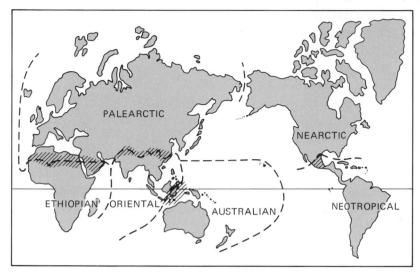

Figure 4.14. The six major biogeographical realms.

masses (terrestrial environment). These were initially defined (by A. R. Wallace, a pioneer in zoogeography) on the basis of their distinctive constituent mammalian faunas, but have since proven to be generally valid for other faunas as well as floras.

1. *Nearctic realm*—includes all of North America down to the Mexican plateau in Central America, and Greenland. Typical mammals include the bison (buffalo), prong-horned antelope, bear, mountain lion, caribou, elk (wapiti), muskrat, mountain goat, prairie dogs, opossums, skunks, and racoon.

2. *Palearctic realm*—includes Asia north of the Himalaya Mountains, Europe, and Africa north of the Sahara Desert. Typical mammals include deer (European red, fallow, and roe), antelope, bear, wolf, rodents, moles, oxen, sheep, goats, camel, hedgehogs, and wild pig. Because Europe separated comparatively recently from North America, and because Asia was linked to North America via a land bridge across the Bering Sea in the late Tertiary period, it is not surprising to find that the flora and fauna of these two realms appear very similar. For this reason the Nearctic and Palearctic realms are sometimes grouped together as the *Holarctic realm*.

3. *Neotropical realm*—includes Central America south of the Mexican plateau, all of South America, and the West Indies. Native to this area are numerous edentates (sloths, anteaters, armadillos), the peccary, guinea pig, alpaca, llama, New World monkeys, jaguar, tapir, and blood-feeding (vampire) bats. There are no cattle, sheep, or antelope native to this realm.

4. *Ethiopian realm*—includes Africa south of the Sahara Desert, and the Malagasy Republic (formerly Madagascar). This region contains the richest mammalian fauna including many ungulates (hoofed animals) such as wildebeest, zebras, gazelles, rhinoceros, hippopotamus, and giraffe; also rodents, carnivores (lion, cheetah, leopard), insectivores, primates (gorilla, chimpanzee, baboons, Old World monkeys),

aardvark, and the African elephant. The Malagasy Republic has no large animals and no snakes. Lemurs are endemic. The fauna is quite different from that of Africa, and some zoogeographers prefer not to include the Malagasy Republic in the Ethiopian realm.

5. *Oriental realm*—includes Asia south of the Himalaya and Ming Ling Mountains, India, Ceylon, the Malay Peninsula, Sumatra, the Philippines, Borneo, Java, and Bali. Among the endemic species are the orangutan, gibbons, tarsier, tiger, Old World monkeys, Indian elephant, rhinoceros, antelope, deer, sheep, and goats.

6. *Australian realm*—includes Australia, New Guinea, Tasmania, and New Zealand. The fauna here is rich in marsupial (pouched) mammals (kangaroos, wombat, koala bear). Monotremes (egg-laying mammals) such as the echidna and platypus are endemic. Among placental mammals native to this region are many bats and rodents (some of which may have been introduced by primitive man).

An imaginary boundary through the East Indies called *Wallace's line* (after A. R. Wallace, codiscoverer of the natural selection theory) separates the Australian and the Oriental realms (Figure 4.15). The line passes between Bali and Lombok, through the straits of Makassar between Borneo and the Celebes, and continues east of the Philippines. Although some of the channels along Wallace's line are only a few tens of miles wide, they contain some of the deepest trenches known in the oceans. On either side of Wallace's line, the flora and fauna are quite distinctive even though the topographies and climates are very similar. When the seas retreated from the continental shelves (as they appear to have done several times in the past), the deep ocean trenches persisted as water barriers against migration between the Australian and Oriental realms.

By ascending a mountain, one may observe changes in the vegetation similar to the biomic transitions that would be seen in traveling from the equator to the arctic zone. Many alpine species are identical with or closely resemble lowland species much farther north. Darwin noted that many plants of the White Mountains of New England were nearly the same as those on the highest peaks in Europe. During the Pleistocene ice ages, the advance of glaciers from the north caused the retreat of temperate plants toward the tropics. The only plants to survive in the region of glaciation would be the arctic types. As the ice age came to a close and the glaciers retreated, the only arctic plants to survive at the more southern latitudes would be restricted to the peaks of high mountains where the climatic conditions resemble those of the colder northern latitudes. Thus, it is not surprising to find alpine (high mountain) and lowland arctic types that are identical or very nearly so. A species that was widely distributed in the northern hemisphere before the advance of the ice sheets may have survived only in localized milder regions of their former range. After the retreat of the glaciers, the few surviving relict populations may be separated by great distances and may sometimes be even continents apart.

> At present, magnolias only occur naturally in southeastern China and in the southeastern United States. Many regions within the Oriental and Neotropical realms have ecological conditions in which magnolias would thrive. Indeed their distribution was continuous during the Tertiary period. Climatic changes caused widespread extinction

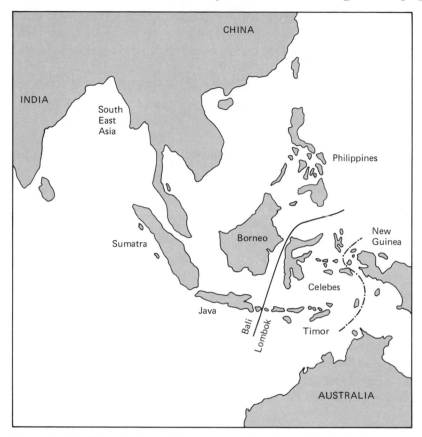

Figure 4.15. Wallace's line (solid) runs along the edge of the South-East Asian continental shelf. The edge of the Australian continental shelf is marked by the broken line.

of the intermediate populations during the Pleistocene ice ages, leaving only these remnant populations isolated on different continents. The fact that they are now found only in the Holarctic realm is thus difficult to understand on the basis of independent creations of the same species. The same can be said for the alligator and several species of amphibians, which today are found only in the United States and China.

Darwin understood that populations of the same species go their own evolutionary way only when isolated from each other. All other factors being equal, the greatest differences are expected between organisms that have been isolated for the longest time. Inhabitants of similar regions in different biogeographical realms (e.g., the savannas of Africa and the Argentine pampas) would thus be expected to be differentiated sharply, whereas those within different ecological regions of the same realm (e.g., the mountains and the plains of North America) would be expected to show greater resemblances. When new mountain ranges form, they become colonized by representatives from the surrounding lowlands. Selection for characters adaptive to mountain environments may thus cause, in time, the evolution of new species. Yet they would still be expected to retain many resemblances to the lowland types from which they were derived. Because of the longer isolation, resemblances between related members of different biogeographical realms would be expected to remain only at the level of higher taxa.

Deep ocean basins are barriers to the dispersal of inhabitants of the continental shelves. Thus, the littoral and neritic flora and fauna are quite different along the eastern and western shores of the Pacific basin. Specific ecological requirements for temperature, salinity, light, currents, nutrients, pressure, substrate, oxygen, etc. can further limit dispersal of a species to relatively small zones within the ocean habitat. Salt water is a barrier to most fresh-water organisms. Salmon are exceptional; they are spawned in fresh water, but migrate to the oceans where they spend most of their lives. They almost invariably return to the same fresh water system to breed and die. This homing instinct is a barrier to colonizing new rivers or streams. The organisms in ponds, lakes, streams, or rivers might be expected to show considerable differentiation because of their isolation. Quite the opposite is true. Many fresh-water organisms are widely distributed. One way by which inhabitants of different fresh-water systems may be interchanged is mediated by floods that periodically join rivers, lakes, etc. Waterfowl may be very effective in the dispersal of plants and animals. Seeds or spores of plants and the eggs or larvae of animals may be carried in mud clinging to the feet and legs of these birds. Some seeds may even pass intact through the digestive tract of birds.

Oceanic islands are those located beyond the continental shelf and are almost invariably of volcanic origin. These islands usually are inhabited by relatively few species and were originally colonized via a sweepstakes route. However, a large percentage of such species are typically *endemic* (i.e., found nowhere else). On the Galapagos Islands, Darwin recognized twenty-six species of land birds, twenty-one of which were endemic. Among the eleven species of marine birds, however, only two were endemic. Darwin noted that the flora and fauna of the Galapagos Islands resembled very closely those of South America, some 500 miles to the east. Yet more than half the angiosperms were endemic, some of them even restricted to a single island of the archipelago.

Darwin postulated that these islands were initially colonized by a few migrants from the mainland. Natural selection for different kinds of traits, in time, caused the evolution of distinctive species. Yet they obviously retain many characteristics indicating their close relationship to the South American species from which they were derived. The climate and geological characters of these islands are quite different from those of South America. Therefore it is difficult to explain the distribution of these related species by special creation of similar organisms to inhabit similar environments. Darwin's explanation is much more reasonable. Amphibians and terrestrial mammals are generally not native to oceanic islands. Many of the islands do have bats, as expected, because the flight of bats and birds allows relatively long-range dispersal. Consider the yearly migrations of many birds between the arctic and South America. The Bermuda Islands, about 700 miles east of North Carolina, are located on one of these major bird migration routes. The fact that the inhabitants of the Bermudas all resemble those of the Nearctic realm rather than those of the Neotropical realm is an expectation of Darwinian theory, but not of special creation theory.

The overwhelming majority of facts concerning biogeographical distributions are only understandable in terms of continental drift, migration, isolation, and evolution.

Essay Questions

1. Outline the conditions required for the optimum preservation of an organism as a fossil.

2. Cite as least two facts from paleontology that support the theory of evolution.

3. Discuss the ramifications that uniformitarian theory may have on estimates of the earth's age.

4. What are the major lines of evidence that support the theory of continental drift?

5. How do the facts of biogeography relate to the theory of organic evolution?

FURTHER EVIDENCE
OF EVOLUTION

Two of the major lines of evidence supporting the evolutionary theory, namely paleontology and biogeography, were presented in the last chapter. This chapter presents further supporting evidence drawn from the disciplines of taxonomy, comparative embryology, comparative anatomy, comparative biochemistry, and physiology.

Taxonomy

Taxonomy is the theory and practice of naming, describing, and classifying organisms. *Systematics* is the scientific study of the kinds, diversity, and similarities of organisms and of the relationships among them. These two terms are often incorrectly used as synonyms. Attempts at classification have been of practical use to naturalists since the time of Aristotle, who introduced the concept of a "ladder of nature." According to this idea, a human is higher on the ladder (more complex) than an insect, which in turn is higher than a worm. The ladder was merely an aid to classification and should not be interpreted (as many have done) as an evolutionary lineage. Systematics is a relatively new scientific discipline, emerging after the Darwinian revolution. Today some systematists are even engaged in genetic research in attempts to understand evolutionary relationships.

Our present system of classification is largely based on a scheme developed by Swedish naturalist Carolus Linnaeus (1707–1778; Figure 5.1). A century before Darwin, virtually every naturalist believed that all the various kinds of organisms now called species originated by special acts of creation. Linnaeus also believed in the immutability of species and developed his classification scheme biased by this philosophical position. Recognizing that certain species resembled each other very closely, he grouped similar species together in a category called the *genus*. The terms genus and species are derived from the Greek words *genos* and *eidos* used by Aristotle. Groups of genera seemed to cluster into larger assemblages sharing more fundamental characteristics, which he named *orders;* the orders with only elementary traits in common were grouped into *classes*. Linnaeus divided living things into the two *kingdoms* Plantae and Animalia. Thus, within a kingdom organic diversity appeared to Linnaeus as a hierarchy, i.e., groups within groups. About 1730 Linnaeus began to defend the then revolutionary theory (developed by Rudolph Camerarius, 1694) that plants reproduce sexually. The first edition of his *Systema Naturae* (1735) contributed a new concept to botanical classification. There were twelve classes based on the number of stamens in the flower; the orders were based on the number of pistils. The tenth edition of this book (1758) popularized *binomial nomenclature,* i.e., the assignment of two Latin words for each species.

The binomial system was actually developed by Swiss naturalist Kaspar Bauhin (1560–1624), but it received wide attention only after it had been coupled with the Linnaean system of classification. In naming each species, the first word is the genus (capitalized) to which it belongs, and the second word (uncapitalized) is the species category. All such names are either underlined or italicized.

Figure 5.1. Carolus Linnaeus (1707–1778). (From *Linnaeus and the Linnaeans* by Frans A. Stafleu, 1971. Published by A. Oosthoek's Uitgeversmaatschappij N. V., Utrecht, Netherlands; with permission of Stechert Hafner Service Agency, New York)

For example, the fruit fly of genetic fame is *Drosophila melanogaster* (Figure 5.2). *Dros* means dew or honey; *phila* is love; *melano* refers to black pigment; and *gaster* implies stomach or belly. Hence, the name means a "black-bellied honey-lover." The abdomen of the male is black and they are fond of ripe fruit, so the Latin description in the binomial is quite informative.

Formerly, each species was identified by several lengthy descriptive sentences. Moreover, though a common name was given to a species, it would be written differently in each language. Even within the same country, a species was known by several common names. For example, the mountain lion of North America is also known as puma, catamount, panther, and cougar. All this confusion, both within and between nations, disappears with the use of the single scientific name *Felis concolor*.

About one-hundred years after Linnaeus, German embryologist Ernst Haeckel introduced two other taxonomic categories. The *phylum* was to be the largest grouping within a kingdom; the *family* was an assemblage between the genus and order. Today the complete hierarchy appears as follows (with two examples for the domestic dog and white clover, respectively):

Kingdom	Animalia	Plantae
Phylum	Chordata	Pterophyta
Class	Mammalia	Angiospermae
Order	Carnivora	Rosales
Family	Canidae	Leguminosae
Genus	*Canis*	*Trifolium*
Species	*familiaris*	*repens*

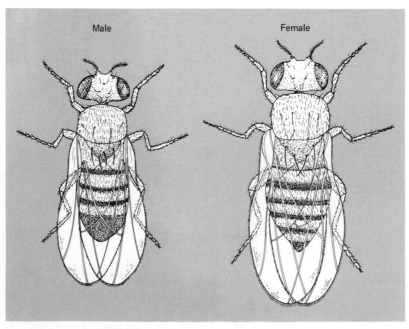

Figure 5.2. *Drosophila melanogaster.*

Other intermediate categories are available for use by taxonomists, but they are not officially recognized by the International Congress of Zoology.

Modern taxonomic schemes attempt to show genealogical relationships and, hence, are *natural systems.* The Linnaean system was essentially an artificial one, as it was based solely on various degrees of morphological similarity. Furthermore, under the philosophical outlook of special creation, there was no reason to look for biological meaning in the taxonomic hierarchy. Yet it was disconcerting to early taxonomists that the simplest and most logical schemes for classifying organisms always seemed to be that of a hierarchy. Why some other geometrical configuration could not be justified remained a puzzle until taxonomy became philosophically reoriented toward an evolutionary approach. Linnaeus tried without success to develop map-like diagrams that would always group similar organisms together and separate dissimilar groups. Others, including Lamarck, attempted to fit organic diversity into a ladder-like diagram on the assumption that more complex and/or more specialized forms evolved from less complex and/or less specialized types. This may be possible for certain segments of the hierarchy, but cannot be applied universally. A branching tree-like system is much more useful. For example, a case could be made for amphibians being more advanced than fishes, and reptiles being more advanced than amphibians. However, the ladder-like series cannot be continued because birds cannot be considered more advanced than mammals, or vice versa. The ladder must have a fork leading from reptiles to both birds and mammals. This kind of reasoning forced taxonomists to diagram biological variation in the form of a tree with the more specialized groups near the ends of the branches, and the similar yet more generalized groups near the bases of these branches. It is generally observed that groups in different twigs near the base of a common fork tend to resemble each other more strongly than do the basal and terminal members of the same twig. This fact is inexplicable on the philosophical basis of a special creation for each species, for the common characteristics of similar groups would then be expected to be distributed independently of the degree of specialization. According to the theory of evolution, however, a tree-like taxonomic hierarchy is to be expected. The more unspecialized (generalized; primitive) groups near the base of a common fork are those that have undergone only relatively minor modifications from some ancestral form; the more specialized (advanced; derived) groups have undergone much more extensive changes.

Although a *dendritic* (tree-like) diagram serves to outline the major features of taxonomy, a configuration more in keeping with evolutionary theory would be partly *reticulate* (net-like). The criteria by which a group of organisms is recognized as a species will receive a detailed examination in Chapter 17. For now, we can simply define a *species* as an interbreeding group of organisms sharing most of their traits in common and reproductively isolated (at least in part) from other such groups. Two populations of a given species may, in isolation, begin to diverge from each other and may even eventually fail to interbreed freely with each other when brought together (i.e., they have evolved into different species). Through hybridization, one (or possibly both) species can acquire characteristics derived from the other. The evolutionary significance of even small amounts of interspecific matings could be vastly out of proportion to the incidence of such hybridizations. Therefore, if all organic changes were thoroughly known, it is highly probable that numerous cases of hybridization would be found to convert the dendritic classification scheme into a partly reticulate evolutionary progression.

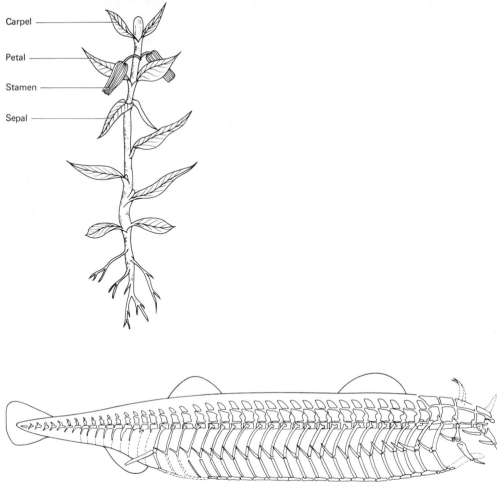

Carpel

Petal

Stamen

Sepal

Figure 5.3. The archetype concept prompted M. J. Schleiden to propose a primitive plant whose floral parts are aggregates of transformed leaves (top). The same stimulus prompted Richard Owen to propose an archetype of the vertebrate skeleton (below) as a series of vertebrae with rib-like attachments; jaws developed from the first two segments, limbs from others.

At the time of Linnaeus, it was generally conceded that all life was the result of divine creative acts as outlined in the first chapter of the Bible, Genesis. All of Creation was interpreted as initially perfect, including humans. There would be no need for any biological changes in such a world. With this philosophical outlook, evolution appeared to be superfluous; if changes did occur after the perfect Creation they would have to be harmful. Linnaeus attempted to explain the taxonomic hierarchy in terms of *archetypes*. It was assumed that the Creator had in mind a conceptual set of ideal (perfect) forms, archetypes. Each Linnaean class represented a major archetype, each Linnaean order represented a lesser archetype, and so on down the hierarchy. Thus, the species common to a genus were viewed as imperfect copies of the archetype at this level. How imperfection came to characterize the biological world was not explained. Somewhat later, J. W. Goethe speculated upon the archetypes (ideal plans or Platonic ideas) that would best fit organisms to survive in nature. According to his view, all organisms had deviated from the original archetype by the forces of the environment. By strictly intuitive methodology, he proposed that all parts of a flower were modified leaves and that the skull was composed of modified vertebrae (Figure 5.3). These ideas were incorrect, but they eventually stimulated work in comparative anatomy by Geoffroy St. Hilaire, Georges Cuvier, and others, which later was to form part of the body of evidence for the theory of evolution.

It might be anticipated that drastic changes would have been necessary to convert the artificial taxonomic scheme (based on morphological similarities) of the creationists to the natural system (based on degrees of genetic relationship) of the evolutionists. The fact that only relatively minor changes were required to make this transition is evidence of the close harmony between the theory of evolution and objective observations of organic diversity.

Embryology

The science of *embryology* is concerned with the cause and effect relationships that lead to differentiation of tissues, organs, and organ systems. Sperm cells were first seen by Leeuwenhoek in the seventeenth century, but the concept of a cellular structure for all living organisms (the cell theory) was not formalized until the nineteenth century. During this interval, several embryological theories were popular. A Dutch microscopist, Jan Swammerdam, proposed that an organism develops by the enlargement of preformed miniature parts in the germ cells (sperm and egg). This came to be known as the *preformation theory*. Charles Bonnet extended Swammerdam's idea into an *encasement theory*, according to which the embryo has within its sex cells other preformed embryos like boxes within boxes. This theory was formerly favored by creationists, because it was compatible with the idea of total creation, immortality of the soul, and the ultimate fate of humanity. Furthermore, it gave support to the dogma of original sin and the doctrine of predestination. If God had placed a finite number of embryos within the gametes of Adam and/or Eve, then when the last box had been opened (development of the ultimate embryo) the continuity of humanity's existence would terminate. This theory had such a profound influence that some microscopists thought they saw a preformed human embryo inside a sperm cell; it was even given the name *homunculus*. One school, called the *animalculists*, consisted of those who believed that the embryos were within the sperm cells; the other school, called *ovists*, comprised those who thought the embryos were within the egg cells. During the last half of the eighteenth century, Caspar Wolff became convinced, by his microscopical observations of chick embryos, that the amorphous zygote becomes gradually organized into the various tissues and organs of the embryo. This concept was known as the *epigenesis theory* and became the forerunner of our modern embryological concepts. After overwhelming scientific evidence had been amassed in support of epigenesis, creationists were forced to accept this new doctrine.

August Weismann was probably the first to propose a coherent embryological theory in the 1890s. He suggested that cellular differentiation attending embryological development was a consequence of sorting out from the fertilized egg certain hereditary "determiners" (the chromosomes) in successive cell divisions. In other words, the loss of some chromosomes (or portions thereof) would produce one cell type, whereas the loss of other determinants would cause a different type of cell to develop. Only in cells of the germ line (gonads) would the entire set of genetic instructions (genome) be preserved. Weismann's germ-line theory was finally laid to rest in 1964, when J. B. Gurdon performed a crucial experiment (Figure 5.4). Gurdon destroyed the nucleus of a frog egg by ultraviolet radiation. Using a micropipet, he extracted the nucleus from an intestinal cell of a tadpole (a highly differentiated cell type) and introduced it into the enucleated frog egg. Most of the embryos made in this fashion either failed to divide, or became abnormal and aborted development. But the fact that he did obtain some mature adult frogs by this process indicates that all the "determiners" (chromosomes and genes) are still present in highly differentiated cells. Therefore, embryonic development must ultimately be explained in terms of differential gene action, i.e., different genes becoming active at different times and in different cells to produce the proteins (enzymes) peculiar to each cell type.

In 1828, Karl Ernst von Baer condensed his embryological observations into a set of four basic principles: (1) during development, the more generalized characters appear before the specialized traits; (2) specialized characters always develop from the more generalized characters; (3) each animal becomes progressively more unlike other animals as development proceeds; and (4) early embryos of a species resemble the early embryos of animals of lower taxonomic status. After the publication of Darwin's *Origin of Species,* some embryologists apparently became overzealous in their evolutionary interpretations of von Baer's dicta. One of these was Ernst Haeckel who developed the *recapitulation theory* (also called the *biogenetic law*). According to this theory, "ontogeny recapitulates phylogeny." This simply means that the embryological development (*ontogeny*) of an organism repeats (recapitulates) the evolutionary history (*phylogeny*) of its species.

Haeckel thought that higher animals evolved by the addition of extra embryonic stages to the end of a preexisting ancestral path of development. According to his theory, the single-celled zygote of a mammalian animal corresponded to a protozoan ancestor. Cleavage produced the coeloblastula (hollow ball of cells stage), which corresponded to a primitive ancestor, similar to the colonial alga *Volvox.* The process of *gastrulation* produces an embryo with two germ layers: an outer *ectoderm* and an inner *endoderm.* He saw the early gastrula stage as representative of the coelenterate stage of evolution (modern coelenterates include sea anemones, jellyfish, corals, and hydra). Soon after gastrulation, a third embryonic (germ) layer called *mesoderm* appears between ectoderm and endoderm. This trilaminar (*triploblastic*) stage corresponded to the flatworm (platyhelminth) level of development. The basic chordate characters of hollow dorsal nerve cord, notochord, and pharyngeal gill slits (or pouches) develop later. The embryo first resembles a fish in that gill slits and six aortic arches are present, and the early kidney passes through the pronephros and mesonephros stages. As limb buds appear, the embryo becomes recognized as a member of the *tetrapods* (four-legged vertebrates). The kidney becomes metanephric, typical of *amniotes* (animals that develop an extraembryonic membrane called the amnion; including reptiles, birds, and mammals).

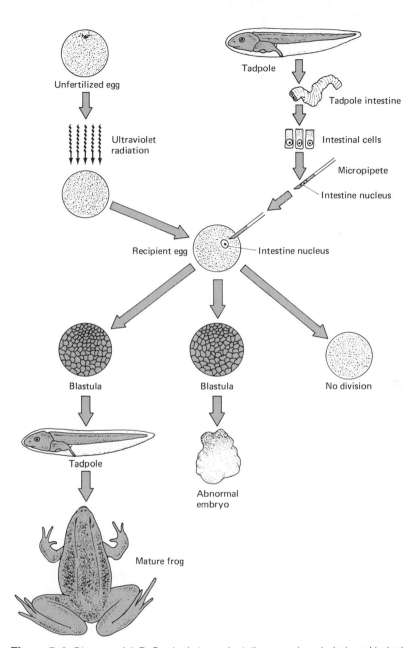

Unfertilized egg

Ultraviolet radiation

Recipient egg

Tadpole

Tadpole intestine

Intestinal cells

Micropipete

Intestine nucleus

Intestine nucleus

Blastula

Blastula

No division

Tadpole

Abnormal embryo

Mature frog

Figure 5.4. Diagram of J. B. Gurdon's transplantation experiment, designed to test the hypothesis that no essential loss of genetic information occurs in tissue differentiation.

Finally the truly mammalian traits begin to appear; a muscular diaphragm separates thorax from abdomen, three bones develop in the middle ear, hair is produced, etc.

> Development of the mammalian kidney begins with a *pronephros* consisting of a series of anterior tubules collecting fluid directly from the body cavity (coelom) and conducting it to the outside via a pair of pronephric ducts. This kind of kidney is functional only in primitive vertebrates such as the lamprey. A different kind of kidney then develops (the *mesonephros*) posterior to the pronephros. The mesonephros extracts wastes from the blood rather than from the coelom. Two Wolffian ducts communicate with the outside. Most fishes and amphibians develop both a pronephros and mesonephros, but only the latter functions in adults. The mesonephros is functional in embryos of *amniotes* (reptiles, birds, and mammals). Adult amniotes posteriorly develop a third kind of kidney, the *metanephros,* which carries wastes to the outside by way of two ureters (Figure 5.5). Although amniotes have both a pronephros and a mesonephros during embryonic development, only the metanephros remains in adults.
>
> If an all-powerful Creator wished to devise an embryological pathway for the production of the mammalian kidney, it would seem both logical and energetically more efficient to cause it to develop directly from primary mesoderm. The theory of special creation is unable to explain why a mammalian embryo must first develop two other more primitive kidney types and eventually discard them both. These facts are explainable by the theory of evolution as epigenetic modifications of a developmental pattern that functioned in remote ancestors.

The recapitulation theory was shown to be unsound even in Haeckel's day, but it all seemed so tidy that the fundamental inconsistencies were largely ignored. It was pointed out that most modern coelenterates do not form coeloblastulae (hollow balls of cells), but solid *morulae* or *steroblastulae*. The method of gastrulation according to Haeckel's scheme was quite different from that of most modern coelenterates. Furthermore, most of them are triploblastic rather than diploblastic (two-layed) as Haeckel envisaged.

The biogenetic "law" implies that embryos of higher animals pass through developmen-

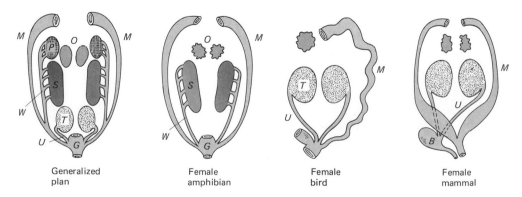

Generalized plan · Female amphibian · Female bird · Female mammal

B = urinary bladder; *G* = gut; *M* = Müllerian duct (oviduct); *U* = ureter; *W* = Wolffian duct;

O = gonad; *P* = pronephros; *S* = mesonephros; *T* = metanephros.

Figure 5.5. Diagrams of adult female urogenital systems in selected vertebrates. Notice in the generalized plan the anterior, middle, and posterior positions of the pronephros, mesonephros, and metanephros, respectively.

tal stages comparable to the adult forms of lower animals. The development of gill pouches in a mammalian embryo does not mean that the embryo is in an adult fish stage. The gill pouches of mammalian embryos never function in respiration as they do in adult fishes. Resemblance between the mammalian embryo and the adult fish is never extensive at any stage of its development. Thus, we cannot expect to find the complete evolutionary history of any species to be revealed in the developmental sequences that constitute its embryology. The biogenetic "law" also implies that embryos are unable to experience evolutionary adaptations (or are much less able to do so than adults). This is not true. Many insects pass through an embryonic larval stage. For example, the caterpillar is the larval stage of adult butterflies. Larvae are often adapted for feeding on sources completely different from the adult stage. For example, caterpillars usually feed on leaves, whereas adult butterflies have mouth parts highly modified into a tube for sipping nectar from flowers.

> The classification of barnacles remained unsettled until it was discovered that they developed from six-legged nauplius larvae characteristic of the crustaceans. Adult barnacles attach to rocks or other animals by their heads and elaborate calcareous shells. The larval stage, however, is motile and can evolve various adaptations for enhancing dispersal, location of suitable substrate for the adult stage, etc. The closest relatives of primitive chordates (hemichordates) are the echinoderms (e.g., starfish) as revealed by similarities in their larval forms (Figure 5.6).

A relatively close resemblance is expected to exist between the larval or embryonic stages of related organisms, because fewer developmental steps have occurred to differentiate them. The longer two organisms share a common (similar) developmental pathway, the closer they resemble one another. For example, the developmental pathways of humans and apes diverge relatively late, whereas the paths of fish and humans diverge much earlier (Figure 5.7). Embryos are more easily modified early in development, because most tissues are relatively undifferentiated at this stage (developmentally plastic). Many tissues of later embryos have become specialized and can no longer be altered. Therefore, Haeckel's idea of adding extra steps to the end of an ancestral developmental pathway runs counter to present knowledge of embryology. The branching pattern of modified embryological pathways is consistent with the bush-like diagrams used to represent evolutionary divergence. The ladder-like end-addition sequences of the recapitulation theory is inconsistent with this evolutionary scheme. Haeckel's "biogenetic law" is now thoroughly discredited. Von Baer's principles, however, still remain as fundamentals of modern embryology.

> Another classical example of relationships easily explained by evolutionary modifications of a primitive construction is found in the aortic arches of vertebrates (Figure 5.8). Most all vertebrates initially develop six branches of the ventral aorta called aortic arches. In the primitive chordate subphylum Cephalochordata, each branch is continuous with the dorsal aorta. Fishes have modified this primitive construction by forming in each arch a ventral (afferent) and dorsal (efferent) branchial artery, the two being connected by a capillary network in the gills. Sharks have completely lost the first (anteriormost) arch. Most bony fishes (teleosts) have lost the first two arches. The lung fish *Protopterus* shows a curious mixture of primitive and advanced arches, some being continuous, others being broken by a capillary bed. The first arch is lost, and a branch of the sixth dorsal branchial artery serves the lung. Amphibians and higher classes have

A

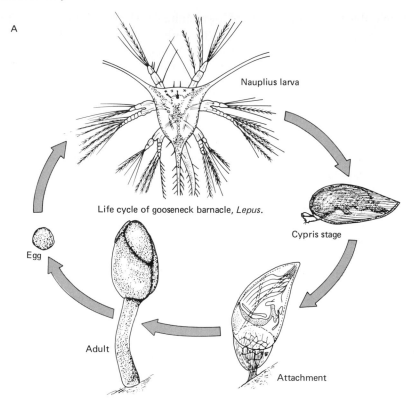

Nauplius larva

Life cycle of gooseneck barnacle, *Lepus*.

Cypris stage

Egg

Adult

Attachment

B

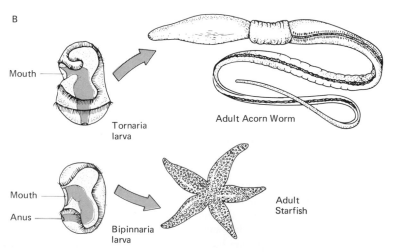

Mouth

Tornaria larva

Adult Acorn Worm

Mouth

Anus

Bipinnaria larva

Adult Starfish

Figure 5.6. Larval evidences of evolutionary relationships. (A) Life cycle of gooseneck barnacle, *Lepas,* reveals its relationship to the crustaceans through its nauplius larva; (B) similarity of tornaria larva of the acorn worm (a hemichordate) to the bipinnaria larva of the starfish (an echinoderm) indicates their evolutionary relatedness.

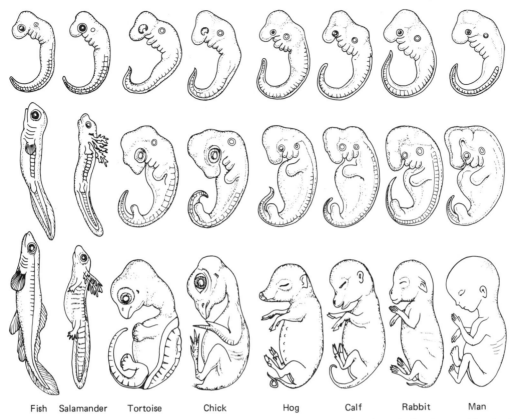

Fish   Salamander   Tortoise   Chick   Hog   Calf   Rabbit   Man

Figure 5.7. Comparison of vertebrate embryos at three different stages of embryonic development (early at top through late at bottom).

lost the branchial (gill) capillaries. Tailed amphibians (urodeles) have lost the first two arches; the third arch serves the head region as the carotid arteries. The dorsal connection between the third and fourth arches is also lost. The fifth arch is small and may be lost. A branch from the sixth arch serves the lung. In reptiles, arches 1, 2, and 5 are lost. The sixth arch serves the lung as the pulmonary artery. The only dorsal connection is between the third and fourth arch. In birds, all dorsal connections are lost as is the left fourth arch; otherwise the pattern is similar to that of reptiles. Mammals are similar to birds except the right fourth arch is lost.

Occasionally a baby is born suffering from oxygen deprivation due to a patent ductus arteriosus (the dorsal branch of the sixth aortic arch) or to a patent foramen ovale (opening between the upper chambers in the heart). Both open structures are reminders of the more primitive circulatory patterns of our ancestors.

Gill arches and gill slits form in mammalian embryos, but they never function in respiration as they do in fishes. Only one gill slit has any function in mammals; it forms a tube connecting the pharynx with the middle ear (Eustachian tube).

The primitive jawless fish, the lamprey, has seven gill arches. One of the most important modifications in the evolution of the vertebrates was the conversion of the first two gill arches into jaw components and their articulative and supportive structures (Figure 5.9). Remnants

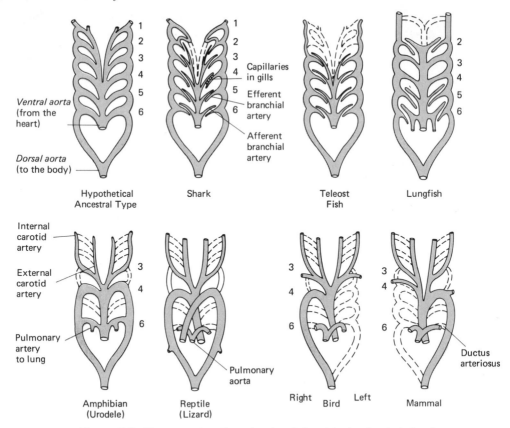

Figure 5.8. Diagrams of aortic arches in adult vertebrates (ventral views).

of gill arches in adult mammals can be found in the three middle ear bones (functioning to transmit sound from the tympanic membrane to the inner ear) and in the throat region where they form the hyoid apparatus and other structures such as the thyroid and cricoid cartilages that support the larynx and trachea. Why should the mammalian embryo have to pass through a stage in which it forms gill arches and gill slits if these structures are never to function as such? The most logical answer is that mammals have retained some genetic information in common with their distant fish-like ancestors. This information is used during early embryological development to form primitive structures capable of subsequent modification by the acquisition of the uniquely mammalian genetic information obtained during the long course of evolution.

Some examples are known of sexual maturity occurring in essentially larval (embryological) bodies, a phenomenon known as *paedogenesis (neoteny)*. The axolotl salamanders (genus *Ambystoma;* Figure 5.10) can attain sexual maturity while retaining functional gill respiration in their larval aquatic habitat. By injecting a thyroid hormone, *Ambystoma* can be induced to metamorphose into a lung-breathing terrestrial adult form. Among the simplest chordates are the sea squirts or tunicates (Figure 5.11). Adult tunicates are sessile, filter-feeding, soft-bodied animals that move water in and out like a clam. The larval tunicate appears superficially like a tadpole; it has a dorsal hollow nerve tube, a rod-like notochord, and a pharynx with gill slits. These are the three major characteristics of the phylum Chordata (the group to which humans belong). The most specialized tunicates are neotenous, permanently tailed forms that become sexually mature in the tadpole stage. Their development is direct, i.e., their embryos hatch as miniature adults. It is thought that ancestors, like these permanent larvae, represented the stock that eventually gave rise to a new chordate subphylum, *viz.* the vertebrates. Subsequent evolution from these neotenous tunicates may have produced an intermediate form similar to the modern lancelet or *Amphioxus* (subphylum Cephalochordata). Segmentation of the body (especially in the tail) in this animal is a very important adaptation for active swimming (instead of being at the mercy of water currents). The first vertebrates to appear were jawless fishes, heavily armored with dermal bony plates (Figure 5.12). Cartilage was present primitively as strengthening material around the notochord. Later, calcium was deposited in replacement bones of the axial and appendicular

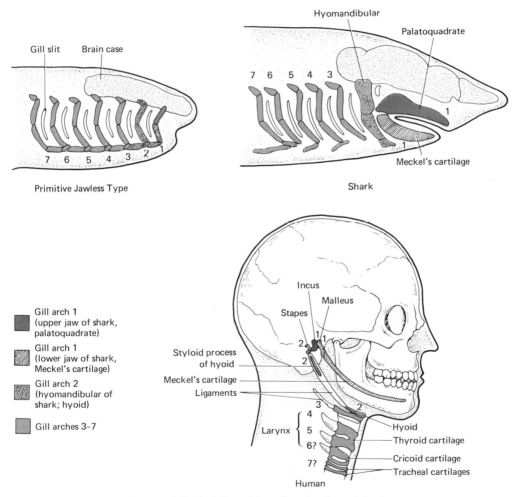

Figure 5.9. Evolution of the gill arches in vertebrates.

Figure 5.10. The axolotl salamander *Ambystoma tigrinum.* (Courtesy Carolina Biological Supply Company)

Tunicate

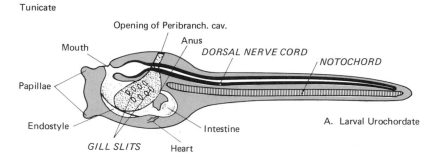

Opening of Peribranch. cav.

Mouth

Anus

DORSAL NERVE CORD

NOTOCHORD

Papillae

Endostyle

Intestine

A. Larval Urochordate

GILL SLITS

Heart

Amphioxus

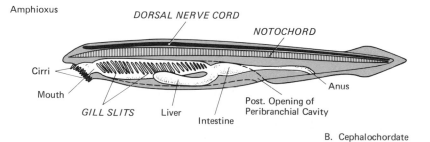

DORSAL NERVE CORD

NOTOCHORD

Cirri

Mouth

Anus

GILL SLITS

Liver

Post. Opening of
Peribranchial Cavity

Intestine

B. Cephalochordate

Lamprey (a modern jawless fish)

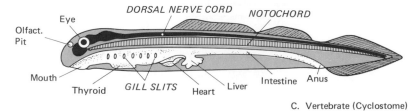

DORSAL NERVE CORD

NOTOCHORD

Eye

Olfact.
Pit

Mouth

Intestine

Anus

Thyroid

GILL SLITS

Heart

Liver

C. Vertebrate (Cyclostome)

Figure 5.11. Three modern kinds of primitive chordates.

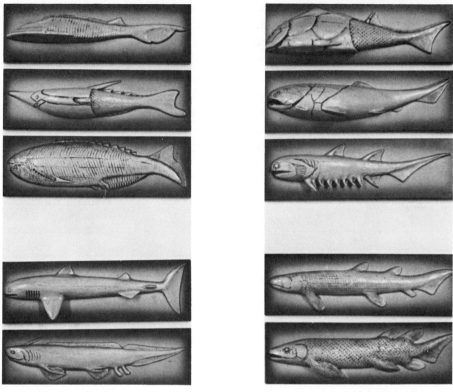

Figure 5.12. Models of primitive and rare fishes. Group of three at top left: Class Ostracodermi (extinct, armored, jawless fishes). Group of three at top right: Class Placodermi (extinct, primitive, jawed fishes). Group of two at bottom left: Class Chondrichthyes includes modern fishes with a cartilagenous skeleton as well as extinct, shark-like forms as shown here. Group of two at bottom right: Class Osteichthyes includes most modern fishes with bony skeletons; shown are a lung fish (top) and a Coelacanth (bottom). (Courtesy of Ward's Natural Science Establishment, Inc.)

skeletons. Jaws developed from visceral arches, as described earlier. The vertebrate line was well on its way to domination of the waters and the land, and it all began with neoteny.

Darwin's definition of evolution as "descent with modification" becomes the basis for our understanding of van Baer's principles. From embryology we gain one of the most impressive pictures of how great that modification can be.

Comparative Anatomy

In *comparative anatomy* we try to discern genetic affinities as expressed in adult structures rather than in embryological structures. The two disciplines of comparative anatomy and comparative embryology are very closely related and could be treated under a general heading of "comparative development," but because of historic and traditional separation they have been treated accordingly herein.

Two major components of biological structure are heredity and environment. The streamlined body contours common to the penguin, porpoise, shark, and the extinct *Ichthyosaurus* are obvious adaptations to a pelagic swimming niche (Figure 5.13). These animals, representing the vertebrate groups, bird, mammal, fish, and reptile, respectively, are not closely related. The superficial resemblance of these animals is probably the result of natural selection accumulating different changes that modified their conformations in the direction of reducing the resistance to the movement of water over their surfaces. The *analogous* biological features of different organisms are those displaying a common function or common mode of niche exploitation rather than those resulting from common ancestry or common heredity

Figure 5.13. Convergent evolution through adaptation to a common marine environment (swimming niche). From top to bottom, penguin, porpoise, shark, and *Ichthyosaurus.*

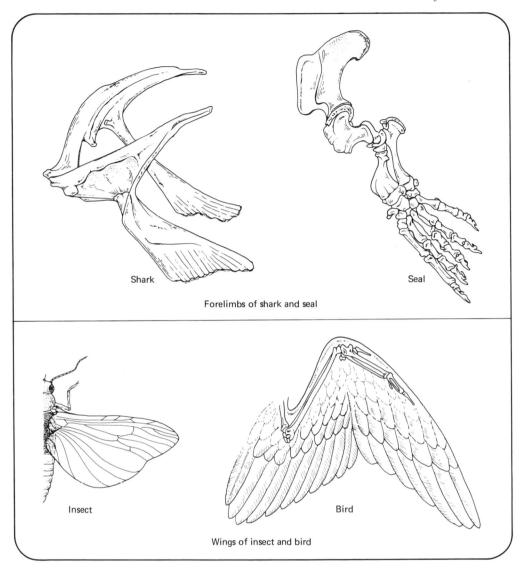

Shark

Seal

Forelimbs of shark and seal

Insect

Bird

Wings of insect and bird

Figure 5.14. Analogous structures perform a similar function but are built on different basic plans.

(Figure 5.14). The streamlined swimming forms referred to above are thus analogous structures. When widely divergent groups such as these come to resemble each other in one or more traits, the process is termed *convergent evolution*. If the similarity occurs among closely related organisms, the term *parallel evolution* is sometimes employed. There is no clear-cut distinction between these two terms.

> Limbs of the horse (odd-toed order Perissodactyla) and antelope (even-toed order Artiodactyla) are similarly adapted for running, but they are not as closely related as are deer and antelope (both artiodactyls). Relatively then, horse and antelope might be considered cases of convergent evolution, whereas deer and antelope could be consid-

ered parallel evolution. The use of the terms *convergent* and *parallel* thus depends on the diversity of the groups under consideration.

Elongated, lanceolate leaf blades (willow-like) appear in some members of a large percentage of the dicotyledonous orders of angiosperms, apparently without regard to the degree of relatedness. This kind of leaf is also found in the monocotyledonous plants, in gymnosperms, and in ferns. Within a major group, development of willow-like leaves could be called parallel evolution; between major groups (such as angiosperms and gymnosperms) it would probably be called convergent evolution.

Many succulent species of the family Euphorbiaceae (spurge order Euphorbiales) and of the milkweed family Asclepiadaceae (dogbane order Apocynales) in the Eastern Hemisphere superficially resemble members of the cactus family (Cactaceae; cactus order Cactales) of the Western Hemisphere. Apparently both groups have developed spines (all derived from different structures) and succulent growth forms in response to similar hot, dry climates through convergent evolution (Figure 5.15).

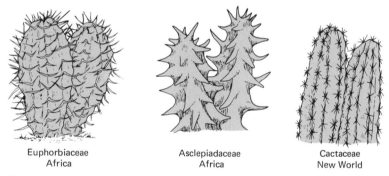

| Euphorbiaceae | Asclepiadaceae | Cactaceae |
| Africa | Africa | New World |

Figure 5.15. Convergent evolution in desert plants belonging to different families.

Within the mammalian class, we can find animals that have modified limbs for swimming (e.g., seal), for digging (e.g., mole), for running (e.g., horse), for flying (e.g., bat), for brachiating (e.g., monkey), for jumping (e.g., kangaroo), etc. (Figure 5.16). There is good evidence that all these highly modified organisms have been derived from primitive ancestral insectivores with unspecialized, five-toed (pentadactyl) feet. This divergent evolution from a generalized primitive type into diverse specialized species, each adapted to a distinct mode of life, is termed an *adaptive radiation*. Although the limbs of the aforementioned species appear superficially quite different because of adaptations to different niches, they are actually very similar in basic construction.

For example, the upper arm of a human consists of a single bone, the humerus. The forearm has two bones (radius and ulna). There are a number of small bones in the wrist (carpals), five bones in the hand (metacarpals), and two or three finger bones in each digit (phalanges). These same bones can usually be found in exactly the same order in each forelimb of all mammals (Figure 5.17). In some species, such as the horse, loss of some bones has occurred, but the general plan is the same.

Homologous structures are those that display a common construction because of common

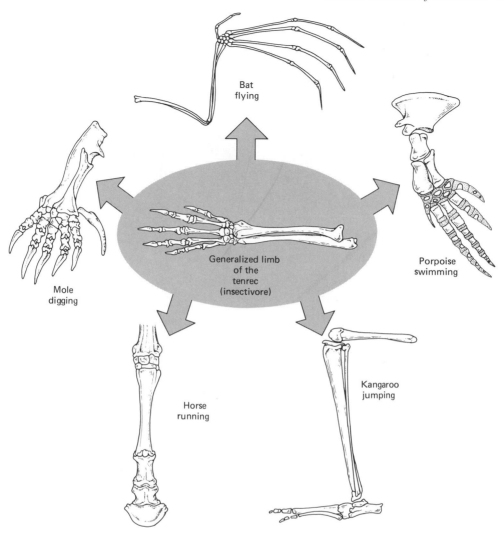

Figure 5.16. Adaptive radiation in the forelimbs of mammals.

ancestry, regardless of the use to which these structures may be put. Thus, the forelimbs of mammals are homologous structures that have undergone an adaptive radiation allowing the mammals to occupy all the major life zones. Within the pea family (Leguminosae), there is considerable variation in the structure of the stipules (leaf-like structures at the base of the leaf). In some members, it may be green and photosynthetic; in others it is nongreen and either glandular or reduced to functionless scales; in still others it may be sharp and spine-like. Although they do not perform the same functions, they are nonetheless modifications of structurally related parts, and therefore homologous.

Some structures may be considered analogous or homologous, depending on one's point of view. The forelimbs of birds, bats, and certain extinct reptiles (genus *Pteranodon*) are all adapted for flight (Figure 5.18). These wings may be considered homologous structures from

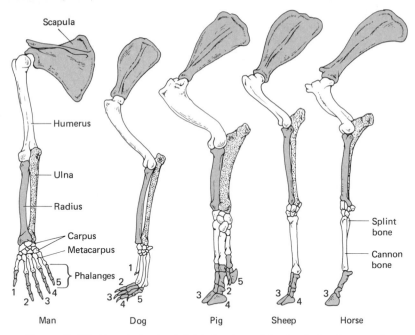

Scapula

Humerus

Ulna

Radius

Carpus
Metacarpus

Phalanges

Splint bone

Cannon bone

Man   Dog   Pig   Sheep   Horse

Figure 5.17. Homologies in skeletal structure of the forelimbs of mammals.

the view that they are all modifications of a basic tetrapod forelimb bone structure. All five digits are present in the bat wing. Birds have lost the fourth and fifth digits. The *Pteranodon* had only four digits, and the fourth was elongated to support the wing membrane. Furthermore, only birds have feathers for the flight surface. Flight arose independently in these three classes of vertebrates. Although there are common adaptations for flight, the constructions of the airfoils are so different in these three groups that they might best be viewed as analogous structures.

Within an individual, we sometimes find two or more structures that are basically similar in construction, but modified to perform different functions. Such features exhibit *serial homology*. Those who believe in the classical (Goethean) *foliar theory* of flower evolution view the various parts of a flower (carpels, stamens, petals, sepals) as serially homologous with leaves lower on the stem. This concept is not universally accepted, however. The origin(s) of angiosperms (flowering plants) is still shrouded in mystery.

Crabs, shrimp, and crayfish (Figure 5.19) belong to the Class Crustacea in the Phylum Arthropoda (jointed-footed animals). These animals typically bear a pair of appendages on each body segment. These appendages are serially homologous in that they usually have two segments in a basal portion (protopodite) that supports two parallel distal portions (a medial endopodite and a lateral exopodite), each consisting of a series of segments. In the head region of the crayfish, the first two pairs of appendages serve as sensory antennae. The third pair are the biting mandibles. The fourth and fifth pairs are maxillae specialized for food handling. In addition, the fifth appendages have a dorsal outgrowth of the exopodite, called the epipodite, which functions to deflect a current of

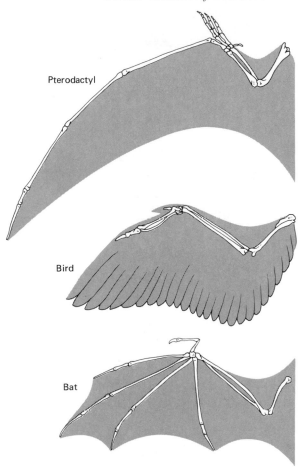

Figure 5.18. Homologies obviously exist in the skeletal elements of these vertebrate wings. However, the softer portions that form the airfoils are so different that they might be considered only analogous structures.

water over the gills. In the thoracic region of the crayfish, the appendages of segments 6, 7, and 8 form maxillipeds modified for touch, taste, and food handling. The ninth segment bears the largest appendages called pincers (or chelae) used in food gathering and defense. The tenth and eleventh pairs of appendages bear small pincers and serve as walking and grasping legs. The next pairs of walking legs do not have pincers. Appendages in segments 7 through 12 also bear gills. The abdominal region of the crayfish has six segments. The first two abdominal appendages (on segments 14 and 15) are modified in the male for the transport of sperm to the female; in the female these may be reduced or serve as swimmerets. The next three pairs of appendages are also swimmerets, which function in water circulation in both sexes and for carrying eggs and young in the female. The last body segment has a pair of broad, flattened appendages that form the uropod used in swimming (and for egg protection in the female). If each appendage was created to serve the functions it now performs, it is difficult to understand why they are all constructed on the same basic plan. Each function is carried out in nonsegmented animals by structures with a completely different design. According to the theory of evolution, however, it is easy to envisage how natural selection gradually accumulated adaptive modifications, which enhanced a division of labor among the nearly identical swimmeret-like appendages of the presumptive primitive ancestor.

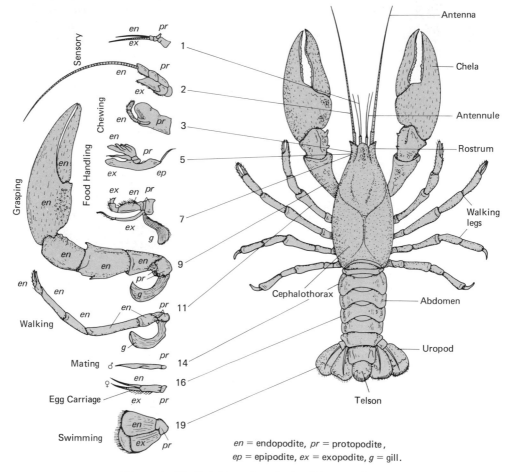

Figure 5.19. Serial homology in crayfish appendages.

Evolution need not always result in the production of greater complexity. Much like water following the path of least resistance to the sea, evolution is opportunistic. Whatever variation (modification) is adaptive to any degree will be favored by natural selection so that greater numbers of individuals in subsequent generations will come to possess the trait, until it becomes virtually universal within the species. Even the loss of complexity may be advantageous.

Crabs are among the most highly modified members of the crustaceans. There has been a great deal of loss in complexity in their evolution. The abdomen is small, recurved, and held tightly against the ventral part of the thorax. The first two abdominal appendages serve as copulatory organs, but the rest of them are rudimentary or missing entirely.

Barnacles are so highly modified that at one time they were mistaken for mollusks. The cypris larva of the gooseneck barnacle (Figure 5.6) attaches to the substrate by its head, losing its antennae and compound eyes in the process. The abdomen is usually reduced or absent. Calcareous plates are secreted around the animal. It uses its thoracic appendages, which typically are plume-like cirri, to kick food into its mouth. It is difficult to see in the greatly modified adult structure an affinity to the crustaceans, but it is quite evident in the larvae.

Sacculina is a barnacle parasitic on crabs (Figure 5.20). The early nauplius larva of this organism resembles that of other crustaceans, but lacks a digestive tract. The later cypris larva burrows into the host and degenerates into a mere mass of cells that eventually invades all parts of the crab's body in a root-like growth, absorbing nourishment from the tissues of its host.

All the above are examples of organisms in which loss of complexity has been advantageous to the exploitation of a particular ecological niche. They are still very successful creatures in the economy of nature.

Probably one of the most dramatic lines of evidence for evolution is seen in *vestigial* or *rudimentary structures*. These nonfunctional structures and organs are easily explained by the theory of evolution as the now useless remnants of functional structures in ancestral stock.

The finding of a rudimentary pelvic girdle in whales suggests that they have evolved from

Figure 5.20. The adult barnacle *Sacculina* parasitizes crabs as a root-like growth of cells. Its nauplius larva reveals its crustacean relationships.

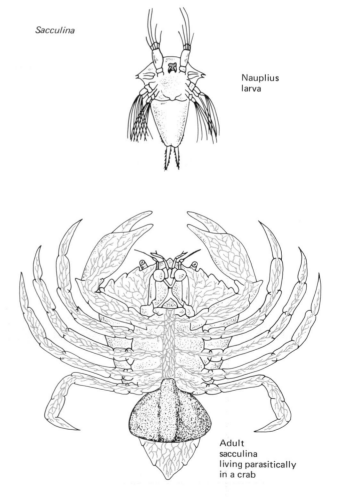

Sacculina

Nauplius larva

Adult sacculina living parasitically in a crab

a four-legged (tetrapod) ancestor. Even some snakes such as pythons and boas have remnants of hindlimbs. Hoofed mammals (ungulates), such as the horse, display a small vestige of one of the toes called a "splint bone" (Figure 5.21). These are unmistakable traces of evolution from an ancestor with a more typical mammalian limb construction.

Primitive members of the snapdragon family Scrophulariaceae (such as the mullein;

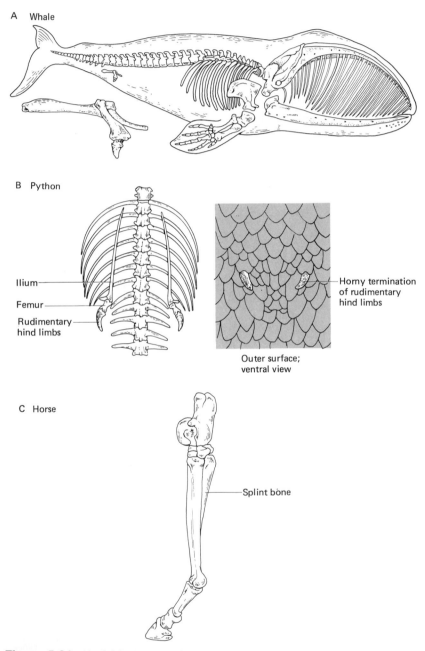

Figure 5.21. Vestigial structures in vertebrates. (A) rudimentary pelvis in whale; (B) vestigial hind limbs in python; (C) splint bone on forelimb of modern horse.

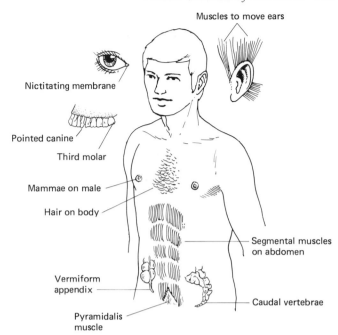

Figure 5.22. Some vestigial structures in humans.

Verbascum) have five stamens in the flower whereas the more advanced members of this family (such as the bush monkey-flower, *Mimulus longiflorus*) have only four stamens. Among other genera of this family, rudiments of the fifth stamen are present. In some cases this rudiment has been converted into a nectar gland; in others it has been modified into a structure resembling a toothbrush.

The pairs of lateral glands that appear on the petioles of apricot leaves (*Prunus armeniaca*) apparently represent modified rudimentary leaflets corresponding to those of roses in the same family (Rosaceae). Sometimes an "abnormal" apricot leaf will have the lateral glands replaced by small green leaflets.

Nearly a hundred vestigial structures have been cataloged in modern humans (Figure 5.22). The best known is the vermiform appendix, a blind sac at the juncture between the small and large intestines. In certain other mammals, such as the guinea pig and the horse, the homologue of this organ is a large caecum in which bacterial digestion of food occurs. Presumably the appendix functioned similarly in the distant ancestors of humans.

Many land vertebrates have, in the medial corner of the eye, a transparent nictitating membrane, which can be used to wipe the eyeball clean by a horizontal movement akin to the vertical movement in the blinking of our eyelids. Frogs move their clear nictitating membranes into place when they dive, where they function as underwater goggles. Humans retain a vestige of this membrane as a small crescentic fold in the corner of the eye.

Humans have a complete, albeit rudimentary, set of muscles for moving both ears and a nonexistent tail. Presumably our ancestors had a tail, and it was an adaptive advantage for them to move their ears for the detection of sounds.

Comparative Biochemistry and Physiology

The morphological traits that characterize a species ultimately depend on the constellation of enzymes, hormones, and other biochemicals involved in the process of differentiation. As a matter of fact, homologies at the molecular level are even more convincing than those at the morphological level, because the former are either closer to immediate gene products or are themselves gene products. It is in terms of the degree of genetic relationship that evolutionary lineages are optimally expressed.

Hemoglobin is the oxygen-carrying pigment of red blood cells. It is a conjugated protein (molecular weight approximately 64,500) composed of four polypeptide chains. The normal adult human hemoglobin molecule consists of two identical alpha chains and two identical beta chains. Each alpha chain has 141 amino acids, and each beta chain has 146 amino acids. The genetic information for specifying these two polypeptide chains resides in two sequences of nucleotides in DNA; i.e., there is one gene for the alpha chain and one gene for the beta chain. A comparison of the amino acid sequences in the beta chains of humans and gorillas shows them to be identical except at one position. The beta chain of the pig differs at about ten sites and the horse at 26 sites from the human beta chain. As a general rule, the number of differences between homologous polypeptide chains of a certain type found in different species is roughly proportional to the degrees of difference in genetic relationships as established by standard methods of phylogenetic classification. In comparing the structures of many homologous proteins, the inescapable conclusions are that one of man's nearest living relatives is the gorilla and that these two species have evolved from a common ancestor in fairly recent times.

Another protein studied in great detail is *cytochrome c,* a respiratory pigment essential for oxidative metabolism. There is more information currently available on the evolution of this molecule than on that of any other protein. The cytochrome c molecules are identical in chimpanzees and humans, consisting of 104 amino acids in exactly the same order. The cytochrome c of humans differs from that of the red bread mold *Neurospora* in 44 of the 104 amino acid positions, yet the three-dimensional structures of these two molecules are essentially alike. The functional similarity of cytochrome c molecules from all eukaryotic (truly nucleated) cells is demonstrated by the fact that, regardless of its source, the cytochrome c can react in the test tube with cytochrome oxidase (a multienzyme complex that normally accepts electrons from cytochrome c) from any other species. Although functionally the same, the amino acid sequences of cytochrome c from different species may be quite different. The degree of difference appears to correspond quite well to the distance between species on the evolutionary tree. Computer analyses of these differences for the more than forty species whose primary cytochrome c structures are known have produced elaborate family trees of living organisms entirely without recourse to the more traditional morphological data (Figure 5.23). These trees agree remarkably well with those obtained by the more traditional method. It is encouraging to know that morphological similarities, in most cases, are a reflection of molecular (and hence genetic) similarity.

Cytochrome c appears to have evolved much more slowly than hemoglobin. One explanation for this phenomenon is that cytochrome c is a relatively small protein that

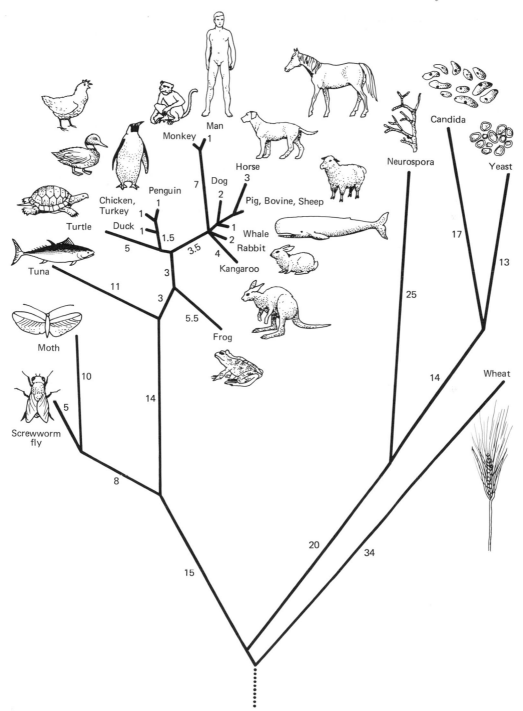

Figure 5.23. Phylogenetic tree constructed by computer analysis of homologies between cytochrome c molecules in various organisms as shown. Each node in the tree represents a common ancestral protein; the length of each branch (indicated by numbers thereon) represents the most likely number of point mutations that occurred along the way.

interacts over a large portion of its surface with molecular complexes that are larger than itself. Therefore, natural selection has operated to conserve specific amino acids at many key positions on the molecule where it mates with its "substrates" (the reductase and oxidase complexes). On the other hand, hemoglobin interacts at many fewer loci with oxygen, which is smaller than itself. Therefore, more "nonessential" amino acid positions could be changed in hemoglobin and still retain its functional integrity.

The primary structure (amino acid sequence) of the protein called α-melanocyte stimulating hormone (MSH) is a string of thirteen amino acids identical in domestic horses, cattle, and pigs. This hormone is produced by the intermediate lobe of the pituitary gland attached to the base of the midbrain. It is probably responsible for pigmentation changes in the skin and/or hair of many mammals. A hormone produced by the anterior lobe of the pituitary gland is called ACTH (adrenocorticotropic hormone) and serves to stimulate the adrenal gland (another endocrine organ). The first thirteen amino acids in ACTH from sheep, cattle, and pigs are identical to those in MSH. There are amino acid sequences unique to each of these species in positions 25 to 33 (from the amino terminal end), but otherwise all 39 amino acid positions of the complete protein hormones are identical. Cushing's disease of humans is characterized by high levels of ACTH. A symptom of the disease is hyperpigmentation of the skin, because of the MSH-like activity of ACTH. The identical biochemical features of these two hormones (Figure 5.24) argues very strongly for the evolution of one from the other (perhaps by partial gene duplication). The minor differences in ACTH structure within the hoofed mammals is a reflection of their relatively close genetic relationship through evolution.

Insulin is a polypeptide hormone produced by the pancreas and serves to regulate blood sugar level. Beef insulin is so similar to human insulin that it has been used for the treatment of human diabetes and normally escapes being recognized as foreign by the very discerning immune system. Occasionally, however, an individual becomes intolerant of the foreign insulin. In these cases, the foreign molecule acts as an antigen, which stimulates the production of antibeef insulin antibodies. These antibodies are thought to neutralize the physiological activity of the hormone in a manner analogous to the neutralization of bacterial toxins by homologous antitoxins. The animal insulins analyzed thus far (including beef, sheep, pig, whale, horse, rabbit) differ only in one to three amino acid positions. Yet a highly discriminating antiserum can tell these molecules apart. This power of serological analysis has been widely employed for the study of genetic relationships between different species. A rabbit injected with beef insulin may produce an antibeef insulin antibody that will not only specifically react with cattle insulin, but will also cross react broadly with many other mammalian insulins. This indicates that the molecular structure of mammalian insulins are very similar. Serum contains a very complex set of proteins. Some of these proteins, such as insulin, may be very similar (or identical) at the molecular level within a relatively large taxonomic group. Other proteins, such as albumins, may differ markedly from one species to another within a relatively small taxonomic group.

The technique most widely used to detect the union of antigen and antibody *in vitro* is the precipitation test. This test can be quantified by measuring the amount of precipitate formed. The more proteins antigenically related to that molecule from which the antiserum was derived, the more precipitate that should be formed. By testing various primate sera against antihuman serum produced by a rabbit, it was found that the sera of the orangutan,

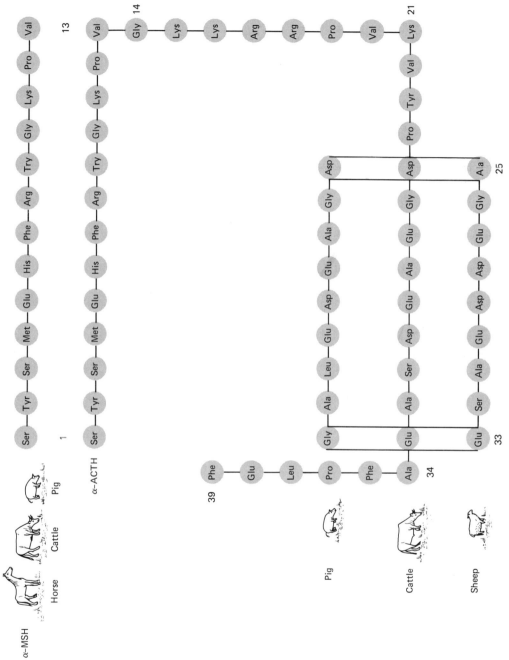

Figure 5.24. Molecular homology between MSH (melanocyte stimulating hormone) and ACTH (adrenocorticotropic hormone) in selected hoofed (ungulate) animals.

chimpanzee, gorilla, and Old World monkeys were more closely related serologically to human serum than were those of the New World monkeys and marmosets. Lemurs (primitive primates) appeared to be almost unrelated by this technique. These results are in agreement with the phylogenetic trees constructed from morphological data. A higher percentage of positive reactions is generally found when antigens and antibodies are derived from animals within the same taxonomic order. For example, antihuman serum antibodies made in the rabbit react only with the blood of primates (Figure 5.25). This phenomenon is the basis of the forensic precipitation test for the identification of blood stains. Antibeef serum has been shown to react positively with 72 per cent of the ungulate (hoofed animals) antigens. Antiwallaby serum gave positive tests with 68 per cent of the marsupial (pouched animals) antigens. Very little, if any, reaction is seen between birds and other vertebrate taxonomic classes. Blood obtained from a frozen Siberian mammoth (now extinct) and antiserum against the Indian elephant produced a detectable precipitation.

An antisheep serum (made by injecting a rabbit with sheep serum) will give the strongest precipitation reaction with homologous sheep serum, a fairly strong reaction with beef serum, and a relatively weak reaction with pig and horse sera. Thus, with one serological reagent we infer that cattle and sheep are more closely related (at the molecular level of their serum proteins, and hence in the genetic endowments that specify these proteins) than are sheep and pigs or horses. This unidimensional view fails to tell us, for example, that pigs and horses have relatively little serological relationship with one another. By preparing antisera against each animal's sera and testing them in all combinations, the systematic serologist can view antigenic relationships from several vantage points. Such data allow the construction of three-dimensional models of serologic relationships as shown in Figure 5.26. There is nothing innate in serological data to indicate primitiveness. Generally, however, there is good agreement

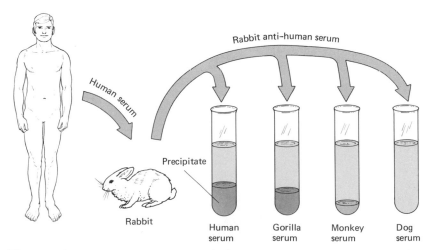

Figure 5.25. Use of the precipitation test for demonstrating serological relationships. Human serum is injected into a rabbit, which responds by making antibodies to the foreign human serum proteins. Testing these antibodies against various vertebrate serums produces precipitates in proportion to the degree of antigenic relationship with humans.

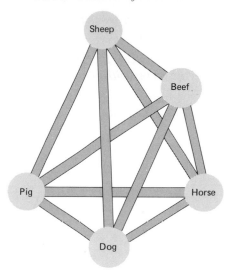

Figure 5.26. Three-dimensional model of the relative serological distances between five mammals as determined by precipitation tests.

between these serological models and phylogenetic trees constructed from conventional morphological data.

Antigenic (molecular) relationships between plant species have also been studied by the precipitation test. Varieties of grapes that could not be successfully grafted were found to be serologically distinctive, whereas varieties that were grafted successfully were serologically similar. Contamination or adulteration of foods can be detected serologically. For example, the presence of horse meat in "pure pork" sausage can be detected through the precipitation test. Likewise the presence of ergot (a fungus that infests cereals) in flour or of potato in wheat flour can be detected serologically.

The tools available to the biologist for demonstrating the phylogenetic origin of modern species have progressed to the point that individual changes at the molecular level are routinely under investigation.

Perhaps the most important source of evidence in support of the theory of evolution comes from the science of genetics and its manifold subdisciplines (including cytogenetics, biometrics, population genetics, etc.). A thorough understanding of the modern theory of evolution can only be gained by those who have a good basic understanding of the principles of heredity. For this reason, a brief review of genetics and its implications for the theory of evolution are presented in the next section.

Essay Questions

1. *According to the principle of natural selection, should functionless structures be perpetuated?*

2. *Why are the embryological stages of any two related species more likely to resemble each other than their adult forms?*

3. *Certain human proteins apparently have evolved at a much faster rate than others. Offer a physiological (biochemical) explanation for this phenomenon.*

4. *How can the degree of geneological relationship be estimated in adult organisms with similar morphologies?*

5. *What kinds of characteristics are of greatest use to the taxonomist? To the systematist?*

part II

THE MECHANISMS OF HEREDITY

REPRODUCTION

Only cells have all the properties required by the current definition of "life." One of the cardinal characteristics of "life" is the ability of a single entity to multiply, retaining its essential "nature" in the process. This is *reproduction*. Viruses are not classified as "living" even though they can reproduce, because they do so only within cells and lack important cellular attributes, such as metabolism and irritability. Viruses probably arose from cellular components and evolved a parasitic reproductive cycle dependent on cells. Therefore, although viruses do not represent a "connecting link" in the evolution of "living" systems from abiotic chemical systems, they do serve to remind us that all the properties we now ascribe to "life" need not have evolved together. Only cells exhibit the properties of living matter, and the mechanisms of *cellular reproduction* therefore are of great importance in the science of evolution.

Life, as we know it today, began with the first cell. Many modern organisms still exist in the primitive unicellular condition (e.g., bacteria, blue-green algae, protozoa). Through cellular reproduction, many "copies" of the original cell could be formed. If these cells clung together into a multicellular colony perhaps some of the cells could become specialized in anatomy (structure) and/or physiology (function) to perform certain tasks ("division of labor"). Obviously a cell specialized to do one or a limited number of tasks could do a better job thereon than a generalized (unspecialized) cell. This is probably the adaptive advantage that was seized on by natural selection early in the evolution of living systems and resulted in numerous kinds of multicellular organisms. Thus, cellular reproduction is responsible not only for the continuity between generations, but also for the elaboration of multicellular plants and animals.

The Prokaryotic Cell

Prokaryotes are organisms lacking membrane-bound subcellular components (*organelles*). There is no nucleus, and chromosomes are not present. There is, however, a "nuclear region" occupied by a single circular molecule of deoxyribonucleic acid (DNA), uncomplexed with proteins. The instructions for making another similar or identical organism are encoded into the DNA molecules. DNA has the ability to not only replicate itself (*autocatalysis*), but also direct the synthesis of other large molecules (*macromolecules*) such as ribonucleic acids (RNA) and proteins through *heterocatalysis*. The segments of DNA that specify the structure of these particular macromolecular products are called *genes*. Ultimately, genes are responsible for the cytological, anatomical, morphological, physiological, and behavioral characteristics of each organism. Understanding evolution necessitates understanding how gene complexes replicate, and how beneficial changes in the system may be introduced and maintained.

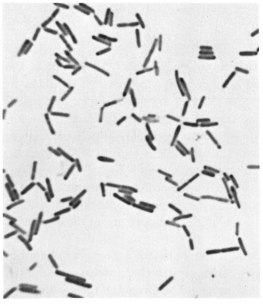

A

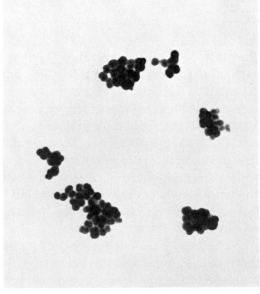

B

The prokaryotic condition is undoubtedly primitive and today is found only in bacteria (Figure 6.1) and blue-green algae (Figure 6.2). Because of their relative simplicity and ease of culturing, bacteria have been used widely in genetic investigations. Much has been learned of the structure and functions of DNA from these lowly organisms. They reproduce almost exclusively by asexual means, but even bacteria may occasionally experience recombinations through partial "sexual" mechanisms. Usually only a relatively small fragment of DNA is transferred from one cell to another (*meromyxis*). This is in contrast to the more complex eukaryotic cells, where sex involves the union of two complete sets of genetic instructions furnished by gametes. Prokaryotes do not make gametes. Bacteria use at least three mechanisms to accomplish genetic recombination:

1. *Conjugation.* A donor ("male") cell may unite with a recipient ("female") cell by a thin cytoplasmic bridge and transfer part of its DNA. Recombination occurs by deletion of a part of the recipient's DNA and incorporation of the donor's DNA (Figure 6.3).
2. *Transformation.* When cells disintegrate, DNA fragments are released. Other living cells may absorb this DNA and incorporate it into their own DNA by a breakage and reunion mechanism (Figure 6.4) similar to the one to be described later in the eukaryotic cell.
3. *Transduction.* Certain viruses (called *bateriophages* or *phages*) attack and lyse bacterial cells (Figure 6.5A). Occasionally a virus may exchange part of its own nucleic acids for that of the bacterium. When the virus is released from the host cell and reinfects another cell, the new host may incorporate some of the genes from the previous host cell (Figure 6.5B).

Figure 6.1. Three kinds of bacterial morphology: (A) bacillus; (B) coccus; and (C) spirillum. (Courtesy Carolina Biological Supply Company)

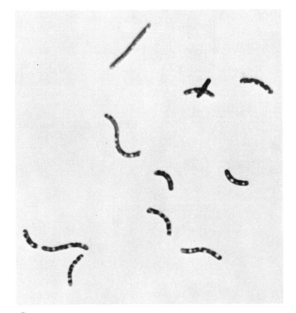

C

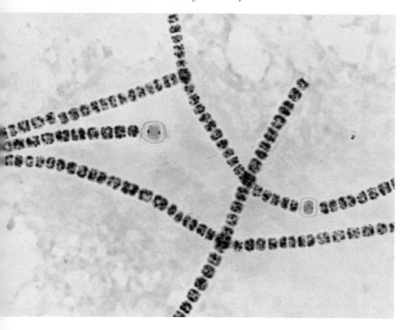

Figure 6.2. A filamentous blue-green alga (*Nostoc*). The darkly staining chromatin material inside these cells is not confined within a nucleus (prokaryotic condition). (Courtesy Carolina Biological Supply Company)

CONJUGATION

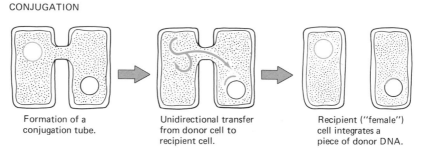

Formation of a
conjugation tube.

Unidirectional transfer
from donor cell to
recipient cell.

Recipient ("female")
cell integrates a
piece of donor DNA.

Figure 6.3. Diagram of conjugation in bacteria.

TRANSFORMATION

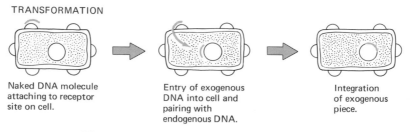

Naked DNA molecule
attaching to receptor
site on cell.

Entry of exogenous
DNA into cell and
pairing with
endogenous DNA.

Integration
of exogenous
piece.

Figure 6.4. Diagram of transformation in bacteria.

A. Lytic life cycle of a bacteriophage.

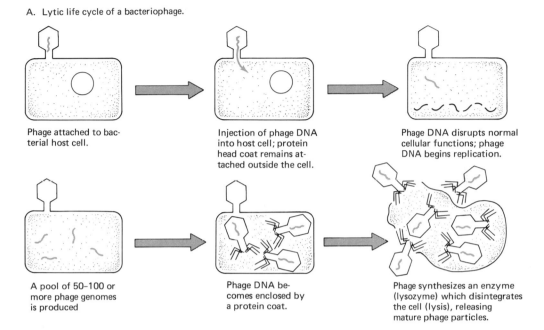

Phage attached to bacterial host cell.

Injection of phage DNA into host cell; protein head coat remains attached outside the cell.

Phage DNA disrupts normal cellular functions; phage DNA begins replication.

A pool of 50–100 or more phage genomes is produced

Phage DNA becomes enclosed by a protein coat.

Phage synthesizes an enzyme (lysozyme) which disintegrates the cell (lysis), releasing mature phage particles.

B. Transduction

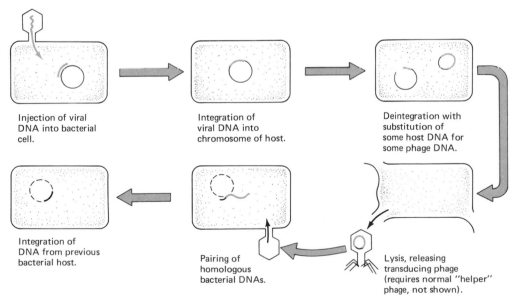

Injection of viral DNA into bacterial cell.

Integration of viral DNA into chromosome of host.

Deintegration with substitution of some host DNA for some phage DNA.

Integration of DNA from previous bacterial host.

Pairing of homologous bacterial DNAs.

Lysis, releasing transducing phage (requires normal "helper" phage, not shown).

Figure 6.5. Two alternative fates of a single phage. (A) Lytic (vegetative) life cycle; (B) nonlytic and transducing event. Sometimes a piece of bacterial DNA becomes substituted for phage DNA. Upon infection of a genetically different host cell, the new host may incorporate one or more genes from the previous host cell.

Even viruses, which are not "free-living" entities, may recombine their genetic information. When two or more genetically different viruses infect the same cell, many replicas of the viral DNA are made. Several rounds of "matings" (pairings) and exchanges may then occur between these different nucleic acid molecules.

The Eukaryotic Cell

Cells of the more highly evolved plants and animals have membrane-bound organelles such as nuclei, mitochondria, and chloroplasts. Here again, there appears to be a division of labor even within the cell. These "truly nucleated" (*eukaryotic*) cells have many similar features even though they are found in such widely different organisms as trees and mammals. The eukaryotic cell (Figure 6.6) typically consists of an outer *plasma membrane* enclosing the *protoplasm* (an organized chemical milieu which *in toto* exhibits the properties of "life"). Two major regions of the protoplasm are the membrane-bound *nucleus* and the *cytoplasm* (substances outside the nucleus). Within the nucleus are the *chromosomes* composed of DNA and protein (*nucleoprotein*).

The cell is analogous to a small factory. The nucleus represents the "supervisor" with the blueprints and gives directions to the "workers" in the cytoplasm. The cytoplasm contains the

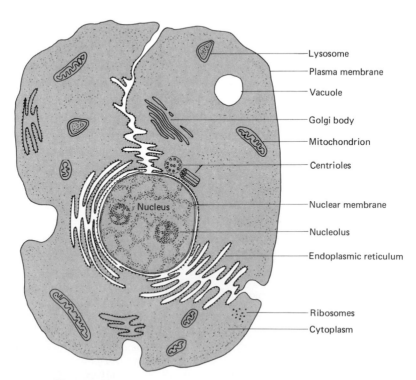

Figure 6.6. Diagram of an eukaryotic cell (animal, generalized).

machinery for executing these instructions. *Chloroplasts* (found in plant cells) act as "traps" for solar energy, converting it by photosynthesis into chemical energy stored in the form of sugars. The *mitochondria* contain enzymes that convert raw materials (nutrients such as sugars) into high-energy compounds such as adenosine triphosphate (ATP), that then serve as common currency wherever energy is required in the cell. *Ribosomes* function as "jigs" to hold RNA molecules and enzymes together in the formation of new proteins. All enzymes are proteins. *Enzymes* are molecules that accelerate chemical reactions but are not themselves consumed in the process. The explanation for how different cells (of a multicellular organism) become specialized to perform different tasks (*differentiation*) lies in the understanding of how specific genes become active (in enzyme synthesis) only in certain cells at critical times during embryonic development. Thus, evolution may involve not only the particular constellations of genes represented in organisms from one generation to the next, but also the numerous control mechanisms that allow these genes to express themselves at various times during the life cycle.

Chromosomes

A typical eukaryotic cell contains two sets of chromosomes. One set is of *maternal* (mother parent) origin; the other is of *paternal* (father parent) origin. Each species has a characteristic number of chromosomes in each cell, called the *diploid number* ($2n$). The diploid numbers of some selected species are shown in Table 6.1. Note that the number of chromosomes bears no correlation to the position of the species in the phylogenetic scheme.

Humans have a diploid number of 46, but domestic cattle have $2n = 60$. Cattle are not considered to be more highly evolved than humans. The range of diploid numbers in

TABLE 6.1. Diploid Chromosome Numbers of Selected Species

Organism	Dipoid Chromosome Number
Ascaris worm	2
Fruit fly (*Drosophila melanogaster*)	8
Garden pea	14
Corn (*Zea mays*)	20
Opossum	22
Bull Frog	26
Salamander (*Ambystoma*)	28
Domestic swine (hog)	38
Mouse	40
Human	46
Tobacco	60
Horse	60
Monkey (*Cebus*)	54
Crayfish	~200
Wild Rhubarb	~200

animals extends from 2 in a parasitic worm (*Ascaris univalens*) to approximately 1,600 in a protozoan (*Aulacantha* species). In plants, the range is from 4 in the flowering plant *Haplopappus gracilis* to about 1,260 in a fern (*Ophioglossum reticulatum*). The normal diploid range for most plants, however, is 14–24, and perhaps slightly higher for most animals.

Nonreproductive cells are termed *somatic* (body) cells; reproductive cells (sperms, eggs) are called *gametes*. The divisional process called *mitosis* replicates cells and maintains the same number of chromosomes in the products (daughter cells) as in the original (parent) cell. In the divisional process called *meiosis,* the two sets of chromosomes of the parent cell are reduced to a single set in the products (gametes), called the *haploid number* (*n*). All the genes in a gamete are collectively termed a *genome*.

Individual chromosomes may be recognized by the following peculiar cytological characteristics.

1. *Length.* Chromosomes of the haploid set are conventionally numbered consecutively from longest to shortest. These measurements are best made when the chromosomes are at their maximal contraction during the division process and therefore most easily visible. All the chromosomes in a somatic cell represent a *karyotype,* typical for each species (Figure 6.7). Chromosomes from a photomicrograph of a somatic cell may be cut out and rearranged as matched pairs in a standard sequence based on size and other characteristics. When the karyotype is represented in this fashion (either as pictures or as diagrams), it is referred to as an *idiogram* (Figures 6.8 and 6.9).

2. *Position of the centromere.* If the centromere (primary constriction) is near the middle, the chromosome will have arms of equal length. Such a chromosome is called *metacentric.* If the centromere is slightly off center, the chromosome will have a long and a short arm (*submetacentric*). The term *acrocentric* is applied to a chromosome

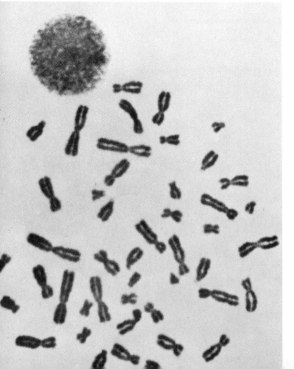

Figure 6.7. Photograph of the chromosomes of a normal male. This cell has been stopped in the middle of cell division and chemically treated to make the chromosomes spread apart for easier counting. [From Redding, A. and Hirschhorn, K.: *Guide to Human Chromosome Defects* In Bergsma, D. (ed.): Birth Defects: Orig. Art. Ser., Vol. IV, No. 4 Published by The National Foundation-March of Dimes, White Plains, N.Y. 1968, p. 92]

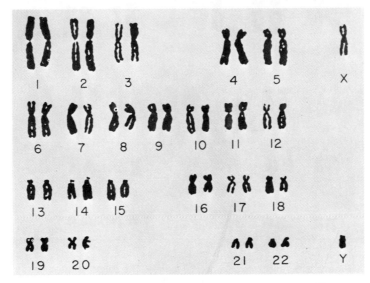

Figure 6.8. Idiogram of the chromosomes of a normal male. The same chromosomes shown in Figure 6.7, cut apart and arranged in pairs for easier analysis. Notice that the 22 paired chromosomes (the autosomes) are arranged in descending order according to size and shape. The X and Y chromosomes, which determine the sex of the individual, are easily recognized as different from the autosomes. [From Redding, A. and Hirschhorn, K.: *Guide to Human Chromosome Defects* In Bergsma, D. (ed.): Birth Defects: Orig. Art. Ser., Vol. IV, No. 4 Pulished by The National Foundation-March of Dimes, White Plains, N.Y. 1968, p. 92]

with a centromere very near one end, creating a very long arm and a stubby arm. *Telocentric* chromosomes have a terminal centromere and hence only a single arm.

3. *Satellites.* Not all chromosomes have satellites, but such structures may be diagnostic when present. A satellite is a small glob of chromatin connected to the end of a chromosome by a thin chromatic filament.

4. *Banding.* For reasons not yet understood, certain regions of the chromosome stain differentially with quinacrine derivatives (producing Q-bands) or with Giemsa stain (producing G-bands). Each chromosome has its own specific banding pattern, allowing identification of individual members of the karyotype (Figure 6.10).

Each kind of chromosome in a sperm cell has its counterpart in an egg cell. These chromosomal "twins," called *homologous chromosomes* or *homologues,* have identical cytological features and are usually genetically very similar in the sequence of their DNA nucleotides. This genetic similarity allows the homologues to *synapse* or form pairs during meiotic division. Pairing is an important step in the formation of gametes. The genetically dissimilar chromosome sets in a hybrid between two species may not have sufficient homology to form pairs. This would lead to hybrid sterility and thus keep the species reproductively isolated from one another (see Chapter 17).

In many animals and a few plants, one of the homologous pairs of chromosomes

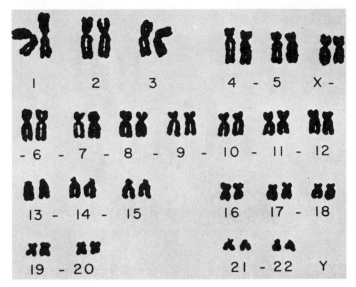

Figure 6.9. Idiogram of a normal female. The paired autosomes are not recognizably different from those of the male. There is, however, no Y chromosome. Instead, there are two X chromosomes, which appear in the upper right-hand corner of the idiogram. [From Redding, A. and Hirschhorn, K.: *Guide to Human Chromosome Defects* In Bergsma, D. (ed.): Birth Defects: Orig. Art. Ser., Vol. IV, No. 4 Published by The National Foundation-March of Dimes, White Plains, N.Y. 1968, p. 93]

1. Length and Position of Centromere

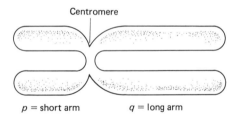

Centromere

p = short arm q = long arm

p/q ratio is useful for characterizing each chromosome

3. Banding

2. Satellites

Satellite

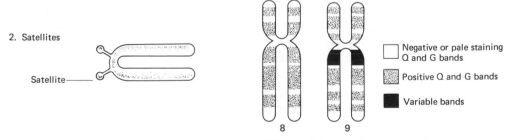

Negative or pale staining Q and G bands

Positive Q and G bands

Variable bands

8   9

Differentiation of morphologically similar human chromosomes (8 and 9) by banding pattern.

Figure 6.10. Characteristics of individual chromosomes.

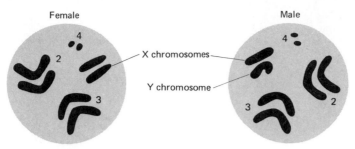

Figure 6.11. Diagram of somatic cells in *Drosophila melanogaster* ($2n = 8$).

determines the sex of the organism. These homologues are often of recognizably different morphologies (i.e., *heteromorphic*). These are called the *sex chromosomes* and are commonly labeled X and Y. For example, male mammals have an X and a Y chromosome; females have two cytologically identical X chromosomes. Although the X and Y chromosomes appear different, they usually have some genetically homologous regions that allow them to form pairs during meiosis. All chromosomes other than these sex chromosomes are called *autosomes* (Figure 6.11).

Mitosis

Plant cells have two major features that generally distinguish them from animal cells: (1) a cell wall of cellulose and other strengthening materials and (2) chloroplasts. Unicellular organisms (either plant or animal) can reproduce by dividing mitotically into two cells (new organisms). A multicellular organism may pinch off a reproductive cell, called a *spore,* by mitosis. The spore multiplies (by repeated rounds of growth and mitotic divisions) into the multicellular adult to complete the life cycle. Mitosis is essentially the same in every eukaryotic cell (Figures 6.12 and 6.13). The stage between mitotic divisions is called *interphase.* In this stage most cells are metabolically active in carrying out their specific functions. DNA molecules also replicate in interphase prior to the next mitotic division cycle. Mitosis is a continuous series of events classified for convenience into four major stages.

1. Prophase. Each chromosome enters mitosis from the previous interphase with two identical DNA molecules forming the backbones of identical sister nucleoprotein strands called *chromatids.* The sister chromatids are held together for a time by a late replicating segment of DNA called the *centromere* region. Each chromosome shortens and thickens as it coils and supercoils on itself in a fashion analogous to the winding of a rubber band connected to the propeller of a toy airplane. The nuclear membrane finally dissolves, giving the chromosomes access to the entire cell. A spindle-shaped apparatus consisting of protein fibers begins to form from each of two polar regions of the cell. Some of these fibers attach to the centromeres.

2. Metaphase. Each chromosome then becomes attracted to a central region of the cell called

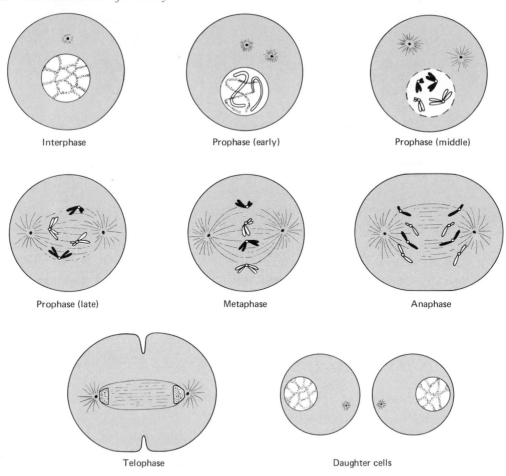

Interphase   Prophase (early)   Prophase (middle)

Prophase (late)   Metaphase   Anaphase

Telophase   Daughter cells

Figure 6.12. Diagram of mitosis in animal cells.

the *equatorial plate* (*plane*). The centromere region replicates its DNA and frees the two sister chromatids. The *spindle* becomes fully formed.

3. Anaphase. Contractile fibers of the spindle apparatus pull the former sister chromatids (now recognized as individual chromosomes) apart toward opposite poles of the cell.

4. Telophase. A group of chromosomes congregate at each of the two polar regions and begin to unwind, thereby returning to the attenuated state of interphase. A nuclear membrane begins to form around each chromosome cluster. The cytoplasm divides what was formerly one cell into two. Cytoplasmic division (*cytokinesis*) occurs in animal cells simply by the cell's constricting into two daughter cells. New cell walls must be deposited to divide the cytoplasm of plant cells in two. Almost all the genetic information is in chromosomal DNA. The mechanics of chromosomal distribution to daughter cells by mitosis is the same for both plants and animals; the continuity of heredity is thereby assured. Mitosis can occur in haploid cells

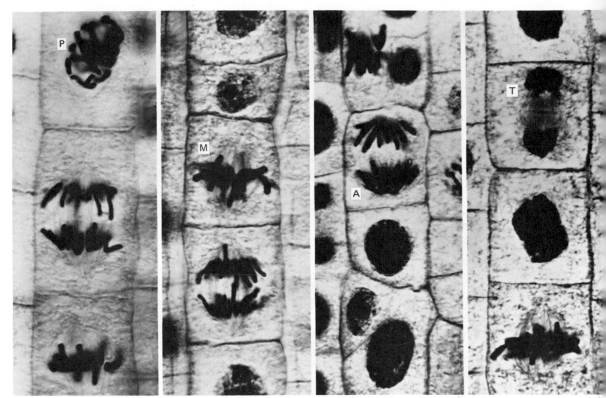

Figure 6.13. Photomicrographs of mitotic figures in onion root tips. From left to right note (1) prophase, (2) metaphase, (3) late anaphase, and (4) telophase (cell plate formation). (Courtesy Carolina Biological Supply Company)

as well as diploid cells. Notice that mitosis is an *equational division,* i.e., one that delivers each member of a pair of sister chromatids into different daughter cells. The daughter cells produced by mitosis, therefore, have identical chromosomal heredity. Mitosis is a conservative process, the antithesis of evolution!

Meiosis

Mitotic (asexual) reproduction is considered to be a more primitive mechanism than meiotic (sexual) reproduction. In most forms of *sexual reproduction,* two *sex cells* or *gametes* unite to form a single cell called a *zygote.* Barring virgin birth (*parthenogenesis*), a gamete does not produce a new organism by itself. Any process that accumulates, in a single cell, genetic contributions from other cells is *sexual.* When two haploid (*n*) gametes unite, they form a diploid (*2n*) zygote. The life cycle is completed when the diploid organism again produces haploid cells. Through meiosis (Figure 6.14), which contains a *reductional division,* a diploid cell becomes two haploid cells. Obviously, a haploid cell cannot undergo meiosis. One meiotic

Figure 6.14. Diagram of meiosis in plant cells.

cycle is characterized by a single replication of DNA followed by two divisions of the nucleus and cytoplasm (referred to as the first and second meiotic divisions). Within each meiotic division, four major stages are recognized.

First Meiotic Division (Meiosis I)

Prophase I. As explained for mitosis, the DNA replicates during interphase, producing chromosomes with two identical sister chromatids. The chromosomes shorten and thicken by coiling, the nuclear membrane dissolves, and the spindle apparatus develops. These events are common to both mitosis and meiosis. However, two important events occur in meiosis that have no counterpart in mitosis: (1) *synapsis*—homologous chromosomes derived from one parent pair up with those from the other parent, and (2) *crossing over*—segments of nonsister chromatids may be exchanged between each pair of chromosomes by breakage and reunion of homologous parts. This causes *genetic recombination* to occur, i.e., new combinations of hereditary instructions are molded into the transmission unit of the chromosome. As we shall soon learn, new gene complexes constitute the raw materials on which the process of natural selection acts to cause evolution in populations.

Metaphase I. The synapsed (paired) chromosomes move to the equatorial plane, but the centromeres do not replicate. The spindle apparatus is fully developed at this stage.

Anaphase I. Homologous chromosomes (each consisting of two sister chromatids) move toward opposite poles. This separation of homologues will, in effect, reduce the number of chromosomes in daughter cells by half (a reductional division).

Telophase I. A haploid set of chromosomes congregate at each pole of the cell and cytokinesis divides the cytoplasm into two haploid cells. (Note: To determine the number of chromosomes at any stage, we count the number of functional centromeres.) In some organisms the chromosomes unwind, the spindle degenerates, and a nuclear membrane reforms. In other organisms there is little change through prophase of the second division.

Second Meiotic Division (Meiosis II)

Prophase II. If the chromosomes uncoiled in telophase I, they recoil again in prophase II. Likewise, the spindle reforms, and the nuclear membrane disappears.

Metaphase II. Each chromosome moves as an independent unit to the metaphase plate. The centromere now replicates, freeing the sister chromatids.

Anaphase II. The sisters (now called chromosomes) move toward opposite poles in an equational division (comparable to a mitotic anaphase).

Telophase II. The chromosomes again become thread like, the spindle degenerates, the nuclear membrane reforms, and cytokinesis divides each cell into two. Thus, from an original cell, one complete meiotic cycle yields four haploid cells, none of which contain identical heredities.

Gametogenesis

Although meiosis is an important aspect of gamete formation (*gametogenesis*), it generally is not the only process involved. Meiotic products of higher plants and animals usually require additional maturation events before they become functional gametes.

Animals (Figure 6.15)

Male Gametogenesis (Spermatogenesis). The original diploid cell in which meiosis is initiated is termed a *primary spermatocyte.* After the first meiotic division is complete, two haploid products (*secondary spermatocytes*) are formed. The four haploid products of the second meiotic division are called *spermatids.* Functional gametes (called *spermatozoa* or *sperm* cells) are often generated by a process called *spermiogenesis,* which reorganizes the spermatid cytoplasm into a long whip-like tail, leaving relatively little cytoplasm around the nucleus. The sperms of some lower animals (e.g., roundworms) do not have flagella for locomotion, but rather exhibit amoeboid movement.

Female Gametogenesis (Oogenesis). Meiosis in most females is characterized by a grossly unequal cytokinesis, such that the bulk of the cytoplasm remains with only one of the meiotic products. The diploid primary oocyte undergoes the first meiotic division, yielding one large haploid secondary oocyte and a tiny haploid polar body. The secondary meiotic division produces a large haploid ootid and a tiny haploid secondary polar body. A primary polar body may also divide to produce two secondary polar bodies, but all polar bodies normally degenerate and do not function in fertilization. The ootid matures into a functional *egg* cell (*ovum*) with little obvious change. Note that only one functional gamete is produced from

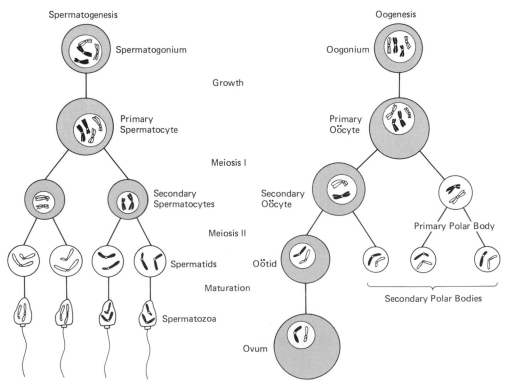

Figure 6.15. Animal gametogenesis.

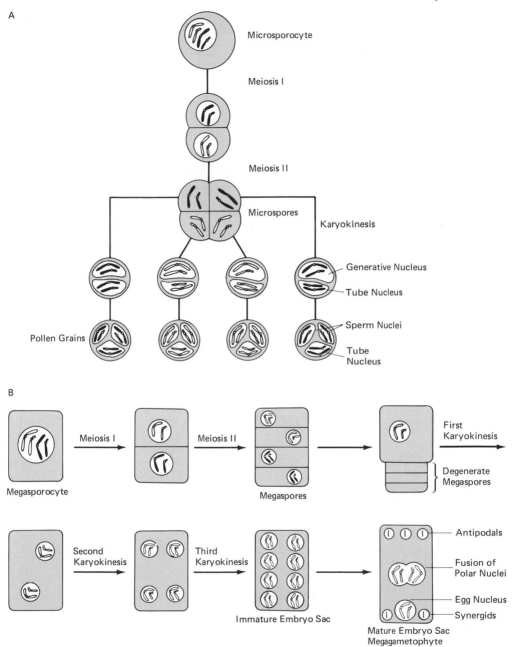

Figure 6.16. Gametogenesis in angiosperms (flowering plants). (A) Pollen production (microsporogenesis); (B) embryo sac production (megasporogenesis).

each primary oocyte. Natural selection apparently has modified the cytoplasmic divisions in this fashion so that the egg will contain ample nutrients for the development of the embryo upon fertilization.

Plants (as represented in many flowering plants; Figure 6.16)

Male Gametogenesis (*Microsporogenesis*). Certain diploid cells in the anthers of flowers, called *microsporocytes* undergo meiosis, producing four haploid *microspores* (immature *pollen grains*). Only the nucleus (chromosomes) of a microspore replicates (*karyokinesis*); the cytoplasm does not divide. One of the two nuclei thus produced undergoes a second karyokinesis. The entire male *gametophyte* (gamete-producing plant) thus consists of a single cell (a mature pollen grain) containing these three haploid nuclei (two of which represent the male gametes).

Female Gametogenesis (*Megasporogenesis*). Certain diploid cells in the ovary of a flower, called *megasporocytes,* undergo meiosis, producing four haploid *megaspores*. Three of the four degenerate. The one remaining megaspore undergoes three karyokinetic (mitotic) divisions producing eight nuclei. Six of these nuclei degenerate. Two nuclei unite into a "fusion nucleus" and the other remains as the "egg nucleus" (the latter representing a female gamete). This is the condition of a mature embryo sac (female gametophyte) ready for fertilization.

Fertilization

Animals

Usually when a sperm meets an egg, only the head (containing the nucleus) enters. The tail (and hence the bulk of the cytoplasm of a sperm) remains outside and degenerates. The haploid sperm nucleus and the haploid egg nucleus fuse (*syngamy*) to form the diploid nucleus of the zygote. Mitotic divisions then generate the multicellular features of the embryo (or larva), and eventually the adult structures emerge through growth and differentiation.

Plants (Figure 6.17)

When a pollen grain lights on a stigma of a flower, a pollen tube is generated (*pollination*). This tube grows down the style into the ovary and enters an ovule. Two haploid sperm nuclei from the pollen tube enter the embryo sac. One unites with the haploid egg nucleus to form the diploid zygote. The other haploid sperm nucleus unites with the maternal diploid "fusion nucleus," forming a *triploid* (three sets of chromosomes) nucleus. Mitotic divisions of the triploid cell produce a tissue called *endosperm,* which functions to nourish the embryo. Mitotic divisions of the zygote produce the embryo seedling, which is the new *sporophyte* (spore-bearing plant). Plants with two seed leaves (*cotyledons*) will usually incorporate the nutrients of the endosperm into these structures. This food serves the young plant until it can

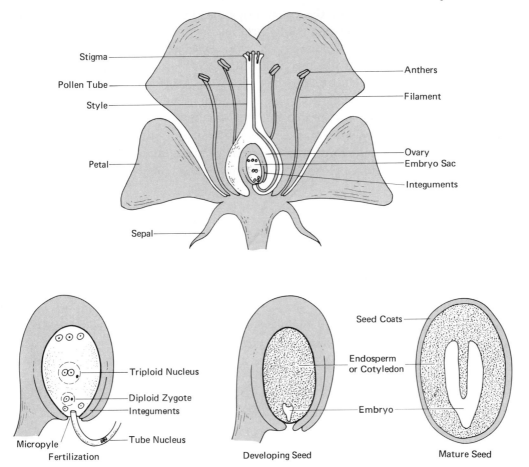

Figure 6.17. Fertilization and development of a seed.

grow up through the soil and begin photosynthesizing in the sunlight. *Double fertilization* is characteristic only of flowering plants (*angiosperms*). One fertilization produces the zygote, and the other produces endosperm.

A seed is represented by the fertilized embryo sac and portions of the surrounding maternal tissues of the ovule. Seeds may remain dormant for prolonged periods of time until suitable conditions exist to induce germination. Many seeds have special modifications that contribute to their dispersal. Some are light or have "wings" for wind pollination; others have barbs or hooks for attachment to animals; still others have hard seed coats that resist digestion within animals. Adaptations for survival and fitness are to be found at every stage in the life cycle.

A *fruit* is defined as a ripened ovary of a flower that encloses the seed(s). However, not all fruits contain seeds. The term *parthenocarpy* applies to the formation of a fruit without seeds. This may result from (1) lack of pollination, (2) lack of fertilization, or (3) death of the embryo at an early stage of development. The edible banana is an example of a fruit without seeds. These plants are dependent upon humans for their propagation through asexual means.

External fertilization (outside the parent organism) requires rather special conditions. Since gametes that participate in external fertilization are not protected from desiccation by thick cell walls, sufficient moisture must be present to allow one or both gametes to "swim" to meet one another. It is not surprising then that natural selection would favor any mechanism in a terrestrial environment that would reduce the risk of gametic loss from a water deficit. The pollen tube of flowering plants is an adaptation that accomplishes internal fertilization. Pollen grains, however, do have a thick coating that protects them during their travels (either by air currents or by attachment to animals) from their sites of production (anthers) to their sites of germination (stigmas). Furthermore, internal fertilization allows the embryo to be nourished and protected for various lengths of time by the female parent.

Recombination

The crossover mechanism that recombines *linked* genes (i.e., genes on the same chromosome) has already been discussed. Another mechanism exists within the meiotic process for recombining genes in different *linkage groups* (i.e., different chromosomes). At the first meiotic metaphase, two pairs of homologous chromosomes may be oriented on the equatorial plate in either of two equally likely positions (see Figure 6.18).

Each chromosome pair thus assorts (segregates) members to the poles independent of every other pair. If we label one pair of homologues A and a and a second pair B and b, we see that four chromosomally different types of gametes can be generated from an individual of chromosomal constitution $AaBb$: AB, ab, Ab, and aB. Each gametic type would therefore be expected to occur with a frequency of 25 per cent. In other words, if a very large number of the gametes made by this individual were placed in a bag, the chance of drawing at random any one kind of gamete would be one out of four ($\frac{1}{4}$). If three pairs of chromosomes are involved, an individual of chromosomal type $AaBbCc$ would be expected to produce eight different kinds of gametes with equal frequency.

Chromosomal Combinations			Frequencies	Gametes
$\frac{1}{2}A$	$\frac{1}{2}B$	$\frac{1}{2}C$	$\frac{1}{8}$	ABC
		$\frac{1}{2}c$	$\frac{1}{8}$	ABc
	$\frac{1}{2}b$	$\frac{1}{2}C$	$\frac{1}{8}$	AbC
		$\frac{1}{2}c$	$\frac{1}{8}$	Abc
$\frac{1}{2}a$	$\frac{1}{2}B$	$\frac{1}{2}C$	$\frac{1}{8}$	aBC
		$\frac{1}{2}c$	$\frac{1}{8}$	aBc
	$\frac{1}{2}b$	$\frac{1}{2}C$	$\frac{1}{8}$	abC
		$\frac{1}{2}c$	$\frac{1}{8}$	abc

The above dichotomous system is an easy method for finding all possible types of gametes. There is a 50 per cent chance ($\frac{1}{2}$) of finding either A or a chromosome in a gamete; likewise a $\frac{1}{2}$ chance of finding B or b, etc. Independent probabilities can be multiplied to give joint probabilities.

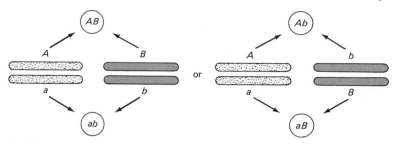

Figure 6.18. Diagram of two possible segregational patterns for two pairs of homologous chromosomes during meiosis.

Given three pairs of chromosomes (A,a; B,b; C,c), the probability of finding A together with b and C in a gamete is $(\frac{1}{2})(\frac{1}{2})(\frac{1}{2}) = \frac{1}{8}$. In general, the number of different chromosomal combinations possible in meiotic products or gametes is represented by the formula 2^n, where n is the number of *pairs* of chromosomes for the species. Humans have 23 pairs of chromosomes. The number of possible different gametic chromosomal combinations is therefore $2^{23} = 8,388,608$. The number of different chromosomal combinations represented in zygotes formed by the union of these gametes is represented by 2^{2n}; for humans this would be 2^{46} (over 70×10^{12}). This by no means represents the full extent of genetic recombination, however, because we have ignored in our calculations the recombination of linked genes by crossing over. Is it any wonder that no two individuals (other than identical twins, etc.) have the same hereditary endowment?

This is part of the explanation for Darwin's observation of "fluctuating variations" in natural populations. Recall that Darwin knew nothing of genes, chromosomes, mitosis, or meiosis. All these discoveries were made after the publication of his *Origin of Species*. His lack of understanding of the origin of variations and the laws of heredity was perhaps his greatest handicap in the development of a more complete theory of evolution. Today we have this knowledge, and its marriage with the natural selection theory has formed the backbone of our modern synthetic evolutionary theories under the banner of *neo-Darwinism*.

Genetic Systems

The efficiency of natural selection, in its creative role of forming better adapted gene complexes, is highly correlated with the amount of genetic variability in a population. There are two major ways by which sexually reproducing organisms can accelerate recombination (and hence increase genetic variability). One way is to increase crossing over so that linked gene complexes become less stable. The other way is to distribute the genome over a larger number of chromosomes so that more recombinant types are produced by independent assortment from metaphase orientations. Both mechanisms theoretically could function so well that the fortuitous formation of a highly adaptive set of genes would have little chance to produce offspring like itself. On the other hand, an organism that has its genome organized into a single linkage group (with little or no crossing over) would generate few recombinant progeny.

Hence, the problem each species must solve is how best to organize its genetic material in terms of numbers of chromosomes, blocks of genes that need to be protected from crossing over, other regions that should be open to liberal recombination, etc. Most organisms have evolved to a middling solution by using a moderate number of chromosomes with variable amounts of crossing over within each linkage group. The *genetic system* of a species encompasses all the biological factors controlling its variability, including the number of chromosomes, the amount of crossing over, the life cycle, the breeding or pollination system, the dispersal potential, the generation time, the barriers to outcrossing, etc. The *population structure* of a species refers to the way in which the individuals are grouped into local breeding populations in terms of size and distance or overlap between groups, the amount of migration (*gene flow*) between groups, etc. (see Chapter 15). Together, the genetic system and the population structure of a species are features of paramount importance in the production of genetic variability and hence in the evolutionary potential of the species.

Life Cycles

Animals

Virtually all animals (but relatively few plants) have a life cycle characterized by a multicellular diploid stage that then produces gametes directly by meiosis. A zygote is formed by the union of the haploid gametes in fertilization. Mitosis then reproduces the multicellular adult (Figure 6.19). Dispersal among sessile animals is usually accomplished by free-swimming embryos or larvae, or in some cases by asexual (*vegetative*) reproductive units.

Sea anemones and some flatworms may split into two or more pieces (*fragmentation*), each piece able to regenerate a whole new organism. Sponges and hydras may grow new organisms from the surface of a parent organism by a process called *budding*. An earthworm or a starfish may reconstruct an entire organism from a severed piece by *regenerative reproduction*. Even higher animals retain some regenerative capacity. For example, a salamander may regenerate an entire limb but cannot regenerate an entire new organism from a severed limb. In most vertebrates the regenerative capacity is limited to repairing relatively small wounds. Most protozoans simply divide mitotically into new single-celled organisms. Early separation of vertebrate embryonic cells (after one or a few mitotic divisions from the zygote stage) may result in two or more "identical" *siblings* (brothers and/or sisters). The production of identical (*monozygotic* = one zygote) twins in humans is therefore an example of vegetative reproduction (Figure 6.20).

Plants

Some primitive plants have a multicellular haploid stage that produces haploid gametes by mitosis. Union of these haploid gametes in fertilization produces a diploid zygote, which soon undergoes meiosis to yield haploid spores. The spores reproduce mitotically to form the multicellular haploid plant (Figure 6.21). Many plants have two conspicuous free-living stages in their life cycle. The multicellular haploid phase produces haploid gametes by mitosis.

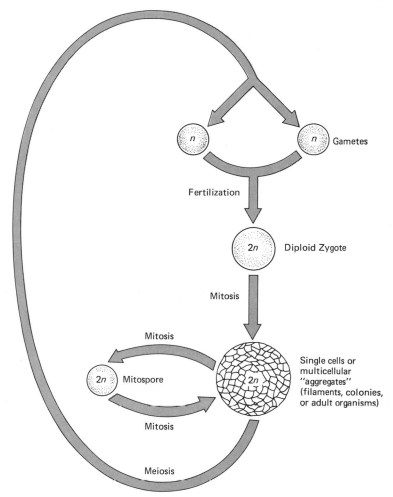

Figure 6.19. A generalized diplontic life cycle. The only haploid stage is in the gametes; found in some algae and fungi, many protozoa, and all higher animals.

Fertilization unites the gametes into a diploid zygote. Mitosis also generates the multicellular diploid phase. Meiosis in the diploid yields haploid spores, which then reproduce the multicellular haploid phase by mitosis. This form of reproduction is characterized by an *alternation of generations* in which a gametophyte (gamete-bearing plant) and a sporophyte (spore-bearing plant) phase follow sequentially (Figure 6.22). An alternation of generations is rarely encountered in animals.

Some coelenterates have a sessile polyp phase and a free-swimming medusa ("jellyfish") dispersal phase in which sex organs develop. However, in contrast to plants, both phases are diploid. This type of alternation of generations is known as *metagenesis* (Figure 6.23).

Regeneration from arm of starfish (Phylum Echinodermata)

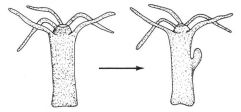

Budding in *Hydra* (Phylum Coelenterata)

Binary fission in *Paramecium* (Phylum Protozoa)

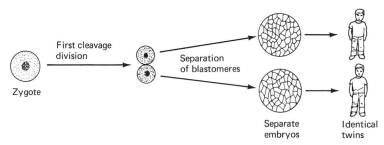

Production of identical human twins (clone of size two); Phylum Chordata

Figure 6.20. Examples of asexual reproduction mechanisms.

In the evolution of flowering plants, the haploid spore phase and the multicellular haploid gametophyte phase have been greatly reduced to tiny multinucleate structures (pollen grain and embryo sac), which are completely dependent on the multicellular diploid sporophyte.

Reproductive cells are thus of two major types: (1) spores and (2) gametes. Spores produced by terrestrial plants have the adaptive advantage of being enclosed in a thick wall and thereby protected from desiccation during dispersal by wind or animals. Aquatic plants may produce flagellated spores, which can accomplish dispersal by "swimming." Spore production is therefore a favorite mode of reproduction among plants immobilized by attachment to a substrate. Spores can develop directly into a new organism. Gametes cannot usually do this.

A few cases are known wherein an animal gamete can directly produce a new organism

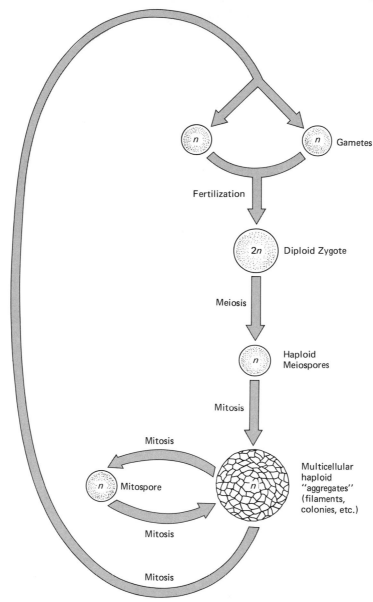

Figure 6.21. Primitive haplontic life cycle characterized by meiosis in the zygote stage. The only diploid stage is in the zygote; found in all primitive and many advanced groups of algae and in many fungi.

without fertilization. This process is termed *parthenogenesis* ("virginal development"). Usually only eggs have enough stored nutrients to undergo parthenogenetic development. Rotifers and bees are examples of animals that use this process naturally. Parthenogenesis has been artificially stimulated in both invertebrates and vertebrates (notably the frog) by pricking the surface of an egg with a needle or by exposing the egg to certain chemicals. A parthenogenetic gamete is thus the functional equivalent of a

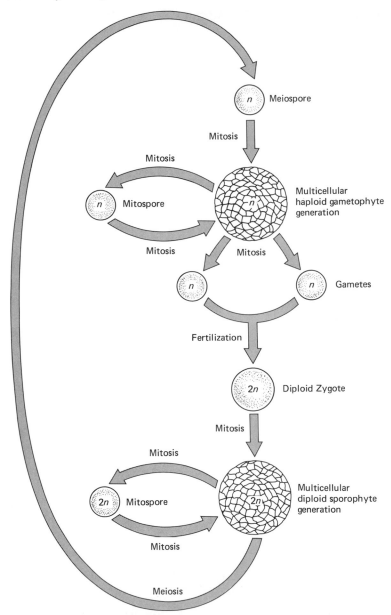

Figure 6.22. Generalized diplohaplontic life cycle characterized by spores derived from meiosis (meiospores) and alternation of haploid gametophytic and diploid sporophytic generations.

spore. *Thelytoky* is a rare type of parthenogenesis in which females develop from unfertilized eggs. The ova that develop in this way are usually diploid, so that the offspring are genetically identical with the parent.

Gametes unite into a zygote, and then the zygote may give rise directly to the new organism. Sexual (*gametic*) reproduction is often stimulated in plants by unfavorable envi-

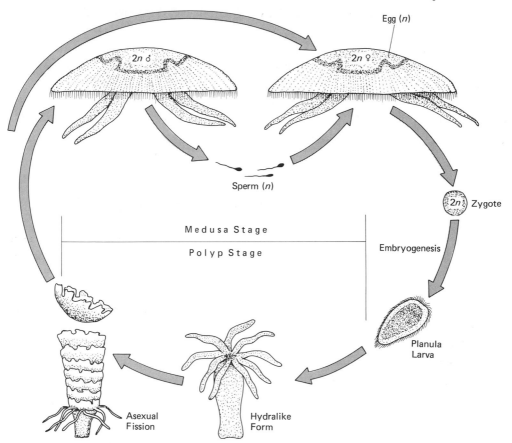

Figure 6.23. Alternation of diploid generations in the jellyfish *Aurelia* (medusa stage dominant).

ronmental conditions. A zygote, bearing two different sets of genetic instructions, apparently can generate more biochemically adaptive responses to stress conditions than can the haploid phase of a plant's life cycle.

Some individuals have both male and female sex organs. Many flowers contain both male (anthers) and female (ovaries) parts. These are called "perfect flowers." Other plants have male (staminate) flowers and female (pistillate) flowers at different locations on the same individual (e.g., corn, *Zea mays*). These are "imperfect flowers."

Plants with both sex organs are classified as *monoecious* ("one house"). The date palm is an example of a plant with male and female flowers residing in different individuals. Such plants are called *dioecious* ("two houses"). *Self-fertilization* is possible in monoecious species, i.e., pollen from a given plant may fertilize ovules of the same plant. Some perfect flowers are *cleistogamous*, i.e., they do not open. Hence, self-fertilization is virtually assured. This appears to be one way to increase the opportunity for gametes to unite. However, as we shall see in a later chapter, self-fertilization commonly has some undesirable effects and the population may pay a price for the privilege of using such a mechanism. Many plants have evolved mechanisms to prevent self-fertilization and thereby to enhance *cross fertilization*, i.e., the union of genetically unrelated gametes. This may produce hybrid vigor and increase fitness among the progeny.

Relatively few animal species have both male and female gamete-producing organs (*gonads*). Such bisexual animals are called *hermaphrodites* (earthworms and land snails are examples). Often these hermaphrodites cannot fertilize themselves, but may participate in mutual cross fertilization. The advantage here is that fertilization can occur whenever any two individuals meet. This is not true of species with separate sexes. Hermaphroditism is thus an adaptation for enhancing the union of gametes in animals with sluggish motility.

Asexual vs. Sexual Reproduction

The collection of genes from two cells into a single cell is the essence of *sex*. Sex is not a reproductive process, quite the contrary. Reproduction produces two or more cells from a single cell. Sex, on the other hand, frequently involves the fusion of two cells (gametes) into a single cell (zygote). Strictly speaking therefore, the commonly used terms "sexual reproduction" and "asexual reproduction" are misnomers. Bearing this important distinction in mind, we will use the terms "asexual" and "vegetative" as interchangeable, and similarly equate the terms "sexual" and "gametic" reproduction.

Most plants, and even some animals, reproduce without sex (asexual or vegetative), i.e., without the production and union of gametes. Bacteria and protozoa commonly reproduce by binary *fission* or splitting in two. The mechanism of genetic reproduction here is essentially equational. All the cells reproduced by fission from a single original cell form a *clone*. Every member of a clone has identical heredity. Some plants (e.g., fungi), can pinch off spore cells, which then grow into new plants. The cycle may repeat endlessly. They may also have a sexual cycle, but this is not essential to propagation. Vascular plants sometimes reproduce by *runners*. These are procumbent stems that take root, becoming a new plant when the connection with the old plant decays. Artificial asexual propagation by humans can be accomplished by several means. A *slip, cutting,* or *scion* separated from a plant may be rooted in soil to form a new plant. Alternatively, a scion may be grafted onto a root *stock* of another plant. The resulting plant is a *chimera* or *mosaic,* consisting of two hereditarily different parts. Natural grafts are rare, but have been known to occur between closely related species. In general, the closer the genetic relationship between the stock and scion the better is the chance of a graft succeeding. The degree of success of artificial grafts can be used as a measure of *"evolutionary distance"* or genetic difference between related species. The same general principle applies in animals. Skin or organ (heart, kidney) grafts between identical human twins usually are successful. Grafts between *dizygotic* (two zygote) siblings will usually be rejected sooner or later by the immune system. As the degree of genetic relationship between donor and recipient decreases, the chances of graft survival decrease correspondingly.

The main point to be made concerning vegetative reproduction is its conservative nature. No new gene complexes are produced by asexual mechanisms. If a species were forced to reproduce exclusively by vegetative means, the long-term prognosis would be extinction. The total range of hereditary variability in the population would not be expected to change from one generation to the next (barring mutation). The environment, however, is notoriously unstable, and over time would very likely change so radically that none of the gene combinations in the population would any longer be adaptive. When that point is reached, the species expires. Although it is true that some organisms (such as bacteria) that reproduce almost exclusively by asexual means have apparently survived without noticeable change over long periods of geological time, even they occasionally have escaped from the monotony of asexual reproduction by some form of "sexuality." If a certain heredity specifies a broad range of adaptive physiological responses to environmental variables, there would be little pressure exerted by natural selection to change it. In such a case, it would be advantageous for that organism to reproduce vegetatively because all members of the resulting clone would be equally adaptively endowed. The problem that each species must solve, therefore, is how to generate the greatest number of gene combinations specifying high fitness for present conditions, while retaining sufficient genetic heterogeneity to meet the exigencies evoked by radical environmental changes.

The only way for a closed (isolated) population to obtain new gene combinations (other than by mutation) is to use a gametic mechanism of reproduction. Sex usually generates genetic variability, and this in turn produces the variability in anatomical and physiological characteristics required by the sieve of natural selection. However, the population pays a price for the privilege of having sex, because a number of ill-adaptive gene combinations will usually be generated through the recombination process. Much gametic wastage is therefore inevitable. The situation is further compounded by the problem of getting sperms and eggs together for fertilization. Consider the tremendous quantities of pollen that must be shed by most flowering plants to ensure adequate pollination, or the millions of sperm cells released by fish and aquatic invertebrates to ensure fertilization of a relatively small number of eggs. This is gametic wastage on a grand scale! For all the disadvantages accompanying sexual reproduction, the advantages of this process are so great that the overwhelming majority of organisms (both plant and animal) reproduce thereby almost exclusively. Remember that the alternative of asexual reproduction may predispose to extinction. If one considers this alternative, one sees that the cost of sex is a relatively small one for the species to pay.

Essay Questions

1. *Contrast the genetic consequences of mitosis with those of meiosis.*

2. *Discuss the advantages and disadvantages of asexual vs. sexual modes of reproduction from the standpoint of evolution.*

3. *Describe two recombinational mechanisms inherent in meiosis.*

4. *Why is the genetic system of a species so important to its evolutionary potential?*

5. *In what respects do eukaryotic cells represent an advance over prokaryotic cells?*

QUALITATIVE TRAITS

Qualitative vs. Quantitative Traits

Johannsen applied the term *genotype* to the hereditary endowment of the organism and the term *phenotype* to any characteristic or trait that the organism exhibits. There are two major classes of traits: (1) *qualitative* and (2) *quantitative*. By *qualitative* we mean traits of "kind," i.e., traits that exhibit *discontinuous variability*. These characters are rather easily classified (or pigeonholed) into unequivocally separate groups. An example of a qualitative trait is the human ABO blood groups. A person is either blood group, A, B, AB, or O, not a mixture of these groups. "Classical" qualitative traits of this kind are usually produced by the action of one (or a very few) gene, and the environment has little or no effect on modifying the phenotype. Thus, a single *genetic locus* (i.e., a specific sequence of DNA nucleotides at a particular position on a chromosome) is totally responsible for the ABO blood groups, and no environmental conditions are known to be capable of changing this genetically determined phenotype.

At the other extreme, we find *quantitative traits* or characters of "degree." These phenotypes cannot be pigeonholed into discrete groups. Rather, if we examine the variability within a population, we find it to be *continuous* (i.e., devoid of gaps). As an example let us consider body height. If the population is large enough, we are likely to find every measurement represented from the very smallest to the very largest. Most individuals are about average in size, however, with progressively fewer individuals exhibiting deviations from the norm. This is a typical quantitative trait. They are often called *metric* traits, because the only way to specify the phenotype is to measure it and assign some numerical value to it (in terms of inches, pounds, whatever). Most quantitative traits are governed by numerous genes (*polygenes*); consequently, quantitative characters are sometimes called *polygenic traits*. In addition, these traits may be subject to considerable modifications by environmental variables. In such cases, the genotype may be considered as determining the physiological or developmental limits within which the environment may modify the phenotype. In the example of body height, a certain genotype may represent a "blueprint" for a tall phenotype, but under a poor level of nutrition only one of moderate stature is realized.

Because numerous genetic changes are generally required for the transformation of one species into another, the fact that two populations differ in the incidence of a qualitative trait is usually of trivial importance in understanding the evolutionary history of the groups. Differences in quantitative traits are of much greater significance, because this reflects genetic divergence involving large systems of polygenes spread throughout the genome. The genetics of quantitative traits is rather complex and further discussion on the subject will be deferred to Chapter 8.

Returning to qualitative traits, one must mention that the environment sometimes does contribute to the phenotypic variability even when only one gene is involved.

> An example of environmental modification of a *monofactorial* (i.e., single gene) trait is found in the Himalayan rabbit. This domestic breed is characterized by a white body and pigmented extremities (tip of nose, tail, ears, and feet). Obviously, the rabbit has the genotype for making pigment, but why is it expressed only in certain parts of the pelage? The answer lies in the temperature-sensitive nature of the gene product. This enzyme, which acts in a biochemical pathway converting a colorless precursor substance into melanin pigment, becomes inactive (denatured) at the normal body temperature of the rabbit. At the extremities, however, where the temperature is slightly lower, the enzyme remains in its active conformation and pigment develops. If the white hair from the side of a Himalayan rabbit is shorn and an ice pack is applied to that area while the new hair grows in, the new growth will be pigmented. Conversely, if a Himalayan rabbit is raised in a room with a temperature kept about 100°F, even the hair on the ears, etc. will blanch with time. This example shows us how the genotype and the environment can cooperate in the production of a phenotype.

Many northern animals (birds and mammals especially) change color with the seasons. During summer months they are usually of darker colors, which may be an adaptation that allows them to blend in with their environment (*cryptic coloration*) and thus escape detection by predators. In winter months, when snow covers the ground, these same animals may become white, again as an adaptive response that allows them to blend in with their background. Even predators have used this device, allowing them to sneak up on their prey without being noticed. Remember that when a prey species evolves a mechanism to escape being eaten, the predator species must evolve compensatory mechanisms in order to survive.

A simple genotype (one or a few genes) under one set of environmental conditions may result in one phenotype, but under a different set of conditions yield a different phenotype. Such a genotype is said to lack *penetrance*. A completely "penetrant" genotype would give a uniform phenotype in all environments. The degree of penetrance is calculated as the proportion of organisms (bearing a common genotype) that exhibits the expected corresponding phenotype regardless of known or unknown environmental variables. Furthermore, although a gene may be penetrant, it may not be uniformly expressed in all individuals or even in all parts of an organism where it is capable of being so.

> Gout is a painful disease caused by the overproduction of uric acid and its subsequent accumulation as crystals of sodium urate in joints (especially those of the feet and toes). It is primarily an affliction of men and older women. After menopause, the levels of female sex hormones are usually disturbed, predisposing older women to the disease. Hyperuricemia (high levels of uric acid in the blood) is postulated to be conditioned by a dominant autosomal gene with 80 per cent penetrance in men and 12 per cent penetrance in women. The levels of sex hormones are viewed as primary components of the "internal environment" affecting the penetrance of this trait.
>
> Marfan's syndrome (arachnodactyly = "spider fingerness") is a genetic disease characterized by excessively long fingers and toes. Among individuals that possess the gene for arachnodactyly, the proportionate extension of the digits is quite variable; hence, the condition is said to exhibit variable *expressivity*. Again, factors of the internal or external environment probably are involved.

The kinds of traits that are easiest to study genetically are those that are governed by only one or a very few genes with very little or no environmental variability. These are the "classical" qualitative traits (100 per cent penetrant and with uniform expressivity) that will be discussed in detail in this chapter.

Mendelian Principles

Dominance vs. Recessiveness

Gregor Mendel is credited as being the "father of genetics." In 1866, he published the results of eight years of breeding studies on the garden pea (*Pisum sativum*). In this publication he set forth some principles of heredity that were later to prove of essentially universal application among sexually reproducing diploid organisms.

Let us reexamine some of Mendel's work. One of the reasons for Mendel's success was his choice of genetically *pure* stock. Inbreeding leads to genetic uniformity. Peas have cleistogamous flowers and hence are obligate self-pollinators (the most severe form of inbreeding). Thus, each strain of pea that Mendel originally selected for his experiments was of uniform genotype. One trait he investigated was the color of the seed coat. He had a yellow strain (which bred true to type because it was a pure line) and a green strain (which did likewise). When he artificially cross-pollinated these two strains, the F_1 (first filial generation of offspring) all possessed yellow seeds. When he allowed the F_1 plants to self-fertilize, the F_2

(second filial generation, or second generation of offspring) had some yellow and some green seeds. The fact that the green trait disappeared in the F_1 and reappeared again in the F_2 suggested to Mendel that the hereditary information was probably particulate rather than fluid. These particles apparently did not contaminate one another when present together in the F_1, because the corresponding traits reappeared in full strength in F_2. Furthermore, the F_1 numerical ratio of yellow to green statistically approximated a 3:1 ratio (actual data: 6,022 yellow:2,001 green; 3.01:1 ratio). This again was Mendel's genius, because many hybridizers had noticed splitting progenies such as this, but they did not bother to count them or formulate hypotheses to explain why this should be so. Mendel reasoned that the simplest explanation involved the postulation of two hypothetical particulate "factors" for each trait. Each individual in the pure lines had two factors: the yellow strain had two factors for "yellowness," the green strain had two factors for "greenness." During gametogenesis, only one representative of each pair of factors was apportioned to a sex cell. Hence, the F_1 progeny possessed one yellow and one green factor. Because the yellow factor (or trait) was the only one to be expressed in the hybrid F_1, Mendel called it the *dominant* factor (or trait). The green factor (or trait) apparently was hidden in the F_1 hybrid, and Mendel referred to it as the *recessive* factor (or trait). The F_1 plants made gametes of two kinds: (1) those bearing a yellow factor and (2) those bearing a green factor. Furthermore, these two kinds of gametes were made with equal frequency. Random union of the F_1 gametes produced an F_2 with a theoretical expected frequency of 3 yellow:1 green. In order to facilitate an understanding of this phenomenon, we shall now use modern terminology. Mendel's factors are now called *genes*. They occur in pairs in diploid organisms. The alternative forms of a gene that reside in homologous DNA nucleotide sequences of homologous chromosomes are called *alleles* (formerly allelomorphs). Hence, the F_1 had an allele for yellow and an allele for green. Mendel used a capital letter to represent the dominant allele, and a corresponding lower case letter to represent the recessive allele. R. C. Punnett developed a "checkerboard" method for displaying the random union of gametes, which now bears his name (Punnett square). We can diagram Mendel's cross as follows:

Parents:	Phenotypes:	Yellow $\times$ Green	
	Genotypes:	GG	gg
Gametes:		$Ⓖ$	$Ⓖ$
F_1:	Phenotype:	Yellow	
	Genotype:	Gg	

Punnett Square:

F_1 Male Gametes	F_1 Female Gametes	
	$Ⓖ$	$Ⓖ$
$Ⓖ$	GG Yellow	Gg Yellow
$Ⓖ$	gG Yellow	gg Green

F_2:   Phenotypes: 3 Yellow: 1 Green

Genotypes: 1 GG:2 Gg:1 gg

Segregation of Alleles

A genotype consisting of only one kind of allele is now called a *homozygous* genotype (e.g., *GG* or *gg*). A genotype consisting of different alleles is now called a *heterozygous* genotype (e.g., *Gg*). Mendel did not use these terms, but he did use similar reasoning in coming to his conclusions. Remember that Mendel knew nothing of chromosomes, meiosis, or mitosis. Yet his work predicts that the pea is a "diploid" and the alleles separate or segregate from one another during gametogenesis (the principle of *segregation of alleles*) to produce "haploid" gametes (having only one of the two "alleles"); that the union of gametes is essentially a random process; and that a certain "allele" can hide or mask the expression of another in the "heterozygous" condition (i.e., exhibit dominance). These principles now form the basis of the science of *genetics* (name coined in 1906). Mendel investigated seven different traits of the pea plant and found each of them explicable in the same manner.

Mendel's principle of "segregation of alleles" can be explained in physical terms now that we know how meiosis operates in the formation of gametes. If we associate the gene for seed-coat color with a *locus* (position) on a pair of homologous chromosomes, then the first anaphase of meiosis pulls homologous chromosomes apart and hence separates (or "segregates") the alleles into different cells (gametes; Figure 7.1).

Independent Assortment

Mendel also investigated the joint behavior of two traits. For example, he investigated the shape of the seed (wrinkled vs. smooth) simultaneously with seed color (yellow vs. green). He knew from previous experiments that smooth seed shape was dominant to wrinkled form. When he crossed a smooth, yellow strain with a wrinkled, green strain, the F_1 exhibited only the two dominant traits (smooth, yellow). The F_2 (produced by self-fertilization of the F_1) yielded a ratio of approximately 9 smooth, yellow:3 smooth, green:3 wrinkled, yellow:1 wrinkled, green. Using *W* to represent the dominant allele for smooth seed shape and *w* to represent the recessive allele for wrinkled seed shape, we can outline this cross as shown in Figure 7.2.

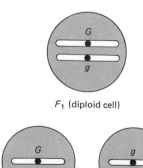

F_1 (diploid cell)

Gametes (haploid cells)

Figure 7.1. Segregation of alleles in a heterozygous genotype (*Gg*) produces equal numbers of haploid gametes bearing either *G* or *g*.

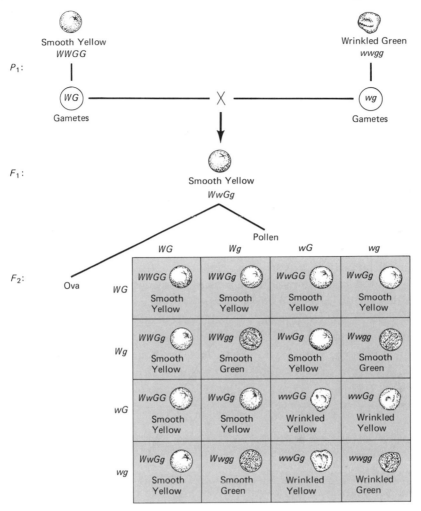

Figure 7.2. Example of independent assortment of two contrasting pairs of seed traits in peas.

The explanation for the 9:3:3:1 phenotypic ratio is found in the fact that each locus is on a different chromosome pair. Each pair of homologues lines up on the first meiotic metaphase plate as an independent unit. As explained in Chapter 6, four different kinds of gametes can be formed by a *dihybrid* individual, i.e., one with two heterozygous loci. This principle of *"independent assortment"* was demonstrated for all seven loci investigated by Mendel.

The number of genetically different gametes that can be formed by a *multihybrid* (heterozygous at more than one locus) is given by the formula 2^n, where n represents the number of segregating loci (i.e., each locus with a pair of alleles). The number of different phenotypes that can appear in the offspring generation from multihybrid parents is also represented by 2^n, where n now is the number of segregating loci with dominant and recessive allelic pairs. The number of different genotypes that can be present among the progeny from multihybrid parents is 3^n, where n represents the number of segregating loci (each with a pair of alleles).

As an example of the recombinational possibilities generated by a multihybrid, consider a genotype with six heterozygous loci (e.g., *AaBbCcDdEeFf*).

Number of genetically different gametes $= 2^n = 2^6 = 32$
Number of phenotypes in offspring $= 2^n = 2^6 = 32$
Number of genotypes in offspring $= 3^n = 3^6 = 729$

Note the great genetic variability that remains hidden under a relatively small amount of phenotypic variability. Natural selection works only on phenotypic variability. Hence, many genes (especially recessive alleles) are protected from the direct effects of selection and may accumulate in considerable numbers in a population in heterozygotes even though they may be somewhat deleterious when in the homozygous condition.

A dominant phenotype may be homozygous (e.g., *GG*) or heterozygous (*Gg*). The most efficient way to discern the genotype, in such a case, is to cross (mate) the dominant phenotype with one exhibiting the recessive characteristic. This is called a *testcross*. If the dominant individual is heterozygous, approximately half the testcross offspring should be of dominant type and half of recessive type. All offspring from a homozygous dominant parent will show the dominant phenotype regardless of the genotype (homozygous or heterozygous) of the other parent. This principle was employed by Mendel in proving the expected theoretical genotypic ratio (2:1) in dominant offspring from monohybrid parents.

Parents: *Gg* × *Gg*
             Yellow   Yellow
F_1:   1 *GG*:2 *Gg*:1 *gg*

             Yellow   Green

Note: $\frac{1}{3}$ of the yellow F_1 progeny expected to be *GG*.
$\frac{2}{3}$ of the yellow F_1 progeny expected to be *Gg*.

Testcrosses of Yellow F_1. $\begin{cases} \frac{1}{3} \text{ of all testcrosses of the yellow } F_1 \text{ [}GG \times gg\text{] should give only yellow progeny [all } Gg\text{, all yellow].} \\ \frac{2}{3} \text{ of all testcrosses of the yellow } F_1 \text{ [}Gg \times gg\text{] should segregate yellow and green in approxi-} \\ \quad \text{mately equal numbers (}\frac{1}{2} Gg = \text{yellow, } \frac{1}{2} gg = \text{green).} \end{cases}$

Post-Mendelian Discoveries

Over half a century of genetic studies of numerous diploid plants and animals have revealed many examples of characters governed by the Mendelian "laws." These traits are sometimes called *Mendelian characters*. However, many other genetic principles, unknown to Mendel, have also been discovered during this period. The remainder of this chapter will be devoted to a brief discussion of some of these other qualitative genetic phenomena.

Codominance

Some alleles fail to behave in a completely dominant or completely recessive manner. Sometimes each allele may be able to express itself to some degree in heterozygous combination. This phenomenon is called *codominance* (also termed *incomplete dominance, lacking dominance,* or *semidominance*). Thus, monohybrid parents can produce progeny with three phenotypes.

An example of codominant allelic interaction is known in the control of snapdragon flower color (Figure 7.3). The gene C^r produces red pigment; its allele C^w produces no pigment (white). A plant with two doses of the pigment gene (homozygous C^rC^r) has fully red flowers; similarly a plant with only C^w alleles (C^wC^w, homozygous genotype) has fully white flowers. A heterozygous plant (C^rC^w) has one dose of redness and one dose of whiteness and consequently is of intermediate color (pink). Pink-flowered plants do not breed true to type, because when pinks are intercrossed they segregate red, pink, and white, in the ratio 1:2:1, respectively. Since each genotype in this case has its own distinctive phenotype, there is no need to perform testcrosses.

A simplistic model to explain the difference between dominant and codominant gene action may be useful. A dominant gene makes an active protein product (e.g., an enzyme), whereas its recessive allele makes no product (or if one is made, it has no biological activity). The amount of enzyme produced by a single dose of the dominant gene is sufficient for the development of a "normal" phenotype. In the case of codominance, both alleles make active products of different kinds. Alternatively, if only one allele makes a product, it is present in insufficient quantity to produce a completely "normal" phenotype.

Dominance or recessiveness (or any other form of allelic interaction) has no intrinsic

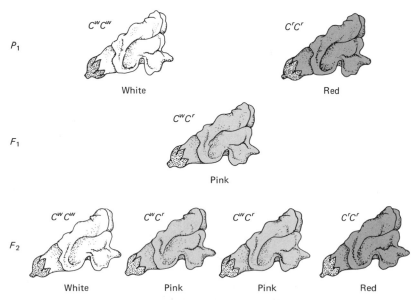

Figure 7.3. An example of codominant allelic interaction in the colors of snapdragon flowers.

relationship to fitness (or adaptive value). Many examples could be cited of highly adaptive alleles exhibiting dominant or recessive characteristics. However, natural selection will usually be much more efficient in changing the incidence of dominant traits, because both homozygous and heterozygous genotypes display the same phenotype. Many recessive alleles can "hide" from natural selection in the heterozygous condition.

There is no way for a harmful codominant allele to escape being culled from the population by natural selection. Homozygotes with two deleterious codominant alleles would be expected to have a lower fitness (adaptive) value than the heterozygote (bearing only a single such allele). The heterozygote, in turn, would be less fit than the homozygote for the alternative normal allele. This codominant form of allelic interaction does not allow the accumulation of undesirable alleles in the population the way that recessive allelic interaction does. Even though some of these alleles may not be advantageous under present environmental conditions, they might be beneficial in some other environment. The evolutionary flexibility of a species is highly dependent on its store of genetic variability and on the genetic system of the species to recombine this genetic variability into new adaptive phenotypes.

A codominant allele that is deleterious in double dose (homozygous) may actually be beneficial when in single dose with the alternative "normal" allele (heterozygous). *Over-dominance* is the name given to the form of allelic interaction in which the heterozygote has greater fitness than either homozygous genotype.

> Sickle cell anemia is an example of overdominant allelic interaction in humans. The "normal" gene S^1 specifies the amino acid sequence of the β chain of normal adult hemoglobin (HbA). The "abnormal" gene S^2 specifies an altered form of the β chain. Homozygotes for the S^1 allele have completely normal hemoglobin and hence have normal capacity for the transport of oxygen from the lungs to the tissues. Homozygotes for the S^2 allele are anemic, because a cell with the abnormal hemoglobin (HbS) cannot transport oxygen efficiently. In many cases, the anemia is so severe that the individual dies, usually before sexual maturity. The phenotype produced by the heterozygous genotype (S^1S^2) is a mixture of HbA and HbS (codominant allelic situation). These individuals generally appear to be normal unless they are subjected to severe oxygen deprivation (such as during athletic endeavors and rapid ascent to high altitudes). Under these stressful conditions, many of the red blood cells curl into a sickle shape (Figure 7.4). They then have a tendency to clump together (predisposing to rapid destruction) and are unable to move easily through capillaries. The patient goes into a painful "sickling crisis" and may die. In most environments this kind of gene would certainly be considered undesirable. Yet in certain parts of Africa where malaria is endemic, the S^2 gene is found at high frequency. The explanation for this phenomenon apparently lies in the fact that *Plasmodium falciparum* (the protozoan parasite responsible for this disease) cannot feed efficiently on erythrocytes containing HbS because of the increased viscosity of the cytoplasm produced by insoluble HbS (HbA is soluble). Therefore the heterozygote (S^1S^2) usually escapes death by malaria and also usually escapes death caused by severe anemia. Such an individual tends to produce more offspring than either homozygous genotype, and this tends to keep both alleles in the population.

The above example of sickle cell anemia should serve to remind us that a gene has no intrinsic value of its own. We cannot say that a gene is "good" or "bad" without specifying the environmental conditions under which that gene is allowed to express itself in the form of a phenotype.

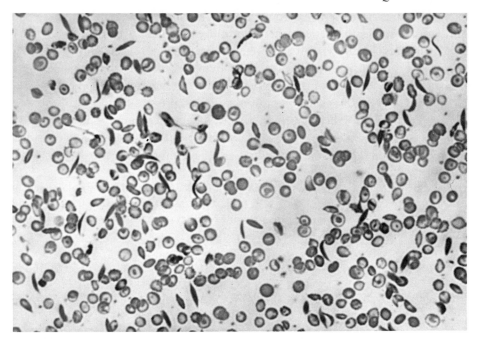

Figure 7.4. Red blood cells (erythrocytes) of a patient with sickle cell anemia in crisis. Many cells are crescentic (sickle-shaped), and some cells are immature (possessing nuclei or chromatin material). (Courtesy Carolina Biological Supply Company)

A multihybrid genotype often exhibits *developmental homeostasis,* i.e., the development of the organism appears to be *canalized* by alternative developmental pathways so that a standard phenotype is produced in spite of genetic and/or environmental variables. A population of organisms may also possess *buffering* mechanisms that tend to prevent change in the system by outside forces. Natural selection organizes the genetic composition of a population in a way that tends to produce the greatest number of highly adaptive phenotypes. To the extent that overdominance contributes to *population fitness,* this usually necessitates segregation of some less fit homozygotes. A highly integrated system of genes that interact harmoniously in forming coadapted genetic complexes is usually refractory to radical changes. Selection acting to change only one component of the system will usually be resisted by the tendency of the population to return (regress) toward the balance of the original composition. Such a population is said to possess *genetic homeostasis,* and the stability of the system is, in large measure, dependent on its store of interacting genetic variability. A well-buffered genetic system is not expected to evolve in response to a changing environment as rapidly as one deficient in the quality of genetic homeostasis.

Multiple Alleles

A gene may be considered a sequence of DNA nucleotides that encodes the information for making a particular polypeptide chain (protein). Through mutation, one or more of these nucleotide "building blocks" may be altered, creating a new allelic form of the gene. Since an "average-sized" gene may consist of about a thousand nucleotides, a change at any one of these numerous positions theoretically could produce an allele specifying an altered protein and hence a different phenotype. It is not surprising, therefore, that most genetic loci are revealed to possess, upon intensive study, more than just two alternative alleles. Whenever more than two alleles are known to exist at a locus, we have a *multiple allelic system*. As examples, there are over 200 self-incompatibility alleles (preventing self-fertilization) known in red clover, and over 300 alleles need to be postulated to explain the inheritance of the B blood gouup system in cattle. Regardless of how many alleles of a genetic locus are represented in a population, no diploid individual can have more than two.

> A simple multiple allelic system in the rabbit is represented by the dominance hierarchy of a coat color locus: $C > c^h > c$ ($>$ = dominant to), where C is the gene for full color, c^h is the Himalayan allele (as explained earlier in the chapter), and c is the allele for albino (no pigment = white; Figure 7.5).

The number of genotypes (g) generated by a locus with n alleles is given by the formula

$$g = \frac{n(n + 1)}{2}$$

If alleles are segregating at more than one locus (each locus with identical numbers of alleles), the number of possible recombinant genotypes is represented by the formula

$$\left[\frac{n(n + 1)}{2}\right]^r,$$

where r is the number of segregating loci. Higher plants and animals are estimated to have an average of about 10^4 loci/genome. If each locus has only three alleles (an extremely conservative example), the theoretical number of genotypes possible in the system is $6^{(10^4)}$ ($6^{10,000}$, an astronomically large figure to be compared with an estimate of the number of stars in the universe, 10^{20}). These kinds of calculations dramatically reveal the tremendous recombinational capabilities inherent in the sexual mode of reproduction. So many genotypes are possible that it is very unlikely that even a small fraction has ever been produced and tested

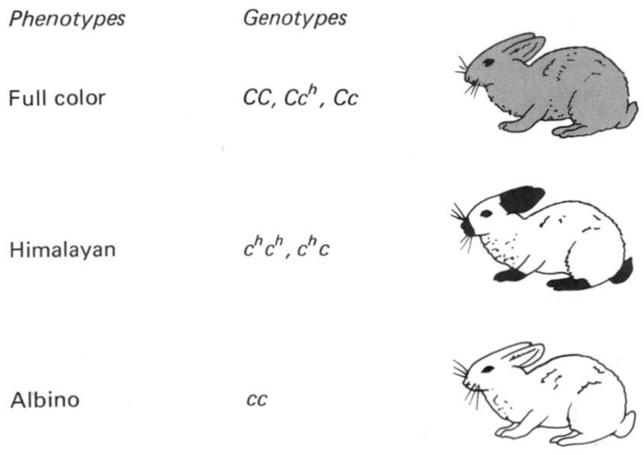

Phenotypes	Genotypes	
Full color	CC, Cc^h, Cc	
Himalayan	$c^h c^h, c^h c$	
Albino	cc	

Figure 7.5. Coat colors of rabbits governed by a multiple allelic system.

against some environment for survival value. Furthermore, for any given genotype that has existed it is highly improbable that it has been tested in all possible environments for survival value. The two factors of genetic variability and environmental heterogeneity are of utmost importance to the evolutionary process.

Dominance hierarchies in multiple allelic systems may be quite complex. When codominant allelic interaction is involved, more phenotypes are produced, thereby allowing natural selection to be more efficient in removing nonadaptive alleles from the population.

> One or more alleles in a multiple allelic series may exhibit codominant interaction rather than dominant-recessive interaction. An example of this type of dominance hierarchy is found in the human ABO blood group system. The allele I^A produces the A antigen on the red blood cell, I^B produces the B antigen, and i is a *"null" allele* (amorph) producing no detectable product. The dominance relationships can then be summarized as follows: $(I^A = I^B) > i$; i.e., I^A and I^B are equal in dominance (codominant), but both are dominant to recessive i.

Blood Group Phenotypes	Genotypes
A	$I^A I^A$, $I^A i$
B	$I^B I^B$, $I^B i$
AB	$I^A I^B$
O	$i i$

Wild Type

Mendel's experimental materials (peas) were cultivated crops. If the wild (natural) ancestors from which crops evolved are unknown, it is difficult to know which traits were common (*wild type*) and which were rare (*mutant type*). It is permissable to use Mendelian gene symbolism in designating genotypes of natural species, but often a different symbolism is employed. Gene symbols are usually assigned to reflect the name of the mutant type. A capital letter represents a dominant mutant gene, and a lower case letter represents a recessive mutant gene. Once a symbol has been assigned to the mutant gene, its wild-type allele is given the same base symbol with a superscript "+."

> Mutant black-bodied fruit flies are produced by a recessive genotype. The lower case letter b is assigned to the recessive mutant gene. Normal body color is the wild type governed by its dominant allele. The same base designator is used with a superscript "+" to designate the dominant allele (*viz. b^+*). After the symbol b had been assigned to the "black locus," a recessive eye color mutant was found that was named "brown." Since no two loci in the same organism can have the same gene symbols, the base designator bw was assigned to the mutant brown gene. Its dominant wild-type allele (contributing to red eye color) is therefore bw^+.
>
> Oval eye shape is the common (wild-type) phenotype in *Drosophila*. A dominant mutant eye shape, called "lobe," was found. The dominant mutant gene was therefore given the capital base letter L. Its recessive wild-type allele therefore uses the same base letter with a superscript "+" (*viz. L^+*).

Lethals

Certain gene combinations apparently fail to provide sufficient instructions for the development of the organism. Perhaps in some cases the gene products are defective enzymes; in other cases these products might interact antagonistically. The end result, however, is nonsurvival (death, lethality). Deaths important to evolutionary theory may occur during the embryonic phase or any time postnatally up to sexual maturity. Some genotypes shorten the life span considerably, yet occur so late in life that the individual's reproductive period is already essentially completed. Natural selection cannot act to reduce the incidence of such genes. On the other hand, a genotype that produces a *sterile* phenotype (unable to breed and/or make functional, viable gametes) may actually live to a "ripe old age," yet die a *genetic death* because it leaves no progeny. Such an organism is never part of the breeding segment of a population. From the standpoint of fitness, it was as though this individual never existed.

An example of a gene that has a lethal effect early in embryological development is known in the mouse. The gene A^y is lethal when homozygous, but interacts with its allele (A) in single dose to produce a viable yellow mouse. The genotype AA is also viable and produces the typical "mousy" color (called *agouti*) found in the wild. When yellow mice are intercrossed, only two phenotypes emerge in the offspring in a ratio of 2 yellow : 1 agouti. The homozygous $A^y A^y$ zygotes die early and are aborted (reabsorbed) and never appear in a litter (Figure 7.6).

$$\begin{array}{ccc} \text{Parents:} & A^y A \quad \male \times \quad A^y A \quad \female \\ & \text{agouti male} \quad \text{agouti female} \\ F_1: & 1\ A^y A^y\ :\ 2\ A^y A\ :\ 1\ AA \\ & \text{inviable} \quad \text{yellow} \quad \text{agouti} \\ & \underbrace{\qquad\qquad\qquad}_{\text{viable}} \end{array}$$

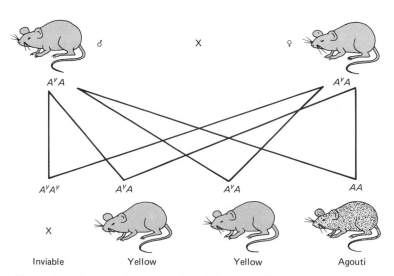

Figure 7.6. An example of a genetic lethal system. Mice homozygous for the yellow allele die *in utero.*

A gene in the snapdragon (*Antirrhinum*) produces its lethal effect only after germination of the seedling. The codominant gene A^1 allows chlorophyll to be synthesized, whereas its alternative allele (A^2) cannot make this photosynthetic pigment. Homozygous seedlings for the A^1 allele appear green, heterozygotes (A^1A^2) appear pale green (because only half of the normal amount of chlorophyll is being produced), but A^2A^2 homozygotes die as soon as the supply of food in the seed is exhausted (because they cannot photosynthesize sugars). Seeds from crosses between pale green plants germinate seedlings in the ratio of 1 green : 2 pale green : 1 white. Among the later survivors we find $\frac{2}{3}$ pale green : $\frac{1}{3}$ green.

$$
\begin{array}{cccc}
\text{Parents:} & A^1A^2 & \times & A^1A^2 \\
& \text{Pale green} & & \text{Pale green} \\
F_1: & 1\ A^1A^1 : & 2\ A^1A^2 & : 1\ A^2A^2 \\
& \text{green} & \text{pale green} & \text{white;} \\
& & & \text{inviable} \\
& \underbrace{} & & \\
& \text{viable} & &
\end{array}
$$

Huntington's chorea is a genetic disease of humans characterized by progressive mental deterioration. It is caused by a dominant autosomal gene that usually begins to show its destructive effects relatively late in life. If the dominant genotype becomes *penetrant* (i.e., shows the corresponding phenotype) during the late 40s, the individual has probably already produced all of the children he or she is likely to have. Consequently, though the afflicted parent may die prematurely, the incidence of this disease is not likely to decrease in subsequent generations (unless the offspring of an afflicted parent volunteers to refrain from having children for fear of passing on the deleterious gene).

A completely lethal gene (from the evolutionary standpoint) is one that prevents the individual from reproducing. *Semilethal* (or *sublethal*) genes are those that kill (or prevent reproduction) of more than 50 per cent, but less than 100 per cent of those with the mutant genotype. *Subvital* genes lower viability, yet cause death (or prevent reproduction) in less than 50 per cent of those with the mutant genotype. Sometimes the combined action of different semilethal and/or subvital genes can result in complete lethality. Populations may harbor many such deleterious recessive genes in heterozygous genotypes. Their *lethal equivalent value* in the population is calculated as follows:

$$
\left[
\begin{array}{l}
\text{Average number of harmful} \\
\text{recessive genes carried} \\
\text{by each member of the} \\
\text{population.}
\end{array}
\right]
\times
\left[
\begin{array}{l}
\text{Average probability that} \\
\text{each gene will cause death} \\
\text{(before reproducing) when} \\
\text{homozygous.}
\end{array}
\right]
$$

A population's store of lethal equivalents is sometimes called its *genetic load*. As an example, a population wherein every individual has an average of eight recessive semilethals, each of which (when homozygous) has a 50 per cent probability of premature death, would have a "genetic load" of four "lethal equivalents."

Epistasis and Other Gene Interactions

Whether or not the allelic variability at a certain genetic locus in a population can be expressed in the form of different phenotypes frequently depends on the genetic constitution of some other locus (or loci) in the genome. In such a case, the locus that hides or masks one or more other genes is termed *epistatic;* the term *hypostatic* is applied to the concealed member (or members) of the system. The phenomenon of *epistasis* is another mechanism that allows genetic variability to accumulate in the population, protected from the culling effects of natural selection (analogous to the protection afforded a recessive gene under its dominant allele).

A pair of alleles (*C* and *c*) in the mouse is responsible for colored and noncolored (albino) pelage, respectively. At an independently assorting locus (on another chromosome) is the agouti locus; allele *A* determines wild type (agouti = pigmented hairs tipped with yellow) and its recessive allele (*a*) produces solid hair color (e.g., solid black). Whenever the genotype contains *cc*, the mouse will be albino, regardless of the genotype at the other locus. Only when the *C* allele is in the genotype can *A* or *a* become expressed. The *C* locus therefore is acting epistatically, and the *A* locus is behaving hypostatically. When dihybrid parents are mated, the classical 9:3:3:1 phenotypic ratio becomes modified by this form of epistasis into a 9:3:4 ratio (Figure 7.7).

The hypothetical biochemical scheme shown below explains how this form of epistasis operates. Gene *C* makes enzyme 1, which in turn catalyzes the reaction from a colorless compound (Q) to the black pigment (R). When gene *C* is absent (as in the recessive homozygote, *cc*), there is no enzyme made and the biochemical pathway is blocked at compound Q. Gene *A* specifies enzyme 2, which in turn catalyzes the conversion of R into S (yellow pigment of the agouti pattern). The recessive allele (*a*), when homozygous, makes no enzyme and the pathway is blocked at compound R.

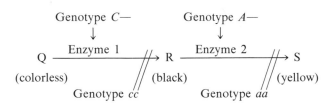

Note: The dash (−) in a genotype represents lack of knowledge of the homologous allele on one of the chromosomes. It makes no difference whether the genotype is *CC* or *Cc*, both make some enzyme 1.

Nonallelic genes may interact without behaving epistatically. Numerous examples of *complementary interaction* are known. When the dominant alleles of two (or more) loci are required for the expression of a unique phenotype, "dominant complementarity" is involved. When the homozygous recessive genotype at two (or more) loci is required for the expression of a unique phenotype, "recessive complementarity" is involved.

In 1897, a plant called Shepherd's purse (*Capsella bursa-pastoris*) was thought to be monomorphic (single form or phenotype) with regard to seed-pod shape (heart-shaped). In that year, G. Heeger discovered in Germany a few *Capsella* plants with ellipsoidal pods. It was considered a new species of Shepherd's purse and was named *Capsella*

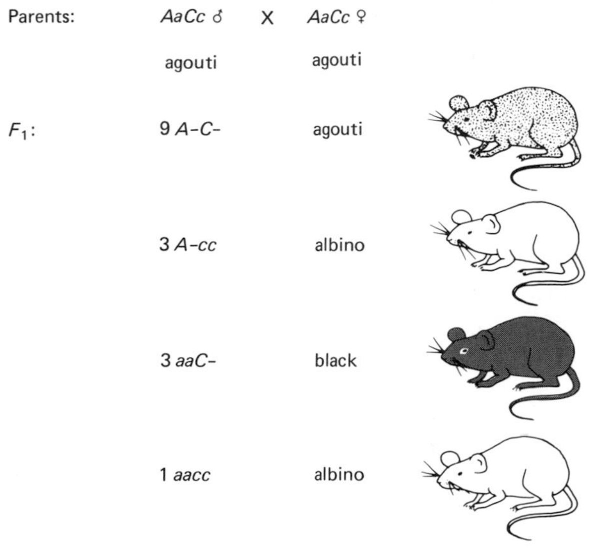

Parents: *AaCc* ♂   X   *AaCc* ♀

           agouti   agouti

F_1:   9 *A–C–*   agouti

         3 *A–cc*   albino

         3 *aaC–*   black

         1 *aacc*   albino

Figure 7.7. Example of an epistatic interaction of coat color genes in the house mouse. Blanks in the genotypes indicate that it makes no difference to the phenotype which allele is present.

heegeri in honor of its discoverer. Hugo De Vries attributed the appearance of this new "species" to mutation. Recall that De Vries favored the "mutation theory" and discounted the "natural selection theory" as the prime-mover in the evolutionary process. The two "species" were artificially crossed by G. H. Shull. The F_1 all had heart-shaped seed pods and were fully fertile. The F_2 (produced in the conventional way by allowing the F_1 to freely intercross, including selfing) segregated heart-shaped and ellipsoidal phenotypes in a ratio approximating 15:1, respectively. Shull interpreted the results in terms of recessive complementarity involving two independently assorting loci (Figure 7.8).

It was concluded from these results that both plant forms belonged to one species, because the F_1 hybrids were fully fertile and the morphological difference in seed capsules could be explained by segregation and interaction of alleles (already in the population) at two independently assorting loci.

Figure 7.8. Recessive complementarity in the inheritance of seed-pod shape in *Capsella*.

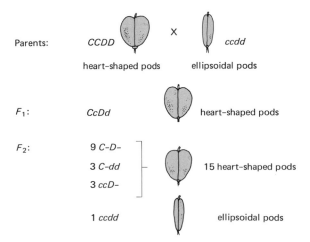

Parents: *CCDD*   X   *ccdd*

      heart-shaped pods   ellipsoidal pods

F_1:   *CcDd*   heart-shaped pods

F_2:   9 *C–D–*
        3 *C–dd*   15 heart-shaped pods
        3 *ccD–*

        1 *ccdd*   ellipsoidal pods

When other kinds of epistatic interactions are operative, dihybrid parents can produce progeny in ratios such as 12:3:1, 9:6:1, 13:3, and 9:7.

Sex Determination

Many plants and some animals have the ability to make both sperms and eggs (gametes). Such individuals are called *monoecious* and *hermaphroditic,* respectively. The differentiation of two kinds of gamete-producing tissues in the same individual is no more surprising than the differentiation of any other two kinds of tissues from cells of identical genotype. Recall that most angiosperms have complete flowers with staminate and pistillate components. Differences in the internal environment must be responsible for the bisexual nature of these organisms because all cells of the body presumably contain identical heredity.

Sometimes external environmental factors can be instrumental in sex determination. Eggs of the marine worm *Bonellia* (Figure 7.9) will develop into females in the absence of adult females in the immediate environment. They develop into males in the presence of females or extracts from the proboscis of females. The presence of a female chemical messenger in the external environment, then, seems to be responsible for the development of any egg (regardless of genotype) into a functional male.

A segregating genetic basis for sex is found in the unicellular alga *Chlamydomonas.* Different sexes in some of the lower forms of plants and animals cannot be distinguished morphologically. Physiologically (and genetically), however, there must be a difference because members of the same clone cannot conjugate together, but may be able to do so with members of a genetically different clone. These morphologically indistinguishable sexes are

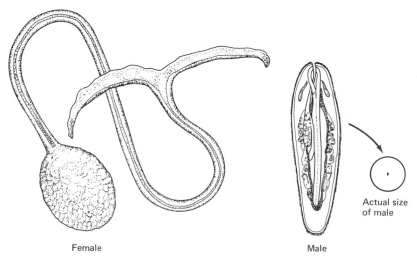

Female   Male

Figure 7.9. A marine worm, *Bonellia viridis,* in which maleness develops in larvae reared in the presence of females (or extracts thereof). The proboscis of the female (left) is about an inch long. The degenerate male (enlarged at right) lives parasitically in the genital ducts of adult females.

designated (+) and (−) *mating types*. Soon after (+) and (−) *Chlamydomas* cells unite to form a zygote, meiosis produces four haploid cells (two are (+) and two are (−) mating types). This 1:1 ratio among the meiotic products can be explained by the segregation of a pair of sex alleles.

Higher animals and dioecious plants usually have a chromosomal mechanism (therefore polygenic) of sex determination (rather than a single-locus mechanism). All mammals probably have a pair of sex chromosomes, called X and Y. The presence of a Y chromosome is male determining. Thus, normal diploid males are XY and females are XX. Males are called the *heterogametic sex* because they make two kinds of gametes (with equal frequency). Females make only one kind of gamete and are called the *homogametic sex*. Equal numbers of male and female progeny are expected from each mating using the X-Y system (Figure 7.10).

Some male insects (e.g., grasshoppers and roaches) have only a single X chromosome and no Y chromosome. This XO condition (O = no homologue) predisposes to male development, whereas the XX condition predisposes to female development. Males make X-bearing sperms and sperms without a sex chromosome in a 50:50 ratio; hence, equal numbers of males and females are expected each generation.

In some animals (e.g., butterflies and moths), the female is the heterogametic sex (making chromosomally different kinds of gametes); males are homogametic. To emphasize this difference the sex chromosomes are labeled Z and W (corresponding to X and Y, respectively). Females are ZW, and males are ZZ. Females make Z-bearing and W-bearing eggs with equal frequency. All sperms contain a Z chromosome. A 50:50 sex ratio is thereby expected in every generation.

In chickens and some other birds, the female is also the heterogametic sex and the male is the homogametic sex. But whereas the male has two Z chromosomes, the female has only one Z chromosome. Therefore, she makes two kinds of eggs with equal frequency; those bearing a Z chromosome and those without a Z chromosome. This ZO (O = no homologous chromosome) method also produces a 50:50 sex ratio in the progeny (Figure 7.11).

A peculiar form of sex determination occurs in *Drosophila*. On rare occasions, a fly is found that has a mixture of male and female characteristics (*intersex*). These intersex flies have three sets of autosomes and two X chromosomes (*AAAXX*). Failure of the autosomes to

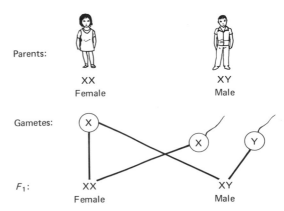

Figure 7.10. The X-Y sex determination mechanism of humans and other mammals.

Parents:

XX
Female

XY
Male

Gametes:

X   X   Y

F_1:   XX   XY
Female   Male

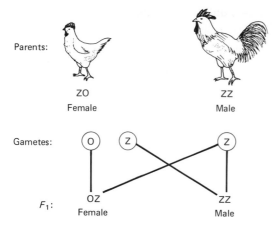

Parents:

ZO
Female

ZZ
Male

Gametes:

F_1:

OZ
Female

ZZ
Male

Figure 7.11. The Z-O sex determination mechanism of chickens.

separate during meiotic anaphase (*nondisjunction*) in one of the parents produced an unusual gamete with two sets of autosomes and a single X chromosome. For example, *D. melanogaster* has $2n = 8$; $n = 4$ (3 autosomes and 1 sex chromosome). An intersex could have $3 \times 3 = 9$ autosomes and 2 X chromosomes (total 11 chromosomes). It appears as though a balance has been struck between the female determiners on the two X chromosomes and male determiners on the autosomes. If we arbitrarily assign a male-determining value of 1.0 to a set of autosomes, then the female-determining value of genes on an X chromosome would be 1.5 ($2 \times 1.5 = 3 \times 1$). This mode of sex determination is called *genic balance*. The Y chromosome apparently contributes nothing to sex determination, because *AAXXY* flies are females and *AAAXXY* flies are intersexes. The fact that *AAX* flies are sterile males indicates that genes essential for fertility are on the Y chromosome. Thus, the normal diploid situation (*AAXX*♀♀ and *AAXY*♂♂) behaves as though sex was determined by the X-Y heterogametic male mechanism.

Sex in bees and other social insects is determined by a diploid or haploid chromosomal constitution (*arrhenotoky*). Queen bees are fertile diploid females; worker bees are infertile diploid females. The difference between fertility and sterility here is determined by the kind of food the diploid larvae receive. The somatic cells of male bees (drones) are haploid. Therefore, gametogenesis in males must involve a mitotic process rather than meiosis (because having only one set of chromosomes, it cannot be further reduced). If the queen lays unfertilized eggs, they will develop parthenogenically into males; if she lays fertilized eggs, they will develop into females. This type of sex determination only allows genetic recombination in the formation of female gametes and hence is not as efficient in generating genotypic variability in the offspring as the XY or ZO methods. Furthermore, arrhenotoky does not inherently generate a 1:1 sex ratio because the female usually chooses to fertilize more eggs than not. Workers are productive hive members. Only one or a few males are needed for the singular insemination of the queen. All other males are useless. But even here natural selection may be operative, because presumably only the strongest (swiftest) male(s) will be able to catch the female during her nuptial flight.

Sex-Linked Inheritance

Genes located on the X chromosome are said to be *sex linked*. Most sex-linked loci have no homologous counterparts on the Y chromosome. The Y may be thought of as genetically inert (for the most part). A sex-linked gene in the male is only represented once (on the singular X chromosome) and hence is *hemizygous;* females (with two X chromosomes) may be either homozygous or heterozygous, as usual. *Reciprocal crosses* involving autosomal genes are expected to yield comparable results; not so for sex-linked genes.

Red eyes in the fruit fly (*Drosophila*) is governed by a dominant gene (W) and white eyes by its recessive allele (w). White-eyed females crossed to red-eyed males yield an F_1 in which all males are white eyed and all females are red eyed.

Parents: ww ♀ × WY ♂
white female   red male

F_1:   1 Ww ♀♀ : 1 wY ♂♂
red females   white males

F_2:

	Males	Females
Red	½ WY	½ Ww
White	½ wY	½ ww

The reciprocal cross involves white-eyed males and red-eyed females as parents. If the female is homozygous, the F_1 will have only red-eyed progeny.

Parents: WW ♀ × wY ♂
red female   white male

F_1:   1 Ww ♀♀ : 1 WY ♂♂
red females   red males

F_2:

	Males	Females
Red	½ WY	½ WW : ½ Ww
White	½ wY	—

Since males have only one X chromosome, recessive genes can be expressed in hemizygous condition. Thus, sex-linked recessive mutations tend to be exposed to the sieving action of natural selection earlier and more frequently than autosomal recessive mutations. This is so because a recessive sex-linked mutation becomes immediately expressed in a male (with a single X chromosome), but an autosomal recessive mutation would be sheltered from the immediate scrutiny of natural selection under its dominant allele. Many generations may elapse between the time a recessive autosomal mutation occurs and the time a homozygous recessive genotype (by chance mating of two heterozygous parents) appears. If the sex-linked recessive mutation occurs in a female, it would have a 25 per cent chance of being expressed in the very next offspring (assuming that the probability of producing a male is $\frac{1}{2}$, and the probability of the mother transmitting the mutant X chromosome is $\frac{1}{2}$). If half the population is male, then one third of all X chromosomes of the population would have their recessive sex-linked genes exposed to natural selection in the hemizygous condition of males. Thus, sex-linked recessive mutations tend to be more efficiently selected than their autosomal counterparts.

Holandric Inheritance

Genes located in the differential segment of the Y chromosome (i.e., the region that has no homology with the X) are said to be *holandric* ("totally male"). Such genes are normally hemizygous and are transmitted from father to son exclusively. Natural selection can operate with maximal efficiency on Y-linked traits, because the single Y chromosome of males exposes all its holandric genes in hemizygous condition. There is no opportunity for a holandric gene to "hide" under an alternative allele (as can a recessive X-linked gene in females). The effect is essentially the same as if selection were operating directly upon the haploid genome of the gametes.

> A suspected case of holandric inheritance in humans involves a gene for hairy pinna (the external flaps of skin commonly called "ears"). This trait has been traced through several family pedigrees and found to appear in all the sons of a hairy-eared man but none of his daughters for several generations.
>
> Normal male *Drosophila* have an XY Chromosome constitution. Occasionally an abnormal male is found with an XO condition (O = no homologue). Such males are invariably sterile, indicating that the Y chromosome carries essential holandric genes for male fertility.

Sex-Limited Traits

Certain phenotypes under the control of autosomal or sex-linked genes may only be expressed in one of two sexes. This is often the result of anatomical differences between the sexes, but may sometimes be attributed to differences of the internal environment (e.g., hormones). Obvious examples are the fact that bulls do not give milk and cocks do not lay eggs. Variability in these traits is limited to one sex. Nonetheless bulls do carry genes that influence the milk production capacity of their daughters. A method for determining the genetic value of a bull for dairy production is presented in the next chapter.

Sex-Influenced Traits

A gene may behave as a dominant factor in one sex but as a recessive determinant in the other sex. Characters governed by these kinds of genes are said to be *sex-influenced* (or *sex-conditioned*).

> A classical example of a sex-influenced trait is that of pattern baldness in humans. The autosomal gene for baldness (B^1) exhibits dominance in men and recessiveness in

women. The alternative allele for nonbaldness (B^2) behaves as a dominant in women and as a recessive in men.

Genotypes	Phenotypes	
	Men	Women
B^1B^1	bald	bald
B^1B^2	bald	non-bald
B^2B^2	non-bald	non-bald

Dominance or recessiveness is not an intrinsic property of a gene, as demonstrated in the above example. The environment may be an important factor in determining the phenotypic expression of a genotype. In the case of baldness, the normal levels of sex hormones probably are the most influential factors of the internal environment.

Linkage

When two or more genes reside in the same chromosome they are said to be *linked*. Linked genes do not assort independently of each other. Recombination of linked genes occurs only when crossover events happen to occur between them. Two loci, each with a pair of dominant and recessive alleles, may be linked in either of two ways:

1. When the two dominant genes are on one chromosome and the alternative recessive alleles are on the homologous chromosome, the linkage relationship is called *coupling* phase.
2. When a dominant at one locus and a recessive at another locus are together on one chromosome and the alternative recessive and dominant alleles are together on the homologue, the genes are said to be in *repulsion* phase.

A majority of the gametes is expected to carry genes linked in the same way as they existed in somatic cells of the parents. Occasionally, crossing over will recombine linked genes into new arrangements. Crossing over occurs during prophase I when each chromosome consists of two identical sister chromatids. At any one position on a synapsed pair of chromosomes only two nonsister chromatids are usually involved in a crossover event. From a synapsed pair of chromosomes, a single crossover event yields two *parental* (*nonrecombinant* or *noncrossover*) products and two *recombinant* (or *crossover*) products.

Figure 7.12 illustrates recombination of linked genes in a dihybrid (*AaBb*) by crossing over.

In general, the farther apart two genes are located in a chromosome, the greater the chance that a crossover will occur between them; the more tightly linked (i.e., the closer they are together), the less opportunity for recombination. Geneticists have used this principle to construct *linkage maps* based on the *map unit* of distance between genes. One map unit is equivalent to one per cent crossing over.

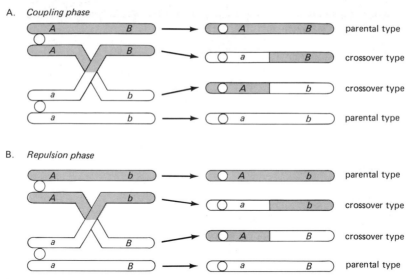

Figure 7.12. Diagrams of the results of crossing over in a dihybrid; (A) coupling phase, (B) repulsion phase.

The most efficient way to determine the gametes a dihybrid (*AaBb*) is forming (and the frequencies thereof) is to perform a testcross. Suppose that among 100 testcross progeny from a coupling phase dihybrid, we find 40 *AB/ab*, 40 *ab/ab*, 10 *Ab/ab*, and 10 *aB/ab* (the slash line separates groups of genes on homologous chromosomes). We diagram the cross as follows:

Parents: *AB/ab* × *ab/ab*

F_1:

(AB)	*AB/ab* 40%	parental types
(ab)	*ab/ab* 40%	
(Ab)	*Ab/ab* 10%	recombinant types
(aB)	*aB/ab* 10%	

The total amount of crossing over is 20 per cent. Therefore, the *A* and *B* loci are estimated to be 20 map units apart.

Recombination of sex-linked genes occurs in females, but not in males because males have only a single X chromosome. Any mechanism that retards recombination tends to restrict the genotypic variability of a population and hence may restrict its evolutionary potential. In this sense, sex-linked genes have a lower evolutionary potential than autosomally linked genes because recombination of the latter occurs in both sexes.

When two highly adaptive genes occur on different chromosomes, they will often become separated by independent assortment during meiosis. A higher percentage of gametes bearing both genes could be produced if the organism could rearrange its structural components (chromosomes) to bring both genes into the same linkage group. *Translocations* are able to do this. Similarly, two highly adaptive genes in the same linkage group can be transmitted together with greater frequency if they could be moved closer together in the chromosome. *Inversions* can accomplish this. Structural heterozygosity involving translocations and inversions may also be useful for maintaining genetic heterozygosity and for preventing the breakup (through crossing over) of chromosome regions containing highly adaptive gene complexes. The roll of translocations and inversions as *crossover suppressors* will be discussed in detail in Chapter 9.

An adaptively neutral trait (neither harmful nor beneficial) may appear in relatively high frequency in the population if the corresponding gene happens to be closely linked to a highly adaptive block of genes. Therefore, it is unwarranted to assume that a high-frequency trait must have a high fitness value. On the other hand, just because a trait appears to be trivial does not mean that the corresponding genotype makes no contribution to fitness.

> The ability of people to taste a chemical called PTC (phenylthiocarbamide) is dependent on a dominant autosomal gene; nontasters of PTC are homozygous for the recessive allele. Whether or not a person can taste this chemical would seem to be an adaptively neutral characteristic. However, in the endemic area of Brazil, the incidence of nodular goiter has been found to be significantly higher among nontasters of PTC. Obviously, this locus may have effects important for fitness other than the "major" effect that first drew attention to its study.

Because gene products often are active in complex, interconnected metabolic and developmental pathways, it is not surprising that most loci, when well studied, reveal multiple subsidiary effects that may seem (on first encounter) to be completely unrelated to the original one. All such ancillary manifestations of a single gene are called *pleiotropic effects*.

> Aside from their obvious importance during blood transfusions, the antigens of the ABO blood group system do not appear to have important pleiotropic effects. One study revealed that group O individuals had 17 per cent greater incidence of duodenal ulcers than controls, and group A individuals had 10–13 per cent more gastric carcinomas or pernicious anemias. These associations may have relatively little selective value, however, because these diseases usually do not strike until relatively advanced age (in the postreproductive period). On the other hand, in regions in which malaria is endemic, the frequency of individuals lacking the Rh antigen is extremely low or zero. This strongly suggests that natural selection may be operative at this blood group locus. About 90 per cent of West African Negroes do not have the Duffy blood group antigens a or b (in Northern European Caucasians only about 3 per cent, and about 10 per cent of Mongoloids of China do not have these antigens). Most West Africans are resistant to the protozoan malarial parasite *Plasmodium vivax*. Human blood cells were exposed to a parasite known to cause malaria in monkeys (*P. knowlesi*); *P. vivax* has not yet been successfully cultured outside the human body. About 80 percent of the cells possessing the Duffy antigens were invaded by this parasite; only about 2 per cent of the cells lacking the Duffy antigens were infected. This indicates that these antigens (and

possibly others) are probably receptor sites by which malarial parasites (and perhaps other pathogenic organisms) gain entry into cells.

Extrachromosomal Inheritance

Small amounts of DNA are known to reside in such organelles as chloroplasts and mitochondria. This extranuclear or cytoplasmic genetic information, in the form of *plasmagenes,* usually is transmitted only from the female parent to her offspring. The phenomenon of *maternal transmission* is probably due to the fact that very little if any cytoplasm of the male gamete is involved in fertilization. Cytoplasmic elements are transmitted to daughter cells in an unpredictable fashion, because there is no mechanism to insure their regular segregation (as there is for chromosomes).

> Three kinds of leaves may be found in the "four-o'clock" plant (*Mirabilis jalapa*): dark green, pale green, and variegated. The dark green leaves contain only normal chloroplasts; pale green tissues contain abnormal chloroplasts; variegation results from a mixture of normal and abnormal chloroplasts. Plants grown from seeds that developed on branches with only dark green leaves all develop dark green leaves; likewise, those from branches with only pale green leaves bear only pale green leaves; but those from variegated branches segregate dark green, pale green, and variegated leaves in unpredictable ratios. The phenotype of the branch from which the pollen was derived is of no consequence in this regard.

The genetic information of very primitive cells probably was not organized into "chromosomes" initially. Perhaps it became advantageous for normal development to ensure that groups of plasmagenes were inherited as a unit. This could be accomplished by joining several plasmagenes into a linkage group. With the evolution of the mitotic mechanism, each daughter cell was assured of receiving a full set of "chromosomal" genetic instructions. Chromosomal replication is now geared to the division of the cell. Perhaps for some genetic elements it is advantageous that they remain plasmagenes, to be replicated at various times and rates during the life of a given cell according to its needs. Some nuclear genes and plasmagenes are known to interact in the production of a normal phenotype. For example, normal chromosomal genes and normal plasmagenes must be present for normal chlorophyll production. All phenotypes are the result of gene-environment interactions. For every gene, part of its internal environment is the residual genotype (be it chromosomal or extrachromosomal).

Essay Questions

1. Explain how qualitative traits differ from quantitative traits. Which kind of trait is more important for the differentiation of populations into new species?

2. Discuss three major principles of genetics discovered by G. Mendel. How does each principle operate in mechanistic terms?

3. Discuss five genetic principles unknown to Mendel and their significance to modern evolutionary theory.

4. Under what conditions would gene linkage be of adaptive value? How can the "strength" of linkage be increased?

5. How may "adaptively neutral traits" become high-frequency characteristics in a population through selective mechanisms?

QUANTITATIVE TRAITS

Conflict: Mendelians vs. Biometricians

As explained in Chapter 2, the Mendelian approach to solving the riddles of heredity was considered useless for most of the practical problems faced by plant and animal breeders. Although Mendelism might be applied to qualitative traits (discontinuous variability), these same methods could not be used to investigate characters of degree exhibiting continuous variability. How can one make counts and ratios where no gaps exist in the phenotypic variability? The biometricians of the late nineteenth century were vitally concerned with quantitative traits and could not embrace the Mendelian principles after their rediscovery in 1900.

Bridging the Gap

One of the first experimental discoveries that helped to bridge the gap between the Mendelians and the biometricians was that of Swedish geneticist Nilsson-Ehle (Figure 8.1). He was studying the inheritance of kernel color in wheat. Some pure line crosses between white and red strains yielded F_2 ratios of about 3 red:1 white; other crosses gave F_2 ratios approximating 15 red:1 white; still others were consistent with a 63 red:1 white ratio. Nilsson-Ehle explained these results by segregation of alleles at one, two, and three loci, respectively.

> In those crosses that gave 15:1 F_2 ratios, various shades of red appeared to be due to the equal and cumulative effects of additive alleles (lacking dominance) at two independently assorting loci. Let us designate the alleles that contribute to redness with capital letters, and the null alleles (contributing nothing to color) with small letters.

$$\text{Parents:} \quad R_1R_1R_2R_2 \quad \times \quad r_1r_1r_2r_2$$
$$\text{dark red} \qquad\qquad \text{white}$$

$$F_1: \qquad\qquad R_1r_1R_2r_2$$
$$\text{medium red}$$

$$F_2: \quad \tfrac{1}{16}\, R_1R_1R_2R_2 \qquad\qquad \text{dark red}$$

$$\left.\begin{array}{l} \tfrac{2}{16}\, R_1R_1R_2r_2 \\ \tfrac{2}{16}\, R_1r_1R_2R_2 \end{array}\right\} \tfrac{4}{16} \quad \text{red}$$

$$\left.\begin{array}{l} \tfrac{1}{16}\, R_1R_1r_2r_2 \\ \tfrac{4}{16}\, R_1r_1R_2r_2 \\ \tfrac{1}{16}\, r_1r_1R_2R_2 \end{array}\right\} \tfrac{6}{16} \quad \text{medium red}$$

$$\left.\begin{array}{l} \tfrac{2}{16}\, R_1r_1r_2r_2 \\ \tfrac{2}{16}\, r_1r_1R_2r_2 \end{array}\right\} \tfrac{4}{16} \quad \text{light red}$$

$$\tfrac{1}{16}\, r_1r_1r_2r_2 \qquad\qquad \text{white}$$

Three segregating loci were postulated to explain the 63:1 F_2 ratio.

$$\text{Parents:} \quad R_1R_1R_2R_2R_3H_3 \quad \times \quad r_1r_1r_2r_2r_3r_3$$
$$\text{dark red} \qquad\qquad \text{white}$$

$$F_1: \qquad\qquad R_1r_1R_2r_2R_3r_3$$
$$\text{medium red}$$

The F_2 would be expected to have six different shades of red as well as one class lacking color, in frequencies represented in Figure 8.2 as a histogram (bar graph).
It begins to be difficult to determine accurately in which phenotypic class a given shade of red belongs. With just one or two more loci segregating in this fashion, it would no longer be possible to unambiguously assign each phenotype to its own class. The trait has lost its character of discontinuous variation and has become essentially quantitative in nature.

The genetic models used by Nilsson-Ehle to explain wheat kernel color are obvious oversimplifications for most quantitative traits. For one thing, these models have assumed that all phenotypic variability is due to gene effects and that the environment makes no contribution. Furthermore, it assumes that only two alleles exist at each locus, all plus alleles contributing equally and cumulatively to the phenotype, all loci are segregating independently (not linked), etc. The models worked well for the quasi-quantitative material investigated by Nilsson-Ehle. However, most traits governed by *multiple factors* (polygenes) are not so easily analyzed because of the complicating realities of many more genes, sizable environmental effects, linkage, dominance, epistasis, etc. Nonetheless these quasi-quantitative traits show us that Mendelian genes still form the hereditary basis for continuous phenotypic variability.

Figure 8.1. Nils Herman Nilsson-Ehle (1873–1949). [From *Genetics,* **18**: Frontispiece (1933)]

The major differences between qualitative and quantitative traits is that the latter are regulated by many genes, each with effects too small to be noticeable individually, and that the environment often makes substantial contributions to the phenotypic variability. The multiple factor theory has greatly facilitated our understanding of the genetic basis of Darwin's "fluctuating variability."

Darwinian evolution was proposed to operate by the accumulation of many small (perhaps imperceptibly small) variations over long periods of time. Qualitative traits of the kind that Mendel studied hardly qualify as the kinds of variations Darwin had in mind. Furthermore, most qualitative traits known at that time involved superficial features viewed trivial for purposes of defining differences between species and higher taxonomic groups. Morphological characters such as size and shape of various parts of the organism and physiological traits such as growth rate, age of maturity, and productivity are the more important ones for taxonomic purposes, for the improvement of production in domestic crops and livestock, and for the evolutionary divergence of populations into new species. These are quantitative traits characterized by continuous variability and governed by many different genes (polygenes) scattered throughout the genome. The exact numbers of elements in a polygenic system are seldom known, but the relatively few estimates for quantitative traits

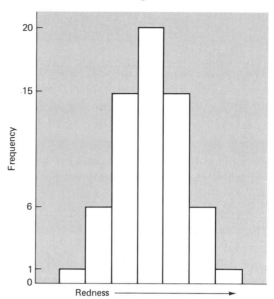

Figure 8.2. Histogram of wheat kernel colors expected in offspring of trihybrid parents.

made to date range from ten to perhaps a hundred (or more). Each locus of a polygenic system contributes such a small amount to the phenotype that the effects of its segregation cannot be studied through the classical Mendelian methodology of making counts and ratios. The system must be studied as a whole in terms of population measurements (*parameters*). It is rarely possible (in nature) to measure every individual in a given population to determine a population parameter such as the mean (average) phenotypic value. We therefore must depend on a sample of the population to estimate the corresponding parameter. Measurements derived from samples are called *statistics*. Sampling should be done in such a way that the various genotypes are represented in the sample proportionate to their frequencies in the population. Random sampling may bias the data if the distribution of the organisms in the sample area is not at random. Given an unbiased sample, the larger the sample size, the more accurate should be our estimate of the corresponding parameter.

A single gene with a major effect on the phenotype (*oligogene*) may have seemingly unrelated minor pleiotropic effects on a completely different quantitative trait. Other instances are known in which a quantitative trait has at least one (and perhaps several) oligogene directly involved in the genetic system. For example, Mendel studied the height of pea plants (a typically quantitative trait). He crossed a tall pure line with a short (dwarf) pure line. All the F_1 plants were tall; the F_2 segregated 3 tall:1 dwarf. This indicates that a single pair of alleles is involved, with tall dominant to dwarf. Undoubtedly an extensive system of polygenes in the residual genotype contributes in a comparatively minor way to plant height, but the total effect was so small that segregation in the polygenic system did not obscure the "talls" from the "dwarfs" in the F_2. The penetrance and expressivity of oligogenes (i.e., genes involved in typical qualitative traits) is undoubtedly due to alterations produced by "internal environmental" systems of quantitative polygenes (*modifier genes*), as well as to exogenous environmental factors.

Environmental Effects

Highly inbred (pure) lines or vegetative clones have been used to estimate the relative contributions that the genotype and environment make to various quantitative traits. Phenotypic variability within a pure line is entirely environmental because the genetic variability is zero. Different estimates of the environmental contribution may be obtained from different, genetically uniform populations reared under the same conditions. The F_1 hybrids from crossing two pure lines are usually highly heterozygous, but genetically uniform (genetic variability $= 0$). Corn hybrids tested under identical conditions displayed about 30 per cent less environmental variance than did the pure lines from which they were derived. Homozygous genotypes appear to be more "plastic," i.e., more subject to environmental modification than heterozygous genotypes. The latter are said to be better "buffered" against environmental factors.

Among animals, *identical (monozygotic* or *single egg) twins* can also be used for determining environmental effects, but their rarity limits widespread use. When groups of monozygotic twins reared under the same environmental conditions are compared with groups of twins whose members were reared in different environments, the effects of environment are revealed. Groups of monozygotic twins are often compared with groups of like-sex *fraternal* (*dizygotic,* or *two egg) twins,* because the differences between them reflect the genetic component of the trait under study. Fraternal twins are no more closely related than siblings born at different times. However fraternal twins contemporaneously share the same intrauterine environment and would be expected to share a more common postnatal environment than siblings reared at different times. Investigations of human twins reveal that certain quantitative traits (such as body height) are much less subject to environmental modifications than other traits (such as body weight).

It is a well-known fact that environmental stimuli many sometimes induce nongenetic changes in the phenotype that resemble the effects of specific gene changes. *Phenocopies,* as they are called, are relatively easy to detect for qualitative traits, but their presence usually adds major complications to the analysis of quantitative traits.

> When *Drosophila* pupae from normal parents are subjected to shocks of temperatures higher than normal (35°C) for short intervals during this stage of development, some abnormal phenotypes thus produced are indistinguishable from the vestigial-winged flies (Figure 8.3) produced by the homozygous recessive genotype of the "vestigial locus" at normal rearing temperature (25°C).
>
> Certain people, homozygous recessive for an autosomal gene, are vitamin-D resistant, i.e., they do not respond to the dose that satisfies the normal physiological requirements. The vitamin-D requirement for the mutant homozygotes is about one hundred times normal. Vitamin D is required for the intestinal transport of calcium and phosphate. Without sufficient vitamin D, children's bones develop the abnormal condition called rickets (Figure 8.4). The genetic disease of "vitamin-D resistance" is indistinguishable from the phenocopy produced in an individual reared on a diet deficient in vitamin D (or in an environment with too little sunlight for conversion of precursors in the skin to vitamin D).

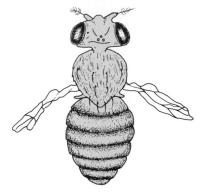

Figure 8.3. Vestigial wings in *Drosophila*.

A major deficiency in Darwinian theory was its failure to clearly distinguish between the genetic and environmental components of phenotypic variability. If we recall from Chapter 6 the example of how the environment and the genotype cooperate in the production of the pelage coloration peculiar to the Himalayan rabbit, it allows us to understand more easily how the environment can play such an important role in the development of a quantitative trait. The polygenic nature of continuous variability presents many opportunities for environmental factors to become involved in multiple metabolic, physiological, and developmental pathways. In some cases, the environment produces phenocopies that mimic genetically determined phenotypes. As a result, the phenotypic spectrum of a quantitative trait commonly has a sizable component of environmentally induced variability together with a certain amount of genetically determined variability. Consequently, a "superior" genotype (i.e., one with the

Figure 8.4. Normal leg bone development in child at left compared with a child afflicted with rickets (right).

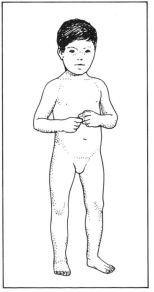

Normal

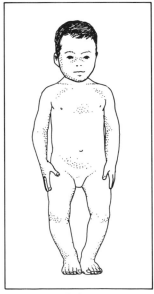

Rickets

capacity to produce a highly adaptive phenotype, given a favorable environment) may sometimes be hidden under a mediocre or poor phenotype (because of unfavorable environmental factors). Similarly, a relatively "poor" genotype may present a superior phenotype, because the individual was lucky enough to be reared under nearly ideal environmental conditions for the realization of its full genetic potential.

Natural selection operates only at the phenotypic level, not at the genetic level. Therefore, selection often makes "mistakes" in allowing certain phenotypes to survive (and reproduce) while culling others (by early death or disallowing reproduction). These mistakes are directly attributable to the nongenetic components of phenotypic variability. Traits with a large amount of environmental variability will cause natural selection to be relatively ineffective in changing the genetic composition of the population from one generation to the next. A highly *plastic* trait is one in which the environment makes a large contribution to the phenotypic variability. Plastic traits are not expected to evolve under changed environmental conditions as quickly as phenotypes determined almost exclusively by genes. After all, there is really no point in modifying the genetic composition of a population that adapts easily to changing conditions by environmentally induced modifications.

The water crowfoot (*Ranunculus aquatilis*) provides an excellent example of phenotypic *plasticity*. (Figure 8.5). The shape of the leaves on a given plant is determined by the environment acting on a genetically uniform background. Leaves above water are broad, flat, multilobed structures adapted to receiving high-intensity sunlight and of sufficient structural strength to withstand buffeting from disturbances in the atmosphere. Leaves below water level are much different. They are delicate filamentous structures, adapted to using the diffused underwater rays of light in photosynthesis. They can be very thin without danger of excessive water loss through transpiration because they are

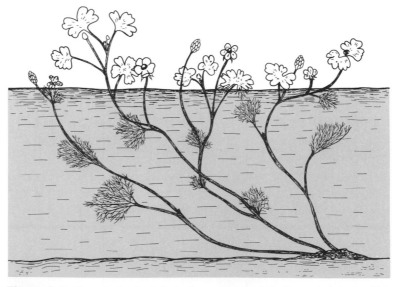

Figure 8.5. Phenotypic plasticity in the water crowfoot (*Ranunculus aquatilis*). Leaves which develop above water are much different than those which develop under water.

surrounded by water. Furthermore, these delicate leaves are buoyed to receive support and protection from the elements by the surrounding water. The ability of a singular genotype to respond in its various parts to different environments demonstrates a high degree of phenotypic plasticity. Environmental changes would be expected to cause less evolution in such a species than in one unable to respond with environmentally induced adaptive alterations.

If this is true, then it prompts the logical question, "Why haven't all organisms evolved a highly plastic phenotype?" The answer to this question appears to lie in the fact that a genotype allowing much environmental modification of the phenotype (and therefore exploitation of different environments) usually cannot compete effectively in any one environment with a genotype allowing little plasticity through canalized development of specific adaptations to these special conditions. Plasticity appears to be more common among plants than animals, probably because plants cannot move to more favorable locations when the environment changes.

Artificial Selection

Darwin was well aware of the progress that could be made in changing the heredity of domestic crops and livestock through the purposeful selection by humans of characters useful to them. All the various breeds of dogs can be traced back to a few original "mongrel-like" ancestors. *Artificial selection* by man has produced modern breeds of dogs so different that a biologist encountering only the extremes of this species for the first time would almost certainly classify them as members of different species. Darwin reasoned that if artificial selection could cause such profound changes to occur in a species during the relatively short time that crops and livestock have been under domestication, then natural selection operating over eons of geological time could explain the evolution of new species and higher taxonomic groups.

About 1902 Danish botanist W. L. Johannsen (Figure 8.6) began selection experiments using the garden bean (*Phaseolus vulgaris*). As with Mendel's peas, the bean is almost totally self-fertilized. Prolonged selfing produces homozygosity at all loci within a given lineage. He obtained a commercial mixture of the Princess variety of bean; each seed in the mixture was the product of a long history of selfing and hence was homozygous (genetically "pure"). From this mixture of different homozygous genotypes, Johannsen selected the largest and the smallest beans, planted them, and harvested the seeds from his selected plants. He found that plants grown from large beans tended to produce large beans; similarly, plants grown from small beans tended to produce small beans. This confirmed that selection was effective in separating different genotypes from a mixture. He then chose 19 seeds, each from a different plant and began to propagate each of them as a separate line. Within each line, he selected the heaviest and lightest beans for continuing the line in the next generation. This process was continued for six years. At the end of this time, it was obvious from the data that selection had not been effective. In each generation, the average weight of progeny seed from "light" parents was approximately the same as the average weight of progeny seed from "heavy"

Figure 8.6. Wilhelm L. Johannsen (1857–1927). [From *Genetics,* **8**: Frontispiece (1923)]

parents (Figure 8.7). Johannsen concluded that selection was of no consequence within a pure line. In other words, acquired characteristics are not inherited; all the phenotypic variability within a pure line is due to environmental modifications. The fact that Johannsen's selection had been successful only initially was interpreted to mean that selection could only isolate certain genotypes already present in the original mixture and that it had no "creative" potential. This interpretation made it difficult to accept Darwin's theory of evolution of new types through natural selection. The error of this interpretation is in the assumption that all populations consist of a mixture of homozygous selfing genotypes. Later it became clear that most natural populations are outcrossing (cross pollinating) and that each individual is highly heterozygous.

> The "creative" roll of selection can be appreciated from the following model. Consider a quantitative trait under genetic control of 40 loci. Each locus has a (+) allele contributing positively to the trait and a (−) allele that makes no contribution to the phenotype. In a population in which each individual is heterozygous (+/−) at all 40 loci, the probability of forming the most extreme phenotypes (40 loci homozygous for + alleles or 40 loci homozygous for − alleles) by chance is $(1/2)^{80}$. This probability is infinitesimally smaller than one divided by the total number of electrons in the universe. Under the guidance of natural selection, however, it might be possible for an extreme phenotype to be synthesized in about ten generations. This is the "big gun" of evolution theory; *this is creativity!!*

Much of the genetic variability of an outcrossing (cross pollinating) population usually lies concealed under a relatively narrow range of phenotypic variability. Because of the complexities of gene product interactions in metabolism, growth, and differentiation, epigenetic compensations have ample opportunity to canalize the development of many different

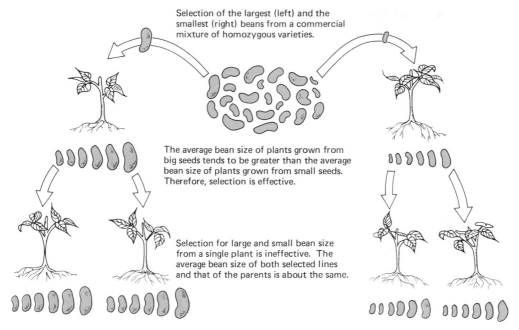

Selection of the largest (left) and the smallest (right) beans from a commercial mixture of homozygous varieties.

The average bean size of plants grown from big seeds tends to be greater than the average bean size of plants grown from small seeds. Therefore, selection is effective.

Selection for large and small bean size from a single plant is ineffective. The average bean size of both selected lines and that of the parents is about the same.

Figure 8.7. Diagram of Johannsen's selection experiment for large and small bean size.

genotypes into a relatively uniform phenotype referred to as *wild type* or "normal type." Artificial selection can reveal this hidden genetic variability, moving the population far beyond its normal limits of phenotypic expression. These attempts by humans to propagate extreme phenotypes are usually successful, but often as soon as artificial selection is relaxed, natural selection causes the population to revert (or regress) toward the original composition and re-establishment of the wild-type norm of reaction. Domestic animals that have escaped confinement and live in the wild are called *feral* animals. Feral stock tend to regress toward the original types from which man began his selective breeding. This regression phenomenon is caused by genetic (population) homeostasis and the "wild" phenotype is a result of developmental (individual) homeostasis (as explained in the previous chapter). Humans often select for traits that are nonadaptive in natural populations. Domestic stock can survive because man provides an environment in which they can flourish. When domesticated individuals return to the wild, however, natural selection begins to move the population back to its *adaptive norm* (i.e., the array of genotypes which, as an interacting system, maximizes fitness relative to the environment in which they live).

The Normal Distribution

Many quantitative biological characteristics follow what is called a *normal* (or *Gaussian*) *distribution,* i.e., they form a symmetrical bell-shaped curve when the phenotypic measurements are plotted on the abscissa and the frequencies of individuals at each metric value are plotted on the ordinate. Body height is an example of a trait that commonly follows a normal distribution. If we examine a large number of individuals in a given population, we find that the most frequent phenotype is one of average size; progressively fewer individuals are found in the taller and shorter categories as we move away from the mean value (Figure 8.8).

Population parameters are usually designated by letters of the Greek alphabet. Sample statistics are represented by letters of the English alphabet. Thus μ (mu) represents the population mean; the corresponding statistic is $\bar{X}$. To find the mean of a sample, we simply add the individual measurements and divide the total by the number of individuals in the sample. This is expressed in statistical form as

$$\bar{X} = \frac{\sum_{i=1}^{n} X_i}{n},$$

where n = number of individuals sampled, X_i = the measurement of the ith individual, and the capital Greek letter sigma (Σ) directs the statistician to sum (add) each item following this symbol beginning with individual number one ($i = 1$) inclusive through the last (nth) individual.

One other parameter is necessary to define the variability of a normally distributed population. This is the *standard deviation,* symbolized by the lower case Greek letter sigma (σ) and estimated from samples by the corresponding sample standard deviation (s). The statistical formula for calculating the sample standard deviation is

$$s = \sqrt{\frac{\sum_{i=1}^{n}(X_i - \bar{X})^2}{n - 1}}.$$

To calculate this statistic, we subtract each individual measurement from the mean, square this deviation, add all the squared deviations, divide the total by the number of individuals in

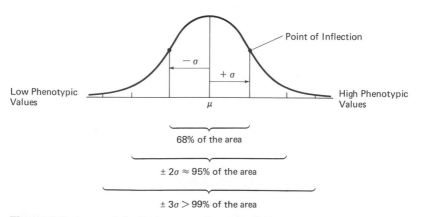

Figure 8.8. A normal distribution curve (μ = population mean; σ = standard deviation).

the sample minus one, and take the square root of this value. It is not the purpose of this book to require a detailed knowledge of how these calculations are performed. Rather the purpose is to become acquainted with statistical concepts so we can understand the action of selection in population models of quantitative traits.

By a method of the calculus, the point of maximal slope of the normal curve can be calculated. If we construct a line perpendicular to the base line passing through this point of inflection of the curve, then the distance from the mean to this perpendicular line represents one standard deviation unit. For every normally distributed trait, approximately two thirds (68 per cent) of the measurements will lie within plus or minus one standard deviation of the mean. Approximately 95 per cent of the measurements will lie within $\pm 2\sigma$ of the mean; more than 99 per cent of the measurements will lie within $\pm 3\sigma$ of the mean. Now a metric trait can be easily defined for each population by specifying the mean and the standard deviation ($\mu \pm \sigma$).

Samples of identical size drawn from the same population commonly have slightly different means. The distribution of means from an infinite number of such samples would form a normal distribution around the true mean of the population. The standard deviation for this distribution of sample means is called the *standard error of the mean,* represented by $\sigma/\sqrt{n}$, where n is the number of individuals in each sample. Approximately 95 per cent of the area under this curve lies within plus or minus two units of "standard error of the mean" (actually 1.96 units). Most biologists are satisfied with statistical evidence that gives 20:1 (95 per cent) or better odds in support of the conclusions drawn from experiments (or samples). Therefore, to estimate the true value of the population mean (μ) from a given sample, we use the formula

$$\mu = \bar{X} \pm 1.96 \, s/\sqrt{n}.$$

Suppose that the mean hemoglobin concentration in a sample of 500 males of a given population is $\bar{X} = 14.3$ grams/100 ml of blood; the sample standard deviation $s = \pm 1.2$ grams/100 ml of blood. We could say, with 95 per cent probability of being correct, that the mean of the entire population (μ) from which our sample was taken lies within the limits

$$\mu = 14.3 \pm 1.96 \left(\frac{1.2}{\sqrt{500}} \right) = 14.3 \pm 0.107.$$

Therefore, 95 per cent of estimates made by repetitive sampling of this same population should yield means ranging from $(14.3 - 0.107) = 14.2$ to $(14.3 + 0.107) = 14.4$. Now suppose that another investigator samples a group of 500 or more males and finds a mean of 15 grams/100 ml of blood. There is the possibility that the measurement techniques of this investigator were different from our own. Barring this, we would conclude (and 95 per cent of the time would be correct) that this group of males was taken from a population different from the one we first studied. In other words, these two populations are significantly different in the statistical sense. They might represent different races or subspecies. This example represents the kind of statistical tools that evolutionists can use to determine the extent to which two or more populations of a given species have diverged.

Many statistical procedures use the square of the standard deviation (σ^2 or s^2), called *variance,* because the variance can be partitioned or fractionated into additive components by a process called *analysis of variance.* It is far beyond the scope of this book to even outline how an analysis of variance is performed. All we need concern ourselves with now is the fact that the phenotypic variance ($\sigma_P{}^2$) of a given quantitative trait in a population is the result of different genotypic effects and different environmental contributions. Analysis of variance can yield estimates of the genetic variance ($\sigma_G{}^2$) and environmental (nongenetic) variance ($\sigma_E{}^2$).

$$\sigma_P{}^2 = \sigma_G{}^2 + \sigma_E{}^2$$

The genetic variance can be further partitioned into components of additive variance ($\sigma_A{}^2$), dominance variance ($\sigma_D{}^2$), and epistatic (interaction) variance ($\sigma_I{}^2$).

Likewise, the environmental variance may be partitioned into various components that can reveal such effects as those due to climatic, edaphic, density, age, and maternal effects. The sum of all of the environmental components of variance is the total nongenetic (environmental) variance ($\sigma_E{}^2$).

Gene Action

Different alleles may interact with one another in a variety of ways. If an allele contributes a constant increment to the phenotype whenever it is present in the genotype, it is said to act *"additively."* The heterozygote in this case would have a phenotype exactly intermediate between the two homozygous genotypes.

Gene Action	Arbitrary Phenotypic Value/Genotype		
Additive	0 *aa*	1 *Aa*	2 *AA*
Partial Dominance	0 *aa*	1.4 *Aa*	2 *AA*
Complete Dominance	0 *aa*		2 *Aa* *AA*
Overdominance	0 *aa*		2 2.2 *AA* *Aa*

One of the alleles may exhibit partial or complete dominance over the recessive allele when in heterozygous combination. Overdominance exists when the heterozygote has a more extreme phenotype than either homozygote. The subject of overdominant gene action will be discussed further with reference to hybridization in Chapter 18.

The linearity between additive alleles and their corresponding phenotypic values can be extracted by an analysis of variance. An underlying component of additivity also exists in genes showing various degrees of dominance or epistasis (Figure 8.9). This linear relationship is also extracted by analysis of variance and appears together with the additive gene effects in the *additive variance*.

In the case of dominant gene action, the deviations (shown as dotted lines in Figure 8.9) of each genotype from the line of best fit (additive component) would appear as the *dominance variance*. In a much more complicated way, the nonadditive interaction between different loci can be extracted as the epistatic or *interaction variance*.

A fortunate epistatic gene combination occurring in one or both of the parents will often be broken up by recombination and therefore not recovered in the progeny. Similarly, a deleterious recessive allele contributed to the offspring by a homozygous recessive parent may fortunately be hidden by the completely dominant normal allele from the other parent. This gives the offspring a better phenotype than one of its parents. In another case, however, the offspring may receive dominant alleles from both parents and therefore not be any better than either parent. Furthermore, if we have a multiple allelic series with the dominance hierarchy $A^1 > A^2 > A^3 > A^4$, the gene A^3, which behaves as a dominant in one parent (genotype A^3A^4), may be combined in the offspring with an allele (A^2) that was a recessive in the other parent (genotype A^1A^2). In the offspring (genotype A^2A^3) the A^2 allele now appears as a dominant gene. In other words, when dominance or epistasis is involved, one cannot predict the contribution a parent will make to the improvement of its offspring. This lack of predictive value associated with the dominance and epistatic components of genetic variability prohibits their use in the formulation of breeding plans for the improvement of crops and livestock. The additive component, however, does have predictive value, because every time an additive gene appears in the genotype we can count on it contributing a certain increment to the desired phenotype regardless of what other alleles may be present. Therefore, an individual's "additive genotype" represents its *breeding value*.

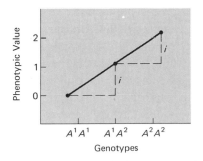 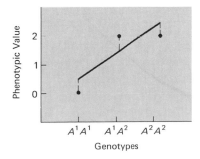

Figure 8.9. Model of additive gene action (left); each substitution of an A^2 allele for an A^1 allele adds a constant increment (i) to the phenotypic value. Model of dominant gene action (right) of the A^2 allele. Line of best fit to the three points is an underlying component of additivity; deviations from this line would appear in the dominance variance.

Heritability

What fraction of the phenotypic variability for a specific trait in a given population is due to gene effects? This parameter is called *heritability* (h^2) and can be expressed mathematically as

$$h^2 = \frac{\sigma_G{}^2}{\sigma_P{}^2},$$

(i.e., the ratio of the genetic variance to the phenotypic variance). Notice that the numerator of this "broad" estimate of heritability contains all kinds of genetic factors. As explained in the section on gene action, only the additive variance has predictive value. Therefore, breeders prefer to work with the "narrow" estimate of heritability expressed as

$$h^2 = \frac{\sigma_A{}^2}{\sigma_P{}^2}.$$

Unless specified otherwise, further references in this book to heritabilities will be used in the narrow sense. Heritability is an extremely important attribute of a quantitative trait in a population. The effectiveness of selection is directly correlated with the heritability of the trait. If all the phenotypic variability for a given trait is environmentally induced, heritability is zero, and selection will not be able to move the population mean. For example, all the phenotypic variability Mendel observed within a pure line of peas had to be due to nongenetic effects because each individual in a pure line has the same homozygous genotype. On the other hand, when all the phenotypic variability of a given trait is of genetic origin ($h^2 = 1$), no mistakes will be made in evaluating the breeding value of a particular genotype on the basis of its phenotype. As an example, if our serological tests tell us that an individual is blood group AB, then we know that person's genotype is heterozygous at the ABO blood group locus. In such a case, selection would infallibly pick out the superior genotypes and the population could evolve quite rapidly if need be. Most quantitative traits in outcrossing populations have heritabilities between these two extremes.

Let us now develop a model that allows us to perceive the relationship between heritability and the efficiency of selection. Given a very large population, we measure each individual for a particular quantitative trait (e.g., body length of the Yorkshire breed of swine at eight months of age). Suppose that the average body length of this group ($\bar{P}_1$) is 40 inches. If we wish to increase body length within this population, we will select for breeding purposes those individuals with exceptional length; perhaps we decide to use only those measuring 50 inches or longer. Suppose that the average length in this breeding group ($\bar{P}_p$) is 52 inches. We breed this selected parental group and obtain a large progeny with a mean ($\bar{P}_2$) of 46 inches. The difference between the $\bar{P}_p$ and $\bar{P}_1$ is called the *selection differential* (ΔP), and the difference between $\bar{P}_2$ and $\bar{P}_1$ is called *genetic gain* (ΔG). It is assumed that, in a very large population, the plus and minus effects of nongenetic factors within a generation will tend to cancel each other out in the mean, so that the mean phenotype of a population is a reflection of its mean genotype. If the environment has not changed from one generation to the next, the average phenotype of the progeny is a reflection of the average genotype of the selected parental breeding group from which it was derived. The realized (effective) heritability for body length in this population (i.e., in the genetic background represented by this breed and in the specific environmental or management conditions under which this experiment was conducted) is estimated as

$$h^2 = \frac{\Delta G}{\Delta P},$$

i.e., the ratio of the genetic gain to the selection differential. In other words, we "reached" out from the original mean 12 inches ($\Delta P = 52 - 40$) in our selection of the breeding group. However, we did not recover in the progeny all that we reached for. We realized a genetic gain of only 6 inches ($\Delta G = 46 - 40$). Therefore, the heritability estimate for body length in this population is 50 per cent ($h^2 = \frac{6}{12} = 0.5$). That is to say, we made about 50 per cent error in selecting genotypes for long body based on phenotypic measurements (Figure 8.10). On the average, only half the phenotypic variability is due to gene effects; the other half is nongenetic (environmental).

Suppose that there were twenty loci segregating [(+) alleles contribute a standard amount to body length; (−) alleles contribute nothing]. Selection generation after generation for long body length should eventually accumulate + alleles in homozygous condition at each of the 20 loci. When all the loci are "fixed" for the + alleles, there no longer is any genetic variability for this trait. There will still very likely be some phenotypic variability, but all of it must be due to nongenetic (environmental) causes. At this point, heritability of body length is zero and selection, no matter how rigorous (in terms of magnitude of the selection differential), will no longer effect a shift in the mean value from one generation to the next. Thus, the heritability of a quantitative trait may change as the population evolves under selection.

Most economically important traits in our domesticated stock have relatively low herita-

Figure 8.10. Hypothetical selection experiment for eight month body length in the Yorkshire breed of swine. The dark area at top (with mean = $\bar{P}_p$) represents the selected parental group from which the next generation (below) is derived.

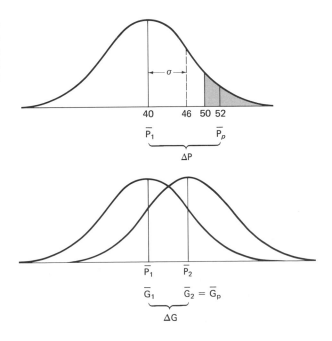

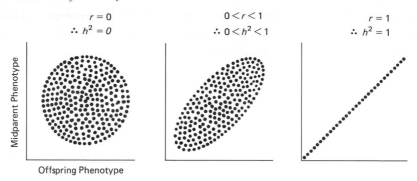

Figure 8.11. Heritability (h^2) of a metric trait as estimated by the correlation (r) between offspring phenotypes and that of their midparent.

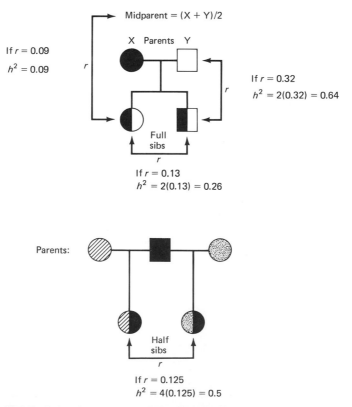

Figure 8.12. Pedigrees depicting the genetic relationships between various members of a family; also the relationships between various phenotypic correlations (r) and the heritability (h^2) of a metric trait. Circles = females; squares = males.

bility estimates. Long-term selection experiments in organisms with short generation times (e.g., *Drosophila*) have revealed that much genetic variability remains hidden under a standard "wild" phenotype. This genetic variability can be exploited by artificial selection to move the population far from its original adaptive norm. When selection for a given trait has been applied in opposite directions, different estimates of heritability are often obtained. For example, the heritability estimate when selecting for longer bodies may be different from the estimate when selecting for shorter bodies. The reasons for these differences are poorly understood.

Other methods of estimating heritability depend on measuring the phenotypic similarities between relatives. The correlation between phenotypic measurements in parents and their offspring can be used for this purpose, if we assume that there has been no essential change in the environment between generations. A *correlation coefficient* (r) is a measure of how closely two things tend to vary in the same direction. This parameter has the limits ± 1. When two factors are so closely related that a 1 unit increment of positive change in one factor always produces a specific increment of positive change in the other factor, they are 100 per cent positively correlated ($r = 1$). Since offspring receive a sample half of genes from each parent, heritability for a metric trait can be estimated by the correlation between offspring and the midparent value (i.e., the average of the two parents; Figure 8.11); or it can be estimated by twice the correlation between offspring and one parent ($h^2 = 2r$). The same relationship is true for contemporary full sibs, because litter mates share approximately 50 per cent of their genes in common. Half sibs (e.g., same mother, but different father) have only 25 per cent genetic relationship; therefore $h^2 = 4r$ between half sibs (Figure 8.12).

Selection Indexes

Trucation selection for a single trait (as presented in the discussion of realized heritability) is the simplest form of artificial selection. In practice, however, one cannot simply select for one character and ignore the multitude of other economically important ones. For while a breeder may make progress in selecting for one trait (e.g., disease resistance), other desirable characteristics (e.g., productivity as pounds of grain per acre or number of eggs laid per year) may be lost. Therefore, the most efficient way to improve the economic value of domestic stock is to select several important characters simultaneously. The breeder must consider the relative economic importance, the heritabilities, and the genetic correlations of the characters in which improvement is desired. The breeder then constructs a *selection index* in which each parameter receives appropriate weighting. For example, all other things being equal, a trait with a high heritability is likely to yield the greatest genetic gain from selection and therefore deserves greater "weight" in the index. The single score of the index is then treated as a single character. Some individuals can be of high merit in one component and low in another. Therefore, when one tries to improve a number of traits simultaneously (by using an index) the amount of selection pressure that can be applied to any one trait is considerably below that of selection for a single character (Figure 8.13). This is essentially the problem faced by natural populations in selection of the extremely complex "trait" called fitness. However,

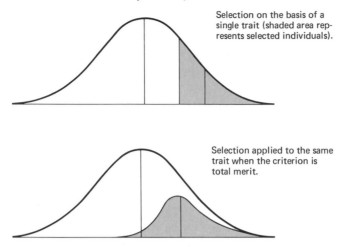

Selection on the basis of a single trait (shaded area represents selected individuals).

Selection applied to the same trait when the criterion is total merit.

Figure 8.13. Selection on single trait vs. total merit.

natural selection is mindless of heritabilities and genetic correlations. Given two traits that make equal contributions to fitness, one with high heritability and one with low heritability, natural selection does not put more emphasis on selecting the trait with the higher heritability. For this reason, natural selection is less effective in improving fitness than artificial selection is in improving economic value through use of a selection index.

Breeding Principles

Once the principles of quantitative inheritance have been mastered, practical breeders can do only two things to improve the genetic value of cultivated crops and domestic livestock. First, they can use several methods to choose those individuals they wish to be members of the breeding population (selection). Second, they can then employ the most efficient breeding scheme(s) to accomplish their specific purposes. If heritability of a quantitative trait is high (above 50 per cent), then much of what we see in the individual's phenotype (or record of performance) is the effects of genes (on average). Under such conditions, *individual selection* (based on the individual's phenotype) should be emphasized. The term *mass selection* is sometimes used synonomously with individual selection (especially when the selected individuals are bulked together *en masse* for subsequent matings). Most economically important traits commonly have medium to low heritabilities, and therefore not much reliance can be placed on individual selection. In such cases, perhaps at least some attention should be paid to the merits of the ancestors, a process called *pedigree selection*. It is also useful for characteristics that can only be manifested in one sex (e.g., milk or egg production) or when selection must be made when the organism is still too young to show the trait. If the merits of the immediate ancestors (parents) are well known, then proportionately more emphasis can be placed on pedigree selection. Little is to be gained by considering remote ancestors when the merits of the intervening ones are equally well known.

Some traits that take only a two-fold classification can nonetheless be governed by multiple factors (e.g., vitality or disease resistance can be measured on the basis of whether the organism lived or died). For these kinds of traits, as well as for ones subject to considerable chance variation, *family selection* can be useful. Under family selection, entire families of siblings that possess one or both parents in common may be considered for reproductive purposes. Full sibs (having both parents in common) share approximately 50 per cent of their segregating genes in common (genetic relationship = 50 per cent); half-sibs (having only one common parent) have a genetic relationship of approximately 25 per cent, etc. Family selection tends to be most effective when the degrees of genetic relationship within a family are high (usually as the result of inbreeding). When the observed resemblance between family members is higher than their genetic resemblance, the individual should be judged partly on its own characteristics and partly on its deviation from its family average. Less attention should be paid to family averages when comparing two individuals from related families than when comparing individuals from unrelated families.

The most definitive method for determining the genetic worth of an individual is by the performance of his/her offspring. This is termed a *progeny test*. Again, this method is most useful for traits that have low heritability, can only be expressed in one sex, or can only be measured after the organism is dead (e.g., carcass characteristics in livestock). Progeny testing is very expensive. Only a relatively few individuals can be selected by this method. Since male germ plasm can be disseminated much more widely than that of females, it is almost exclusively used on males. Suppose a dairy bull is to be progeny tested for its genetic contribution to butterfat production in its daughters. A number of females would be allocated to the bull for his progeny test. These females should represent a random sample of the breed. The environmental conditions should be standardized for all offspring insofar as it is possible. If the average butterfat production records of the daughters exceeds that of the average of the dams, the increase can be attributed to superior germ plasm contributed by the sire. Obviously, the more dams allocated to each sire, the more accurate the estimate of each sire's breeding value. But if more dams are allocated to each sire, then fewer sires can be progeny tested. In any case, those individuals to be progeny tested must have been previously selected for this purpose by some other method (usually by pedigree selection).

Selection by any one or a combination of the above methods can be applied in at least three general ways. One way is to put all the selection pressure on one trait. This is expected to make the most rapid gain in improving that trait. After the desired level of improvement has been realized, a second trait becomes the sole object of selection, etc. This is called *tandem selection*. The problem with this method is that when selection is relaxed on the first trait in order to improve the second trait, the first tends to revert back to the population average. A second method is to establish independent levels of minimal acceptable performance for each trait and select simultaneously on more than one trait. This tends to prevent selected traits from slipping back (regressing) toward the population average. However, it may result in culling some individuals that are very desirable for one trait, but mediocre or poor for another trait. A third method is based on some kind of a total score or selection index to measure net merit as discussed previously in this chapter.

Once individuals have been selected to be parents of the next generation, they can then be mated in one of several ways. One of the most commonly used methods for commercial beef and sheep enterprises is *random mating*. Simply turn enough sires out on the range with the dams to insure that all become inseminated as they come into breeding receptiveness (heat). This is the scheme used earlier in this chapter for estimating realized heritability. Other mating methods are termed *assortative matings*. When matings occur between individuals that are more closely related than the average of the population, they are said to constitute positive assortative matings; when matings occur between individuals that are less alike than expected by chance, then negative assortative matings exist. This subject will be discussed in some detail in Chapter 15.

Correlated Characters

Charles Darwin discusses the problem of "correlated variation" in *The Orgin of Species*. He recognized that white house cats with blue eyes are usually deaf. We now know that blue eyes and deafness are pleiotropic effects of the same dominant autosomal gene controlling white pelage. Suppose that a similar gene was present in a population of arctic hares. The white coat would probably allow hares to blend in with their snowy background and thereby more easily escape being seen by predators. However, the correlated characteristic of deafness would not allow the hare to hear predators nearby. The adaptive feature of camouflage might be largely neutralized by the nonadaptive feature of deafness. Such a gene would not be expected to increase in the population if the overall change in adaptive value was zero. This exemplifies how two traits can be negatively correlated with respect to their effects on fitness even though they are highly positively correlated at the phenotypic level.

One of the anticipated consequences of long-term simultaneous selection for two correlated quantitative traits is that their genetic correlations should become negative. The reason for this is that genes with favorable pleiotropic effects on both traits will be brought to *fixation* (homozygosity) relatively quickly by selection and subsequently will contribute essentially nothing to the genetic variances or covariance of the two traits. Pleiotropic genes that affect one trait favorably and the other character unfavorably will remain relatively unaffected by selection and remain at their original intermediate frequencies. Thus, most of the covariance between these traits would eventually appear negative. Both traits, therefore, might have sizable heritabilities, yet fail to respond to selection when applied to both traits simultaneously. The phenotypic correlation between characters gives no indication of their genetic relationship because environmental variation (unlike genetic variation) tends to affect both traits in the same way. Fitness, determined by the total genotype, is the compound character continuously shaped by natural selection. Thus, we might expect its major components to be negatively genetically correlated. Darwin also cites the associations between beak length and feet size in pigeons. Both traits are polygenic in nature. The correlation is a positive one such that pigeons with longer beaks tend to have larger feet and vice versa. If artificial selection is applied for longer beaks, the breeder might find that the population also responds with larger feet because the genes that favor development of longer beaks are also involved in the growth of larger feet. We refer to the *primary character* as the one to which selection is directly applied, and the *secondary character*(s) as the one(s) that is/are changed through *indirect selection* because of its/their correlation with the primary character. For purposes of improving domestic stock, indirect selection cannot be as useful as direct selection, unless the secondary character has a much higher heritability than the desired primary character, and the two traits are highly correlated genetically; or unless a much greater selection differential (selection intensity) can be applied to the secondary than to the primary character. Indirect selection might be useful when the desired primary character can be measured in only one sex (sex-limited quantitative trait) but the secondary character can be measured in both sexes. In such a case, the most rapid progress (i.e., the greatest genetic gain) would be anticipated by selecting the primary character directly in the sex that expresses it, and by applying indirect selection in the sex that does not express it.

Allometry

The head of a human baby is much larger in proportion to the length of its arms and legs than that found in the adult. This exemplifies the well-known principle of *allometric growth* (*allometry, heterauxesis*). Different parts of the body often grow at different rates relative to each other. If the part in question develops faster than the whole (or another part), it shows *tachyauxesis;* if less rapidly, *bradyauxesis;* if at the same rate, *isauxesis* (Figure 8.14).

The larger breeds of a domestic species have predictable allometric growth differentials for many morphological characteristics compared with the smaller breeds. Allometry is an important principle in modern evolutionary theory and is understandable in terms of genetic correlations.

An extinct member of the deer family known as the Irish elk (*Megaloceros;* Figure 8.15) had such gigantic antlers that one might wonder about their adaptive value. It has been

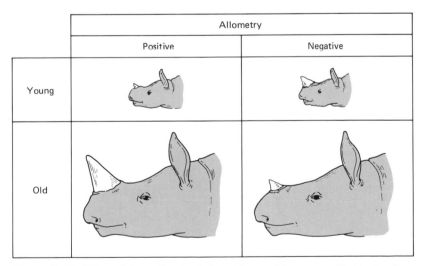

Figure 8.14. Differential growth rates (allometric growth) as exemplified by the face length and the size of nasal horn in a hypothetical rhinoceros-like animal. Under positive allometry, the nasal horn exhibits tachyauxesis with reference to facial length. Under negative allometry, the nasal horn exhibits bradyauxesis with reference to facial length.

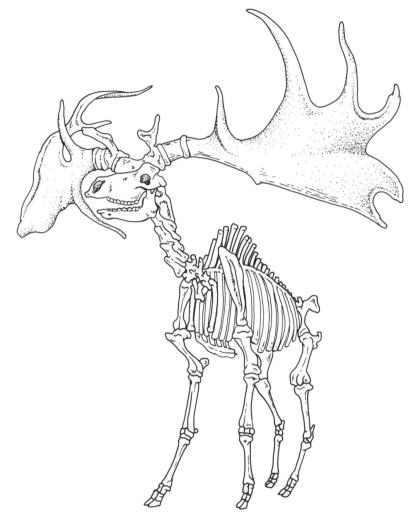

Figure 8.15. The Irish elk (*Megaloceros*).

suggested that the development of progressively larger antlers was adaptive at an early period in the history of this species, but that the "evolutionary momentum" for large antler size got out of hand (so to speak) at a later period and eventually contributed to its extinction. This concept of a built-in, goal-seeking, straight-line evolutionary propensity is called *orthogenesis* and is now thoroughly discredited by modern evolutionists. It is probably true that a large amount of energy was expended by the elk in producing these huge antlers. Their weight and size would appear to be detrimental in fleeing from predators, especially through brush and undergrowth. On the other hand, these oversized antlers could be formidible weapons in defense or in battles with other males for the breeding privileges of herd stag.

Among members of the deer family, the larger the body size, the greater the relative size of the antlers (allometric growth). If selection favored increase in body size (as it commonly appears to have done in many lines of mammalian evolution), the size of the antlers would tend to increase even faster because of the genetic correlation and allometric growth relationships between these two traits. Continuous selection in one direction, operating over long periods of time in this fashion, is termed *orthoselection*. This does not imply that evolution runs in straight lines toward some ultimate goal. Evolution is opportunistic, functioning only to maximize fitness or adaptation to present environmental conditions. A very large-bodied animal might be at a selective disadvantage as the climate becomes progressively warmer (because heat loss is a function of skin surface area relative to volume). In this case, selection could be expected to reverse the trend, producing progressively smaller bodied animals (with smaller antlers) in subsequent generations until the optimum size is reached. If the size of the *Megaloceros* antlers did begin to detract from its adaptive value, then selection (under a stable environment) would be expected to hold the population at an equilibrium in which the advantages of large body size were maximized and counterbalanced by the disadvantages of allometric antler growth. The specific events that led to the extinction of the Irish elk, as for almost any other kind of organism, are not known. *Megaloceros* disappeared during the ice ages and might have been hunted to extinction, along with other large animals, by early humans. The ice ages were characterized by relatively rapid climatic changes. The demise of the Irish elk, along with the wooly mammoth, saber-toothed cat, and many other Pleistocene species was undoubtedly precipitated in part by these rapid environmental changes and the inability of these species to generate new adaptive genotypes at a speed commensurate with the demands of survival.

Essay Questions

1. *How can the multiple factor hypothesis explain continuous variation?*

2. *What effect does phenotypic plasticity have on the effectiveness of selection?*

3. *Explain "heritability" and its importance for genetic gain under selection.*

4. *In what way can selection perform a "creative" function?*

5. *What effects do the genetic correlations between different characters have on their response to selection?*

CYTOGENETICS

A Hybrid Discipline

Recall that Mendel knew nothing of the processes of meiosis or mitosis. Perhaps one reason why Mendel's discoveries were not appreciated by the scientific community of his day was the universal ignorance of any physical basis to which his theoretical elements or hereditary factors could be ascribed. In the years between 1870 and 1900 (Mendel died in the year 1884) rapid advances were made in the study of cells and their components (*cytology*). By 1900, when Mendel's laws of heredity were rediscovered, chromosome splitting had been observed and the major features of mitosis and meiosis had been worked out. It then became a very logical step to propose that the hereditary elements were carried by chromosomes (the *chromosome theory of heredity*). Walter S. Sutton and Theodor Boveri, among others, were champions of this theory. The new science of *genetics* began to place genes (which tend to be transmitted together) into linkage groups. It was found that the number of linkage groups never exceeded the number of chromosome pairs characteristic of the species (and in well-studied organisms was, indeed, equal to the number of chromosome pairs). Mendel postulated that genes occur in pairs in individuals; so do chromosomes. His law of "segregation of alleles" has its counterpart in the separation of homologous chromosomes during meiosis. Also during meiosis, each pair of homologues passes to opposite poles as independent units; this accounts for Mendel's principle of independent assortment. Some genes behave as though only one member of a pair was present in the cell; these can be associated with the singular X chromosome (sex-linked genes) in one sex. Linked genes do not always stay together; likewise chromosome pairs are known to exchange homologous regions (crossover mechanism). When genes seem to behave in unexpected ways, the chromosomes frequently exhibit corresponding abnormalities. These correlations provide overwhelming evidence in support of the "chromosomal theory of heredity." A hybrid discipline called *cytogenetics* attempts to unite the principles of genetics with cytological phenomena (especially those of the chromosomes).

The rate at which genetic recombination can occur in a population is extremely important for evolution. A population that fails to generate new adaptive gene combinations at a rate commensurate with the rate of environmental change is headed for extinction. Therefore, the ways in which genes are structurally related in chromosomes and the mechanisms that regulate their recombination, i.e., the *genetic architecture* or *genetic system* of a population, are significant aspects of a population's evolutionary potential. Relatively little is known about the control systems of the meiotic process. We lack adequate techniques for precise measurement of the degree of chromosome pairing (synapsis) and chiasma formation (associated with genetic crossing over). The fact that two organisms have the same chromosome number does not necessarily indicate that they are closely related phylogenetically. Closely related species would be expected to have few chromosomal differences (either numbers of chromosomes or the arrangement of genes within the genome); distantly related species commonly have distinctive karyotypes. The degree of genetic relationship between two organisms is revealed in a rough way by the extent to which chromosomes form pairing figures during meiosis. Therefore, considerable information about the phylogenetic relationships between different organisms is potentially revealed through cytogenetic investigations.

Giant Chromosomes

Most chromosomes are too small to allow detection of minor structural changes by even the most powerful light microscopes. Almost twenty years after *Drosophila* was first used in genetic experiments, it was rediscovered that cells of the so-called "salivary glands" in the larval stage possessed giant chromosomes—perhaps ten times thicker and a hundred times longer than chromosomes found in other cells of the body (Figure 9.1). These giant chromosomes are probably formed by repeated duplication of chromatids so that about a thousand identical strands come to lie side by side. The name *polytene* chromosome indicates they are formed by "many threads." It is thought that at specific locations on these chromosomes the strands are more tightly coiled than at other sites; thus, some regions stain darker than other regions and produce a "banded" appearance. The pattern of bands and interbands is characteristic for each species in this genus. Each band can be individually identified on a cytological map. Thus, even relatively minor alterations of the chromosomal architecture can be detected in these giant structures. Another unusual feature of these chromosomes is that the two

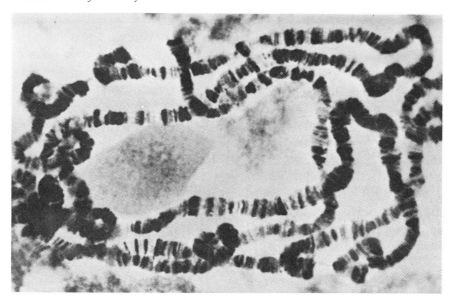

Figure 9.1. Giant chromosomes from the salivary glands of *Drosophila*. (Courtesy of Ward's Natural Science Establishment, Inc.)

homologues are paired so intimately that they cannot be individually identified. Pairing of chromosomes in somatic cells is a highly unusual phenomenon and its biological significance, together with a function for the exaggerated nature of the chromosomes, has yet to be deciphered. The centromeres of the giant chromosomes (and the Y chromosome of males) coalesce into a diffuse *chromocenter* composed largely of heterochromatin. Darkly staining *heterochromatin* is thought to be highly coiled and relatively genetically inert; lighter staining *euchromatin,* forming the bulk of these chromosomes, is thought to be relatively uncoiled and genetically active (e.g., in the formation of mRNA). Heterochromatin is commonly found near centromeres and appears to influence the expression of genes (perhaps even "switching them off") when in close proximity. This may contribute to a phenomenon known as "position effect" to be discussed in the next section. Some chromosomes, e.g., the Y chromosome, appear to be almost entirely composed of heterochromatin.

Structural Changes

Neither the degree of dominance, nor the phenotypic expression, nor the mutation rate is an intrinsic property of a gene. Every gene exhibits these characteristics as a function of the other genes with which it is associated in the genome together with other internal as well as external environmental factors. This is an extremely important point for evolutionary theory. A gene newly arisen by mutation might be adaptively neutral in its original location. If it is moved to a different location in the genome, it may exhibit new properties (perhaps beneficial, perhaps detrimental). These *position effects* are very common. They serve to remind us of the potential for evolving new adaptations inherent in rearrangements of the existing genetic material (quite apart from the introduction of new genes through mutation). The genetic architecture of the population is also subject to natural selection. Fortuitous chromosomal rearrangements, when beneficial, will tend to be perpetuated and eventually stabilized by whatever mechanisms the opportunistic nature of evolution can secure. Position effects are of two general kinds: (1) stable (or S-type) and (2) variegation (V-type). The stable type of position effect is usually associated with euchromatin; the variegated type with heterochromatin. Furthermore, S-type position effects are limited to modification of single phenotypic expressions. The V-type position effect has a "spreading effect" and usually modifies the phenotypic expression of several genes in close proximity to the heterochromatin. Examples of both types of position effect will be given later in this chapter.

Translocations

A *reciprocal translocation* (*segmental interchange*) involves the breakage of two different chromosomes and exchange of the nonhomologous segments (Figure 9.2). Chromosome breakage may be considered a gross mutational event that occurs spontaneously at low frequency, but may be induced with high frequency by exposure of the organism to ionizing radiations (X rays, gamma rays). If both the translocated chromosomes have a single centromere, the organism can form functional gametes. If the interchange produces a dicentric chromosome (one with two centromeres), the organism will probably not be able to form functional gametes. The two centromeres would likely move to opposite poles during meiosis and rupture the connecting strand; the reciprocal chromosome would be *acentric* (without a centromere) and would probably not move to either pole and thereby fail to be incorporated into the nuclei of daughter cells. It is possible for a chromosome to break and the broken ends to rejoin. In this case, the mutational event would probably be of no consequence because the original chromosome is restored. If the broken piece fails to rejoin, however, part of the genetic information of the ruptured chromosome would probably be lost from gametes

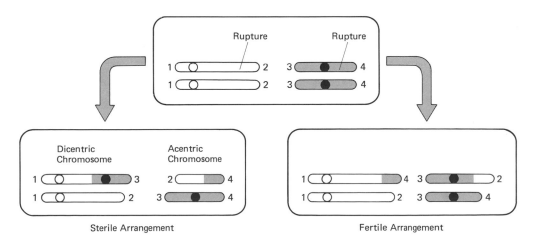

Figure 9.2. Formation of a reciprocal translocation.

formed by that organism. *Deletions* of genetic material are usually harmful. When chromosomes fragment, the broken ends behave as though they were "sticky" and tend to be healed by attachment to other broken ends. An unruptured chromosome has no tendency to join other chromosomes. Hence, a single break in only one chromosome of the genome is not likely to produce a translocation. A reciprocal translocation heterozygote produces *structural hybridity,* i.e., a normal chromosome now has a rearranged homologue with which it can pair only partially. During the formation of gametes, synapsis of homologous regions forms a cross-shaped configuration that later opens into a circle (Figure 9.3).

If adjacent chromosomes in the ring move to the same pole, gametes with repeated segments (duplications) and genetic loss (deletions) will be formed. Such gametes are usually nonfunctional in plants (although they apparently can function in animals, but commonly result in embryonic deaths). The only way for a translocation heterozygote to form genetically balanced gametes is to move alternate members of the ring to the same pole. Therefore a newly formed translocation heterozygote will usually be semisterile (only partly fertile). Fertility will be expressed to the degree that it can segregate (separate) adjacent chromosomes to opposite poles from the ring. If the structural hybridity of a translocation heterozygote has adaptive value, then natural selection would tend to favor the accumulation of genetic variants that could stabilize the segregation of adjacent chromosomes from the ring (remember that the process of meiosis is also under gene control).

Half the functional gametes formed by a translocation heterozygote will carry translocated (T) chromosomes, the other half will carry normal (N) chromosomes. If T and N gametes unite, they form another translocation heterozygote. If two N gametes unite, they form a completely normal karyotype. When two T gametes unite, they form a translocation homozygote, in which each chromosome has a completely homologous counterpart; meiosis then appears normal and produces only T gametes (Figure 9.4).

If the translocation produces adaptive position effects, the structural homozygote would be selected and perhaps become the basis of a new "race" within the species. Obviously, the linkage relationships of genes in these chromosomes would be different from those of the original population from which the translocation homozygote was derived.

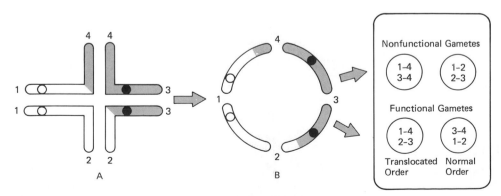

Figure 9.3. Reciprocal translocation heterozygote. (A) pairing figure; (B) circle of chromosomes at metaphase.

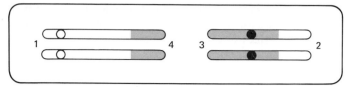

Figure 9.4. Translocation homozygote.

If a block of genes in one chromosome is especially harmonious with a group of genes in a different chromosome, it would be advantageous to merge these two segments into a single chromosome by translocation). This would enhance the chances that the coadapted complex would be transmitted as a unit in gametes. We know that crossing over would tend to break up these fortunate linked combinations. Therefore, another possible adaptive ramification of translocation heterozygotes is that genes in the region between the centromere and the point of translocation are "protected" from recombination. That is, the block of genes from the centromere to the place where the nonhomologous chromosome is attached behaves as a "transmission unit" or "supergene." If a crossover does occur in this region, the crossover gametes will contain duplications and deficiencies and be genetically unbalanced. The only functional gametes will contain parental type (noncrossover) chromosomes. This phenomenon is called *crossover suppression,* obviously a misnomer because crossing over is not prevented, but rather the recovery of funcional crossover gametes is blocked (Figure 9.5).

An example of a position effect induced by a translocation is known in a species of the evening primrose *Oenothera blandina.* In its normal location, the gene P^s produces flower bud sepals with stripes of broad red alternating with narrow yellow-green. At a locus linked with P^s, the dominant genotype S- governs yellow petal color and recessive ss gives sulfur petal color. When the segment of chromosome bearing these loci is moved to a nonhomologous chromosome in a translocation heterozygote, sepals of genotype P^sP^s have irregular patches of red pigment rather than broad stripes. If gene s is in the normal chromosome and its allele S is in the translocated chromosome, the petals show a mosaic of sulfur and yellow colors. The genotype Ss has fully yellow petals if s is in the translocated chromosome and S is in the normal chromosome. By crossing over, the genes P^s and S can be transferred from the translocated chromosome

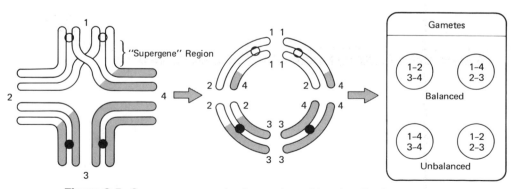

Figure 9.5. Crossover suppression in a reciprocal translocation heterozygote.

to the normal homologue. The genes then exhibit their normal phenotype effects, indicating that the structure of these genes has not been changed by the translocation; only their phenotypic expressions have been modified (i.e., a variegated position effect).

Inversions

An *inversion* is produced when a chromosome breaks at two positions and the intercalated segment is reinserted after turning 180° (Figure 9.6). Genes at opposite ends of the inverted segment now have new neighbors that may influence their phenotypic expression (position effect). The linkage map of an inversion homozygote would have some genes in a different order from the map of its normal counterpart. An inverted region can be detected cytologically in giant chromosomes as a segment in which the order of bands has been reversed. There are two major classes of inversions based on the location of the centromere.

(a) *Pericentric inversions* have the centromere within the inverted segment. *Inversion heterozygotes* (i.e., one normal chromosome and one inverted chromosome) must form loop-like configurations to maximize pairing of homologous regions during gametogenesis. One chromosome must form a loop and the other must twist in the inverted region. A single two-strand crossover within the inversion produces genetically unbalanced crossover gametes (because of duplications and deficiencies). Only the noncrossover gametes will be genetically balanced (Figure 9.7). Inversion heterozygotes, therefore, also act as crossover suppressors and prevent recombination of the block of genes within the inverted region. It would be advantageous to keep a harmonious combination of genes together as a transmission unit ("supergene"). In this case, natural selection would tend to favor the formation of inversion heterozygotes.

Inversion homozygotes do not incur gametic loss as do the heterozygotes and therefore are at a selective advantage in this sense. However, inversion homozygotes do not exhibit crossover suppression and therefore cannot prevent a block of highly coadapted genes from being broken up through crossing over. The position of the centromere is shifted so that the gross appearance of the chromosome may be radically changed in a pericentric inversion homozygote. For example, a metacentric chromosome can be changed to a submetacentric or acrocentric form through a pericentric inversion. Thus, the karyotypes of races within a

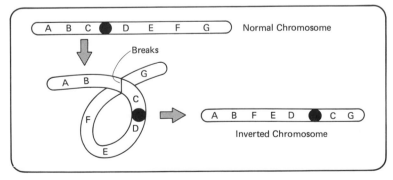

Figure 9.6. Formation of an inversion; various regions of the chromosome are labeled A through G.

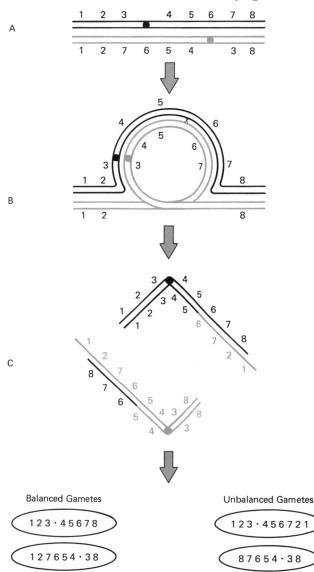

Figure 9.7. Effects of a single two-strand crossover between regions 5 and 6 within the inverted region of a paricentric inversion heterozygote. (A) Unpaired figure; (B) synaptic configuration maximizing pairing of homologous regions; (C) anaphase I of meiosis.

Balanced Gametes

$$1\ 2\ 3\cdot 4\ 5\ 6\ 7\ 8$$

$$1\ 2\ 7\ 6\ 5\ 4\cdot 3\ 8$$

(Noncrossover chromatids)

Unbalanced Gametes

$$1\ 2\ 3\cdot 4\ 5\ 6\ 7\ 2\ 1$$

$$8\ 7\ 6\ 5\ 4\cdot 3\ 8$$

(Crossover chromatids)

species (or of closely related species) may appear quite dissimilar in morphology and yet be genetically the same (or very similar).

(b) *Paracentric inversions* have the centromere located outside the inverted region. A single two-strand crossover within the inverted region again produces genetically unbalanced recombinant gametes.

At first meiotic anaphase, a *bridge* forms in the *dicentric chromosome* (possessing two centromeres), and the *acentric* fragment (without a centromere) usually would not move poleward and fails to be included in the reconstituted nuclei of the daughter cells (Figure 9.8).

Somewhere along its length, the bridge will rupture during first meiotic anaphase,

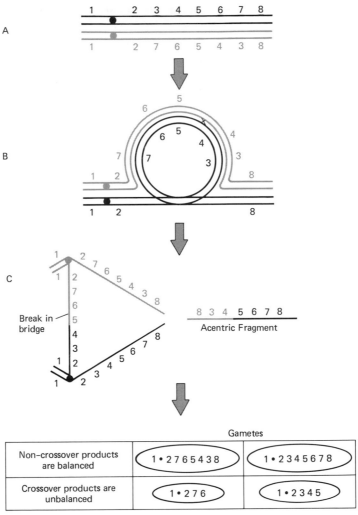

Figure 9.8. Effects of a single two-strand crossover between regions 4 and 5 within the inverted region of a paracentric inversion heterozygote. (A) Unpaired figure; (B) synaptic configuration maximizing pairing of homologous regions; (C) anaphase I figure showing bridge and acentric fragment formation.

creating genetically deficient gametes as crossover products. Here again, the region within the inversion exhibits crossover suppression in the inversion heterozygote and is thus protected from recombination. Paracentric inversion homozygotes have the same centromeric position and hence the same gross chromosomal morphology as the normal chromosomes from which they were derived. Paracentric inversion homozygotes do not have any synaptic problems and do not exhibit semisterility as do the structural heterozygotes, but neither do the homozygotes have a mechanism for protecting chromosomal regions from recombination. A species that has a karyotype consisting entirely of unequal-armed acrocentric chromosomes would be expected to incur paracentric inversions far more frequently than pericentric ones. If the karyotype of a species consists entirely of equal-armed metacentric chromosomes, the frequencies of the two types of inversions should be the same. The frequency of translocations is expected to be greater than that of inversions in species with high chromosome numbers relative to those species with lower chromosome numbers.

It seems as though nothing is "free" in nature; for every advantage gained by a population, nature usually extracts a fee in one way or another. If the population gains a selective advantage by holding a highly coadapted block of genes together by structural hybridity, it usually must "pay the price" of loss of some of its gametes (crossover gametes) and hence reduction in its biotic potential.

A special mechanism in *Drosophila* prevents crossover products from being incorporated into female gametes. Meiotic divisions in the female occur in tandem. Only the meiotic product at one end of the linear arrangement becomes the egg nucleus. When a single two-strand crossover occurs within the rearranged segment of a paracentric inversion heterozygote, a dicentric chromosome bridge is formed at first anaphase. The bridge tends to prevent poleward migration of the crossover chromatids; only the *eucentric* (single centromere) nonrecombinant chromatids move to the outermost two of the four poles at second anaphase. Thus, all eggs of *Drosophila* contain a complete genome and are functional. Reproductive loss is prevented in the female because genetically unbalanced gametes are not produced. Evolution has seized on a mechanism (*viz.* linear meiotic divisions) that allows the organism to reap the benefits of structural hybridity without paying the price of gametic wastage (at least in the female sex). Any mechanism that tends to produce abnormal ratios of gametes is said to produce a *meiotic drive* favoring certain genotypes and acting against other genotypes. Meiotic drive will be discussed in more detail near the end of this chapter.

Deletions (Deficiencies)

Generally speaking, loss of genetic material (*deletion* or *deficiency*) is usually harmful to the organism. Genetically deficient gametes produced by crossing over within a paracentric inversion heterozygote will not function in fertilization or, if they do function, the embryos will probably die. Deletion homozygotes (deficient in both chromosomes) are usually less viable than deletion heterozygotes (one normal, one deficient chromosome). Deletions, therefore, are of little value to our understanding of the broad sweep of evolution. Chromosome deletions are of two major kinds: (1) terminal and (2) interstitial (intercalary). A single break in a chromosome may cause the loss of an acentric end fragment; during synapsis the abnormal chromosome appears shorter than its normal homologue. This would produce a *terminal deletion* (Figure 9.9). An *interstitial (intercalary) deletion,* i.e., the loss of an internal chromosome segment, requires two breaks in the same chromosome and rejoining of the distal ends. During meiosis, pairing of homologous regions in an interstitial deletion heterozygote tends to form a buckle or loop in the normal chromosome (Figure 9.10).

Suppose an individual is of homozygous dominant genotype *Aa*. During meiosis, a deletion occurs involving loss of one chromosome segment bearing the dominant gene. This genetically deficient gamete now unites with a normal gamete from an individual of homozygous recessive genotype *aa*. The resulting zygote is *hemizygous* at this locus (*a-*) and at all loci within the deleted region. Its phenotype will be recessive, a condition called *pseudo-*

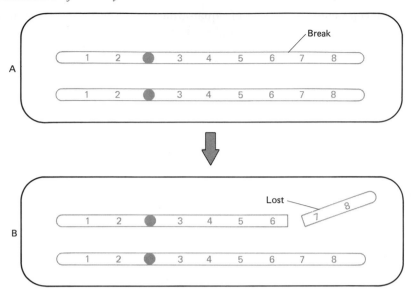

Figure 9.9. (A) Formation of a terminal deletion heterozygote; (B) synapsis in prophase I of meiosis.

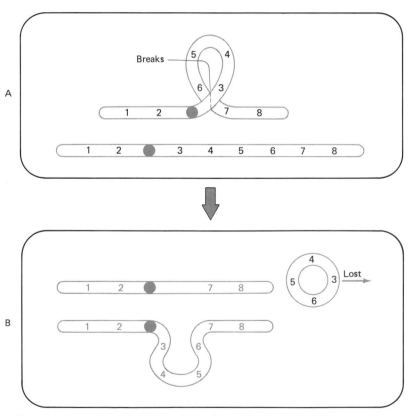

Figure 9.10. (A) Formation of an interstitial deletion heterozygote; (B) synapsis; normal chromosome forms a buckle to maximize pairing of homologous regions.

dominance. This phenomenon mimics the effects of gene mutation $A \to a$. Geneticists have used deletions to locate the residence of genes in chromosomes; they have used overlapping deletions in giant chromosomes to make physical maps. The linear order of genes as deduced from crossover maps is the same as that established for their actual physical locations in the chromosome. Crossover distances between genes, however, are not directly proportional to their physical distance in the chromosome because the frequency of crossing over may vary widely in different regions of a chromosome.

> When a hemizygous white-eyed *Drosophila* male (wY) was mated to a red-eyed female (thought to be homozygous WW) with a "notch" in her wings, the female notch offspring also had white eyes. As explained in Chapter 7, all female offspring should have been heterozygous (Ww) with red eyes. This finding suggested that the notch phenotype was produced by a small deletion, which included the white-eye locus on the X chromosome. A check of the giant salivary gland chromosomes revealed a small deletion in one of the X chromosomes of the female offspring, confirming the notion of pseudodominance for white eyes and structural heterozygosity (one normal and one deletion chromosome) for notch wing (Figure 9.11). Homozygosity for "notch" is lethal in females, and hemizygosity is lethal in males. Here, then, we have an example of a distinctive phenotypic effect associated with a deletion heterozygote because of genetic imbalance or *dosage effect* (quite apart from the phenomenon of psuedodominance).

If a deficiency occurs in a somatic cell, mitosis may produce a clone of cells exhibiting pseudodominance; the organism then becomes a *mosaic (chimera)*, possessing genetically different cells and exhibiting phenotypic variegation. For example, a heterozygous person might develop one normal brown eye and one pseudodominant blue eye. Somatic mutations are not transmitted to offspring and therefore are of no consequence for evolution.

Figure 9.11. (A) Interstitial deletion heterozygote of about 45 bands in the giant X chromosomes of the salivary gland cells in *Drosophila,* associated with the Notch phenotype; the normal chromosome forms a buckle to maximize pairing of homologous regions. (B) Terminal deletion heterozygote at one end of the paired X chromosomes, including the yellow and achaete loci.

Notch female

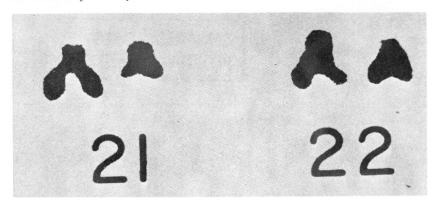

Figure 9.12. Philadelphia chromosome. This excerpt from an otherwise normal karyotype shows a deletion or loss in one of the number 21 chromosomes, which is associated with certain types of leukemia. This deletion is found only in the leukemic cells of the patient and never in normal ones. The Philadelphia chromosome almost never runs in families. [From Redding, A. and Hirschhorn, K.: *Guide to Human Chromosome Defects* In Bergsma, D. (ed.): Birth Defects: Orig. Art. Ser., Vol. IV, No. 4 Published by The National Foundation-March of Dimes, White Plains, N.Y. 1968, p. 106]

About three fourths of patients with a cancer of the white blood cells (called chronic granulocytic leukemia) have lost a terminal segment of the long arm of chromosome 21 (Figure 9.12). Because this condition was discovered in Philadelphia, the deleted chromosome was called the *Philadelphia chromosome* (Ph[1]). Apparently the deletion can occur in stem cells, from which all bone marrow cells originate, at any time during the life of the individual. The genic balance that controls the production and release of certain white blood cells (granulocytic leukocytes) is apparently upset in a stem cell possessing a Ph[1] chromosome. This results in an abnormally high level of whites, and a corresponding lower frequency of red blood cells (which can cause severe anemia and death). The Ph[1] chromosomes show a remarkable uniformity of the deletion (approximately 29 per cent loss). This suggests that there may be some kind of structural "weakness" on chromosome 21 where breaks are more likely to occur.

Duplications

An extra segment of chromosome in the genome is referred to as a *duplication*. Several mechanisms are available for the production of duplications. A *tandem duplication* can be formed by two breaks in the same chromosome, with rejoining of the broken ends delayed until after replication, followed by interstitial insertion of an extra acentric piece (Figure 9.13).

If the extra piece is inserted in an inverted manner, a *reverse tandem duplication* is produced. By translocation, an interstitial piece of one chromosome can be inserted into a nonhomologous chromosome; the resulting *transposition* requires three breaks (two in one chromosome, and one in a nonhomologous chromosome; Figure 9.14). A *shift* requires three breaks in the same chromosome with exchange between the interstitial pieces. Crossing over in the region of a shift heterozygote can then produce a chromosome bearing a *displaced duplication* (Figure 9.15). Duplications may produce *dosage effects,* i.e., the phenotypic consequences of a gene when present in more than double dose.

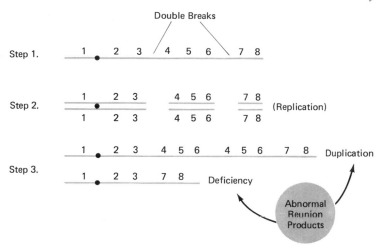

Figure 9.13. Formation of a tandem duplication.

When the X-chromosome region $16A_{1-2}$ bands in the giant salivary gland chromosomes of *Drosophila* are present in single dose, normal eye size is observed. A narrow (bar-shaped) eye is associated with a tandem duplication of this segment. Homozygous stocks of bar-eyed flies do not breed absolutely true, however, because about 1 in 1600 offspring appear with normal (wild-type) eye size. With about equal frequency an extremely narrow-eyed fly is also produced, called ultrabar (double bar; Figure 9.16).

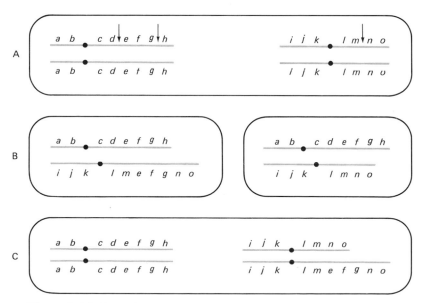

Figure 9.14. Formation of a transposition. (A) Three breaks indicated by arrows (two in one chromosome and one in another; (B) abnormal gamete (left) and normal gamete (right); (C) fertilization produces a zygote with *efg* region in triplicate.

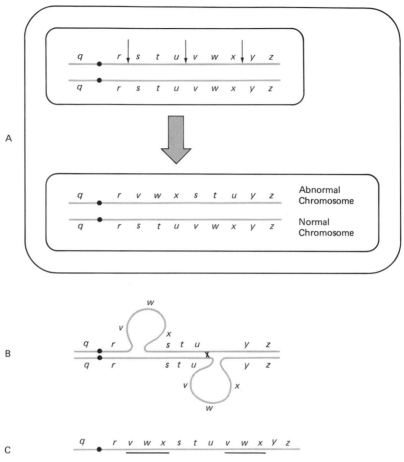

Figure 9.15. (A) Formation of a shift heterozygote by three breaks (arrows) in a single chromosome; (B) pairing and crossing over within the shift heterozygote; (C) resulting recombinant chromatid possessing a displaced duplication of the *vwx* region.

These unexpected phenotypes apparently are produced by rare *oblique synapsis* and *unequal crossing over* within the duplicated region of females (Figure 9.17). The ultrabar phenotypes have three *repeats* of the $16A_{1-2}$ bands on a single chromosome, whereas the normal flies have only a single set of these bands on a chromosome. Notice that a bar-eyed female has the same number of $16A_{1-2}$ bands as an ultrabar female. The ultrabar phenotype, therefore, represents a *stable (S-type) position effect*. The segments when repeated in triplicate, lying side by side on the same chromosome, behave synergistically, yielding a more extreme phenotype than when the same genetic material is distributed equally among the two X chromosomes. Here again, it must be emphasized that there are probably many opportunities to produce evolutionary novelties through reorganization of the existing genetic material of the population, i.e., through structural changes.

Potentially, duplications can be of great importance in the evolution process because they provide genetic redundancy in which mutations may accumulate without necessarily lowering

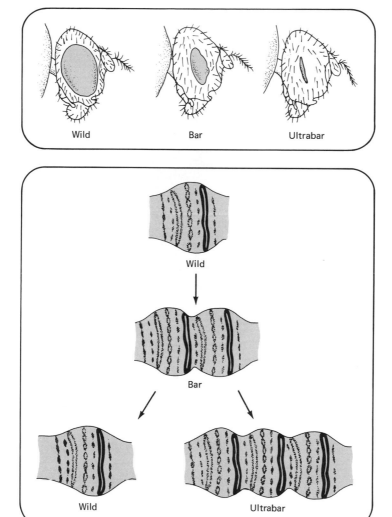

Figure 9.16. Duplication of a specific region of the X chromosome of *Drosophila* produces the bar-eyed and ultrabar-eyed phenotypes.

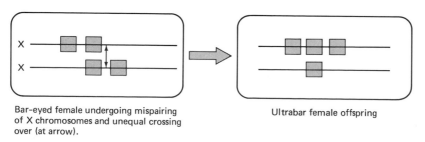

Bar-eyed female undergoing mispairing of X chromosomes and unequal crossing over (at arrow).

Ultrabar female offspring

Figure 9.17. Diagram of oblique synapsis and unequal crossing over in homozygous bar-eyed female *Drosophila*.

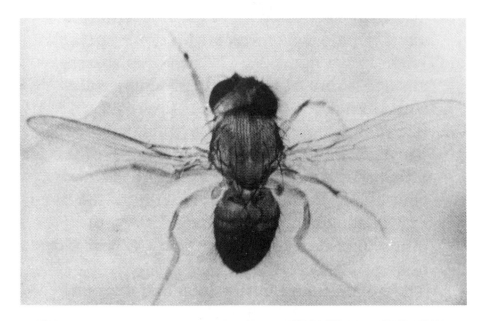

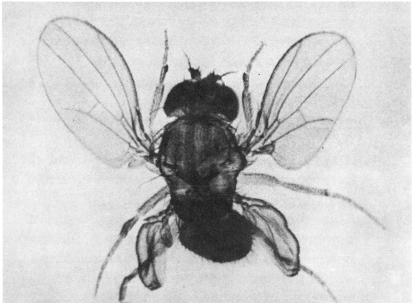

Figure 9.18. Pseudoalleles associated with the development of halteres (balancers) in *Drosophila*. Top left: wild type (normal fly) with club-shaped pair of halteres on the posterior part of thorax. Bottom left: bithorax. Top right: postbithorax. Bottom right: double mutant with fully developed second pair of wings. (From *Study Guide and Workbook for Genetics* by Irwin H. Herskowitz, McGraw-Hill Book Company, 1960)

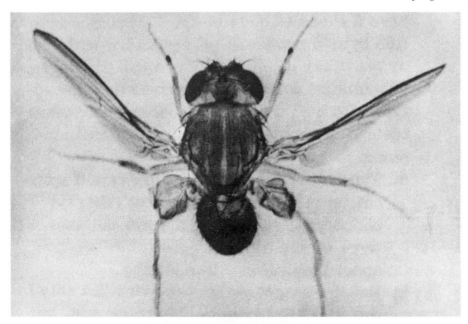

the fitness of the organism to present conditions. Some of these mutations in the duplicated segment might become beneficial if the environment changes. Thus, the population is able to increase its store of genetic variability and thereby enhance its evolutionary potential. Normal genes in the original chromosome segment could provide adaptation to present conditions, thus freeing the duplicated segment to mutate in various ways, some of which may provide a selective advantage under changed conditions. If this is indeed a mechanism whereby greater genetic complexity is evolved, it should be possible to find examples of duplicated genes in a transitional state where one locus is performing a specific function and its modified "twin" is performing a slightly different function.

Pseudoalleles may exemplify these transitional states. Initial studies of a genetic locus may lead to the conclusion that a pair of "alleles" are segregating. Later, when larger samples are obtained, it is found that the "alleles" can be recombined at low frequency by crossing over because they are really very closely linked loci rather than true alleles. Furthermore, when the dominant normal alleles ($+$) are in coupling linkage phase (or *cis* position) of a dihybrid, the resulting phenotype will be different than when the dihybrid has mutants in different homologues, i.e., when linked in repulsion phase (or *trans* position). Therefore, pseudoalleles exhibit a phenomenon termed *cis-trans position effect*. Several examples of pseudoallelic loci in *Drosophila* are correlated with duplicated or triplicated segments in the giant chromosomes. The condition is also suspected in other organisms, but due to the small size of the chromosomes it cannot be supported by cytological data.

> The sex-linked white eye locus of *Drosophila* was originally believed to have at least three alleles (w^+ = normal red eye color, w^a = apricot, and w = white). Later it was found that apricot and white were psuedoallelic as closely linked loci of a duplicated chromosomal segment. The white locus has the alleles w^+ (normal red) and w (white); the apricot locus has the alleles apr^+ (normal red) and apr (apricot). Dihybrids in cis position ($w^+ apr^+/w\ apr$) have normal red eyes, but those in trans position ($w^+ apr^+/w\ apr^+$) have pale (dilute) apricot eye color. Notice that the same genes are present in both the cis and trans heterozygotes, but the phenotypes are not the same. This is an example of a cis-trans position effect. It seems as though the product of the w^+ gene must be produced in close proximity to the product of the apr^+ gene to yield a normal phenotype. It has been proposed that a sequential biochemical pathway is involved, such that the product of gene w^+ acts as substrate for the product of gene apr^+ (or *vice versa*). If these genes are in trans position (i.e., in different chromosomes), their products may be quantitatively too small and too distant to interact efficiently in the production of normal eye pigments. The "white region" in *Drosophila* has subsequently proven to be a nest of five (possibly more) recombinationally separate, closely linked genes with similar phenotypic effects. The simplest explanation is that an original gene has become duplicated and mutation has differentiated the repeated segments into functionally different pseudoalleles.

One characteristic of the order Diptera within the class insecta is the presence of a single pair of wings (the second pair has been reduced to a set of "balancers" called *halteres*). This is the order to which flies belong, including the fruit fly *Drosophila*. A pair of pesudoalleles is involved in the development of the halteres. At one locus, the gene bithorax (*bx*) converts the anterior region of the haltere into a wing-like structure. At the other locus, the gene postbithorax (*pbx*) converts the posterior part of the haltere into a wing-like structure. These genes are separable by recombination, but closely linked (0.02 map units apart). They are also associated with triplicated segments of the giant chromosomes. When the two mutants are in cis position of a dihybrid ($bx^+pbx^+/bx\ pbx$), normal halteres are produced. When the mutants are in trans position ($bx^+pbx/bx\ pbx^+$), a slight postbithorax effect is achieved (cis-trans position effect). Flies that are homozygous for both mutants have a fully developed second pair of wings (Figure 9.18). Most other orders of insects have two pairs of wings. Therefore, two recessive mutations in these pseudoallelic loci can transform a fly into an insect that can no longer be classified in the order Diptera! This is not the way evolutionists perceive the origins of higher taxonomic groups (i.e., groups above the species level such as genera, families, and orders). A more likely explanation is that two pairs of wings is a *primitive* condition from which the dipteran insects evolved; a single pair of wings is the *derived* condition of modern flies. The normal development of halteres is undoubtedly governed by many genes, and this pathway may be interrupted at any of several stages by mutations such as those of the *bx-pbx* pseudoallelic loci. In other words, it is relatively easy for one or a few mutations to interfere with normal development of halteres in a gross way, but it is incongruous to believe that a few mutant genes could provide all of the instructions for the elaboration of an anatomical feature as complicated as a fully developed second pair of wings.

Duplications can produce *dosage effects,* as exemplified by the bar-eyed condition in *Drosophila.* It may be possible for an organism to develop mechanisms that can nullify dosage effects. These mechanisms may be classified under the term *dosage compensation.* Probably one of the best examples of dosage compensation is provided by the sex-linked genes of humans. Males have a single dose of sex-linked genes, whereas females have a double dose. Yet the sex-linked traits of females do not usually appear any more intense than corresponding traits in males. Nuclei of female somatic cells have a darkly staining mass of heterochromatin affixed to the inner border of the nuclear membrane. This structure is called *sex chromatin* or *Barr body* (named for M. L. Barr, its discoverer; Figure 9.19). M. F. Lyon postulated that sex chromatin is a single, highly condensed, genetically inactivated X chro-

Figure 9.19. Positive buccal smear with two Barr bodies. This cell, with two Barr bodies (arrows), was taken from a female patient with three X chromosomes rather than the usual two. [From Redding, A. and Hirschhorn, K.: *Guide to Human Chromosome Defects* In Bergsma, D. (ed.): Birth Defects: Orig. Art. Ser., Vol. IV, No. 4 Published by The National Foundation-March of Dimes, White Plains, N.Y. 1968, p. 95]

mosome. Males do not usually have a Barr body, because the genes on their single X chromosome must remain functional. In other words, females, by "turning off" one of their X chromosomes, can achieve a functional hemizygous condition of their sex-linked genes equivalent to that of males as a dosage-compensation mechanism (the Lyon hypothesis).

> An enzyme known as glucose-6-phosphate dehydrogenase (G-6-PD) is the product of a human sex-linked gene. A recessive mutant allele makes no enzyme. The blood of males (hemizygous for the G-6-PD gene) and females (homozygous for the G-6-PD gene) have essentially the same concentration of this enzyme. It was found that half the cells of a female (heterozygous for the mutant allele) produce the enzyme, the other half does not. Therefore, the female is a functional mosaic for this locus. Random inactivation of one X chromosome in each cell would account for these results. Several other X-linked human loci are also known to exhibit dosage compensation.

Meiotic Drive

In our discussion of genetics in previous chapters, it was assumed that all gametes are equally able to participate in fertilization with those of the opposite sex. This is not always true, however. Selection may act directly on the gametes; those gametes bearing a particular gene (or gene complex) may be much more viable and/or capable of fertilization than other gametes. Genetic analysis of these genes may reveal an apparently abnormal segregation ratio. A heterozygote (A^1/A^2) should produce A^1 and A^2 gametes with equal frequency (1:1 ratio). Some examples are known in which this ratio is grossly distorted. This is a phenomenon called *meiotic drive*. Obviously the allele favored by gametic selection should steadily increase in the population. But this does not always happen, indicating that other selective factors may counterbalance meiotic drive at this locus. Perhaps the allele favored in the gametic (haploid) stage is at a disadvantage in the zygotic (diploid) stage. The contribution meiotic drive has made to organic evolution is unknown. From the relative paucity of known examples, it appears to be a rare phenomenon. We should be cautious, however, not to equate rarity with unimportance, because (as with fortunate mutations) if it happens under the appropriate conditions its importance could be far out of proportion to its frequency of occurrence.

> A dominant mutant gene in *Drosophila* called SD (segregation distorter) on chromosome II causes that chromosome (and the other closely linked genes) to be recovered with high frequency in the progeny relative to its recessive wild-type allele. It has been suggested that in heterozygotes the SD allele often causes its wild-type homologue to fragment and behave irregularly during spermatogenesis. Consequently, few chromosomes bearing the wild-type allele are found in functional gametes.
>
> A tailless condition of the house mouse is produced by the heterozygote involving one of three dominant alleles (*T*) and any one of a number of recessive alleles (*t*). The dominant alleles are lethal when homozygous; some of the recessive alleles are also lethal when homozygous. The wild-type allele (+) allows normal tail development when homozygous (+/+), or when in heterozygous combination with any one of the recessive alleles (+/*t*). In some cases, *t*-bearing gametes appear to constitute 99 per cent

of the functional sperm cells formed by a $+/t$ heterozygote. Meiotic drive in this case tends to maintain the deleterious t alleles against strong selection to remove them from the population via lethal homozygotes.

Maintenance of Structural Hybridity

In some cases, translocation or inversion heterozygotes (one normal chromosome, one altered chromosome) are adaptively superior to either homozygote (two normal chromosomes or two altered chromosomes). That is to say, genetic heterozygosity, often present together with structural heterozygosity, tends to produce *hybrid vigor* (*heterosis*). It would be advantageous under these conditions for the population to stabilize the structural heterozygote, i.e., prevent the formation of the less-fit homozygotes. *Gametic lethals* (an extreme form of meiotic drive) represent one mechanism whereby *structural hybridity* (*structural heterozygosity*) can be maintained in a population.

A species of the evening primrose, *Oenothera muricata* ($2n = 14$), has apparently undergone a series of reciprocal translocations involving the ends of all its chromosomes. During meiosis a ring of fourteen chromosomes is produced through pairing of the homologous regions at the tips of its chromosomes. This extensive structural hybridity has been retained by the appearance of gametic lethals in each of the haploid sets of seven chromosomes. One set is designated the *rigens* (R) complex, and the other set is called the *curvans* (C) complex (so named by Otto Renner, hence dubbed *Renner complexes*). The gametic lethal in the R complex inactivates the pollen; a different gametic lethal in the C complex inhibits the embryo sac. Thus, the only zygotes that form are those from R embryo sacs and C pollen grains (Figure 9.20). Nonallelic recessive lethal genes, each in a different homologous chromosome (or in different Renner complexes), that cause permanent structural hybridity are classified as a *balanced lethal system*.

Structural hybridity can also be maintained by *zygotic lethals,* i.e., genes that prevent the

Figure 9.20. Example of gametic lethals in *Oenothera muricata*. A gametic lethal in the R chromosome complex inactivates pollen; a different gametic lethal in the C chromosome complex inhibits development of the embryo sac. Only R embryo sacs and C pollen are functional in fertilization.

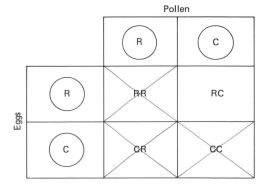

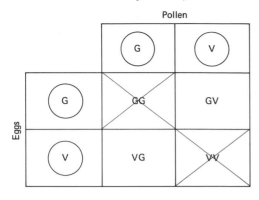

Figure 9.21. Example of zygotic lethals in *Oenothera lamarckiana*. Homozygotes for either the G or V chromosome complexes are lethal. Only the G-V combinations survive.

development of the diploid organism. Meiotic drive operates only at the level of the gametes, not after the formation of the zygote.

Another species of the evening primrose, *Oenothera lamarckiana* ($2n = 14$), also has two sets of Renner translocation complexes designated *gaudens* (G) and *velans* (V). Homozygotes of either complex are lethal; only the heterozygous G-V complexes are viable (Figure 9.21).

Two lethals linked in repulsion phase can be maintained as a balanced lethal system if prevented from crossing over by structural heterozygosity for an inversion. The dominant conditions of curly wings (*C*) and plum eye color (*P*) are on chromosome II of *Drosophila* and within the region of an inversion. Either gene is lethal when homozygous. Crosses between parents in repulsion phase linkage produce viable progeny just like the parents (Figure 9.22). Heterozygous loci within the inverted segment are also maintained in permanent hybrid conditions by the balanced lethal-inversion system. The population pays a high price for the privilege of using a zygotic lethal system—*viz.* loss of half the offspring.

Studies of natural populations of *Drosophila* have revealed that individuals near the center of the range tend to exhibit considerable chromosomal variability in the form of inversion heterozygosity. Individuals near the margins of the range tend to display more uniformity in chromosomal structure (inversion homozygosity). This distribution of structural forms may be explained as follows. Environmental conditions are more likely to be optimal

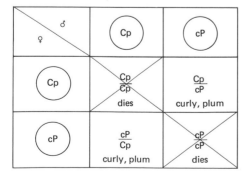

Figure 9.22. Example of a balanced lethal system in *Drosophila*. Dihybrid flies, in repulsion phase linkage for genes C and P, produce only parental type gametes because they are structurally heterozygous for an inversion including these two loci. Homozygotes at either locus are lethal; only heterozygotes survive.

near the center of the range than near the periphery. Selection would tend to be less intense near the center than at the edge of the range where niche selection becomes highly restrictive. Adaptive gene complexes tend to be maintained through structural hybridity; recombination of linked genes is relatively low. This would be advantageous near the center of the range. At the boundary, however, the ability to recombine linked genes becomes an important factor for the evolution of new complexes adapted to the more restrictive conditions in that area. Structural homozygosity allows greater recombination among linked genes and hence becomes advantageous at the borders of the range.

Ploidy

The term *ploidy* refers to "sets" of chromosomes. Multiples of the entire haploid set are classified under the heading *euploidy* ("true or even sets"); multiplication of only parts of a set are grouped under the name *aneuploidy* ("not true or uneven sets").

Aneuploidy

Homologous chromosomes usually separate from each other at anaphase of the first meiotic division. Occasionally a pair of homologues fail to part (*nondisjunction*) and both go to one pole. This creates genetically unbalanced gametes with one extra chromosome ($n + 1$ gamete) or with one less than the normal haploid number ($n - 1$ gamete). The union of a normal gamete (n) with an $n - 1$ gamete produces a *monosomic* ($2n - 1$) individual with one of its chromosomes represented as a singlet (all others as doublets). When an $n + 1$ gamete is fertilized by a normal haploid gamete (n), a *trisomic* ($2n + 1$) individual is produced (i.e., one chromosome is represented three times and all others are in pairs).

> As discussed in Chapter 7, white eyes are produced by a recessive sex-linked gene in *Drosophila*. When a white-eyed female is crossed to a red-eyed male, all the F_1 females should be red eyed and all the F_1 males should be white eyed (criss-cross inheritance). Rarely, some unusual F_1 progeny appear in which males are red eyed and females are white eyed. These unexpected phenotypes are sometimes aneuploids: the red-eyed males are monosomic (single X, no Y; called XO), the white-eyed females are trisomic (XXY). Recall that the Y chromosome of *Drosophila* does not contribute to sex determination the way it does in man. Nondisjunction of the X chromosomes occurred in the white-eyed female parent producing $n + 1$ and $n - 1$ gametes. The union of an abnormal XX egg with a normal X-bearing sperm produces an XXX "superfemale" (metafemale) zygote that usually dies. When an $n - 1$ egg (lacking an X chromosome) is fertilized by a normal sperm bearing the Y chromosome, the YO zygote also fails to develop (lethal). The cross is diagrammed as shown in Figure 9.23.

Loss of chromosomes is usually accompanied by harmful rather than beneficial consequences. Most examples of viable monosomics are in plants, apparently because plants can tolerate genic imbalance better than animals can. Normal development in animals seems to be delicately balanced at the diploid level and any deviations from the diploid condition are

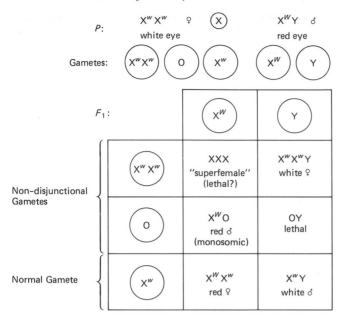

Figure 9.23. Consequences of nondisjunction of X chromosomes in white-eyed female *Drosophila* when crossed with red-eyed males.

usually grossly abnormal. Even in plants, monosomics are generally less viable and less vigorous than their diploid counterparts. Monosomy exposes recessive genes on the single chromosome to immediate selection in the same way as does hemizygosity for sex-linked genes in diploid males. Without a pairing partner, the singlet chromosome of a monosomic individual may not move from the first meiotic metaphase plate, resulting in the loss of that chromosome during gametogenesis.

Chromosomal gain is not usually as harmful as chromosomal loss in plants. Trisomics ($2n + 1$) tend to form a trivalent association (three paired homologues) during first meiotic prophase. If two homologues of the trivalent move to one pole and the third to the other, a trisomic will produce n and $n + 1$ gametes. If the $n + 1$ gametes or the trisomic zygotes lower fitness (reproduction), natural selection will tend to eliminate the aneuploids from the population.

A striking example of trisomy is available in the common Jimson weed, *Datura stramonium*. The diploid number for this species is $2n = 24$. Twelve different trisomics are therefore possible (one for each of the twelve chromosomes in the haploid set). All twelve trisomics have been found and each has distinctive phenotypic effects (Figure 9.24). For example, trisomy for chromosome A produces narrow in-rolled leaves; that for chromosome B produces dark green, glossy leaves; that for chromosome E has large seed capsules with very long spines; that for chromosome K yields globular capsules with thick spines; etc. Half the ovules produced by the globular capsules have n and half have $n + 1$ chromosomes, indicating regular segregation of two members of the trivalent to one pole and the third member to the opposite pole during meiosis. Furthermore the $n + 1$ ovules are functional. But pollen grains with $n + 1$ chromosomes are inviable; hence, only normal (n) pollen grains are functional in this trisomic plant.

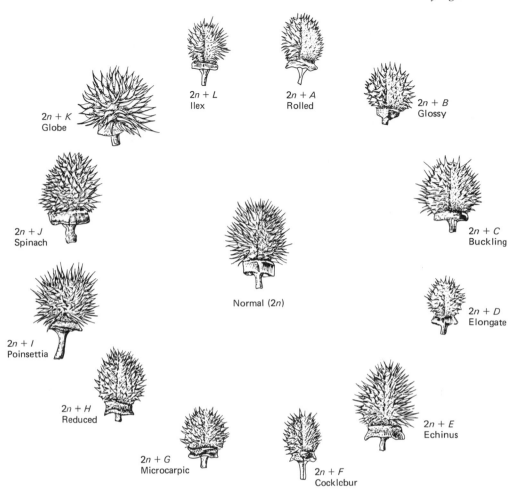

Figure 9.24. *Datura stramonium* seed capsules. Center is normal capsule ($2n = 24$), surrounded by twelve capsules representative of each of the different trisomics.

Double trisomics ($2n + 1 + 1$), tetrasomics ($2n + 2$), and other more complicated aneuploid conditions have been identified in various plant species. In general, the greater the deviation of aneuploidy from the normal diploid level, the greater the deleterious effects on the phenotype. Duplications can arise through the insertion of an acentric segment of the extra chromosome into either a homologue or a nonhomologous chromosome. The fate of the residual centromere would depend on the amount of essential euchromatin still attached.

> In humans, the condition formerly called "mongolian idiocy" (now called *Down's syndrome,* after the English physician Langdon Down) has been associated with a trisomic condition for one of the smallest of the human chromosomes (No. 21). It is thought to be the most common human congenital abnormality (approximately 1:600 births). Symptoms include mental retardation, small stature, broad hand with stubby fingers, oversized tongue cut with furrows, eyelid fold resembling those of Mongolian people, round faces, and loose jointedness. Women over age forty incur about ten times

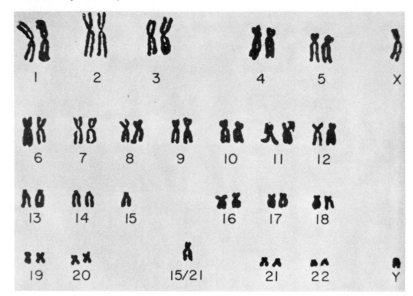

Figure 9.25. Translocation trisomy 15/21. This is due to "centric fusion." The patient has Down's syndrome with only 46 chromosomes. The trisomic chromosome 21 is attached to a normal chromosome 15, so that the patient has sufficient chromosome material for 47 chromosomes. [From Redding, A. and Hirschhorn, K.: *Guide to Human Chromosome Defects* In Bergsma, D. (ed.): Birth Defects: Orig. Art. Ser., Vol. IV, No. 4 Published by The National Foundation-March of Dimes, White Plains, N.Y. 1968, p. 103]

greater risk of giving birth to an afflicted child than younger women. It seems that the prolonged interval from the time eggs begin to form in the ovaries of females until the time they eventually become "ripe" for ovulation allows the accumulation of changes that interfere with the normal meiotic mechanism. No ageing effect is seen in males, because primary spermatocytes are continually being replaced. Approximately 95 per cent of the people with Down's syndrome have $2n + 1 = 47$ chromosomes as a result of nondisjunction in one of the parents. The other 5 per cent have $2n = 46$ chromosomes, but in these patients the extra chromosome No. 21 is translocated onto one of the large autosomes (e.g., No. 15). Karyotypes of the phenotypically normal parents of such a patient reveal that one of them (not necessarily the mother) has $2n - 1 = 45$ chromosomes; one of their No. 21 chromosomes has become attached to a nonhomologous autosome (Figure 9.25). Translocation-type Down's syndrome is uncorrelated with maternal age, i.e., younger women have just as many defective children as older mothers. Children in families in which one parent is a translocation heterozygote and the other parent has no translocation are expected in the ratio of 1 normal: 1 Down's syndrome: 1 normal (carrier of the translocation) as diagrammed in Figure 9.26.

Euploidy

Diploid organisms ($2n$ chromosomes) produce gametes with the haploid (n) number of chromosomes. The term haploid should be reserved for the reduced number of chromosomes found in a gamete. Some organisms have only one set of chromosomes (the n number) in cells

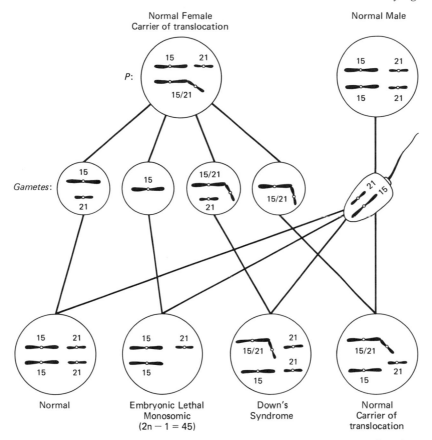

Figure 9.26. Origin of a translocation type Down's offspring from a normal mother (carrying a 15/21 translocation) and a normal father.

other than gametes. For these organisms the term *monoploid* (one set) is proper (e.g., bacteria and fungi, mites, rotifers, and male bees). Newly formed monoploids are usually weak and inviable, and unless they can vegetatively (asexually) reproduce, their fate is almost certainly a genetic death. Meiosis is commonly abortive because singlet chromosomes may not move from the metaphase plate; if they do move, the distribution of chromosomes to the meiotic products is haphazard, producing genetically unbalanced gametes (sterility). The recessive alleles at all loci become exposed to natural selection in monoploids. Monoploids are considered of little or no importance in the evolution of new species and higher taxonomic groups.

Polyploidy is the term applied to all chromosomal sets above the diploid level. Polyploidy is much more common in plants than in animals. One reason for this is that the sex of the individual, as determined by the X-Y chromosome mechanism, would fail to develop properly in the presence of extra sex chromosomes (e.g., XXX superfemale *Drosophila* flies are usually inviable; XXY men are usually sterile, etc.).

Triploids, containing three sets of chromosomes (*3n*), may be produced by the union of an unreduced diploid gamete (*2n*) and a normal haploid gamete (*n*). Triploids are at a reproductive disadvantage, however, and therefore rarely survive in nature. Meiotic pairing in any region of the chromosome is restricted to only two homologues at a time. The extra chromosome set of triploids may fail to pair with their homologues, or if they do pair as *trivalents* (association of three homologous chromosomes), segregation may randomly distribute the extra chromosomes to the meiotic products. This again results in genetically unbalanced gametes and consequent sterility. Organisms with odd numbers of chromosome sets ($5n$, $7n$, etc.) are usually highly sterile for similar reasons. Polyploids sometimes exhibit a phenomenon known as *gigantism,* i.e., excessively large and showy flowers, fruits, leaves, etc. (Figure 9.27). This may be considered a special kind of dosage effect, whether the extra chromosomes cause an increase in cell size or not. When a desirable triploid is found, man may choose to propagate it vegetatively for food or (in the case of ornamental plants) for its aesthetic qualities. Such has been the case with the common banana, a triploid. Here sterility is advantageous because normal banana seeds are hard and inedible. Several apple varieties are triploids (Winesap, Gravenstein, Baldwin); European pears and Keizer's Kroon tulips are also triploids.

  Tetraploids (four sets of chromosomes, $4n$) are fairly common in nature. Higher degrees of ploidy are well known but not generally of major evolutionary significance. Many of our important crop varieties are tetraploids (e.g., McIntosh apples, pears, cherries, blueberries, blackberries, alfalfa, potato, coffee, and peanuts). If all the chromosomes are derived from the same species, the term *autopolyploid* is applied (*auto,* indicating "self," or same species). Autopolyploids may be produced in several ways. An egg may rarely be fertilized by more than one sperm (*polyspermy*). Mitotic nondisjunction of the entire chromosomal endowment may occur in a cell of a sex organ; meiotic nondisjunction of the entire complex produces extra sets of chromosomes in a gamete. It is possible to induce artificially autopolyploidy with

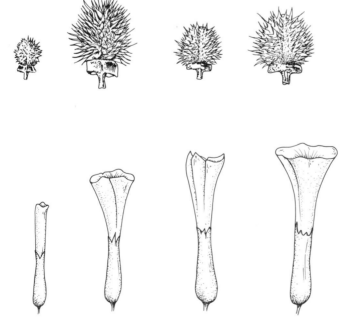

Figure 9.27. At top are the seed capsules of the Jimson weed (*Datura stramonium*) and at the bottom are its flowers. From left to right are haploid (*n*), diploid (2*n*), triploid (3*n*), and tetraploid (4*n*). Notice that flower size tends to increase with the ploidy level, but the size of seed capsules is uncorrelated with the ploidy level.

high frequency by application of the alkaloid *colchicine*. This chemical prevents development of the spindle apparatus and, hence, doubles the chromosome number. Sterility is commonly observed in autotetraploids for the same reason as in triploids; only now there is even greater opportunity for irregular segregation of chromosomes into gametes from a mixture of univalent, bivalent, trivalent, and tetravalent pairing configurations. Occasionally a species may evolve a mechanism that can stabilize segregation of chromosomes by pairs in an autotetraploid. Fertility is restored in the species to the extent that it can do this. For example, autotetraploids in the genus *Datura* are known frequently to separate chromosomes equally during meiosis. This is also accomplished in the tulip and lotus, because only a single chiasma tends to form in any one pair of homologues. Thus, each group of four homologues separates as two bivalents.

Hybrids sometimes are formed between different related species. In many cases, the hybrid is not as well adapted to local conditions as its parents were. Furthermore, if the basic number of chromosomes was not identical in both parental species, meiosis would probably fail to produce genetically balanced gametes in most cases. More importantly, the genetic content of the two sets of chromosomes in the hybrid may be so different that pairing is ineffective, resulting in abortive gametes. Species hybrids are, therefore, usually sterile. However, if the diploid content of hybrid reproductive cells could be doubled, each chromosome set would now have an identical set with which to pair. Fertility could be restored to the *allopolyploid* (*allo*, indicating its origin from different sets of chromosomes) produced by union between the exceptional gametes from the hybrid. The haploid number found in gametes of an allotetraploid would be $2n$. Union of two $2n$ gametes regenerates the tetraploid, and in this sense it breeds true to type.

Allotetraploids tend to be much more successful in nature than autotetraploids, not only because meiosis is more likely to be normal, but also because the hybrids may express novel characteristics that allow them to exploit niches unoccupied by either parent. Furthermore, the allotetraploid is reproductively isolated from the parents because outcrossing of a rare allotetraploid with either parent would produce a triploid (usually sterile). An allotetraploid may therefore represent a new species.

In order for the new species to become established, however, two restrictions usually must be overcome. First, a suitable environment must exist in which the polyploid can survive. In many cases, polyploids are able to exploit the opportunities in disturbed habitats newly opened to colonization by plants (e.g., regions uncovered by retreating glaciers at the end of one of the "ice ages"). This suggests that polyploids may possess a wider range of tolerances than diploids. Second, a tetraploid must usually migrate beyond the range of the parental species in order to avoid being "swamped" by diploid pollen and the consequential production of sterile triploids. The terms *amphiploid*, *amphidiploid*, and *double diploid* are sometimes used synonymously with allotetraploid.

An excellent example of the origin of a new species through allopolyploidy is represented in the experiment of Russian geneticist Karpechenko. He crossed two plants belonging not just to different species, but to different genera: the radish (*Raphanus sativa*) and the cabbage (*Brassica oleracea*). Both species have the same diploid number ($2n = 18$), but the genetic content of the two genomes are quite dissimilar, and the chromosomes fail to pair. The hybrid is, therefore, usually sterile. A very small fraction of his hybrids looked different and, on cytological examination, were found to be allotetraploids ($4n = 36$). The chromosome number apparently had been doubled shortly after the zygote was formed. Each set of chromosomes had an identical set with which to pair in the allotetraploid, and they were therefore fertile. The allotetraploid had many traits that were intermediate between the parental extremes and characteristically showed some degree of gigantism. Backcrosses of the allotetraploid to either parent produced sterile triploids. Hence, a new, reproductively isolated species had arisen. Karpechenko constructed a new genus for his species, *Raphanobrassica*. Unfortunately the allotetraploid had no economic value because its leaves were like those of the radish and its roots like those of the cabbage. The cross is shown in Figure 9.28.

Autopolyploids tend to be very similar to their parents in morphology and ecological requirements. Autopolyploids are produced by doubling the chromosome numbers of normal individuals or *intraspecific* hybrids (e.g., a cross between two varieties within a species). They usually exhibit varying degrees of sterility, because of the tendency of the nearly identical chromosomes to form multivalent pairing associations during meiosis. Allopolyploids, on the other hand, are produced from *interspecific* hybrids and are generally of intermediate character or else combinations of parental characteristics. They can be fully fertile (although this depends on the harmonious interaction of two relatively unrelated genomes) and almost exclusively form bivalent pairing configurations during meiosis. If the genomes of two diploid species that hybridize are fairly closely related, the hybrid may exhibit some bivalent formation during meiosis. Doubling the chromosome number of such a hybrid produces what is called a *segmental allopolyploid* in which both bivalent and multivalent pairing figures are observed. G. L. Stebbins (Figure 9.29) defines a segmental allopolyploid as a polyploid

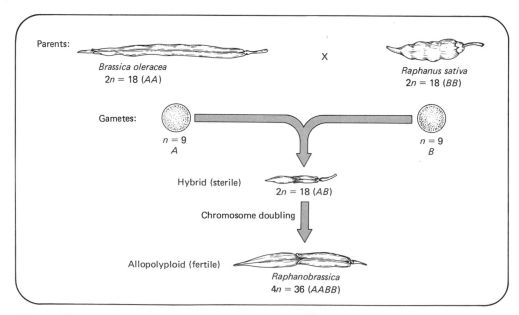

Figure 9.28. Production of a fertile allopolyploid from a sterile hybrid between two species (cabbage = *Brassica oleracea*; radish = *Raphanus sativa*) by doubling the chromosome number. *A* represents one set of chromosomes of the cabbage; *B* represents one set of chromosomes of the radish.

Figure 9.29. George Ledyard Stebbins (born 1906). (From *Study Guide and Workbook for Genetics* by Irwin H. Herskowitz, McGraw-Hill Book Company, 1960)

containing two pairs of genomes that have many homologous chromosomal segments or even whole chromosomes in common, but differ in respect to a sufficiently large number of genes or chromosome segments, so that the different genomes produce sterility when present together at the diploid level. Polyploidy thus forms a continuum from complete autopolyploidy through the intermediate segmental allopolyploids to total allopolyploidy, depending on the degree of genetic and chromosomal differentiation separating the parental species (Figure 9.30). A hybrid between two related species may form many bivalents in meiosis, yet be mostly sterile. Doubling the chromosome number produces a tetraploid that forms bivalents and multivalents in meiosis, but now is mostly fertile. Stebbins explains such odd behavior with his "segmental allopolyploid theory." In the hybrid, the two related sets of chromosomes differ by small inversions that do not interfere with pairing, but result in defective gametes (duplications and/or deficiencies) when crossovers occur within the inverted segments (Figure 9.31). In the segmental allotetraploid, however, each set of chromosomes now has an identical set of true homologues (i.e., chromosomes derived from the same species) with which it preferentially pairs and in which crossing over does not form defective products. One lesson to be learned from segmental allopolyploids is that the absolute degree of homology between chromosomes of different species cannot be adequately evaluated solely by their degree of pairing in hybrids. Segmental allopolyploidy is regarded as an unstable condition. Segregation and chromosomal alteration, guided by natural selection for increased fertility, would be expected to evolve toward either complete auto- or allopolyploidy. Two distinctive properties characterize segmental allopolyploids: (1) they are expected to form at least partly fertile hybrids by backcrossing to autopolyploid derivatives from either of the parental species, and

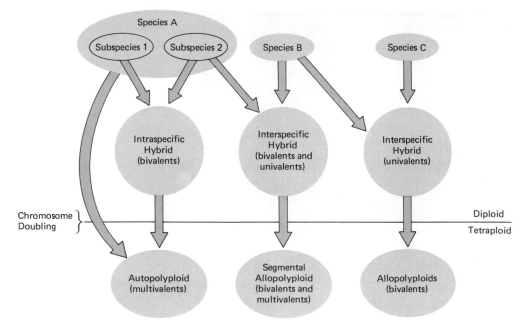

Figure 9.30. Continuum of polyploid types from total autopolyploidy (left) through segmental allopoly-ploidy (middle) to total allopolyploidy (right). (After Solbrig)

(2) they have the capacity to segregate the morphological, ecological, and chromosomal differences that formed the sterility barriers between the parental species.

Polyploidy has apparently been of great evolutionary importance in the plant kingdom. It has been estimated that more than one third of all flowering plants (angiosperms) are polyploid. This is especially true in the grasses, where perhaps 70 per cent of the species are of polyploid origin. A dendritic (tree-like) diagram is commonly used to represent evolutionary divergence through time. Because plant species have so often hybridized and then become new species through allopolyploidy, many branches of the evolutionary "tree" would appear to be fused. The term *reticulate evolution* is applied to the "net-like" lineages of a series of closely related allopolyploid species. Evidence for the polyploid origin of some plants has been provided by breeding the putative ancestors and synthesizing a species essentially identical to the modern form.

The first successful synthesis of a naturally occurring species involved the production of the allotetraploid plant species *Galeopsis tetrahit* by doubling the chromosome number of the hybrid formed from the ancestral diploid species *G. speciosa* and *G. pubescens*. The synthetic amphidiploid is indistinguishable from *G. tetrahit* plants found in nature.

The cultivated, long-staple, New World cotton (*Gossypium hirsutum*) has twenty-six pairs of chromosomes (thirteen long pairs and thirteen short pairs). An Old World (Asiatic) species, *G. arboreum,* has thirteen pairs of long chromosomes; a Central or South American species, *G. thurberi,* has thirteen pairs of short chromosomes. The hybrid between these two species is sterile. Colchicine treatment of the hybrids has produced amphidiploids that resemble the New World cultivated cotton. Crosses of *arboreum* × *hirsutum* or *thurberi* × *hirsutum* produce triploids that form thirteen bivalents and thirteen univalents (single unpaired chromosomes) during meiosis. This indicates that *hirsutum* has genetic homology with the haploid sets in both *thurberi* and *arboreum* and very likely arose as an allotetraploid from the hybrid between these two species.

It is doubtful that polyploidy can lead to the formation of categories higher than that of species, because polyploids do not usually depart radically from their diploid ancestors in either morphology or in ecological requirements. It is generally conceded that most changes that lead to generic, familial, and higher categories occur at the diploid level. Polyploidy is commonly viewed as a conservative mechanism with regard to evolution beyond the species level. There is a drastic reduction in the segregation of recessive traits from polyploids and this, in turn, retards the establishment of divergent lines. As an example, selfing of a diploid dihybrid (*AaBb*) is expected to segregate out the double recessive phenotype in 1/16 of the progeny. The corresponding segregation ratio from a selfed autotetraploid (*AAaaBBbb*) is 1:1296. This means that two recessive characters are expected eighty times more frequently

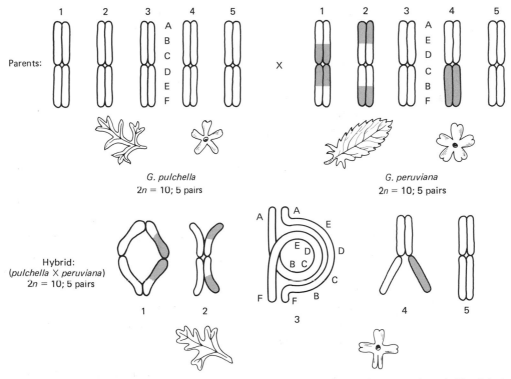

Figure 9.31. Diagram of pairing relationships between chromosomes of two species of *Glandularia* (Verbenaceae). Typical leaves and flowers from each population are also shown. Similarity of shading indicates regions of genetic homology. All chromosomes form bivalents (100 per cent pairing), but fertility of the hybrid is only about 30 per cent, because large segments (shaded) were nonhomologous or contained inversions like the one shown here (and others too small to be seen). Doubling the chromosome number of the hybrid produces an allotetraploid with increased fertility because each chromosome then has an identical partner with which it preferentially pairs during meiosis.

from diploid heterozygotes than from the corresponding autotetraploid heterozygotes. This example considers only two loci; the differences become astronomical when extended to several more loci. The stabilizing influence of autopolyploidy can be nullified through reversion to the diploid condition. *Diploidization* of an autopolyploid might be accomplished by a sequence of chromosomal aberrations that gradually reduces the homology between sets of chromosomes. As mentioned earlier, some polyploids tend to be isolated by their ability to colonize new habitats. This isolation and the increased variation (through recombination) accompanying diploidization provide the key ingredients for evolutionary divergence of higher taxonomic groups. Many woody species that now behave as diploids undoubtedly are evolved from polyploid ancestors through the process of diploidization.

Karyotype Evolution

Closely related species would be expected to have similar karyotypes; distantly related species could have experienced many chromosomal aberrations, with reduction or gain of chromosomes, so their karyotypes might have evolved considerable differences. Thus, comparisons of karyotypes has become an indispensable aid to the systematist in phylogenic interpretations. It has been repeatedly stressed in this chapter that reshuffling chromosome segments may have profound consequences on gene expression. How genes are organized in spatial relationships to one another is therefore of considerable importance for evolution.

The number of chromosomes is usually a constant characteristic for each species. However, there appears to be no correlation between the basic number of chromosomes of a species and its phylogenetic position. In other words, the basic number of chromosomes shows no obvious trends with the degree of "primitiveness" or "specialization" in the higher taxonomic groups such as orders and classes. However, within the much narrower limits of families and genera, some trends are recognized especially in regard to the growth habits of plants. For example, it seems that primitiveness in certain plant genera is more closely correlated to larger chromosome size than is specialization. In the genus *Crepis,* annuals (which are considered to be relatively specialized) tend to have smaller chromosomes than the more primitive perennials. Woody angiosperm species generally possess a higher basic number of chromosomes, smaller chromosomes, and a far greater degree of chromosomal stability than their herbaceous relatives. Woody forms are considered ancestral to the herbaceous types. Chromosomes tend to be larger in monocots and smaller in dicots. But in this case, dicotyledonous plants (those with two seed leaves) are considered ancestral to monocotyledonous plants (those with a single seed leaf). So the trends are not universal, and causative explanations for any of these trends is elusive.

In general, diploid plants tend to be found in the older parts of the total range of a taxonomic group, whereas polyploid plants tend to invade the geologically younger parts of that range. Thus, in such a group, diploids tend to become relics as the more diversified polyploids successively occupy the newly available niches. When conditions become unfavorable for such a group, it is usually the diploids that first become extinct. Therefore, when only one or a few relics of a once prominant group remain, they are more likely to be polyploids than diploids.

Fragmentation of chromosomes also tends to increase the chromosome number of a species. Among some animal groups, an increase of chromosome number by fragmentation appears to be correlated with increasing specialization (similar to polyploidy in plants). For example, the diploid number among lower mammals is about 24; among the higher placental mammals it is commonly 48. In the more specialized placentals, such as ungulates (hoofed mammals), the number is generally 60; and in the highly specialized rodents it may be as high as 84. Moreover, one of the great regularities of evolution is the tendency for similar structural changes to establish themselves in one member of a karyotype after another (the principle of *karyotypic orthoselection*). For example, in a given evolutionary lineage, there may be a tendency for one chromosome after another to incur duplications in the short arms of acrocentric chromosomes; in another lineage, progressive fragmentations may increase the chromosome number; in still another lineage, chromosomal fusions may have successively led to a reduction in the number of chromosome pairs. The reasons for karyoptypic orthoselection remain as one of the great mysteries of evolution.

The quantity and distribution of heterochromatin in the genome is of great importance for the evolution of the karyotype. Heterochromatin has a tendency to break more easily than euchromatin, and generally may be lost with less harmful phenotypic consequences. A reciprocal translocation, involving at least one break in a heterochromatic region, can transfer euchromatin to an *accessory chromosome* (i.e., a centromere to which only heterochromatic arms are attached; also called a *B*-type *chromosome;* Figure 9.32). This process would convert these inert elements to essential chromosomal components. Similarly, transfer of all the euchromatin of one chromosome to another may convert a formerly essential element into an accessory chromosome; loss of the latter component would reduce the basic chromosome number of the karyotype. Reduction in the chromosome numbers of *Drosophila* species appears to have been associated most frequently with *centric fusions* (i.e., a special type of translocation involving linking of euchromatic segments of rod chromosomes by union of their centromeres; Figure 9.33).

An example of centric fusion is known in *Drosophila melanogaster,* where the two X chromosomes of the female appear to be joined by union of their centromeres. This

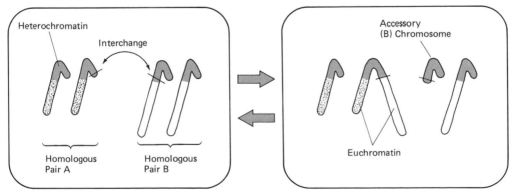

Figure 9.32. Formation of an accessory (B-type) chromosome through an interchange involving heterochromatin (left to right); conversion of a nonessential accessory chromosome into an essential one by the reverse process (right to left).

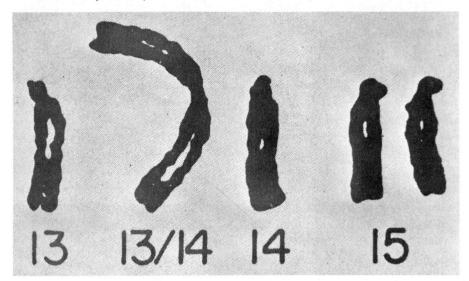

Figure 9.33. "Centric fusion" type of translocation occurring in a clinically normal individual. There are only 45 chromosomes in this karyotype. One each from pairs 13 and 14 have fused together. This translocation carrier is clinically normal, because he neither lacks nor has extra chromosome material, but he has a high risk of having abnormal offspring. [From Redding, A. and Hirschhorn, K.: *Guide to Human Chromosome Defects* In Bergsma, D. (ed.): Birth Defects: Orig. Art. Ser., Vol. IV, No. 4 Published by The National Foundation-March of Dimes, White Plains, N.Y. 1968, p. 103]

"attached-X" condition changes the karyotype from $2n = 8$ to $2n = 7$ in these unusual females. The two rod-shaped X chromosomes now appear as a single metacentric chromosome. An *isochromosome* is thus formed, with arms that are mutually homologous. If a female, homozygous for the recessive allele specifying yellow body color, is crossed with a normal male, the male offspring will appear like the male parent with regard to their sex-linked traits. This cross is diagrammed in Figure 9.34. Geneticists

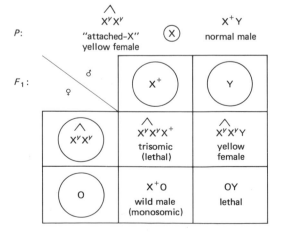

Figure 9.34. Diagram of a cross between an "attached-X" yellow-bodied female *Drosophila* and a normal male.

have used the attached-X condition to study the effects of various agents on mutation rates in males, because mutations may be observed in the F_1 males (normally another generation would be required to segregate out the parental male X chromosome in hemizygous condition).

Centric fusions are made possible when large blocks of inert heterochromatin surround the centromere. Thus, a certain amount of heterochromatin in the genome would appear to provide flexibility for evolution of the karyotype. Species with little or no heterochromatin would have little opportunity for karyotype evolution. This seems to be the case in the North American plant genus *Tradescantia,* which lacks detectable heterochromatin and whose species exhibit a highly uniform karyotype.

Occasionally a V-shaped chromosome breaks to form two rod chromosomes. Perhaps the original centromere may fragment to yield two functional centromeres; alternatively one of the centromeres may be gained by attachement to an accesory chromosome. In any event, this increases the basic chromosome number of the karyotype. Pericentric inversions may shift the centromere, converting a symmetrical metacentric chromosome into an asymmetric sub-metacentric or acrocentric chromosome. The more primitive species in the genus *Drosophila* tend to have asymmetrical chromosomes; the more specialized species tend to exhibit symmetry. Among plants, however, just the reverse seems to be the case. Chromosomal asymmetry appears to be correlated with specialization. For example, in the genus *Crepis,* asymmetry is associated with reduction in the number as well as specialization of floral and vegetative parts. If centric fusions are more common in animals, and unequal translocations more frequent in plants, this could account for the observed trends toward symmetry in animal and asymmetry in plant karyotypes.

Chromosomal *polymorphisms* have been recognized in both plants and animals. This indicates that structurally different chromosomes are represented in the population at relatively high frequencies. For example, seven different types of Y chromosome (all heterochromatic) are known in geographically separated populations of the species *Drosophila pseudoobscura.* Among grasshoppers (and other members of the insect order *Orthoptera*), *heteromorphic* homologues (i.e., morphologically distinctive bivalents) indicate that considerable chromosomal polymorphism exists both within and between species. It is assumed that the structural hybridity represented by chromosomal polymorphism, contributes to the fitness of the populations in which it is maintained. Further discussion of this topic is presented in Chapter 16.

The Mutation Theory

As mentioned in Chapter 2, Hugo De Vries' mutation theory was a major challenge to the Darwinian concept of evolution. De Vries had come to his conclusions largely through the study of an escaped American weedy plant called the evening primrose (genus *Oenothera*). The various species of *Oenothera* appeared to breed true, but occasionally a new form would appear and it also would immediately breed true. He proposed that new kinds typically arise in just this way. *Saltations* or major mutations were proposed to account for the instantaneous evolution of new species. Darwin, of course, envisioned that new species arose as a gradual accumulation of small hereditary changes over many generations.

When Mendel crossed pure tall with pure dwarf pea plants, the F_1 was uniformly tall (the dominant character), and the F_2 segregated approximately 3 tall:1 dwarf. When pure breeding *Oenothera* species were crossed, the hybrids were not uniform. Furthermore, selfing of the F_1 revealed that each kind of hybrid bred true (mimicking homozygosity; Figure 9.35). What odd behavior! This problem continued to vex geneticists for almost thirty years. Finally, through the combined efforts of Blakeslee, Cleland, Renner, and others, the problem was largely resolved. Here is what they found. When a reciprocal translocation heterozygote is formed between two nonhomologous chromosomes, a ring of four chromosomes is displayed at metaphase I. If a second reciprocal translocation occurs between any chromosome of the ring of four and a chromosome outside the ring, an additional chromosome pair will be incorporated into the ring (now size 6). Additional reciprocal translocations can eventually bring all the chromosomes into the ring.

This apparently is what happened in the evolution of *Oenothera* species (Figure 9.36). Almost all the *Oenothera* species found in nature have seven pairs of chromosomes, and many species form a circle of fourteen chromosomes during meiosis. Pairing is limited to the homologous ends of the chromosomes; therefore, it is inferred that the arms of the chromosomes have been shuffled in many ways by reciprocal translocations. Different translocations have occurred in each species. Furthermore, at first anaphase, alternate members of the ring move toward the same pole. If maternal and paternal chromosomes alternate in the ring, then only two kinds of gametes would be formed ($\frac{1}{2}$ with all parental chromosomes: $\frac{1}{2}$ with all maternal chromosomes). No satisfactory control mechanism is available to explain why only alternate members of the ring move toward the same pole. With the acquisition of a balanced lethal system in each of the two chromosome complexes, structural hybridity could be maintained. Different balanced lethal systems are operative in different *Oenothera* species. Self-fertilization allows the structural heterozygote to breed true.

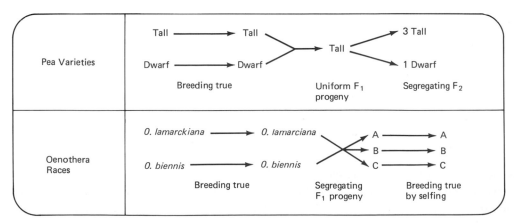

Figure 9.35. Comparison between the behavior of Mendel's pea varieties and De Vries' *Oenothera* "species" (now recognized as structurally hybrid races within a species).

O. hookeri *O. biennis*

Figure 9.36. Two of the many species of the *Oenothera* complex.

New forms of *Oenothera* can arise by crossover events between different Renner complexes. As an example, in *O. lamarckiana,* the *gaudens* (G) and *velans* (V) complexes pair as shown in Figure 9.37 (ends of chromosomes are numbered as reference points). If a crossover occurred in the interstitial region near the centromere between the G chromosome 11-7 and the V chromosome 13-14, a new complex called *deserens* (D) would be formed consisting of three V chromosomes (3-4, 10-9, 12-11), and two G chromosomes (14-8, 5-6) in addition to a new chromosome 7-13. The D and V complexes are compatible, producing a ring of six chromosomes and four bivalents during meiosis. A D/V complex segregates 1 D/D (seven bivalents) : 2 D/V : 1 V/V (lethal). The origin of D/V complexes from G/V complexes occurs with a frequency of about 6 per 10,000 plants. This is too high a frequency to be ascribed to mutation. Furthermore, single gene mutations do not result in the simultaneous appearance of the numberous character changes that sometimes accompany these chromosomal rearrangements. This unusual reshuffling of genes to form new Renner complexes is apparently the explanation for De Vries "mutations."

It must be emphasized that this is a highly unusual mechanism for the origin of species. The lesson to be learned from the cytogenetics of *Oenothera* is that there is more to evolutionary theory than Mendelism and Darwinism. The parallels between *Oenothera's* unusual genetic behavior and its equally aberrant chromosomal behavior offer striking support for the chromosomal theory of heredity. *Oenothera* has combined three factors, each individually considered generally harmful, into a successful coadapted system of heredity: (1) multiple reciprocal translocations, (2) balanced lethals, and (3) self-fertilization. The 50 per cent

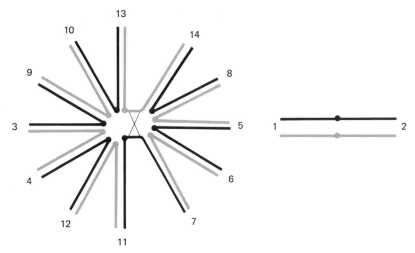

Figure 9.37. Synaptic configuration in *Oenothera lamarckiana* ($2n = 14$). *Velans* complex is light colored; *gaudens* complex is dark colored.

mortality engendered by the balanced lethal system is somewhat compensated by lowering the stigma to the level of the anthers. This results in a much heavier pollination than would be accomplished by insects. Reciprocal translocations, meiotic drive that separates maternal from paternal chromosomes, and balanced lethals all function to prevent the homozygosity inherent in selfing. The enforced heterozygosity produces hybrid vigor (heterosis). *Oenothera's* success is demonstrated by its widespread distribution in nature; it is found from Northern Canada to the tip of South America and from coast to coast on both continents. One of the "giant" forms of *Oenothera*, named *O. gigas* by De Vries, later proved to be a triploid ($3n = 21$).

Essay Questions

1. *What mechanisms can maintain structural hybridity in a population? What are the advantages and disadvantages of such a system?*

2. *Explain the difference between S-type and V-type position effects, giving examples of each.*

3. *Of what evolutionary significance are duplications?*

4. *Explain the origin of new species through polyploidy.*

5. *What light has been shed on the problem of speciation by a cytogenic study of De Vries' Oenothera "mutants"?*

MOLECULAR GENETICS

Introduction

Purely on theoretical grounds, Mendel postulated that hereditary factors (we now call them "genes") occur in pairs. After the acceptance of the chromosomal theory of heredity, the paired nature of genes was attributed to their residence in the paired sets of chromosomes in diploid organisms. The extreme complexity of life processes necessitated that genes also be extremely diversified. For approximately forty years after the rediscovery of Mendelism, most biologists believed that the only kinds of chemicals complex enough to act as genetic material were the *proteins*. Proteins were known to consist of twenty kinds of basic building blocks called *amino acids*. The kinds of amino acids and their specific linear order in polypeptide chains determine the biological activity of proteins. An astronomical number of different kinds of proteins are theoretically possible even for relative short proteins of fifty or one hundred amino acid units. No other chemicals were known to have this much flexibility of structure, and hence it was thought that genes must be proteins.

Nucleic acid had been discovered in 1868 by Fredrich Miescher, only two years after Mendel published the results of his hybridization experiments in peas. Miescher originally

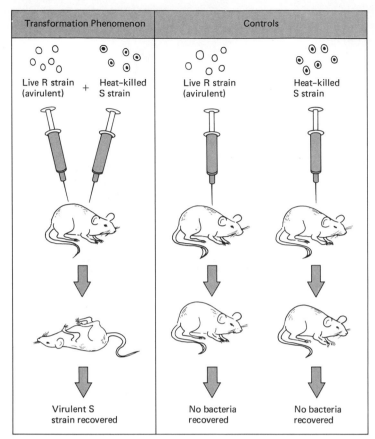

Transformation Phenomenon	Controls

Live R strain (avirulent) + Heat-killed S strain

Live R strain (avirulent)

Heat-killed S strain

Virulent S strain recovered

No bacteria recovered

No bacteria recovered

Figure 10.1. Griffith's transformation experiment.

called this new class of chemicals *nuclein,* because it was obviously associated with the cell nucleus. Much of the chemistry of this class of chemicals had been deciphered by the 1930s. For example, nucleic acids were known to consist of phosphate (acid) groups, pentose (5-carbon) sugars, and two classes of organic bases (*purines* and *pyrimidines*). One kind of nucleic acid, called *deoxyribonucleic acid* (DNA), was associated with the nucleus, and a related class, called *ribonucleic acid* (*RNA*), was found predominantly in the cytoplasm. DNA was known to consist of two kinds of purines (*adenine* [A] and *guanine* [G]) and two kinds of pyrimidines (*cytosine* [C] and *thymine* [T]). Nuclear heredity was well established by the 1930s, but DNA was thought to lack the hetereogeneity of structure required for hereditary material. At that time, most biologists believed in a tetranucleotide structure of DNA. It was thought that DNA was simply a monotonously repetitive sequence of these four interconnected bases. This was the intellectual climate in which the chemistry of genetic material was perceived as late as 1944. In that year, O. T. Avery and his colleagues C. M. MacLeod and M. J. McCarty published a research paper that provided evidence that DNA was the hereditary material of bacteria, not protein.

Avery was following a clue discovered by Frederick Griffith (1928) in experiments with

Figure 10.2. Oswald T. Avery (1877–1955).
[From *Genetics,* 51: Frontispiece (1965)]

different strains of the bacterium *Streptococcus pneumoniae* ("pncumococcus"). Thc S strain was pathogenic in the mouse, but the R strain was not. Virulence of the S strain was associated with the presence of a polysaccharide capsule; the avirulent R strain was unencapsulated. Griffith injected mice with heat-killed S-strain pneumococci and they survived. But when he injected living R strain together with heat-killed S strain, the mice died (Figure 10.1). Biologists were very disconcerted by this result, because for years the science of immunology had taught that protection against disease could be afforded by immunization using live avirulent strains or dead virulent strains. Mice injected with both live avirulent and dead virulent strains of the pneumococcus should have been doubly protected; instead, the dead strain seemed to come back to "life"! Sixteen years later, Avery set out to find whether the protein or the DNA of the pneumococcus was genetic material. The S strain produced glistening, smooth-bordered colonies on agar plates; the R strain had rough margins. Avery (Figure 10.2) discovered that live R bacteria could be converted *in vitro* into living S-type pneumococci in the presence of an extract from the dead S strain (Figure 10.3). In other words, mice were not essential to demonstrate this phenomenon. He called this process *transformation* and referred to the responsible chemical as the *transforming principle*. By separating the protein from the DNA of the S extract, Avery was able to demonstrate that

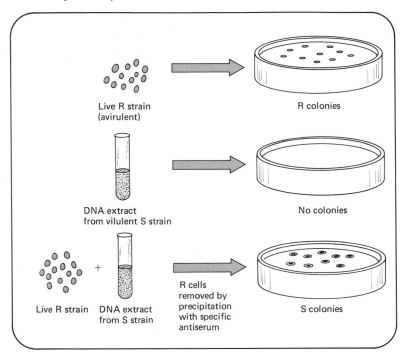

Live R strain
(avirulent)

R colonies

DNA extract
from vilulent S strain

No colonies

Live R strain   DNA extract
from S strain

R cells
removed by
precipitation
with specific
antiserum

S colonies

Figure 10.3. *In vitro* transformation. (After Avery, MacLeod, and McCarty)

only the DNA fraction had transforming activity. Treatment of the DNA fraction with protein-splitting enzymes removed any trace of proteins in the extract, but did not diminish its transforming activity. Similar treatment with a degrading enzyme specific for DNA (*deoxyribonuclease* or *DNase*) completely destroyed its transforming ability. Hence, DNA was concluded to be the transforming principle—i.e., the hereditary material of this kind of bacteria. As other researchers followed Avery's lead, it was found that purified DNA from one bacterium (*exogenous* DNA) can attach to receptor sites on an intact bacterium, enter the cell, and replace a segment of host DNA (*endogeneous* DNA) by a process akin to crossing over in higher organisms (*integration*). Transformation is usually accomplished with low efficiency even under ideal *in vitro* conditions; i.e., less than 5 per cent of exogeneous DNA molecules carrying a genetic marker would succeed in replacing the corresponding host gene. Soon, transformation was demonstrated for many traits in different kinds of bacteria.

By 1950 Erwin Chargaff had exploded the tetranucleotide theory of DNA structure by his discovery that the four organic bases (A,T,G,C) were not always present in equal molar proportions, although the content of purines was always equal to that of pyrimidines. Different species had characteristic base ratios, which could vary widely from one species to another. DNA must not have a monotonous structure after all. It was then logical to assume that the sequence of bases in DNA somehow held the genetic code. Three teams of researchers went to work on the three-dimensional structure of DNA: (1) Linus Pauling (Figure 10.4) at Cal Tech in California, (2) Maurice Wilkins (Figure 10.5) and Rosalind Franklin at King's College, London, and (3) James Watson and Francis Crick (Figure 10.6) at Cambridge,

Figure 10.4. Linus C. Pauling (born 1901). (Courtesy of The Linus Pauling Institute of Science and Medicine)

Figure 10.5. Maurice H. F. Wilkins (born 1916); Nobel Prize laureate 1962.

Figure 10.6. Morning coffee in the Cavendish just after the publication of the manuscript on the double helix. Francis Crick (seated) and James D. Watson (standing). (From *The Double Helix* by James D. Watson, a Signet Book published by The New American Library, Inc. New York, 1969)

Figure 10.7. X-ray diffraction photographs of DNA. The pattern at left is obtained from the sodium salt, and the one at the right from the lithium salt of DNA. (Courtesy of Biophysics Research Unit, Medical Research Council, King's College, London)

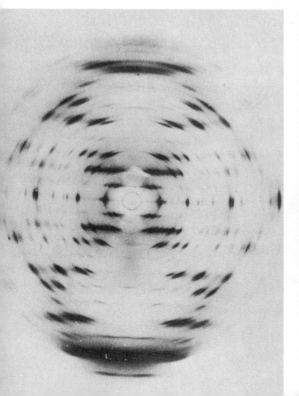 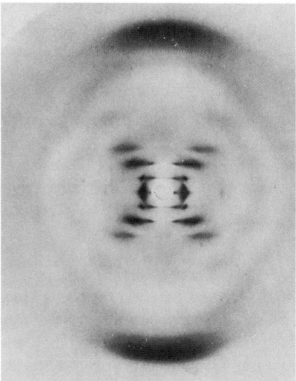

England. The X-ray crystallographic pictures of DNA (Figure 10.7) from the London team provided clues to the helical nature of DNA. Chargaff's finding of approximately equal ratios of purines to pyrimidines in all DNAs provided a clue to the pairing relationships of the four bases. Pauling had probably been unsuccessful, because he did not have access to the crystallographic data. Watson's initial attempts at model building were unfruitful, partly because he was using the rare tautomeric enol forms of guanine and thymine rather than the more common keto forms. American crystallographer Jerry Donohue informed Watson that his source for the structures of these bases, J. N. Davidson's book, *The Biochemistry of Nucleic Acids,* (and many other books) depicted the bases in the enol forms without explanation of their rarity. Watson attributed his success to Jerry Donohue, for he was the only man other than Pauling likely to make the right choice of the keto forms. Watson and Crick capitalized on all this information and in 1953 devised a structure for DNA that has since been proven essentially correct. The Watson-Crick model of DNA immediately opened the door to a veritable flood of investigations on genetics at the molecular level that has not yet begun to recede.

Genetic information must be stored in a highly stable molecule, that rarely permits changes (mutations) to occur. Furthermore, this molecule must be capable of self-replication (*autocatalytic* function) and of directing the synthesis of proteins (*heterocatalytic* function) through which it controls the metabolism and development of the organism. Ultraviolet light of wavelength 260 mμ is *mutagenic* (i.e., causes mutations) in bacteria. Genetic material would therefore be expected to absorb UV light maximally at this wavelength. We would expect diploid cells to contain twice as much of this chemical as haploid cells. DNA meets all these requirements. Indeed, all evidence to date indicates that DNA is the hereditary material of all cellular organisms. However, some viruses use DNA and others use RNA as genetic information. By 1957, M. Meselson and F. W. Stahl had provided experimental evidence for the semiconservative theory of DNA replication (initially proposed by Watson and Crick). In 1961, Marshall Nirenberg (Figure 10.8) cracked the genetic coding problem by his discovery that a sequence of uracil bases (uracil being the counterpart in RNA of thymine in DNA) coded for the amino acid phenylalanine. By 1964, Nirenberg and Khorana had devised ways of deciphering the entire code. Now it became possible to correlate genetic changes (mutations) with changes in the structure of their protein products. The biological activities of proteins may vary widely, depending on their structure and ultimately on the structure of their corresponding genes. One gene might produce a protein that functions best under certain conditions (e.g., at low pH), whereas an alternative allele might produce a protein with advantages in a different environment (e.g., retains stability at higher temperatures). Thus, it becomes clear how a heterozygote might be more adaptive than a homozygote; the heterozygote produces both kinds of proteins and therefore could survive in a wider range of habitats than either homozygote. This is part of the knowledge that molecular genetics has brought to bear on the problems of evolution.

The perpetual riddle of biology has been an explanation for the phenomenon of *differentiation.* How is it that a multicellular diploid individual can develop the morphological and physiological cellular differences that typify the tissues of higher organisms? Presumably, all the cells of an individual contain exactly the same genetic instructions, because mitosis has reproduced them all from the original zygote. Modern theories of differentiation revolve

Figure 10.8. Marshall W. Nirenberg (born 1927); Nobel Prize laureate 1968.

around the central idea that mechanisms exist to control *gene activity,* i.e., the timing at which specific genes are "turned on or off." By "turning on" is meant that DNA becomes active in making products such as proteins; "turning off" indicates that genes cease their hetero-catalytic function. By 1953, François Jacob (Figure 10.9) and Jacques Monod (Figure 10.10) had proposed a model to explain the phenomenon of *enzyme adaptation* in bacteria. Many examples were known in which catalytic enzymes were not produced unless their substrates were present in the growth medium. It appeared as though contact with the substrate caused specific genetic changes to occur, and the bacteria responded adaptively by producing the corresponding enzyme. Does this sound like the old Lamarckian idea of acquired character-istics? Bacteriology was, indeed, the last stronghold of Lamarckism. The Jacob-Monod *operon model* (to be discussed later in this chapter) proposes a mechanism whereby substrates specifically activate the genes required for production of the corresponding enzymes. This is a physiological regulation, not directed mutation! It is not known whether the operon concept (or other models for regulation of gene activity) is applicable to higher organisms. But the implications are clear that cellular differentiation will probably be explained by mechanisms that control the timing at which genes are turned on or off. It now seems highly probable that much of the DNA of higher organisms is involved in elaborate regulatory roles rather than in the synthesis of more enzymes. The adaptations of organisms that allow them to exploit particular environmental niches must eventually be explained by differentiation of morpho-

Figure 10.9. François Jacob (born 1920); Nobel Prize laureate 1965.

Figure 10.10. Jacques Monod (1910–1976); Nobel Prize laureate 1965.

logical and physiological properties in cells, tissues, and organs of individuals. This is where molecular genetics is leading us. We will not fully comprehend evolution until we understand how the constellation of genes of the individual function as a coordinate unit in the development of specific attributes we call the total phenotype. This is why the information in this chapter is critical for the modern synthetic theory of evolution.

DNA Structure and Replication

DNA is constructed like a spiral staircase, i.e., a *double helix* (Figure 10.11). Two outer backbones of alternating sugar-phosphate groups make one complete turn every 34 Å along its length. The rungs of the ladder are represented by two bases paired in specific combinations by maximization of hydrogen bonding. Adenine pairs with thymine by two bonds, guanine with cytosine by three bonds (Figure 10.12); this explains Chargaff's rule of equiva-

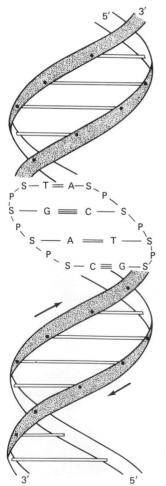

Figure 10.11. Diagram of the Watson-Crick model of DNA.

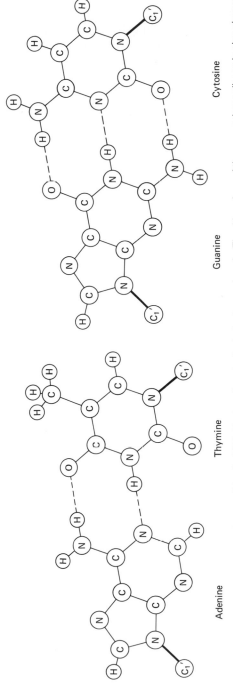

Adenine   Thymine   Guanine   Cytosine

Figure 10.12. Complementary base pairing in DNA. Broken lines indicate hydrogen bonds. The 1'-carbon of the sugar deoxyribose is also shown.

lency between purine and pyrimidine bases. Each base is attached to a sugar, forming a unit called a *nucleoside*. The plane of the sugars is perpendicular to that of the bases. Sugars are interconnected by phosphate groups. A *nucleotide* is a nucleoside plus a phosphate group. Nucleotides are considered the basic building blocks (*monomers*), which when strung together form a long linear molecule of similar units (*polymer*) called a nucleic acid (Figure 10.13). The diameter of the double helix is uniformly 20 Å. Each pair of bases is separated by 3.4 Å, and

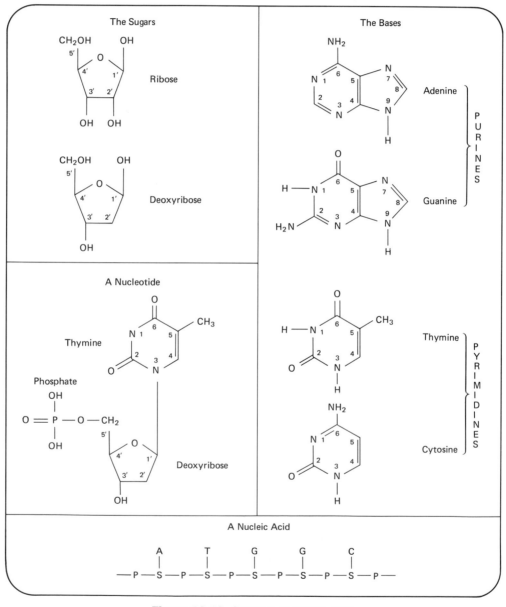

Figure 10.13. Structure of nucleic acids.

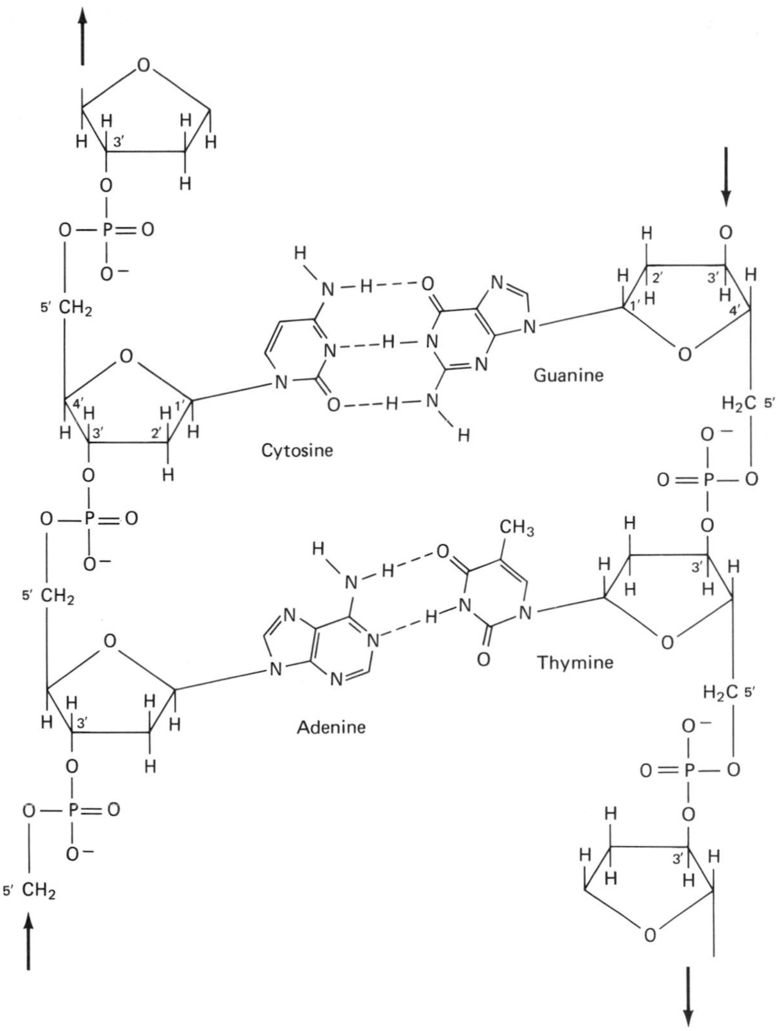

Figure 10.14. A segment of DNA showing the antiparallel orientation of complementary chains.

one complete turn of the helix contains ten nucleotide pairs. Carbon atoms of the sugars receive primed numbers to distinguish them from those of the bases. Reading along one chain from top to bottom of the molecule, we find 5′ at the top and 3′ at the bottom; on the complementary chain (reading as before), we find 3′ at top and 5′ at the bottom. In other words, the chains are *antiparallel* in direction (Figure 10.14). Pauling did not like the idea of antiparallel chains when he first read the Watson-Crick paper, indicating that his concept of DNA structure was wide of the mark.

The structure of DNA immediately suggests several possible modes of replication. In one model, the intact DNA double helix serves as a template for the formation of another complete molecule; this is called *conservative replication*. Another model allows the two strands of the DNA to separate along the hydrogen bonds holding base pairs together. Each strand of parental DNA then acts as a template for the formation of a complementary daughter strand; this is called *semiconservative replication.*

In 1957, M. Meselson and F. W. Stahl performed an experiment that resolved the problem (Figure 10.15). They grew the bacterium *Escherichia coli* in a medium in which the only nitrogen source was the heavy isotope ^{15}N. After many generations of growth in such medium, virtually all the nitrogen atoms in DNA were heavy. They extracted the DNA and measured its density by use of an ultracentrifuge. A 6 molar solution of cesium chloride (CsCl) has almost the same density as DNA. By prolonged spinning in the ultracentrifuge, this solution will form a density gradient, heaviest on the bottom and lightest at the top. When ^{15}N-DNA is spun in the CsCl gradient, it tends to form a band of molecules near the bottom of the tube. DNA from a bacterial culture grown in ordinary (light) nitrogen (^{14}N) tends to form a band near the top of the tube. Meselson and Stahl added a great excess of ^{14}N compounds to the "heavy" culture and thereafter virtually all newly synthesized DNA should incorporate ^{14}N. After one generation in the ^{14}N medium, density gradient ultracentrifugation revealed that the DNA had formed a band intermediate between totally light and totally heavy DNA. In other words, one replication cycle of DNA had created a hybrid molecule, half

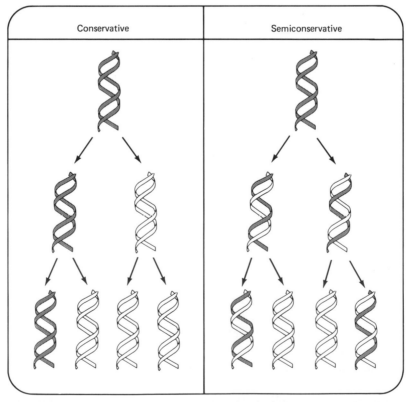

Figure 10.15. Diagram of two major hypotheses of DNA replication. Dark strands contain heavy nitrogen (^{15}N) and light strands contain normal nitrogen (^{14}N) used in the Meselson and Stahl experiment to determine which hypothesis was correct.

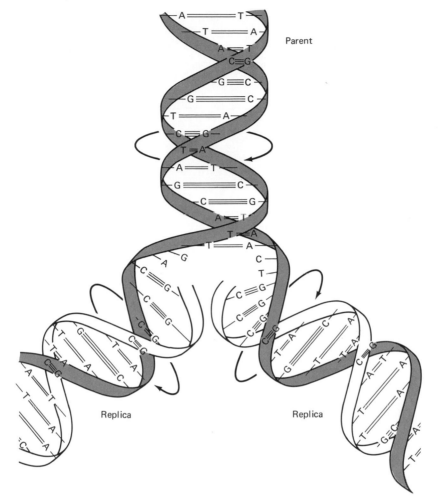

Figure 10.16. Diagram of DNA replication. Parent strands are dark; replica strands are light.

heavy and half light. After a second generation in the ^{14}N medium, half the DNA was hybrid and half was totally light.

The results of the Meselson-Stahl experiment are consonant only with the semiconservative mode of replication. In other words, DNA probably unzips into two strands, and each strand serves as a template for the formation of a complementary stand (Figure 10.16). Free nucleotides become attracted to their complementary bases in the template strand. An enzyme covalently links adjacent nucleotides together. When this process is completed, we have two identical DNA molecules, each consisting of one original strand and one newly synthesized strand.

Arthur Kornberg purified an enzyme called *DNA polymerase* from *E. coli* and found that it was active only in attaching free nucleotides by their 5′ ends to the 3′ end of the growing complementary chain. It is therefore difficult to understand how chain growth proceeds in the 3′ to 5′ direction. There is evidence that *E. coli* DNA daughter chains are synthesized as short

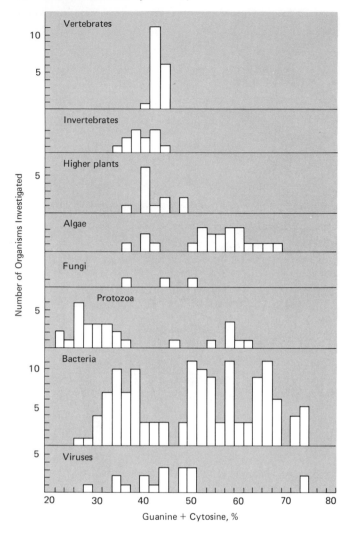

Figure 10.17. Guanine + Cytosine content in DNA of various groups of organisms.

segments rather than sequentially from one end to the other. DNA polymerase could replicate short segments by moving along each of the parental template chains in opposite directions. Another enzyme known as *DNA ligase* catalyzes the formation of an internucleotide phosphate diester bond between a 5'-phosphate and a 3'-hydroxyl group of two adjacent nucleotides at any position along a DNA double helix. Therefore DNA ligase could seal the gaps between adjacent nucleotides left by discontinuous replication of DNA polymerase. One theory of bacterial replication proposes that DNA consists of a number of units capable of independent replication, called *replicons.* Dipteran polytene chromosomes show simultaneous labeling of several discontinuous segments when exposed to short pulses of tritiated (tritium is a radioactive isotope of hydrogen) thymidine (the deoxyribonucleoside of thymine). Perhaps replicons also exist in organisms other than bacteria.

Probably the most ideal measure of genetic relationship between two organisms is the percentage of DNA nucleotide sequences held in common. These determinations are far beyond the ability of present technology to distinguish between closely related species. The

DNA of all vertebrates has a narrow range for the mole fraction [G] + [C] between 40 and 44 per cent (Figure 10.17). The DNA of higher plants has a wider range between 35 and 45 per cent. The widest range is found in bacteria from 26 to 74 per cent. The similar base composition of *Escherichia coli* (52 per cent), *Salmonella typhimurium* (50 per cent), and *Shigella dysenteriae* (54 per cent) provides evidence for a relatively close taxonomic relationship between these three intestinal bacteria. On the other hand, *Streptococcus griseus* (74 per cent) and *Streptococcus pneumoniae* (38 per cent) cannot be considered related except in the very broadest sense despite their common generic name. It is obvious that bacterial taxonomy can profit greatly from the newer knowledge provided by base composition studies.

A much more discriminating technique for determining the degree of genetic relationship is available from *in vitro* DNA hybridization experiments. In 1960, J. Marmur found that when a purified DNA solution is denatured by being heated to the boiling point (causing separation of the two chains of double helix) and then cooled slowly to 60°C over a period of about 80 minutes, a considerable amount of renaturation may occur. If we extract DNA from two different organisms, melt the double helices, put them all together, and allow them to cool slowly, some hybrid double helices may form (one chain from species A and the other chain from species B). The more nucleotide sequences held in common by the two species, the more hybrid double helices should form. Special techniques using this principle have yielded quantitative estimates of genetic relationships between distantly related species of mammals (Figure 10.18). For example, it was found that human DNA and mouse DNA share approximately 21 per cent of their nucleotide sequences in common. Rat DNA and hamster DNA are more closely related to that of the mouse; rhesus monkey DNA is closer to humans than it is to the mouse, etc. This method offers promise of more precisely defining genetic relationships within groups whose taxonomic affinities have been refractory to more conventional methods.

Figure 10.18. Degree of genetic relationship of various animals relative to humans as indicated by the DNA hybridization technique. The amount of hybridization of human DNA with itself is taken as 100 per cent.

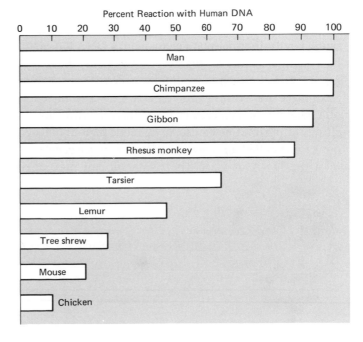

The Central Dogma

DNA has both autocatalytic and heterocatalytic functions. Normally, DNA does not serve as the direct template for the formation of proteins. Rather, the information for making proteins is first *transcribed* into RNA, and then the ribonucleotide sequence of RNA is *translated* into the amino acid sequence of proteins. This is the so-called "central dogma" of molecular genetics (Figure 10.19).

Gene activity is expressed by transcription and/or translation of its coded information into RNA and/or proteins, respectively. In order to understand the origin of any phenotype that an organism presents to the selective forces of the environment, we must ultimately comprehend gene activity.

All cellular ribonucleic acid is made from a DNA template; some RNA viruses can replicate their RNA without a DNA template. RNA differs from DNA in three important respects:

1. The pentose sugar in RNA, called *ribose,* has one more oxygen than the sugar deoxyribose found in DNA.
2. The base *uracil* is substituted for thymine in RNA and preferentially pairs with adenine.
3. RNA molecules are much shorter than DNA molecules and generally are single stranded.

The exact mechanism whereby RNA is transcribed from DNA is not yet known. The DNA double helix is presumed to open locally to allow one or more genes to be transcribed. For any one gene, only one of the two DNA strands makes "sense"; the corresponding nucleotide sequence on the complementary chain would be "nonsense." The mechanism that allows only the sense strand to be transcribed into RNA is unknown at present. The enzyme *RNA polymerase* catalyzes the formation of internucleotide 3′-5′ phosphodiester bonds in synthesizing a polyribonucleotide chain from a DNA template. This enzyme consists of five different polypeptide chains aggregated by secondary bonds; four of them constitute the "core enzyme," and the fifth polypeptide (called *sigma factor,* σ) is easily dissociated. Sigma factor recognizes the "start signals" on DNA, i.e., the nucleotide sequence that signals the beginning of a gene. Some evidence is available from viral chromosomes that the start signal may be a cluster of pyrimidine bases. Any genetic region that signals the beginning of a transcriptional message is called a *promoter region* (*promoter gene*). Once the RNA polymerase has begun to read the DNA message, the sigma subunit may dissociate and recycle to other free core

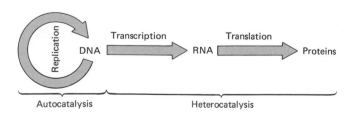

Figure 10.19. The central dogma of molecular biology.

enzymes. DNA is always transcribed beginning at the 3′ end of the message. Because of antipolarity in paired nucleotide chains, the RNA chain made from a DNA template will begin its synthesis at its 5′ end. Its synthesis will be terminated at the 3′ end by a specific protein called the *rho factor* (ρ). Presumably the rho factor recognizes "stop signals" at the end of DNA gene messages and blocks further transcription by RNA polymerase.

There are three major classes of RNA: (1) messenger RNA (mRNA), (2) transfer RNA (tRNA), and (3) ribosomal RNA (rRNA). Messenger RNA carries the coded information for amino acid sequences of proteins. Transfer RNAs form adaptor molecules with the capacity to become specifically attached to amino acids and also to complementary base pair with mRNA molecules. Ribosomal RNAs complex with proteins to form ribosomes; the latter serve as "jigs" for supporting mRNA and tRNA during protein synthesis.

The Genetic Code

Based on mathematical calculations, the simplest genetic code was probably a triplet, i.e., the smallest number of deoxyribonucleotides that could code for an amino acid was three. This conclusion stems from the fact that there are only four bases in DNA (A,T,G,C), but there are twenty amino acids commonly found in proteins. Obviously, a singlet code would not do, because if each base corresponded to an amino acid, there is information for only four amino acids. A doublet code would code for only $4 \times 4 = 16$ amino acids. A *triplet code,* however, has $4 \times 4 \times 4 = 64$ possible combinations, much more than required for the twenty amino acids.

In 1955, the enzyme *polynucleotide phosphorylase* was discovered. It was found to catalyze the polymerization of ribonucleoside diphosphates without any DNA template. Nirenberg capitalized on this discovery. He disrupted bacterial cells, creating a cell-free protein-synthesizing system, i.e., all the machinery (enzymes, ribosomes, energy sources, etc.) was available. He destroyed the endogenous mRNA, making the system dependent on the addition of exogenous mRNA for protein synthesis to proceed. Nirenberg then used the polynucleotide phosphorylase enzyme to make an artificial mRNA consisting of repeating uracil ribonucleotides. Upon addition of the poly-U messenger to the cell-free system, polypeptides were made consisting only of a monotonous string of phenylalanine amino acids. This was the first step in cracking the genetic code. A sequence of uracils (probably three) obviously coded for the amino acid phenylalanine. Following this success, other *homopolymers* (polymers of a single kind of monomer) of mRNA were synthesized: poly-C coded for polyproline; poly-A coded for polylysine, etc. *Heteropolymers,* consisting of two or more different nucleotides, were then made by Ochoa and his colleagues.

By using widely different ratios of the bases, the random formation of various triplet combinations could be calculated and compared with the frequencies of various amino acids in the polypeptide products. This established the empirical formulas for most of the amino acids (e.g., 2C:1A = glutamine, 2A:1C = asparagine, etc.). Khorana (Figure 10.20) found a way to construct regular alternating copolymers such as —UGUGUGUG— which produced an alternating polypeptide such as —valine—cysteine—valine—cysteine—. This permitted

Figure 10.20. H. Gobind Khorana (born 1922); Nobel Prize laureate 1968.

determination of the exact order of nucleotides represented by the empirical formulas. For example, since one of the empirical formulas for valine was known to be 2G:1U, the exact order must be GUG; likewise the code for cysteine must be UGU. Khorana also prepared synthetic-mRNAs with repeating trinucleotide sequences such as —UUGUUGUUG—. Three polypeptides were produced therefrom, each being a monotonous polypeptide, *viz.* polyleucine, polycysteine, or polyvaline. Since there was no "start signal" in the synthetic mRNA, reading of the message could be initiated at any one of three places (UUG = leucine, UGU = cysteine, GUU = valine). This provided proof of the triplet nature of the genetic code. In 1964, Nirenberg discovered that trinucleotides would bind to ribosomes at the position normally occupied by mRNA molecules and would also cause the binding of the corresponding tRNA-amino acid complex. By radioactively labeling the different amino acids, he was able to identify the amino acid corresponding to each of the sixty-four possible triplet combinations (*codons*).

The complete mRNA code is shown in Table 10.1. Notice that some amino acids (e.g., methionine and tryptophan) have a single codon. Others have 2,3,4, or 6 "synonymous" codons. Because more than one codon exists for some amino acids, the code is said to be *degenerate*. Three of the sixty-four codons do not specify any amino acid; instead they appear

TABLE 10.1. mRNA Codons*

		Second Letter				
		U	C	A	G	
First Letter	U	UUU UUC }Phe UUA UUG }Leu	UCU UCC UCA UCG }Ser	UAU UAC }Tyr UAA UAG }Nonsense	UGU UGC }Cys UGA Nonsense UGG Trp	U C A G	Third Letter
	C	CUU CUC CUA CUG }Leu	CCU CCC CCA CCG }Pro	CAU CAC }His CAA CAG }Gln	CGU CGC CGA CGG }Arg	U C A G	
	A	AUU AUC }Ile AUA AUG Met	ACU ACC ACA ACG }Thr	AAU AAC }Asn AAA AAG }Lys	AGU AGC }Ser AGA AGG }Arg	U C A G	
	G	GUU GUC GUA GUG }Val	GCU GCC GCA GCG }Ala	GAU GAC }Asp GAA GAG }Glu	GGU GGC GGA GGG }Gly	U C A G	

* The abbreviated names of amino acids are as follows: ala = alainine, arg = arginine, asn = asparagine, asp = aspartic acid, cys = cysteine, gln = glutamine, glu = glutamic acid, gly = glycine, his = histidine, ile = isoleucine, leu = leucine, lys = lysine, met = methionine, phe = phenylalanine, pro = proline, ser = serine, thr = threonine, trp = tryptophan, tyr = tyrosine, val = valine.

to serve as "stop signals" at the end of a message. An *anticodon* is the group of three adjacent nucleotides on a tRNA molecule, which form complementary base pairs with the codon nucleotides in the mRNA. Crick proposed a "wobble hypothesis" to explain why, in so many cases, an amino acid has multiple synonymous codons that differ only in the third position. According to this theory, the first two bases of a mRNA codon are constrained to pair with the corresponding two bases in the tRNA anticodon in standard fashion (A with U and G with C). The third position, however, is less restricted, i.e., there is some "wobble" or flexibility in the codon-anticodon interaction. For example, U in the third position of an anticodon may pair with either purine (A or G) in the mRNA codon, etc.

Initiation of polypeptide chain growth in *E. coli* appears to be dependent on a special kind of tRNA called formylmethionine-tRNA (f-met-tRNA). The formyl group (CHO) on the amino acid methionine prevents one end from participation in *peptide bond* formation (the kind of bonds that connect adjacent amino acids together into a polypeptide chain). It may, however, initiate polypeptide chain growth in response to the AUG methionine codon or the GUG valine codon and even weakly responds to the UUG or CUG leucine codons when at the 5′ end of the mRNA molecule. All bacterial proteins do not have formylmethionine at their 5′ ends, suggesting that perhaps this chain-initiating residue is later enzymatically removed. This mechanism for initiation of polypeptide chain growth seems to be a general phenomenon involving the kind of ribosomes peculiar to prokaryotes and the self-reproducing cytoplasmic organelles of eukaryotes (e.g., mitochondria). It is doubtful that this mechanism is operative on the different class of ribosomes found in the cytoplasm of higher eukaryotes.

Several other interesting parallels exist between prokaryotes and eukaryotic mitochondria. Bacterial ribosomes sediment in the ultracentrifuge at 70 S (S = Svedberg unit; larger molecules have larger S values). They consist of two subunits: one at 50 S and the other at 30 S. Mitochondrial ribosomes have the same general construction. Eukaryotic cytoplasmic ribosomes sediment at 80 S; their two subunits have values of 60 and 40 S (sedimentation velocities are not simply additive). Protein synthesis on mitochondrial ribosomes is inhibited by the drug chloramphenicol; the same is true for prokaryotic ribosomes. Both bacteria and mitochondria contain circular DNA molecules. These facts suggest that the eukaryotic cells became parasitized by bacteria at some early stage in their evolution. The association evolved into a symbiosis and eventually into a mutualism so that now mitochondria are regular features of eukaryotic cells. Perhaps the early prokaryotes were very efficient in supplying energy (in the form of adenosine triphosphate, ATP) to their hosts and intensive selection secured their obligatory presence. A similar origin for chloroplasts may have evolved from symbiosis with photosynthetic bacteria, because they too possess DNA and prokaryotic-type ribosomes. Both mitochondria and chloroplasts increase in length and divide by fission similar to that seen in bacteria. However these organelles cannot be made to grow or divide outside of cells, indicating that certain vital constituents are produced by nuclear genes. Perhaps some of the genetic information originally in the parasitizing prokaryote has been transferred to the nuclear chromosomes of its eukaryote host. This evolutionary scheme allows us to understand how extranuclear heredity operates. Chloroplasts are found in virtually all plant cells except the male sex cells. This accounts for the maternal transmission of chloroplast traits. Many examples are known of interactions between nuclear genes and cytoplasmic genes (i.e., genes within organelles such as chloroplasts and mitochondria, called *plasmagenes* or *plasmids*).

> In the red bread mold *Neurospora,* a plasmid mutation in mitochondrial DNA is thought to be responsible for a slow rate of growth called "poky" phenotype. The poky trait is transmitted only through the maternal parent. A nuclear gene *F* interacts with the poky mutant to enhance the growth rate, but does not restore complete normality (a condition called "fast poky"). Gene *F* has no demonstrable effect on the phenotype when normal mitochondrial DNA is present. The alternative allele F^1, however, does not correct the abnormal respiratory enzyme system of mutant mitochondrial DNA. Hence, with F^1 in the nucleus, mutant mitochondria produce the poky trait.

Protein Synthesis and Structure

With the wobble hypothesis in mind, there are at least as many kinds of tRNA molecules as there are different amino acids. All tRNAs appear to be single chains of about 75–80 ribonucleotides each. The single chain folds back on itself and tends to form in local regions double helices that stabilize its three-dimensional structure (see cloverleaf model in Figure 10.21). Several unusual bases have been found in tRNAs (inosine, ribothymidine, pseudouridine, etc.), which cannot form convential base pairs. This causes formation of loops of unpaired bases. One of these loops probably contains the anticodon. A second loop is very similar in all tRNAs and is thought to perform a common function such as binding to

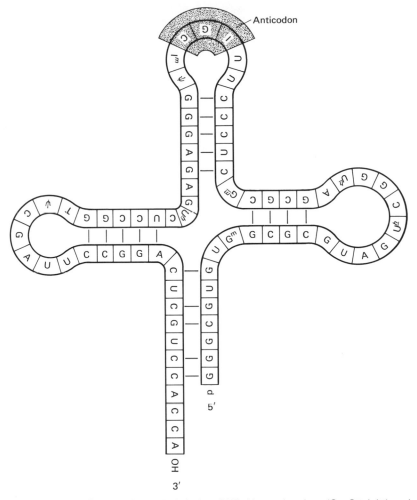

Figure 10.21. Cloverleaf model of alanine tRNA. Unusual purines (G^m, $G^{\underline{m}}$, I, I^m) and unusual pyrimidines (T, ψ, $U^{\underline{h}}$) tend to cluster near bends and unpaired regions of the molecule. Inosinic acid (I) can pair with either U or C. Therefore, the anticodon $^{3'}CGI^{5'}$ can base pair with two of the mRNA codons for alanine ($^{5'}GCU^{3'}$ or $^{5'}GCC^{3'}$).

ribosomes. A third loop may serve as the recognition site for an enzyme that specifically attaches ("loads") one of the amino acids to the 3′ end of the tRNA. A different enzyme is required to join each kind of adaptor molecule (tRNA) to its own species of amino acid.

The 30 S ribosomal subunit attaches to the 5′ end of a mRNA molecule (Figure 10.22). A loaded tRNA base pairs through complementary anticodon-codon interaction. The 50 S ribosomal subunit becomes attached to form an initiation complex. Each 70 S ribosome has two tRNA binding sites: a peptidyl (P) and an amino-acyl (A) site. Once the first loaded tRNA is in the P site, a second loaded tRNA enters the A site. An enzyme called *peptidyl transferase*, one of the proteins of the 50 S subunit, catalyzes the formation of a peptide bond between the two amino acids held by adjacent tRNA molecules on the ribosome. Concomitant

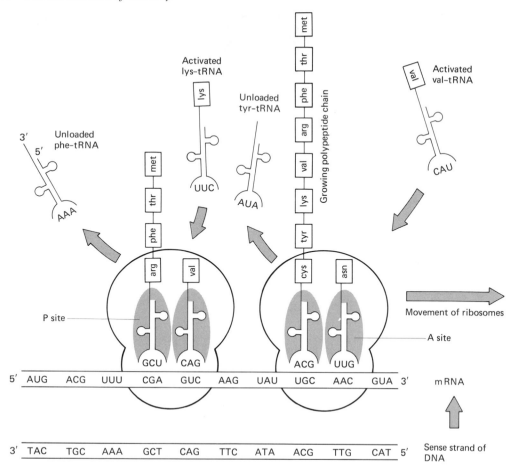

Figure 10.22. Diagram of protein synthesis.

with peptide bond formation, the tRNA in the P site releases its amino acid and departs from the ribosome. It may now recycle by becoming again loaded with its species of amino acid. Meanwhile, the tRNA in the A site moves to the P site, while the mRNA shifts to bring the next codon into register. The vacant A site is now ready to receive the loaded tRNA specific for the newly displayed codon. The process continues over and over again until the stop (nonsense) codon is reached at the 3′ end of the message. A releasing factor (protein) recognizes the stop signal and specifically binds there. It is suggested that the releasing factor may be an enzyme that splits off the terminal tRNA. At the 3′ end of the mRNA message, the 70 S ribosome separates into its 50 S and 30 S subunits for recycling. Many ribosomes may simultaneously read a mRNA strand, forming *polyribosomes*. It is thought that the life of a mRNA molecule is prolonged when covered by polyribosomes; unoccupied mRNA molecules are subject to enzymatic degradation. All mRNA molecules are relatively short lived and must be constantly replaced. The initiating 5′ end of mRNA specifies the free amino end (NH_2) of the polypeptide chain, and the terminal 3′ end of mRNA corresponds to the free carboxyl group (COOH) of the polypeptide chain.

Proteins may have three or four levels of structural complexity (Figure 10.23). The *primary structure* is the sequence of amino acids in the polypeptide chain as determined by the genetic code. Certain regions of proteins may be twisted into a helical *secondary structure* named an α-helix by Pauling. There are 3.6 amino acids per turn of the helix. It is stabilized by weak hydrogen bonds between the carbonyl group of one residue and the imino group of the fourth residue further down the polypeptide chain. Proteins seldom (if ever) exist as simple helices, because the side groups of amino acids tend to interact and bend the polypeptide chain to form the energetically most favorable secondary bonds (*tertiary structure*). In addition, the amino acid proline is a secondary amine (containing no available H) and this also interrupts α-helical regions. Disulfide (S—S) bridges tend to form between cysteine

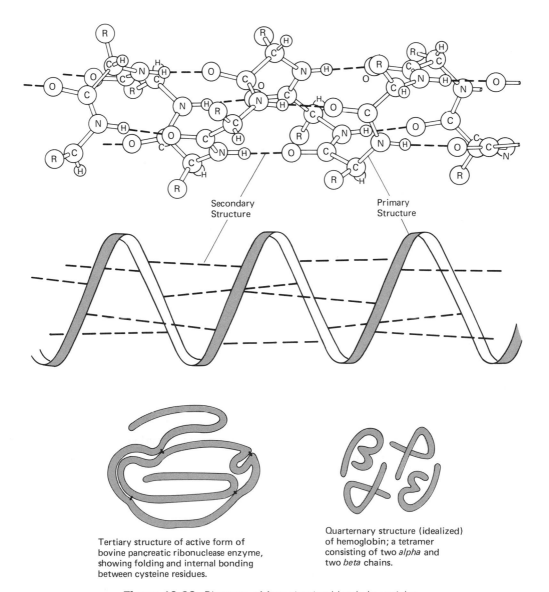

Secondary Structure

Primary Structure

Tertiary structure of active form of bovine pancreatic ribonuclease enzyme, showing folding and internal bonding between cysteine residues.

Quaternary structure (idealized) of hemoglobin; a tetramer consisting of two *alpha* and two *beta* chains.

Figure 10.23. Diagrams of four structural levels in proteins.

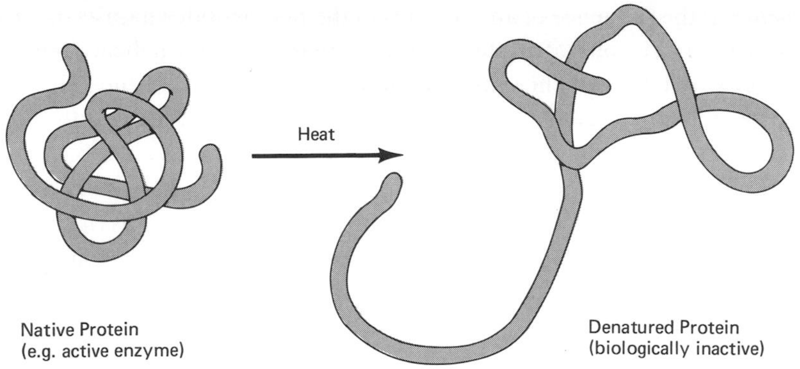

Figure 10.24. Diagram of heat-denatured protein and accompanying loss of biological activity.

residues, and this also distorts otherwise helical regions and aids in stability of the tertiary structure.

The three-dimensional conformation of a single polypeptide chain is thus determined by its primary structure and ultimately by the gene from which it was made. Many functional proteins consist of more than one polypeptide chain. Each chain of a complex protein is referred to as a *protomer*. Several identical or different chains (protomers) may interact to form a *quarternary structure* with physiological activity. For example, the enzyme β-galactosidase from *E. coli* consists of four identical polypeptide chains (*homotetramer*), each chain being 1173 amino acids in length. Another enzyme of this bacterium, tryptophan synthetase, consists of two pairs of dissimilar chains (two A-chains with 267 units each, and two B-chains with about 450 units each; a *heterotetramer*). The polypeptide chains of these quarternary protein structures spontaneously assemble themselves into the correct relationship for functional activity based solely on the constraints of maximal thermodynamic stability.

Some proteins are easily *denatured* (loss of tertiary or quarternary structure) by a moderate increase in temperature, which disrupts the weak secondary bonds in secondary and higher structural levels of proteins (Figure 10.24). If an essential enzyme is temperature sensitive, an organism may survive in the normal temperature range but die at higher restrictive temperatures (*conditional lethal*). Pigment distribution in the Himalayan breed of rabbit was cited in Chapter 7 as an example of a nonessential enzyme denaturation. Hydrangea plants of identical genotype will produce blue flowers if grown in acid soil or pinkish flowers if grown in alkaline soil. This probably represents another example of enzyme modification by environmental factors. Obviously, selection for nongenetic variations (plasticity) such as this would be futile.

Gene Definitions

For a long time, the "classical gene" was assumed to be a transmission unit recognized by its ability to (1) mutate to alternative forms, (2) recombine with other similar units, and (3) endow the organism with a particular phenotypic characteristic. It is now clear, however, that the smallest unit of mutation is a nucleotide, that the smallest unit of recombination involves adjacent nucleotides, and that the unit of function is often a sequence of hundreds or thousands of nucleotides required to specify the amino acid sequence of a complete polypeptide chain. Seymour Benzer coined the terms *muton, recon,* and *cistron,* respectively, for each of these three kinds of genes. Chapter 12 will discuss the problems of mutation.

In order to understand Benzer's concepts of recon and cistron, it would be wise to review the life cycle of the virus from which he obtained these ideas. Benzer used a *bacteriophage* (i.e., a bacterial virus or simply *phage*) called T4, which infects the bacterium *E. coli* (Figure 10.25). Mature infective virus particles consist of a DNA core surrounded by a protein head capsule connected to a tail consisting of a cylindrical sheath, and a base plate with six tail fibers. The phage attaches its tail fibers to specific receptor sites on the host cell. The contractile proteins of the tail sheath aid in injecting the phage DNA into the host cell. Once inside, the phage DNA commandeers cellular resources to manufacture many replicas of phage DNA, and later synthesizes all of its protein components. The DNA and protein subunits spontaneously assemble themselves into mature phage particles. A phage enzyme (lysozyme) is one of the last proteins to be made. It causes dissolution (lysis) of the cell and releases the hundred or so infective phage particles to start the process anew.

The spreading lytic action of an initial single phage particle can be observed as a cleared area called a *plaque* in a confluent growth of bacteria on a petri dish culture. Some phage mutants are known to accelerate the life cycle and result in larger plaque size. These rapid lysing mutants, designated *r*, are thus easily distinguished from the normal (wild-type) phage, designated r^+. Viruses have only a single set of genetic instructions (monoploid or haploid). A special class of *r* mutants (called the *r*II group) fail to grow on *E. coli* strain K; wild-type r^+

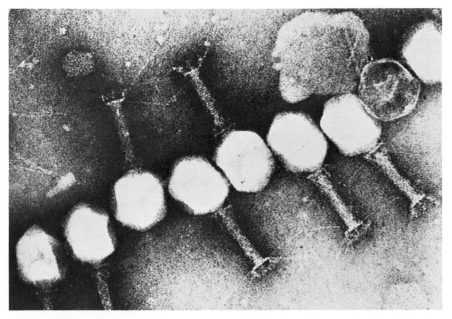

Figure 10.25. Electron micrograph of bacteriophage T4. (Courtesy Carolina Biological Supply Company)

mutants, however, do grow on strain K. If a strain K bacterium is simultaneously infected with an *r*II mutant and a wild-type *r*⁺ phage, both kinds of progeny phage particles are released. Therefore, the *r*⁺ phage particles are providing some factor(s) essential for the growth of *r*II mutants. A large number of *r*II *point mutants* had been discovered, i.e., mutations involving single nucleotide changes. Benzer decided to infect K-strain cells with different combinations of pairs of these *r*II mutants. Any pair of *r*II point mutants that did not cooperate to lyse the cells could be assumed to reside in the same functional unit (cistron) of the phage genome. Any pair of *r*II mutants that could cooperate to yield a normal number of progeny phage were considered to belong to different functional units (cistrons). This was the basis of the "*cis-trans*" or *complementation test* (Figure 10.26). It was found that all the *r*II mutants fell into two functional groups (called cistrons A and B). Mutants within the same functional group fail to complement each other (i.e., fail to grow in a K-strain bacterium when present in the same cell). Any pair of mutants in different cistrons were able to lyse the cells. From this it is inferred that two mutants in the same cistron fail to produce an essential protein and therefore cannot complete the life cycle. If an *r*IIA mutant makes B protein and an *r*IIB mutant makes A protein, the two essential products complement each other and allow the phage particles to complete their life cycle (lyse the cell).

Benzer also found that he could detect very low levels of recombination (0.0001 per cent) between point mutants within the same *r*II cistron. He crossed mutants belonging to the same cistron (noncomplementing mutants) in *E. coli* strain B (on which *r* mutants can grow) and plated the progeny on *E. coli* strain K (on which only wild-type phage grows). The occasional appearance of a plaque on strain K indicated that recombination had occurred to "rescue" a completely normal wild-type cistron. The frequency of recombination between point mutants allows them to be mapped in linear order similar to the mapping of functionally different loci

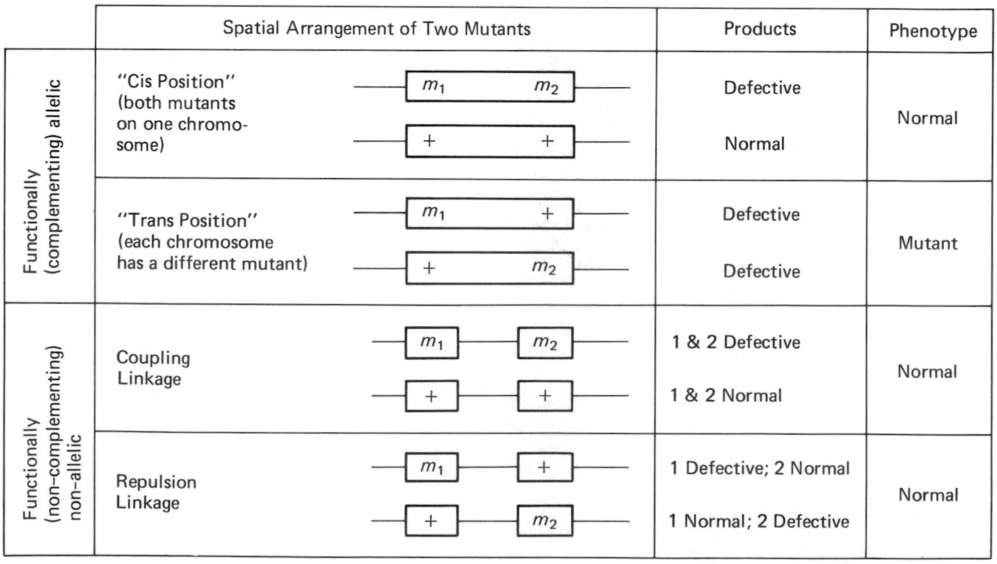

		Spatial Arrangement of Two Mutants	Products	Phenotype
Functionally (complementing) allelic	"Cis Position" (both mutants on one chromosome)	m_1 — m_2 + — +	Defective Normal	Normal
	"Trans Position" (each chromosome has a different mutant)	m_1 — + + — m_2	Defective Defective	Mutant
Functionally (non-complementing) non-allelic	Coupling Linkage	m_1 — m_2 + — +	1 & 2 Defective 1 & 2 Normal	Normal
	Repulsion Linkage	m_1 — + + — m_2	1 Defective; 2 Normal 1 Normal; 2 Defective	Normal

Figure 10.26. The "cis-trans" or complementation test. Each box indicates a functionally different segment of DNA; i.e. a cistron responsible for one complete polypeptide chain.

in higher organisms. The technique was so sensitive that it was virtually able to detect recombination between adjacent nucleotides. Benzer's work indicates that the gene of function (cistron) consists of hundreds or thousands of nucleotides, each one capable of independent mutation (mutons), and recombination may occur between any two adjacent nucleotides (recon). The appearance of a distinctive phenotype (e.g., lysis of K strain) may depend on more than one normal cistron (i.e., more than one polypeptide chain). As explained earlier, some enzymes are now known to consist of two or more different polypeptide chains. Thus, the previous concept of "one gene-one enzyme" had to be modified to the present concept of "one cistron-one complete polypeptide chain."

Cistrons are usually considered as genes of heterocatalytic function, i.e., units specifying the structure of other macromolecules (therefore called *structural genes*). Most of the products of structural genes are probably proteins. All enzymes are proteins. Enzymes are the biological catalysts responsible for the major features of metabolism, growth, and development.

> The contractile muscle protein actomyosin is a major component of many higher animals. The scleroproteins keratin and collagen are products of fibroblast cells and constitute a large fraction of integumentary tissues (skin) and their derivatives (feathers, hair, horns, nails, etc.). Hemoglobin in red blood cells and myoglobin in muscle cells function in the transport, storage, and release of oxygen. Many other proteins also have transport roles (e.g., transferrins carry iron to the bone marrow and tissue storage areas, copper is transported bound to the protein ceruloplasmin, lipids are transported bound to specific proteins as β-lipoproteins, etc.). Albumins in blood create an oncotic pressure that aids in restoring fluid lost from circulation by the leaky endothelium of capillary beds. Cytochromes and flavoproteins function in reversible oxidation-reduction reactions for energy production. The 50 S subunit of bacterial ribosomes alone contains about forty different proteins, and the 30 S subunit has about twenty proteins. Some hormones are entirely protein (e.g., insulin), and other proteins are important components of hormones (e.g., thyroglobulin). Many important "self-recognition signals" are components of the cell membrane (e.g., antigens). Essential components of the immune system are also proteins (immunoglobulins or antibodies, complement components, etc.). Normal blood clotting depends on numerous specific proteins. The list goes on and on. The few examples cited here give only a glimpse of the diversified roles proteins play in the living organism.

Most commonly we think of polypeptide chains (or proteins) as cistron products, but we must remember that not all RNA is mRNA for protein synthesis. Many cistrons are involved in the production of the various tRNAs and rRNAs (the 50 S subunit has two different rRNAs and the 30 S subunit has one distinctive rRNA). In fact, most of the cytoplasmic RNA is complexed with proteins to form ribosomes. There may also be genetic units of autocatalytic function, called *replicons,* whose products are DNA rather than RNA.

Some genetic elements have no product at all, but instead serve as recognition signals (controller elements). For example, a nonsense codon acts as a period at the end of a translational sentence. Promoter genes are initiation segments that bind RNA polymerase. Still other loci function as "translational switches," that may turn certain structural genes "on or off," i.e., determine whether they will be allowed to be transcribed or not.

Genetic Control Mechanisms

The synthesis of certain enzymes is unregulated, i.e., they are continuously produced. Such enzymes are classified as *constitutive enzymes*. An alternative class contains enzymes, the synthesis of which is *regulated* according to the needs of the organism. Mechanisms that regulate the synthesis of proteins are of considerable importance to evolutionary theory, because they control tissue differentiation and hold the key to our understanding the origin of more complex systems from simpler forms.

> There are two major ways of estimating genetic similarities between different species. One way is to investigate differences at the molecular level using electrophoresis, immunological analysis, and gene-sequencing techniques. The other way is to study differences at the organismal level, using anatomical, physiological, behaviorial, and ecological data to estimate genetic distance. Humans and chimpanzees are quite different species at the organismal level, but very similar at the biochemical level. One study involving forty-four proteins and enzymes common to both species found that the average human polypeptide is more than 99 per cent identical to polypeptides in chimps. A question has been raised as to how such great organismal differentiation occurs with so little apparent biochemical differentiation. One theory proposes that subtle changes in regulatory mechanisms by gene mutation or simply by rearrangement of the pattern of existing genes on the chromosomes might be responsible for many major differences at the organismal level. Chimps have forty-eight chromosomes; humans have forty-six. Studies of the chromosomal banding patterns for these species revealed great dissimilarity, indicating that gene orders on the chromosomes are quite different. Geneticists and evolutionists have traditionally believed that gene changes were responsible for the phenotypic changes seen at the organismal level. It may be, however, that the explanation is largely to be found in changes in patterns of gene regulation and gene arrangement.

Two models of regulated enzyme production have been proposed for bacterial systems: (1) inducible systems, found mainly in *catabolic* enzymes (i.e., enzymes that split the substrate), and (2) repressible systems, associated chiefly with *anabolic* enzymes (i.e., enzymes that unite substrates). Jacob and Monod proposed, in 1961, the concept of an *operon* to explain the regulation of enzyme production induced by the sugar lactose. Later, the operon concept was also used to explain repressible systems.

Enzyme Induction

The lactose operon consists of three parts from left to right: (1) a promoter gene recognized by the enzyme RNA polymerase, (2) an *operator* locus, which functions like a switch, and (3) three structural genes coding for the enzymes β-galactosidase, gactoside-permease, and galactoside transacetylase. The disaccharide lactose is cleaved into the monosaccharides glucose and galactose by β-galactosidase (Figure 10.27). Permease enhances the entry of lactose into the cell from the surrounding medium. Transacetylase catalyzes the transfer of an acetyl group from a donor molecule to the galactose component of lactose. The physiological function of transacetylase is unknown at present, because its presence is not

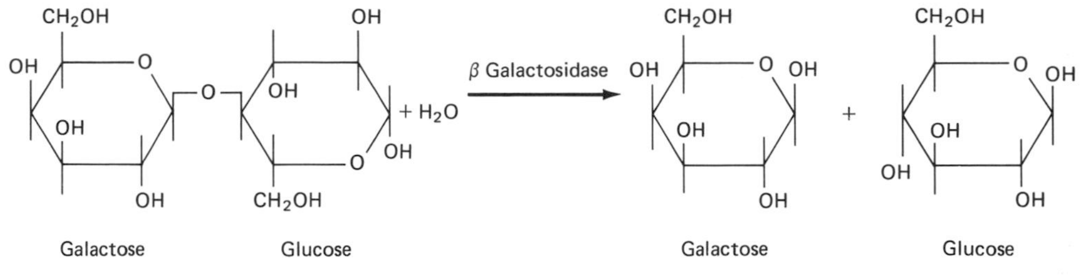

Figure 10.27. Splitting of the disaccharide lactose into two monosaccharides (glucose and galactose) by the enzyme β-galactosidase.

required for normal lactose fermentation. Operon function is dependent on a protein, called *repressor,* synthesized by a noncontiguous constitutive locus called the *regulator gene* (Figure 10.28). In the absence of lactose (noninduced condition), repressor binds specifically with the operator locus of the operon. RNA polymerase (which binds at the promoter locus) cannot move toward the right, if the operator site is occupied by repressor protein. Thus, the "switch is closed" to transcription of the three structural genes in the operon. On the other hand, when lactose is present in the medium (induced condition), a few molecules diffuse into the cells and bind with the repressor protein. This union causes a conformational change in the three-dimensional structure of the protein, altering the site by which it formerly recognized the operator locus.

The phenomenon of a small molecule causing a change in the shape of a protein and the consequent alteration of the interaction of the protein with a third molecule is referred to as an *allosteric effect* (Figure 10.29). The lactose-bound repressor becomes nonfunctional, i.e., can no longer bind to the operator. The "switch" is now opened to allow RNA polymerase to begin the transcription of the structural genes of the operon into mRNA molecules, which will then serve as templates for the synthesis of the three enzymes. When permease appears, the cell begins to pump lactose into the interior with high efficiency. Galactosidase splits the lactose and very shortly the fermentation of lactose is proceeding at maximal rate in the entire culture. When all the lactose is gone, the repressor protein returns to its original shape (reverse allosteric change) and can once again bind to the operator site and "turn off" the operon. The evolution of such control systems was probably in response to the selective advantage gained by organisms that could change constitutive enzyme production into regulated enzyme synthesis, for constitutive enzyme production in the absence of substrate is a waste of energy.

Enzyme Repression

Monod found, in 1953, that one of the bacterial enzymes involved in synthesis of the amino acid tryptophan was repressed by addition of tryptophan to the medium. Later, it was found that four enzymes involved in the biochemical pathway from chorismic acid to tryptophan (see Figure 10.30) were coordinately repressed. Jacob and Monod saw the similarity of this system to that of the inducible lactose system (Figure 10.28). They proposed that the product of the regulator gene was an inactive repressor (*aporepressor*), i.e., unable to

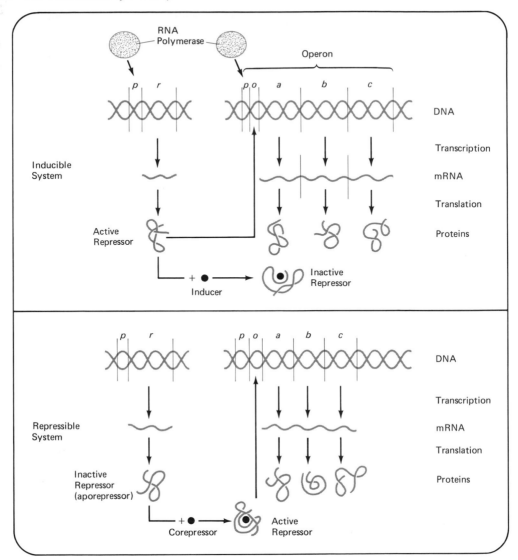

Figure 10.28. Generalized diagrams of inducible and repressible protein synthesis. Symbols: p = promoter; r = regulator gene; o = operator locus; a,b,c = structural genes in the operon.

recognize specifically the nucleotide sequence of the operator locus. When tryptophan synthesis exceeds the requirements of the cell, the excess functions as a *corepressor*. The union of corepressor and aporepressor creates a functional repressor molecule by allosteric transformation. The repressor binds to the operator and shuts off synthesis of all four enzymes in the tryptophan operon. When tryptophan levels return to normal, the corepressor departs from the aporepressor, causing a reverse allosteric change that destroys its ability to bind at the operator locus. The "switch" is then "on," and synthesis of tryptophan is resumed. In a

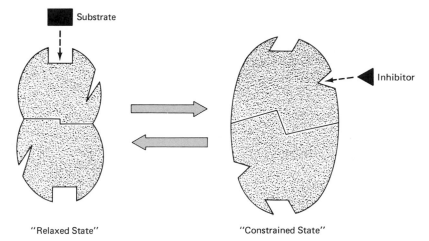

"Relaxed State"   "Constrained State"

Figure 10.29. Model of an allosteric protein. If a protein consists of two identical subunits, it is inferred that an axis of symmetry exists. This model proposes that the protein may alternate between two structural states in which symmetry is preserved. In the "relaxed state," it can bind substrate. In the "constrained state," it can bind inhibitor. Having bound inhibitor, the protein can no longer bind to substrate.

similar fashion, a cluster of ten structural genes, each involved in specifying an enzyme in the biosynthetic pathway of the amino acid histidine, is under the control of a single operator in the bacterium *Salmonella typhimurium*. Inducible and repressible systems can thus be explained by the operon model, differing only in the nature of the regulator gene product (repressor).

At least two kinds of mutations can convert regulated protein production into constitutive synthesis. One group of mutants can alter the regulator gene so that the repressor is abnormal (unable to recognize the operator locus). A second group of mutants can alter the operator locus itself so that a normal repressor is unable to bind to it.

Figure 10.30. Biochemical pathway from chroismic acid to tryptophan.

Substrates and Products

    CDRP = carboxyphenlyaminodeoxyribulosephosphate
    IGP   = indole glycerol phosphate

Enzymes

    AS   = anthranilate synthetase
    PAT   = phosphoribosyl anthranilate transferase
    IGPS = IGP synthetase
    TSA   = tryptophan synthetase (A chain)
    TSB   = tryptophan synthetase (B chain)

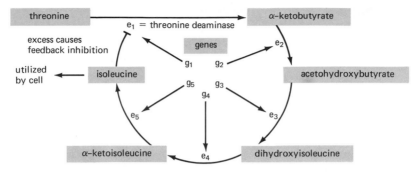

Figure 10.31. An example of feedback inhibition. The end product isoleucine in *E. coli,* when present in high concentration, unites with the first enzyme in its synthetic pathway and thus inhibits the entire pathway until isoleucine returns to normal levels through cellular consumption.

Physiological Controls

Not all regulatory mechanisms operate at the gene level of transcription. Conceptually, regulation can also occur at the translational level or at the level of the protein product. For example, when the concentrations of the amino acid isoleucine accumulate in the cell, the excess isoleucine "feeds back" on one of the first enzymes in the biosynthetic pathway (see Figure 10.31). The union of isoleucine with the enzyme threonine deaminase causes an allosteric transformation that destroys its ability to convert threonine into α-ketobutyrate. The entire pathway then shuts down for lack of this substrate. When isoleucine levels return to normal, threonine deaminase is released from *feedback inhibition,* and the synthetic pathway becomes active again. Feedback inhibition is not nearly as efficient a control mechanism as enzyme induction or repression because enzymes are made constitutively (whether they are needed or not).

Controls in Eukaryotes

Not all enzymes involved in common biochemical pathways are governed by structural genes grouped into operons. However, it would obviously be an advantage to an organism to rearrange its genetic material to bring genes with related biological functions together under the control of a common operator locus. Apparently this has been a general trend in bacterial evolution. Whether or not operons exist in higher organisms remains to be proven.

It is highly probable, however, that many aspects of the differentiation problem will be solved by an operon-type mechanism in eukaryotes. Returning to the *E. coli* lactose operon, we can see how a bacterial cell can exist in alternative, permanently differentiated states. If very low levels of inducer molecules are present in the medium, the internal concentration is not likely to reach levels sufficient to inactivate all the repressor proteins. However, once exposed to high inducer concentrations, the induced cell can be returned to the low levels of lactose and still continue enzyme production. The reason for this is that once permease is induced, it allows the cell to pump in even low levels of lactose with high efficiency. Repressor thus remains bound to lactose, and the operator switch stays open. Transient contact with sufficient inducer causes a physiological change in the functioning of the cell that persists even after the inducer is reduced to levels that formerly had no influence. This phenomenon mimics gene mutation, but obviously is a change in gene activity rather than an alternation of nucleotide sequences in DNA. Lamarckian evolutionary ideas formerly fed on such examples. Now we see the operon system as a possible model for explaining cellular differentiation. As certain gene products become available in early stages of development, they can serve to turn on or off specific genes. This process of sequential gene activity becomes responsible for the orderly unfolding of the physiological and morphogenetic changes that typify embryological development in eukaryotic organisms.

Much evidence now implicates chromosomal proteins in regulatory roles in eukaryotic organisms. *Histones* are chromosomal proteins rich in the basic amino acids arginine and lysine, but completely lacking in tryptophan. Histones increase the stability of the DNA double helix, but at the same time function as repressors of biological activity. Transcription is blocked by histone-bound DNA. *Nonhistone proteins* are a much more heterogeneous group than the histones. Nonhistone chromosomal proteins appear to *derepress* specific genes during certain periods of the cell cycle. One model proposes that a specific nonhistone protein binds to a specific locus on histone-repressed DNA. The nonhistone protein then becomes phosphorylated (negatively charged), forms a strong association with the positively charged histones, and causes the complex to disassociate from the DNA. The "naked" DNA is now open to serve as a template for RNA synthesis.

The Evolution of Dominance

Mutations often tend to destroy the functional activity of their protein products. Hence most mutations behave as recessives to their normal alleles, which produce functional protein products that allow normal (or near normal) development of heterozygotes. Dominance, however, is not an intrinsic property of a gene but subject to modification by other genes as well as external and internal environments. Several theories for the origin of dominance have been proposed. The first of these theories, proposed by Bateson and Punnett, suggested that the recessive condition was caused by the absence of the normal allele. This *presence-absence hypothesis* was abandoned when some deletions were found to have dominant effects. Further evidence against this hypothesis was provided by *reverse mutation* (from abnormal or mutant type to normal or wild type) and by multiple alleles.

Sir Ronald Fisher offered the hypothesis that when a deleterious mutation occurs, the heterozygote would tend to be intermediate in fitness between the two homozygotes. If genes at other loci had pleiotropic effects that could modify the expression of the heterozygote toward that of the homozygous normal (wild-type) condition, the modified heterozygote would gain fitness. Continued selection for such modifying factors would account for the evolution of complete dominance. Several objections to Fisher's "modifying factors theory" have been raised. First, his proposal that a newly arisen mutation leads to a phenotypically intermediate heterozygote begs the question, for this is actually part of the problem to be explained. Another difficulty is the inability of this theory to explain the occasional appearance of mutants that are initially dominant to wild type.

Sewell Wright and J. B. S. Haldane argue that the modifier genes would have major effects of their own, and their frequencies would be much more dependent on the selective advantage associated with their primary function than on their subsidiary pleiotropic effects in enhancing the dominance of some other gene. Thus, it is difficult to envisage how the gratuitous slight advantage of a mutant heterozygote could generate the selection pressure to account for the fixation of those modifiers that enhance the dominance of the normal allele at a given locus. It is well established that selection which alters the genetic background (including "modifiers") can influence the expression of a given gene. An example of the evolution of dominance by selection of modifiers is provided by the "peppered moth" *Biston betalaria* since the industrial revolution in England. Heterozygotes for the melanic (dark color) allele were phenotypically distinct from either homozygote early in this period. After about eighty years of selection favoring the darker forms, the heterozygotes have become almost as dark as the melànic homozygotes. Therefore, some evidence suggests that selection pressure on "modifier genes" may be considerably higher than initially thought possible by Wright and Haldane.

As an alternative to the "modifiers theory," Wright and Haldane have proposed a "physiological theory" of dominance. As pointed out earlier, most mutations tend to be recessive because the products of recessive alleles are commonly less active physiologically (biochemically) than the products of normal (wild-type) alleles. Complete dominance would exist if the heterozygote could produce enough enzyme to allow maximum conversion of substrate. Incomplete dominance would exist if the single normal allele in a heterozygote could not produce enough enzyme for maximal conversion of substrate. The greater the biological activity of a normal allele, the more closely it would approach complete dominance. From what we know of genetic code degeneracy and protein structure, it is easy to visualize how two slightly different alleles could produce slightly different proteins that would carry out the same function. Yet the minor modifications of the amino acid sequences in these two proteins might render them susceptible to differential modifications on different genetic backgrounds or in different environments. Special techniques are usually required to identify these nearly identical alleles, called *isoalleles*.

The wing venation pattern in three populations of wild *Drosophila* is completely normal and identical. Hybrids from crosses between any two of these populations also have normal wings. A recessive mutation ($ci = $ *cubitus interruptus*) causes incomplete development of the cubitus vein (Figure 10.32) when homozygous in all three populations. The normal wild-type alleles in populations 1 and 2 (ci^{+1} and ci^{+2}) behave as complete dominants when in heterozygous condition with the recessive mutant allele (e.g., ci^{+1}/ci). The heterozygote in population 3 (ci^{+3}/ci), however, shows an interrupted cubitus vein and therefore lacks dominance. These data imply that the same wild-type allele exists in populations 1 and 2, but that a different isoallele exists in population 3.

Haldane suggests that dominance may evolve by selection favoring the more efficient

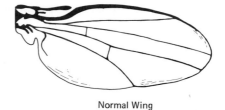

Normal Wing

Figure 10.32. Comparison of a normal wing with a cubitus interruptus wing of *Drosophila melanogaster.*

Cubitus Interruptus

wild-type alleles from a group of isoalleles. Those heterozygotes that contain the more active isoalleles would appear phenotypically closest to wild type and thus be at a selective advantage over those heterozygotes with less-active isoalleles. Selection would tend to establish isoalleles that could provide a safety factor of two in enzyme production so that a single highly active normal allele in a heterozygote could perform the work of two "intermediate" alleles. Being aware of the opportunistic nature of evolution, one can see that it is highly probable that several mechanisms have been employed for the emergence of dominance in different cases and that each theory may hold utilitarian elements.

Evolution of the Genetic Code

The same code appears to function in all modern cells, i.e., the genetic code is *universal.* The evolution of the genetic code is shrouded in mystery, but at least two hypotheses have been entertained. The "frozen accident hypothesis" proposes that the structure of the code evolved by chance. However, once the codon assignments were thus established in a common ancestral cell, they would not be amenable to change in descendants because any mutation that altered codon specificity would produce only missense, and, therefore, usually physiologically inactive proteins. The "steriochemical hypothesis" postulates that, at a very early stage in the evolution of living systems, certain steric relationships existed between triplet codons and the side chains of amino acids they now specify. F. Crick and C. Woese suggest that protein synthesis in primitive systems could have been a far less precise process than we see in operation today. The relatively low degree of functional activity required by proteins of protoorganisms might have allowed considerable substitution of structurally similar amino acids at a given position in the polypeptide chain. It seems highly probable that the codon has always been a triplet, because it would be extremely difficult to align adjacent amino acids on a template formed by a doublet code. Perhaps only the first two nucleotides of the ancient triplet codon were required for recognition of certain classes of amino acids. At some later time, the third position of the codon could have evolved a specific recognition function to eliminate some of the ambiguity. The evolution of specific adaptor molecules (tRNAs) and their specific loading enzymes (aminoacyl-tRNA synthetases) would have also reduced ambiguity. Under the guidance of natural selection relentlessly placing strong pressure on any novelties that reduced ambiguities (enhanced specificities), the structure of proteins became more regular and hence exhibited increased biological activity. Because methionine and tryptophan have only one codon each, they may represent the most recent additions to the code by "invasion" of the AU- and UG- boxes (in Table 10.1), respectively.

Essay Questions

1. *Outline Avery's transformation experiment, including experimental design, results, and interpretation.*

2. *Explain the purpose, design, results, and interpretation of the Meselson and Stahl experiment.*

3. *Diagram the lactose operon of* E. coli *and expound upon its functioning under induced and noninduced conditions.*

4. *How can DNA hybridization experiments estimate the degree of genetic relationships between different organisms?*

5. *What is the significance of gene action for modern evolutionary theory?*

part
III

THE
FORCES OF
CHANGE

THE EQUILIBRIUM POPULATION

The Gene Pool

The *gene pool* of a sexually reproducing population is the sum of all hereditary information resident at any moment in time in the reproductive members of that group (Figure 11.1). The individuals within a population are only temporary repositories of the gene pool, the latter exhibiting continuity from one generation to the next as long as the population exists. All the gametes that could be produced by such a population constitute a theoretical *gametic pool* from which the next generation will arise.

Gene Frequencies

A *gene (allelic) frequency* is that portion of the total number of alleles in the gene pool (or representative sample thereof) at that locus represented by the one under consideration. Because each genotype produced by codominant alleles has a distinctive phenotype, the

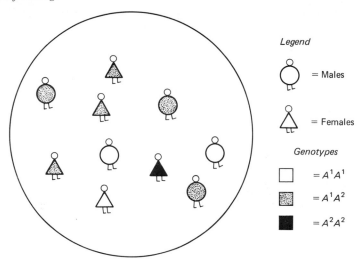

Figure 11.1. Diagram of a population gene pool. Since there are 3 A^1A^1, 6 A^1A^2, and 1 A^2A^2 genotypes, the frequency of the A^1 gene is 0.6, and that of the A^2 allele is 0.4.

number of alleles in homozygous and heterozygous conditions can be ascertained directly from the phenotypes.

> The M-N human blood groups are inherited by a pair of codominant alleles, L^M and L^N. Individuals of group M are homozygous for the L^M allele; those of group N are homozygous for the L^N allele; those of group MN are of heterozygous genotype. Suppose that in a representative sample of 1,000 Caucasians there were 300 of group M, 500 of group MN, and 200 of group N. The frequency of the L^M allele is found by determining the ratio between the number of L^M alleles and the total number of alleles at this locus in the sample. Thus,

$$L^M = \frac{2(300) + 500}{2(1000)} = 0.55.$$

> Similarly,

$$L^N = \frac{2(200) + 500}{2(1000)} = 0.45.$$

> It is therefore estimated that approximately 55 per cent of the gametes in this population carry the L^M gene and 45 per cent carry the L^N gene.

In general terms, the frequency of one allele can be the value p and the gene frequency of the other allele can be the value q, so that $p + q = 1$; that is to say, the sum of the allelic frequencies represents the gene pool unity.

Attributes of entire populations such as mean values and gene frequencies are called *parameters*. These parameters are not known for natural populations, because every member of the population cannot be captured and measured. Instead, a random sample must be used. If the sample is truly representative of the population, the sample measurement should closely approximate the population parameter. All such measurements derived from samples are called *statistics.* Remember that gene frequency calculations derived from samples are only estimates of the parameter. All other considerations being equal, the larger the sample size the more accurately the sample reflects the true gene frequency value for the entire population.

Equilibrium Populations

Many people seem to be confused as to exactly what evolution is. Some statements in modern texts on the subject contribute to this confusion. Take for example the following statement from a currently popular book on evolution: "Evolution results from a change in the hereditary make-up of a species." This implies that evolution is a "thing" rather than a process. In its most basic form, evolution *is* the process whereby changes in the gene pool are affected. Nonscientists are prone to think of evolution only in terms of the appearance of new species or higher categories of organisms. However, much genetic change (evolution) can occur in a population without it splitting into two or more new species. Before embarking on a detailed study of the processes that result in changes in the gene pool, it is essential to grasp thoroughly the concept of a static population, i.e. one that is not changing in its genetic composition from one generation to the next. This concept was developed at the gene pool level by G. H. Hardy (an English mathematician) and independently by W. Weinberg (a German physician; Figure 11.2), both in the year 1908. Their model of a static population has become the cornerstone of the discipline known as *population genetics,* and the formula derived from this model is called the "Hardy-Weinberg Law." This model is one of a hypothetical population conforming to the following assumptions or restrictions:

1. The population is composed of a very large number of sexually reproducing diploid individuals (theoretically infinitely large).
2. Mating is completely at random, including selfing in random amount; this implies that each gamete of one sex has an equal chance of uniting with any gamete of the opposite sex.
3. Both alleles are adaptively neutral, i.e., there is no selection for or against either allele; all genotypes have equal viability and leave progeny directly in proportion to their respective frequencies.
4. The population is closed; no immigration into, nor emigration from the population is allowed.
5. Mutation from one allelic state to another is disallowed.
6. Generation overlap does not exist.
7. All members of the population are of equivalent reproductive age.
8. Meiosis is completely normal so that chance is the only factor operative in gametogenesis and segregation of alleles into functional gametes.
9. Gene frequencies are identical in males and females of the population.
10. Parents make equal contributions to the heredity of their offspring.

As long as a population conforms to these restrictions, neither the allelic frequencies of the gene pool nor the genotypic distribution of the population should change from one generation

Figure 11.2. Wilhelm Weinberg (1862–1937).
[From *Genetics,* 47: Frontispiece (1962)]

to the next. The population is said to be in *genetic equilibrium*. Violation of any of the restrictions on the Hardy-Weinberg model would produce a change in the gene pool (evolution). Since it is virtually impossible for any natural population to conform to all the Hardy-Weinberg restrictions (e.g., no population is infinitely large, completely random mating probably never exists, mutations cannot be prevented, etc.), evolution is almost inevitable.

The remainder of this chapter is devoted to development of the equilibrium equation (Hardy-Weinberg Law) and various applications thereof. The student should always bear in mind the assumptions underlying this law. In subsequent chapters we will be relaxing, one by one, each restriction and investigating the changes that thereby accrue in the gene pool. The forces that contribute to the evolutionary process must be viewed as deviations from the Hardy-Weinberg model. Therefore, this chapter is extremely important because it sets the stage for a thorough understanding of the remainder of this book.

In the simplest case of a single genetic locus with two alleles (A^1 and A^2), *random mating* (*panmixis*) among the three genotypes A^1A^1, A^1A^2, and A^2A^2 is equivalent to the random union of gametes produced by the population. If p and q represent the frequencies of the A^1 and A^2 alleles, respectively, then random union of the gametes of the present generation produces the zygotes of the next generation in the ratio $p^2 (A^1A^1) : 2pq (A^1A^2) : q^2 (A^2A^2)$. This is the Hardy-Weinberg Law (equation).

	♂ Ⓐ	ⓐ
♀	p	q
Ⓐ p	p^2 AA	pq Aa
ⓐ q	pq Aa	q^2 aa

The frequency of the A^1 allele in the gene pool of this generation is

$$p^2 + \tfrac{1}{2}(2pq) = p^2 + pq = p^2 + p(1 - p) = p^2 + p - p^2 = p.$$

Thus, the frequencies of the alleles in the population do not change under random mating, i.e., the population is *not* evolving.

> The serum protein called haptoglobin binds free hemoglobin. In the bound state, hemoglobin is enzymatically degraded. Thus, haptoglobins probably function in the disposal of hemoglobins released by lysis of effete red blood cells. In humans, there are two major electrophoretic variants of haptoglobins produced by two codominant alleles Hp^1 and Hp^2. Let p and q represent the frequencies of the Hp^1 and Hp^2 alleles, respectively. In a Mongoloid population these alleles were estimated to be $p = 0.28$ *and* $q = 0.72$. If this population is in genetic equilibrium the genotypic frequencies should occur in the following ratio:
>
> $$\begin{aligned} p^2 &= (0.28)^2 & &= 0.0784 \text{ of genotype } Hp^1/Hp^1 \\ 2pq &= 2(0.28)(0.72) & &= 0.4032 \text{ of genotype } Hp^1/Hp^2 \\ q^2 &= (0.72)^2 & &= \underline{0.5184} \text{ of genotype } Hp^2/Hp^2 \\ & \text{Total} & &= 1.0000 \end{aligned}$$

For a locus with codominant alleles (as in the previous example), statistical tests can be applied to a population sample to ascertain if it may be considered in genetic equilibrium. These statistical procedures are devised to test the *null hypothesis,* i.e., the assumption that there is no essential difference between the observed numbers of each phenotype and those expected in a sample of the same size according to the Hardy-Weinberg Law. The probability of finding a difference as large as (or greater than) the one actually observed is thereby determined. By convention, most biologists would reject the null hypothesis if this probability was 5 per cent or less. That is, if the difference between the observations and the expectations is so great that 5 per cent or less of an infinite number of samples of the same size from that population would show it by chance alone, the investigator would normally infer that this difference is real (not simply due to chance). Such a large difference is said to be statistically *significant.* Biologists should refrain from using the term *significant* unless they are employing it in a statistical sense, because scientists are prone to think in terms of rejecting the null hypothesis when they hear or see that word. If a statistical test of genetic equilibrium is significant, the evolutionist rejects the null hypothesis and also concludes that one or more of the Hardy-Weinberg restrictions are being violated. In other words, the population is probably not static at this locus. If a human population is sampled for sickle cell anemia (see Chapter 7) and proves to have significantly more heterozygotes than expected according to the Hardy-Weinberg Law, the evolutionist might be led to investigate the factor(s) responsible for the adaptive superiority of that group (relative to the homozygotes). It is not within the scope of this book to present these statistical tests. However, even beginning students of evolution should be aware of the existence of such tests and their general function. A moment's reflection should reveal that statistical tests of goodness-of-fit to the equilibrium expectations can only be applied to loci with codominant alleles or to special cases in which heterozygotes have (by breeding tests) been distinguished from homozygotes.

When an allele exhibits dominance, however, gene frequencies cannot be directly ascertained as in the case of codominant alleles, because the numbers of homozygous dominant and heterozygous genotypes are unknown. The best way to distinguish homozygotes from heterozygotes is by testcrossing the dominant types to recessive types. Testcrossing homozygous dominant genotypes is not expected to produce any progeny with the recessive trait; testcrossing heterozygous genotypes is expected to produce offspring, half of which should have the dominant trait, half the recessive trait. However, testcrossing is not usually feasible in the study of natural populations. If we assume that a population is already in equilibrium, we can estimate the gene frequency of the recessive allele from the square root of the frequency of the recessive trait.

> The ability of people to taste a chemical called phenylthiocarbamide (PTC) is dependent on a dominant gene (T); nontasters are homozygous for the recessive allele (tt). Approximately 70 per cent of the people in the United States can detect PTC. Let p and q be the frequencies of the alleles T and t, respectively. Under the assumption that the population is in equilibrium at this locus, we estimate from $q^2 = 0.3$ that $q = 0.55$. Of the 70 per cent tasters we expect $2pq$ to be heterozygous. Thus, $[2(0.45)(0.55)]/0.7$ is approximately $5/7$; i.e., 5 out of 7 tasters are probably heterozygous.

There is a tendency for the neophyte to assume that a trait governed by a dominant gene should be the most frequent character in the population. A knowledge of population genetics, however, makes it clear that the phenotypic frequencies are functions of the allelic frequencies in the gene pool.

> The presence of horns in cattle is specified by a recessive genotype. Before humans began artificial selection for the dominant hornless (polled) condition in domestic stock, almost all cattle were horned. The presence of horns in range and *feral* (wild) cattle probably is of great value in defense against wolves or other carniverous predators and is therefore subject to natural selection.

When a recessive allele is rare in the population, it is almost invariably found in heterozygotes, not in homozygotes. The phenotypes of such carriers that are presented to the selective forces of the environment are therefore those of the dominant wild (majority) types. If the dominant phenotype is adaptively superior to the recessive phenotype, selection will be very ineffective in removing the deleterious recessive allele from the population. Only in one quarter of the progeny from relatively infrequent matings between carrier parents would the recessive phenotype come into the presence of the selective forces of the environment. When the frequency of one allele is very low in the gene pool, the proportion of heterozygotes is approximately twice the frequency of the recessive allele.

A recessive genetic disease in humans called phenylketonuria occurs in approximately 1 out of 40,000 people. Under the assumption of equilibrium, the frequency of the recessive allele $q = \sqrt{1/40{,}000} = 1/200$. The frequency of heterozygotes is expected to be $2pq = 2(199/200)(1/200) = 398/40{,}000$, which is approximately $1/100$ or twice the frequency of the rare allele.

Two populations that have identical gene frequencies and are in genetic equilibrium will also have identical genotype frequencies. Populations not in equilibrium could have identical gene frequencies (though it is not likely) and yet have unequal genotypic frequencies. After one generation of random mating under Hardy-Weinberg conditions, however, both populations would change their genotypic frequencies to those of the equilibrium population and thereafter remain identical as long as random mating continued.

Examples of Populations with Equal Gene Frequencies, but Unequal Genotype Frequencies

	Allelic Frequencies		Genotypic Frequencies		
	D	*d*	*DD*	*Dd*	*dd*
Population I	0.8	0.2	0.80	0	0.20
Population II	0.8	0.2	0.60	0.40	0
Equilibrium	0.8	0.2	0.64	0.32	0.04

Hardy-Weinberg equilibrium is not a stable equilibrium. Should evolutionary forces disturb the gene frequencies, and the population subsequently return to the Hardy-Weinberg conditions, the new equilibrium established would be dictated by the new gene frequencies, i.e., the population would not return to the previous equilibrium state.

Multiple Alleles

Equilibrium at a genetic locus can be established in any population conforming to Hardy-Weinberg conditions by one generation of random mating regardless of the number of alleles at that locus.

The three electrophoretic variants of the enzyme acid phosphatase in humans are controlled by three major codominant alleles. A sample of 100 Caucasians was found to have the following genotypes: $14 P^a P^a$, $36\ P^b P^b$, $42\ P^a P^b$, $2\ P^a P^c$, and $6\ P^b P^c$ (no $P^c P^c$ genotypes appeared in this limited sample, although they have appeared in larger samples). If we let x, y, and z represent the frequencies of the alleles P^a, P^b, and P^c, respectively, the value for x is calculated as follows. We see that all gametes of the 14 per cent $P^a P^a$ individuals will bear only the allele P^a; half of the gametes of the 42 per cent $P^a P^b$ individuals will bear P^a (therefore 21 per cent of the population's gametes should carry P^a from this source); and half of the gametes of the 2 per cent $P^a P^c$ individuals will bear P^a (therefore 1 per cent of the population's gametes should carry

P^a from this source). In summary,

$$x = (14 + 21 + 1)/100 = 0.36$$

Similarly,
$$y = (36 + 21 + 3)/100 = 0.60$$
$$z = (0 + 1 + 3)/100 \quad = \underline{0.04}$$
$$\text{Total} \qquad\qquad = 1.00$$

The genotypic expectations after one generation of random mating are found from the expansion of $(x + y + z)^2$, which yields x^2 of $P^a P^a$, y^2 of $P^b P^b$, z^2 of $P^c P^c$, $2xy$ of $P^a P^b$, $2xz$ of $P^a P^c$, and $2yz$ of $P^b P^c$. When the observed allelic frequencies are substituted in the equilibrium equation, we expect to find in a population of 100 individuals: $100x^2 = 100(0.36)^2 = 12.96$ or approximately 13 $P^a P^a$, $100(2xy) = 100(2)(0.36)(0.60) = 43.2$ or approximately 43 $P^a P^b$. Similarly we expect approximately 36 $P^b P^b$, 3 $P^a P^c$, 5 $P^b P^c$, but less than 1 $P^c P^c$. The observed frequencies are in close agreement with these expected values. Hence, this population is probably already in genetic equilibrium at this locus and neither the allelic frequencies nor the genotypic frequencies are expected to change in subsequent generations as long as the Hardy-Weinberg conditions are met.

When there are only two alleles at a genetic locus, no more than 50 per cent of the population can be heterozygous under equilibrium conditions. However, the total percentage of heterozygotes can be much greater than 50 per cent when multiple alleles exist.

Let x, y, and z be the frequencies of three alleles; let H be the total proportion of heterozygotes. Then in an equilibrium population $H = 2xy + 2xz + 2yz$. The maximum value for H occurs when $x = y = z = \frac{1}{3}$. In this case, $H = 3(2)(\frac{1}{3})^2 = \frac{2}{3}$.

Sex-Associated Conditions

Autosomal Disequilibrium in the Sexes

When autosomal allelic frequencies are dissimilar in the sexes, they become equilibrated after one generation of random mating, but the genotypic frequencies will not become equilibrated until the second generation of random mating.

Let the frequencies of a pair of autosomal alleles (A,a) be represented by p_m and q_m in males, and by p_f and q_f in females, respectively. Given $q_f = 0.6$ and $q_m = 0.2$, the first generation of panmixis yields

♂ \ ♀	(A) 0.4	(a) 0.6
(A) 0.8	AA 0.32	Aa 0.48
(a) 0.2	Aa 0.08	aa 0.12

from which it can be seen that q has become equilibrated in both sexes at

$$\tfrac{1}{2}(0.48 + 0.08) + 0.12 = 0.40.$$

The equilibrium gene frequency of the a allele, represented by the symbol $\hat{q}$ (pronounced q "hat"), is simply the average of the frequencies in the two sexes, i.e.,

$$\hat{q} = \tfrac{1}{2}(0.6 + 0.2) = 0.4.$$

So in the next generation we have

♂ ╲ ♀	Ⓐ 0.6	ⓐ 0.4
Ⓐ 0.6	AA 0.36	Aa 0.24
ⓐ 0.4	Aa 0.24	aa 0.16

in which the genotypic frequencies have become equilibrated with 36 per cent *AA*, 48 per cent *Aa*, and 16 per cent *aa* genotypes. Full equilibrium has been achieved after two generations of random mating.

Equilibrium for Sex-Linked Genes

If the frequencies of sex-linked alleles on the X chromosomes are unequal in the sexes, the equilibrium values are approached rapidly during successive generations of random mating in a peculiar fashion. The allelic frequencies approach a constant average or equilibrium gene frequency in an oscillatory manner. This phenomenon derives from the fact that females (XX) carry twice as many sex-linked alleles as do males (XY). Females receive their sex-linked heredity equally from both parents, but males receive their sex-linked heredity only from their mothers. The difference between the allelic frequencies in males and females is halved in each generation under random mating. Within each sex, the deviation from equilibrium is halved in each generation, with sign reversed.

Let p_f and p_m be the frequencies of the allele A in females and males, respectively; let $\bar{p}$ be the average allelic frequency in both sexes. Because $\tfrac{2}{3}$ of the sex-linked genes in the population are carried by females $\bar{p} = \tfrac{2}{3}p_f + \tfrac{1}{3}p_m$. Since males get their sex-linked genes only from their mothers, the allelic frequency in males (p_m) is equivalent to the frequency of that allele in females of the previous generation (p'_f). However, the allelic frequency for females (p_f) is the average of the frequencies in males and females of the previous generation (p'_m and p'_f), because females obtain half their sex-linked genes from each parent. Thus, $p_f = \tfrac{1}{2}(p'_m + p'_f)$. The difference between the allelic frequencies in males and females is reduced by half of what existed during the previous generation. Thus,

$$(p_f - p_m) = \tfrac{1}{2}(p'_m + p'_f) - p'_f = -\tfrac{1}{2}(p'_f - p'_m).$$

If we begin with a hypothetical population in which all females are homozygous for the

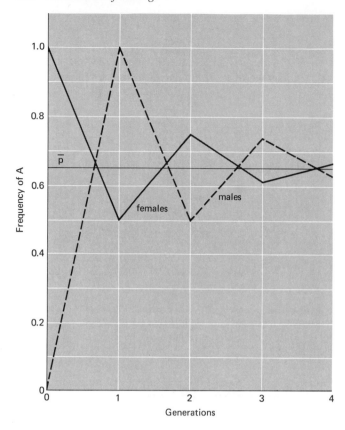

Figure 11.3. Approach to equilibrium at a sex-linked locus in a population initially at maximal disequilibrium.

dominant allele and all males are hemizygous for the recessive allele, $p_f = 1$ and $p_m = 0$. The plot in Figure 11.3 shows the oscillatory manner in which equilibrium is approached by both sexes during the first four generations of random mating.

Generation	p_f	p_m	d_f	d_m
0	1.000	0.000	+0.333	−0.667
1	0.500	1.000	−0.167	+0.333
2	0.750	0.500	+0.083	−0.167
3	0.625	0.750	−0.042	+0.083
4	0.687	0.625	+0.021	−0.042

Notice that the average gene frequency ($\bar{q}$) remains constant through all generations at its initial value

$$\bar{p} = (\tfrac{2}{3})(1) + (\tfrac{1}{3})(0) = \tfrac{2}{3} = 0.666.$$

Also note that the deviation of the gene frequency in each sex from the equilibrium frequency (represented by $d_x = p_x - \bar{p}$) is halved in each generation with sign reversed.

Because males are hemizygous for sex-linked genes, recessive characters are expected to be far more frequent in males than in females. If the frequency of a sex-linked recessive allele is q, then the recessive trait appears in males with the frequency q, but in females with the frequency q^2.

> Color blindness is a sex-linked recessive trait that occurs in Caucasian males with a frequency of approximately 8 per cent. If we let q be the frequency of the colorblind allele, then $q^2 = (0.08)^2 = 0.0064$ or 0.64 per cent of the females are expected to be colorblind.

By similar reasoning, dominant sex-linked traits will always be more frequent in females than in males irrespective of the gene frequencies.

> The human blood cell antigen $Xg(a+)$ is determined by a dominant sex-linked gene (Xg^a). Its frequency in a Negroid population was estimated to be 0.55 (the other 45 per cent of the gene pool being the recessive allele Xg). Letting $p =$ frequency of the Xg^a allele and $q =$ frequency of the Xg allele, females with the $Xg(a+)$ blood group are expected to comprise
>
> $$p^2 + 2pq = (0.55)^2 + 2(0.55)(0.45) = 0.7975$$
>
> or approximately 80 per cent of all female phenotypes. Among males, the frequency of the $Xg(a+)$ blood group is expected to be the same as the frequency of the Xg^a allele, i.e., 55 per cent.

When the dominant gene is rare, females displaying the dominant trait will be approximately twice as numerous as males with the dominant trait.

> Letting $p =$ frequency of a dominant sex-linked gene and $q =$ frequency of its recessive allele, the ratio of dominant female phenotypes to dominant male phenotypes is expressed as
>
> $$p^2 + 2pq:p = p(p + 2q):p = 1 - q + 2q:1 = 1 + q:1.$$
>
> Now if p is rare in the population, q is approximately 1. In this case, the ratio of dominant female types to dominant male types is about $2:1$.

Multilocus Equilibrium

Unlinked Genes

Though the alleles at a single locus reach equilibrium following one generation of random mating, gametic equilibrium involving two independently assorting genes is approached rapidly over a number of generations. Consider one locus with alleles A and a whose frequencies are represented by p and q, respectively, and a second locus with alleles B and b whose frequencies are represented by r and s, respectively. The gametic frequencies at equilibrium can be predicted as the product of the frequencies of the individual alleles in each gamete; thus,

Gametes: (AB) (ab) (Ab) (aB)

Frequencies: pr qs ps qr

Also at equilibrium,

$$(pr)(qs) = (ps)(qr).$$

Consider an initial population with the following genotypic frequencies:

Genotypes	Genotype Frequency	Gametic Frequency			
		AB	ab	Ab	aB
AABB	2	2			
AABb	6	3		3	
AAbb	10			10	
AaBB	18	9			9
AaBb	40	10	10	10	10
Aabb	10		5	5	
aaBB	8				8
aaBb	4		2		2
aabb	2		2		
Totals	100	24	19	28	29

Let pr = frequency of AB gamete, qs = frequency of ab gamete, ps = frequency of Ab gamete, and qr = frequency of aB gamete. This population is not in gametic equilibrium because $(pr)(qs) \neq (ps)(qr)$; i.e., $(0.24 \times 0.19) \neq (0.28 \times 0.29)$. The allelic frequencies now are

for A $p = 0.24 + 0.28 = 0.52$
for a $q = 0.19 + 0.29 = 0.48$
for B $r = 0.24 + 0.29 = 0.53$
for b $s = 0.19 + 0.28 = 0.47$

Therefore, at equilibrium we expect the gametic frequencies to be

for AB $pr = (0.52)(0.53) = 0.2756$
for ab $qs = (0.48)(0.47) = 0.2256$
for Ab $ps = (0.52)(0.47) = 0.2444$
for aB $qr = (0.48)(0.53) = 0.2544$

At equilibrium, the difference

$$d = (pr)(qs) - (ps)(qr) = 0.$$

For independently assorting loci under random mating, the value of d is halved in each generation during the approach to equilibrium, because unlinked genes experience 50 per cent recombination. The maximum value for the absolute value of d is 0.25. This occurs when the population is in either of two conditions:

1. Half the population is *AABB* and the other half is *aabb*, giving a gametic pool in which half the gametes are *AB* and half are *ab*.
2. Half the population is *aaBB* and the other half is *AAbb*, giving a gametic pool in which half the gametes are *aB* and half are *Ab*.

There is a singular exception to the rule that genetic equilibrium at two loci is attained only after a number of generations of random mating. A population that consists initially of only dihybrid genotypes (*AaBb*) should reach genetic equilibrium after a single generation of random mating.

Consider a laboratory population consisting entirely of dihybrid *AaBb* genotypes. These individuals produce four kinds of gametes with equal frequencies: $\frac{1}{4}$ *AB*, $\frac{1}{4}$ *Ab*, $\frac{1}{4}$ *aB*, $\frac{1}{4}$ *ab*, represented by *w, x, y,* and *z*; respectively. If *d* is the departure from equilibrium, then

$$d = wz - xy = (\tfrac{1}{4})(\tfrac{1}{4}) - (\tfrac{1}{4})(\tfrac{1}{4}) = 0.$$

Hence, gametic equilibrium already exists in this initial population. One generation of random mating will produce nine genotypes in the classical F_2 ratio. From then on, the population will be in both gametic and zygotic equilibrium as long as random mating persists.

Linked Genes

The approach to equilibrium for linked genes is slowed because they recombine less frequently than unlinked genes (i.e., less than 50 per cent recombination). The closer the linkage the longer it takes to reach equilibrium. Gametes in repulsion linkage (*Ab* or *aB*) are expected to be equally frequent with those in coupling linkage (*AB* or *ab*) at equilibrium. Therefore, correlations between different phenotypes in an equilibrium population cannot be due to linkage, but rather might be due to the pleiotropic effects of a single locus.

The disequilibrium (d_t) that exists at any generation (*t*) is expressed as

$$d_t = (1 - r)\, d_{t-1},$$

where $r =$ frequency of recombination and $d_{t-1} =$ disequilibrium in the previous generation. If $d = 0.25$ initially and the two loci experience 20 per cent recombination (i.e., the loci are 20 map units apart), the disequilibrium that would still exist after one generation of random mating is

$$d_t = (1 - 0.2)(0.25) = 0.20.$$

This represents $0.20/0.25 = 0.8$ or 80 per cent of the maximum disequilibrium that could exist in the population. The two plots shown in Figure 11.4 are for unlinked genes ($r = 0.5$) and for 20 per cent linkage ($r = 0.2$). Note the slower approach to equilibrium ($d = 0$) as the recombination percentage decreases.

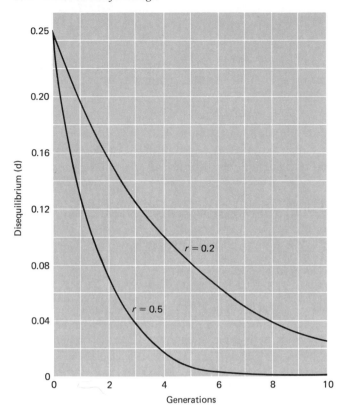

Figure 11.4. Approach to equilibrium at two autosomal loci in two populations (initially at maximal disequilibrium). In one population, the frequency of recombination is 0.2 for two *linked* loci. In the other population, the frequency of recombination is maximal at 0.5 for two *unlinked* loci (independent assortment).

Essay Questions

1. *Why is the equilibrium restriction of random mating seldom, if ever, met in natural populations?*

2. *Why are not dominant alleles always more frequent in the population than recessive alleles?*

3. *If a population is known to be in genetic disequilibrium and associations are observed between different traits, to what causes might these correlations be attributed?*

4. *Hemophilia is a rare recessive sex-linked trait in humans, causing excessive bleeding to occur. In which sex would this trait be most prevalent? Why?*

5. *In reviewing the conditions required for production of the equilibrium state, is it likely that any population is static?*

MUTATION

The term *mutation* has three common meanings: (1) it is the process whereby the structure of DNA is altered, (2) it is a gene modified by this process, and (3) it is an individual manifesting the modified gene. The intended meaning is usually clear from the context in which the word is used. Mutations serve as the ultimate source of new genetic variability on which the long-term process of evolution feeds. Mutations tend to replenish the supply of genetic variability that is constantly being eroded through the elimination of unfavorable variants by natural selection. However, much evolution can occur without new mutations. The immediate source of genetic variability in outcrossing populations resides in the genetic structure of the population and in its recombinational mechanisms. Most sexually reproducing natural populations are highly heterogeneous in their genetic composition, and usually the major limitation on evolution is the facility with which the population can recombine the genetic diversity already present in the gene pool into new adaptive combinations. The origin of the genetic diversity in a gene pool, however, must ultimately be attributed to previous mutations.

Origin of Mutations

All mutations in nature are classified as *spontaneous mutations,* i.e., the specific causal factor(s) is unknown. H. J. Muller (Figure 12.1) and L. J. Stadler (Figure 12.2) independently discovered in 1927 that artificial treatment with X rays can induce mutations at frequencies significantly above the spontaneous level (Figure 12.3). *Induced mutations* are those produced by humans through the use of *mutagens,* i.e., agents that enhance the frequency of mutations over the "background" or "noise" provided by spontaneous mutations. Very high-energy cosmic rays from space are similar to X rays and gamma rays in their mutagenic action. It was initially thought that cosmic rays and decay of radioactive substances in the earth's rocks might be responsible for spontaneous mutations. However, it is now known that these natural sources of radiation play a very minor role in the spontaneous mutation rate. Muller estimated that the spontaneous mutation rate in *Drosophila* was about 1300 times greater than that expected from background radiation sources. It is now believed that spontaneous mutation rates must be explained by mistakes during DNA replication as well as other accidents of nuclear and cytoplasmic metabolism.

If the chromosome serves as a "target" that can be broken by a single "hit" of radiation, the number of single chromosome breaks should be directly proportional to the radiation dosage. The probability of a double hit in the same chromosome would be very small at low levels of radiation, but should increase exponentially with increased dosage of radiation. Investigation of terminal deletions (single hits) and interstitial deletions (double hits), have yielded results congruent with the "target theory." Figure 12.4 shows a linear relationship

Figure 12.1. Hermann Joseph Muller (1890–1967); Nobel Prize laureate 1946. [From *Genetics,* 70: Frontispiece (1972)]

Figure 12.2. Lewis John Stadler (1896–1954). [From *Genetics,* **41**: Frontispiece (1956)]

between radiation dosage and the frequency of terminal deletions; it also shows a curvilinear relationship between dosage and interstitial deletions. The target theory can also be applied at the level of the gene, where a single hit could cause a nucleotide change without necessarily breaking the chromosome.

These high-energy rays collide with atoms, release electrons, and thereby create charged atoms called *ions;* hence, the name *ionizing radiations.* The ionic state is one of high chemical reactivity. It is possible for many ions to be produced along the track of each high-energy ray. In most cases, the same number of mutations results from a given dosage of radiation regardless of whether it is administered all at once or in small amounts over a period of time. An exception to this generalized rule is found in chromosomal aberrations requiring two breaks. If both breaks do not occur within a relatively short time of each other, repair mechanisms begin to heal the earlier break. Radiation may have quite different effects in various tissues. For example, rapidly dividing tissues are highly prone to radiation damage. This knowledge has been used in inhibiting the growth of cancer cells by radiation therapy. The rate of induced chromosomal aberrations in *Drosophila* spermatozoa appears to be independent of radiation intensity, implying that DNA repair mechanisms are inoperative in this highly differentiated cell.

Spermatids usually contain more oxygen than spermatozoa and also exhibit a higher

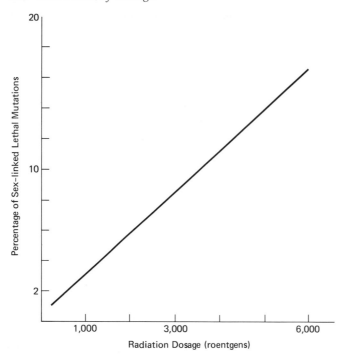

Figure 12.3. Induction of sex-linked lethal mutations in *Drosophila* by various dosages of high-energy radiation.

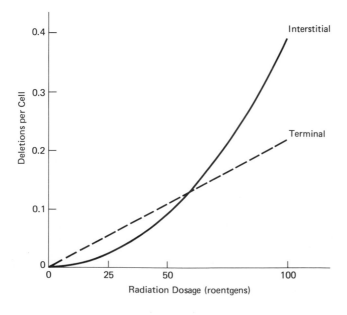

Figure 12.4. Relationship between single-break mutations (terminal deletions) and double-break mutations (interstitial deletions) vs. radiation dosage.

susceptibility to ionizing radiations. Experimental evidence supports the hypothesis that the concentration of oxygen in the cell enhances the mutagenic activity of ionizing radiations. Irradiation of water molecules within the cell creates hydrogen and hydroxyl radicals. These hydrogens can combine with oxygen (if present) to form hydrogen peroxide (H_2O_2). Peroxides are highly reactive chemicals and may be an important source of mutation. Thus, the target theory is inadequate as the sole explanation for radiation effects.

Other known mutagenic factors include extreme temperature shocks and the effects of aging. Nonionizing radiations, such as ultraviolet light (UV), are also known to be mutagenic. DNA absorbs the 2540 Å wavelength maximally, and it is at this frequency that UV light exhibits its greatest mutagenic activity. These wavelengths do not have sufficient energy to produce ions. Instead, they primarily cause covalent linkages between two adjacent thymine bases on the same or on different strands of a DNA molecule. The *thymine dimer* (Figure 12.5) thus produced causes a local distortion of DNA structure and tends to interfere with replication of that region and with transcription of the gene of which it is a part.

The first molecular mechanism for explaining mutations was provided by Watson and Crick in their 1953 paper on the structure of DNA. In an aqueous medium, the deoxyribonucleotide bases exist in an equilibrium state between various possible *tautomeric* forms dependent on where a particular hydrogen atom is attached. For example, almost all adenine nucleotides would normally bear an *amino* (NH_2) group providing a hydrogen atom for bonding to the complementary *keto* (C=O) group of thymine. Very few adenine nucleotides would exist in the rare *imino* (NH) form as a result of a *tautomeric shift* (Figure 12.6) of a hydrogen from the amino group to a nitrogen elsewhere in the same molecule. The imino form of adenine, however, can illegitimately form two hydrogen bonds with cytosine. Similarly the rare *enol* (COH) tautomeric form of thymine can illegitimately form a hydrogen bond with guanine. Thus, the newly synthesized strand of DNA would have replaced one

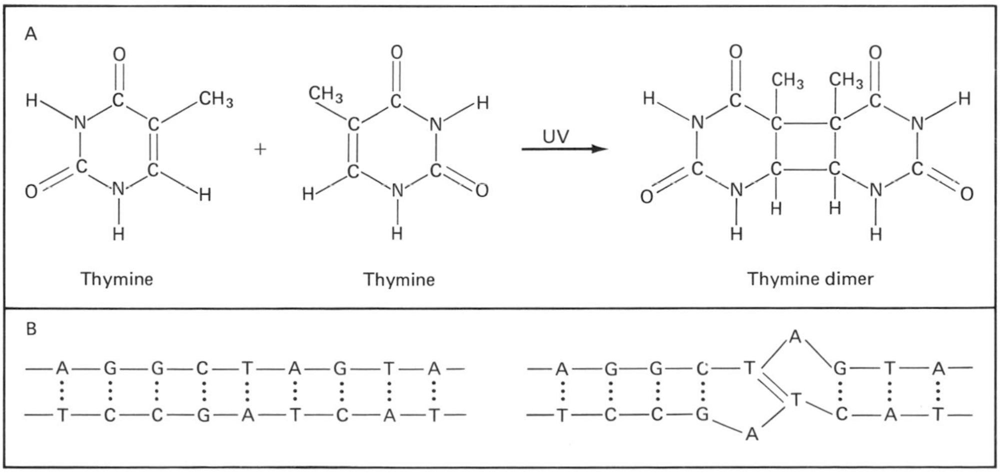

Figure 12.5. (A) Formation of a thymine dimer by ultraviolet radiation; (B) diagram of the distortion a thymine dimer induces by cross-linking bases in different strands of a DNA molecule. Solid lines represent strong covalent bonds; dotted lines indicate weak hydrogen bonds.

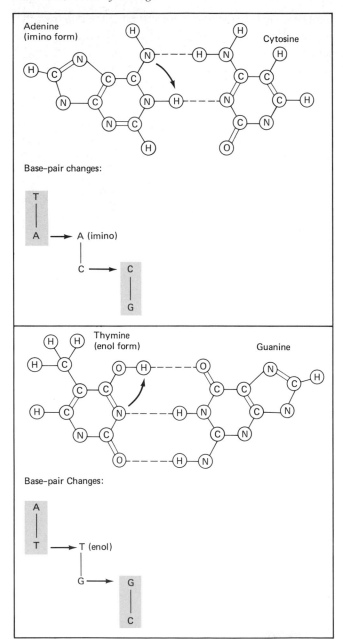

Figure 12.6. Tautomeric shifts of hydrogens (arrows) allow illegitimate base pairings. Replication of the mutant DNA strands produces transitions.

purine by another or one pyrimidine by the other; these kinds of mutations are called *transitions*. Several well-known chemical mutagens cause transition mutations. The base analog 2-aminopurine (Figure 12.7) normally pairs with thymine, but can form one hydrogen bond with cytosine even in its normal (amino) tautomeric state and can form two bonds with cytosine when in its rarer imino form. Similarly, the base analog 5-bromouracil, in its normal keto state, replaces thymine through base pairing with adenine, but can pair with guanine when in its rarer enol form (Figure 12.8). Bromouracil tends to make this error much more

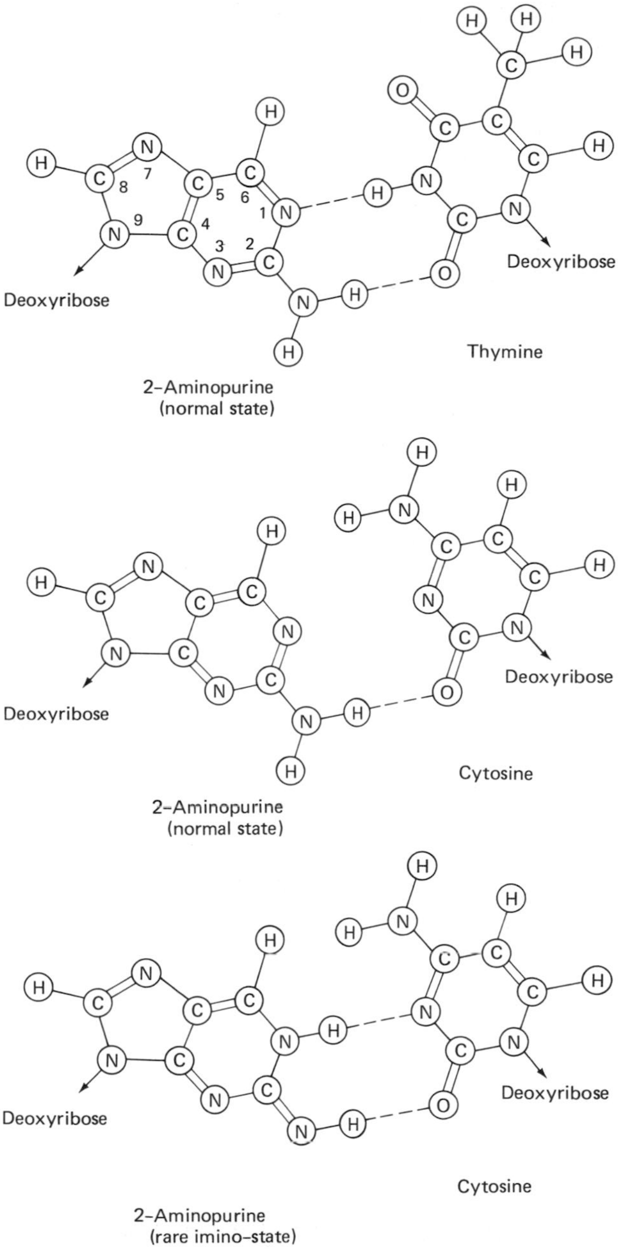

Figure 12.7. Pairing possibilities of the mutagenic base analog 2-aminopurine.

2–Aminopurine
(normal state)

2–Aminopurine
(normal state)

2–Aminopurine
(rare imino–state)

frequently than thymine does. It has been suggested that certain genes may be responsible for synthesis of mutagenic base analogs and thereby contribute to the spontaneous mutation rate. Viruses have also been shown to be mutagenic agents in bacteria. Therefore, we may speak of *mutator genes* of either endogenous or exogenous origin, which enhance mutational events at other loci. Transitions represent one major class of *substitutional mutations,* i.e., exchange of nucleotide bases.

Any kind of mutation that requires replication of DNA for tautomeric shifts to cause a

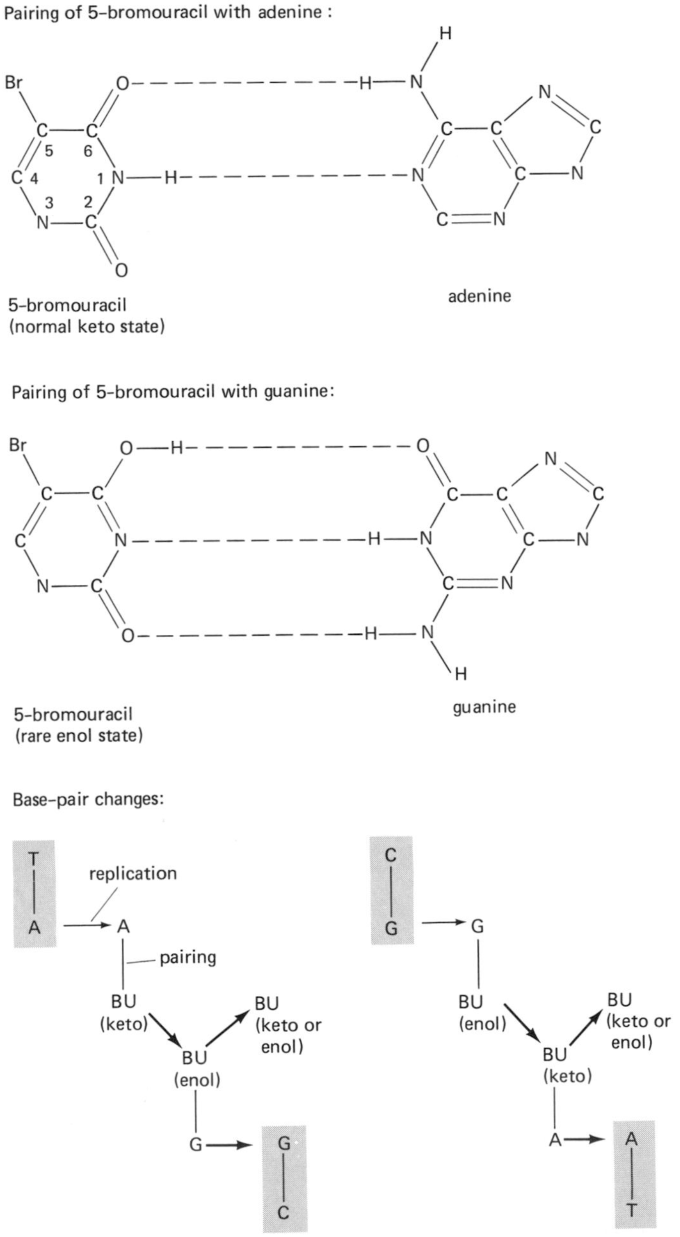

Figure 12.8. The base analog 5-bromouracil pairs with adenine in its normal keto state but can pair with guanine in its rarer enol state.

base substitution is classified as a *copy error*. A second major type of copy error is the *transversion mutations* that substitute a purine base for a pyrimidine base or *vice versa*. When corresponding breaks are induced by radiation between a pair of DNA bases and their sugars, the free doublet (a purine, hydrogen bonded to a pyrimidine) can rotate 180° and become reattached to sugars on the complementary DNA chain. The result of this *rotational base substitution* is classified as a transversion mutation (Figure 12.9).

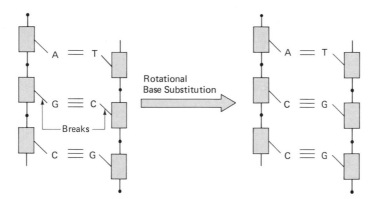

Figure 12.9. Transversion mutation created by a rotational base substitution.

Nitrous acid (Figure 12.10) can cause deamination of cytosine, converting it into uracil *in situ,* i.e., this chemical can cause direct nucleotide changes in nonreplicating DNA. Upon replication, however, the mutated uracil would be expected to base pair with adenine, so that a transition occurs in which the original G-C pair is replaced by an A-T pair.

Acridine dyes such as proflavin and acridine orange belong to a completely different class of mutagens. These chemicals appear to intercalate themselves between the base pairs of DNA and so distort the molecule that upon replication they induce base additions or deletions

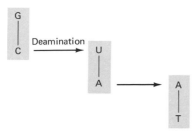

Figure 12.10. Effect of nitrous acid deamination of cytosine causes a transition mutation.

(Figure 12.11). These are called *frame shift mutations,* because the reading frame during translation is shifted out of register. Most of the new codons created by a frame shift mutation are likely to be *missense* codons, i.e., codons that bring the wrong amino acids into the polypeptide chain. These proteins commonly do not function properly, and a mutant phenotype is produced. It is also possible for a frameshift mutation to create a nonsense codon. This would prematurely terminate polypeptide chain growth and would almost certainly produce a mutant phenotype. All changes in DNA structure involving substitution, addition, or deletion of single bases or base pairs are classified as *point mutations.*

Sometimes the substitution of one base for another will occur in a region of DNA that specifies a noncritical amino acid in the protein product; i.e., the amino acid is not part of the catalytically active site of an enzyme, nor is it in a vital position for proper folding of the chain into its normal tertiary shape. Such mutations need not cause mutant phenotypes and would tend to accumulate in the population as part of its genetic variability. Referred to as *silent mutations,* they are one explanation for the origin of isoalleles. Because of degeneracy in the genetic code, it is also possible for a mutation to occur in the third position of a codon without changing the amino acid specificity of the normal and mutant codons.

Forward mutation is a change from the normal nucleotide sequence to the abnormal sequence. *Back* (or *reverse*) *mutation* is a change from abnormal nucleotide sequence to the original normal order. Reverse mutation in a given cistron is commonly at least 10^2 to 10^3 times less frequent than forward mutation. The reason for this is obvious from the structure of the cistron. Change at any one of the hundreds or thousands of nucleotide bases in a normal cistron could potentially destroy the biological activity of its protein product. To put the mutant cistron back into working order, however, usually requires a reverse mutation of precisely the same nucleotide involved in the forward mutational event. Forward mutation rates vary considerably for different traits. There are two major classes of forward mutations: (1) loss of function mutations and (2) change of function mutations.

Loss of function mutations are exemplified by the bacterial changes from lactose fermenting (Lac^+) to lactose nonfermenting (Lac^-) or from prototrophic tryptophan synthesizing (Trp^+) to auxotrophic tryptophan nonsynthesizing (Trp^-). These loci have relatively high mutation rates of about 2×10^{-6} per cell per generation. The reason given above in terms of many potentially mutable sites per cistron is only part of the explanation. Because the synthesis of tryptophan requires many biochemical steps, each step catalyzed by a different

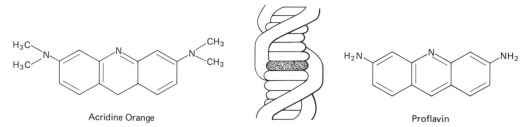

Figure 12.11. Intercalation of acridine dye mutagens (acridine orange, proflavin) into DNA double helix, causing insertion or deletion of nucleotides in replicating DNA chains.

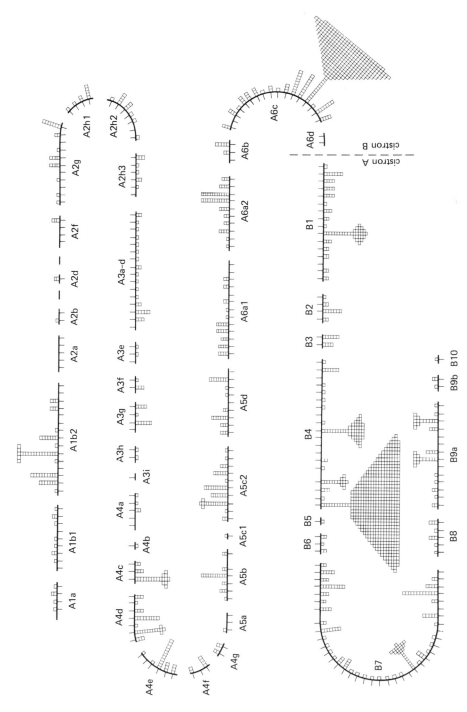

Figure 12.12. Variability in spontaneous mutations of the *rII* region of bacteriophage T4. Each square represents a single observed point mutation. Note that certain loci are more highly mutable than others ("hot spots"). Segments of this map (identified as A1b2, B7, etc.) represent deletions that aided in the analysis of this region. (After Benzer)

317

enzyme, mutations in any of several different cistrons could interfere with tryptophan synthesis. All such mutations would score as Trp⁻ auxotropic phenotypes.

Change of function mutations characteristically have much lower mutation rates than loss of function mutations. For example, the mutation rate in *E. coli* from streptomycin sensitivity to streptomycin resistance (Str^s → Str^r) is very low (about 10^{-10} per cell generation). Streptomycin causes massive misreading of the genetic code in Str^s bacterial cells by binding to normal ribosomes and distorting the protein-synthesizing machinery. The rarity of Str^r mutations is explained by the necessity for maintaining the indispensable function of the ribosomal structure in protein synthesis and yet rendering it immune to the presence of streptomycin. This requires a very special kind of change in probably an extremely limited number of mutons. A similar explanation can be evoked for the rarity of mutants that change the sensitivity of an enzyme to feedback inhibition without destroying the catalytic function of that enzyme.

Regardless of the classifications outlined above, each muton is characterized by a particular mutation rate (either spontaneous or induced) under well-defined environmental (or treatment) conditions. Forward and back mutation rates are commonly of the same order of magnitude for a muton that has incurred single base switches under identical conditions. Exhaustive analysis of the *r*II region of phage T4 has revealed that certain mutons are much more mutable than others ("hot spots"; Figure 12.12). It is now believed that highly mutable nucleotide pairs are correlated with mispairing during crossing over. The rate of mispairing is a function of the number of repeating bases in a given region; the higher the number of repeating bases the higher the frequency of mispairing and consequently of mutation. Thus, the structure of the DNA itself predisposes various loci to characteristic mutation rates.

Suppressor Mutations

Suppressor mutations form another class in which the mutant phenotype caused by one point mutation can be corrected (at least in part) by a second mutation at some other site. For example, a base addition (induced by treatment with acridine dyes) may be "suppressed" by a deletion nearby in the same cistron. The reading frame would be shifted out of register for the nucleotide sequence between the two mutations, and this would usually be reflected as a change in a portion of the amino acid sequence of the corresponding protein (Figure 12.13). If this is only a small segment in a noncritical region, the biological activity of the protein may be normal or nearly so. The first experimental confirmation of the triplet code hypothesis came from deletion and addition mutants. It was found that a sequence of three additions or three deletions could place the reading frame back into register (Figure 12.14). Mutations of this type are classified as *direct intragenic suppressors,* because they directly repair the polypeptide product of a single cistron.

A second type is the *direct intergenic suppressor,* which directly repairs cistronic missense mutations, but it does so by mutation in a different cistron specifying an abnormal tRNA. If the suppressor mutation altered the anticodon of a glycine-loaded tRNA so that it pairs with the mRNA codon for arginine, it would be possible for a missense arginine codon to be

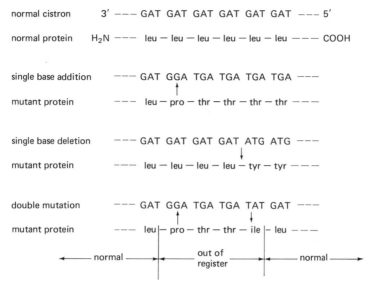

```
normal cistron        3' ——— GAT GAT GAT GAT GAT GAT ——— 5'

normal protein        H₂N ——— leu — leu — leu — leu — leu — leu ——— COOH


single base addition      ——— GAT GGA TGA TGA TGA TGA ———
                                   ↑
mutant protein            ——— leu — pro — thr — thr — thr — thr ———


single base deletion      ——— GAT GAT GAT GAT ATG ATG ———
                                           ↓
mutant protein            ——— leu — leu — leu — leu — tyr — tyr ———


double mutation           ——— GAT GGA TGA TGA TAT GAT ———
                                   ↑                ↓
mutant protein            ——— leu |— pro — thr — thr — ile |— leu ———

            ◄———— normal ————►◄——— out of ———►◄——— normal ———►
                                   register
```

Figure 12.13. Effects of single and double mutants in the regular DNA copolymer GAT.

translated into a normal glycine residue. However, *missense suppression* must not be highly efficient, because it obviously would be lethal if all arginine residues in proteins were replaced by glycine. But if the missense suppressor mutation affected only a minor fraction of the glycine tRNA molecules, occasional conversions of missense to sense could be accomplished without killing the organism. *Nonsense suppressors* also result from mutations in tRNA molecules. If the anticodon of a serine tRNA is modified so that it can base pair with the UAG nonsense codon, the polypeptide chain of a mutant cell would not be terminated prematurely.

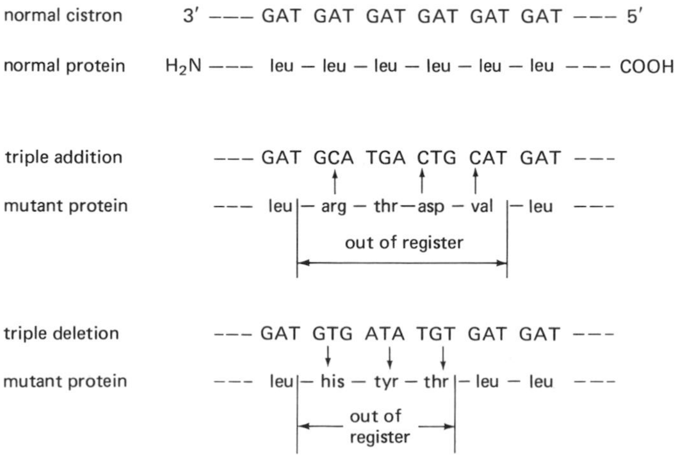

```
normal cistron        3' ——— GAT GAT GAT GAT GAT GAT ——— 5'

normal protein        H₂N ——— leu — leu — leu — leu — leu — leu ——— COOH


triple addition           ——— GAT GCA TGA CTG CAT GAT ———
                                   ↑       ↑       ↑
mutant protein            ——— leu |— arg — thr — asp — val |— leu ———
                                        out of register


triple deletion           ——— GAT GTG ATA TGT GAT GAT ———
                                   ↓       ↓       ↓
mutant protein            ——— leu |— his — tyr — thr |— leu — leu ———
                                        out of
                                        register
```

Figure 12.14. Triplet deletions or triplet additions as mutation suppressors in the regular DNA copolymer GAT.

It may be that the nonsense codon UAA is the natural termination signal, since no highly efficient suppressors of this codon have been found. Sometimes a polycistronic mRNA molecule (coding for several polypeptide chains) is produced from an operon. A nonsense point mutation in the left-hand (early read) portion of the mRNA would prematurely interrupt polypeptide chain synthesis and disallow reading of cistrons further along the message. These are classified as *polar mutants*. Polar mutations can also be corrected by nonsense suppressors.

A third type of suppressor mutation is the *indirect intergenic suppressor*. This class of mutants does not repair the primary genetic lesion, but rather causes a change in some other gene that tends to compensate for the damaged protein product of the primary lesion. For example, a point mutation in the gene specifying the structure of the enzyme anthranilate synthetase could lower the affinity of the mutant enzyme for its substrate chorismic acid (Figure 12.15). This enzyme is in the synthetic pathway of the amino acid tryptophan. Without a functional anthranilate synthetase enzyme, a bacterial cell would not be able to grow on minimal (unsupplemented) medium (*auxotrophic*). Nonmutated (normal) bacteria can synthesize all its amino acids, vitamins, nucleic acids, and other complex organic molecules from simple salts, water, and a carbon source such as a sugar (*protrotrophic*). An auxotrophic cell can be changed into a prototroph by a suppressor mutation that reduces the sensitivity of an enzyme to feedback inhibition involved in the synthesis of the substrate chorismic acid. An overproduction of chorismic acid could compensate for the reduced affinity of the mutant enzyme and thereby restore normal levels of anthranilic acid for use in tryptophan synthesis. All of this discussion concerning suppressor mutations has importance for evolutionary theory. Forward point mutations that produce less-fit mutant phenotypes are not solely at the mercy of rare reverse point mutations to restore normal or near normal functions. Other mutations, either within the same cistron or in completely different cistrons, may compensate for the primary lesion.

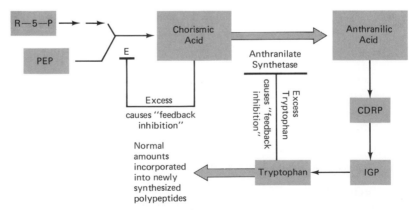

Figure 12.15. Simplified diagram of "feedback inhibition" of excess tryptophan on the enzyme anthranilate synthetase and of excess chorismic acid on a different enzyme (E). Symbols: R-5-P = ribulose-5-phosphate; PEP = phosphoenol pyruvate; CDRP = carboxyphenylamino deoxyribulosephosphate; IGP = indole-glycerol phosphate.

Magnitude of Phenotypic Effect

The magnitude of a mutation's effect on the phenotype would be expected to correlate directly with its fitness value. Since most mutations tend to be harmful, the greater the phenotypic effect the lower the fitness value. Point mutants with small phenotypic effects tend to be detected much more frequently than those with large effects, suggesting that many proteins have noncritical regions where amino acid substitutions do not completely destroy biological activity. Silent mutations are among those classified as *adaptively neutral,* because they do not change the fitness values of their hosts. Some mutations produce effects similar to but with less intensity than their normal alleles and are called *hypomorphs.*

> At the "bobbed" locus in *Drosophila,* the normal dominant wild-type allele (bb^+) produces normal bristles; its recessive allele (bb) when homozygous in double dose, produces shorter, thinner bristles (bobbed). This locus is in the *incompletely sex-linked* portion of the X and Y chromosomes. Genes in this region have corresponding loci in both the X and the Y chromosomes and, therefore, behave just like autosomal loci. Females of XXY chromosomal composition and males with XYY composition have been produced both carrying three mutant bb alleles. Their bristle phenotypes are almost normal in size and shape, indicating that the bb allele is a hypomorph, producing some of the product made by the normal allele, albeit in reduced amounts.

Mutants that produce no phenotypic effect even when present in extra dosage probably make no biologically active products, and are classified as *amorphic* mutants. It is possible for a mutant to be hypomorphic in respect to one trait and yet be amorphic in respect to another trait.

> An example of an amorphic gene in *Drosophila* is the sex-linked recessive allele for white eyes (w). Extra doses of this gene have no effect on changing the phenotype. On the other hand, its pseudoallelic locus, called apricot, has a wild-type allele (apr^+) that produces dull-red eye color and also causes pigmentation of the Malpighian tubules. Its recessive allele (apr) is hypomorphic with respect to eye color (producing a lighter eye color in double dose), but amorphic with respect to its effect on pigmentation of Malphigian tubules (no color).

Hypermorphic mutants are also known with effects similar to, but greater than, that of the normal allele.

> Acute intermittent porphyria is a hereditary disease characterized chemically by the excretion of excessive amounts of porphyrin precursor compounds in the urine, turning it the color of port wine. This was the first "overproduction disease" to be described. The hypermorphic mutant produces too much of the liver enzyme aminolevulinic synthetase. Continued high levels of porphyrins eventually cause neurological damage. King George III, the reigning monarch in Britain when the American colonies fought for their independence, undoubtedly suffered from porphyria. His fits of "madness" so incapacitated him that the Prince of Wales replaced him through the Regency Act of 1811. This disease has been traced back as far as his sixteenth century ancestor Mary, Queen of Scots. It has also been found in his descendants to the present day.

Some point mutations produce a qualitatively new effect that is not produced by the normal allele. Such mutations are called *neomorphs* and can be important sources of ultimately extending the phenotypic variability in a population beyond its former limits. Darwin perceived that variants with small effects would be less likely to disturb radically the normal functioning of the organism. Natural selection could gradually accumulate harmonious combinations of these minor variations over long periods of time into new races and species. This concept is still valid today. Mutations with large phenotypic effects would tend to be very harmful and therefore kept at low frequency by selection.

Genetic Loads

Most "recessive lethal mutations" tend to have a slightly deleterious effect on fitness when heterozygous with the normal allele. These mutations are therefore not completely recessive and contribute to the genetic load of the population in both heterozygous and homozygous genotypes. Most mutants that are detrimental in homozygous condition (but not lethal) also tend to reduce fitness in heterozygous genotypes. If all genotypes have equal fitness values, the population bears no genetic load. Any departure from this model introduces a genetic load into the population and will be accompanied by some genetic deaths. Genotypes that have no fitness suffer a *genetic death*. This need not produce a cadaver, but may simply result in a failure to produce viable, fertile offspring. At least two important factors contribute to the genetic load of populations: (1) mutation and (2) segregation. The term *mutational load* is used for the reduction in population fitness attributed to the recurrence of deleterious mutations. Deleterious mutations are constantly being introduced into the population and must be "weeded out" by the process of natural selection.

The rare occurrence of a beneficial mutation also creates a genetic load, because the old normal allele is now less fit. Natural selection then begins the process of moving the old allele from its original high frequency to a low frequency characteristic of deleterious genes. During this transitional stage, the population pays a price in genetic deaths for replacing one allele by another. This is referred to as the *substitutional (transitional) load*.

In simplistic models of the ideal population, all individuals are homozygous for all the beneficial genes. There are many known examples, however, in which the most adaptive genotypes are heterozygotes rather than homozygotes. The population must maintain both alleles in order to generate the fitter heterozygotes, even though the two homozygous genotypes are at a selective disadvantage. Overdominance with respect to fitness generates a *segregational* or *balanced load*. It should be emphasized that population fitness is not necessarily correlated with genetic load. If a locus exhibits overdominant gene action, the average fitness of a population is increased by retaining both alleles, despite the fact that a genetic load is thereby created. Conversely, a low genetic load does not necessarily indicate a high fitness population. For example, a genetically uniform population would (by definition) have no load, yet it might be so poorly adapted as to be incapable of reproductively maintaining itself.

An *incompatibility load* may also exist in a population if individuals of one genotype exert a detrimental effect on those with a different combination of alleles. The magnitude of the incompatibility load is a function of the frequencies of the alleles involved and the probability that an unfavorable interaction will occur under the appropriate environmental conditions.

> Experiments conducted with various strains of *Drosophila melanogaster* have shed some light on the nature of incompatibility loads. Survival of larvae was observed on culture medium containing either killed larvae of the same or of a different strain; the experiment was controlled by larvae grown on untreated culture medium. In some instances, the survival of one wild-type strain was lowered, when dead larvae of a second wild-type strain were present in the culture medium. Presumably certain metabolic waste products of the dead strain may have an inhibitory effect on the survival of larvae of another strain. In other experiments, treatment of the medium with dead larvae of one strain enhanced the survival of larvae from a different strain. *Facilitation* is the term for interactions between different genotypes that mutually increase their survival. Presumably products of different genotypes may reciprocally supplement each other's needs.

Evolution Rate vs. Mutation Rate

Has the rate of evolution been limited by the appearance of suitable mutations? This is a very important question. The following calculations may help us find an answer. We know that mutation rates vary widely from one locus to another and are influenced by both extrinsic and intrinsic factors. Nevertheless, a conservative average estimate is one mutation per gene locus per 10^5 gametes. A conservative estimate of the number of genetic loci in higher organisms is at least 10^4. Therefore, 10 per cent of all these individuals probably contain a newly mutated gene at one of its loci. Most mutations will be detrimental, but a small fraction (conservatively estimated at one in a thousand) is expected to be beneficial. Thus, in each generation, one out of every ten thousand individuals in the population would have a new mutation of potential evolutionary value. An estimate of 10^8 as the average number of individuals in a population per generation is highly conservative for most plants and many invertebrate animals (especially insects), but undoubtedly this is an overestimate for many higher animals (especially some mammals). Overall then, perhaps 10^8 individuals per generation is a fair general average. The number of generations required to change the original species into at least one new species must surely vary greatly, but perhaps we can use 50,000 generations as a conservative estimate. the total number of individuals that existed during this time span is $(10^8)(5 \times 10^4) = 5 \times 10^{12}$; the number of potentially useful mutations that occurred during this period is $(10^{-4})(5 \times 10^{12}) = 5 \times 10^{-8}$. Again we can only guess how many new mutations are required to transform one species into another, but perhaps 500 is a suitable estimate. From these calculations we discern that using only one in a million of all potentially beneficial mutations (or one in a billion of all mutations) is actually sufficient to provide for the observed rates of evolution.

The answer to our original question than is no; the rate of production of potentially useful mutations does not appear to have been a limiting factor in the rate of evolution. The main restrictions on the supply of genetic variability in a species lie in the recombinational mechanisms of its genetic system and in the rate of gene flow between populations within the species. Within a generation, the contribution made by new mutations to the total genetic variability of the gene pool is miniscule compared to the vast amount of heterogeneity that typifies most natural populations. Increasing the mutation rate several fold would not substantially alter this relationship. Therefore, we should not expect to find any correlations between mutation rates and the rates of evolution.

Mutational Adaptations

As mentioned earlier, bacteriology was the last stronghold of the Lamarckian doctrine. Bacteria, in general, appeared to be highly plastic organisms that easily metamorphosed into different forms on exposure to different environments. This was the doctrine of *pleomorphism* held by the early bacteriologists. Once Koch and Cohn developed pure culture techniques (in the 1870s), "plasticity" was viewed as a function of mixed (unpure) bacterial cultures. The counter doctrine of *monomorphism* proposed that every bacterial type represented an immutable species. This paradigm persisted for the next fifty years despite numerous observations of changes in supposedly pure cultures. Some pathogenic bacteria that grew poorly in a certain host species could be "trained" to grow more efficiently after several serial passages through members of that species. Most bacteria are killed by antibiotics, but a few may survive. These observations were explained by the ability of bacteria to adapt (in the Lamarckian sense) to their new environment. The change from rough to smooth colony morphology in *Streptococcus pneumoniae* (discussed in Chapter 10) was a reversible process. Bacteriologists of the late 1920s entertained the *cyclogenic* or *ontogenic theory* that this "dissociation" process simply represented different phases of the life cycle. Many cases of transient changes induced by the environment were known, as exemplified by the sugar-fermenting enzymes such as β-galactosidase. Obviously, the science of bacterial genetics could not be born until a clear distinction could be made between adaptation and gene mutation.

In 1943, S. E. Luria (Figure 12.16) and M. Delbrück (Figure 12.17) published a paper entitled "Mutations of Bacteria from Virus Resistance to Virus Sensitivity." Their sophisticated experimental design and unambiguous results demonstrated the kind of attack required to solve the riddles of bacterial heredity. The Luria-Delbrück *fluctuation test* was a statistical approach designed to distinguish between induced resistance to the phage T1 and spontaneous mutation to resistance. They reasoned that if mutations were occurring spontaneously in each generation, the numbers of T1-resistant colonies in different bacterial cultures would probably have a large statistical variance. On the other hand, if T1 resistance was acquired only after contact with the phage, the numbers of resistant bacteria in different cultures would be expected to have a much smaller statistical variance.

Figure 12.18 diagrams the results anticipated according to the adaptive mechanism (Part A) and the spontaneous mutation mechanism (Part B). Very few resistant bacteria appear regardless of the mechanism by which they are produced. If the resistant bacteria are distributed randomly among different cultures, they should approximate a *Poisson distribution,* which is a random distribution of rare events. A hallmark of a Poisson distribution is that

Figure 12.16. Salvador E. Luria (born 1912); Nobel Prize laureate 1969.

Figure 12.17. Max Delbrück (born 1906); Nobel Prize laureate 1969.

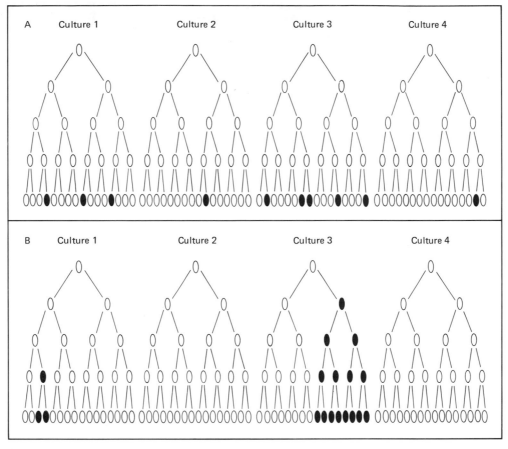

Figure 12.18. Diagram representing a typical distribution of ten *E. coli* mutants resistant to bacteriophage T1 produced according to (A) an adaptive mechanism, and (B) the spontaneous mutation mechanism.

the mean and the variance are equal. Thus, the ratio (variance/mean) = 1 is characteristic of a Poisson distribution. Returning to Figure 12.18, we notice that in both A and B there are ten resistant bacteria in the four culture plates. The average number of resistant bacteria per culture is 10/4 = 2.5. Now let us calculate the variance/average ratios for parts A and B:

$$A = [(2.5 - 3)^2 + (2.5 - 1)^2 + (2.5 - 5)^2 + (2.5 - 1)^2]/4/2.5 = 1.1$$
$$B = [(2.5 - 2)^2 + (2.5 - 0)^2 + (2.5 - 8)^2 + (2.5 - 0)^2]/4/2.5 = 4.3$$

Thus, we see that if bacteria are adapting (Part A) to the presence of phage, the variance/mean ratio should approximate 1.0. If they are spontaneously mutating (Part B) to resistance (whether phage are present or not), the same ratio should be considerably higher than 1.0.

Now we are ready to review the Luria-Delbrück experiment. They used a strain of *E. coli* sensitive to phage T1. Twenty individual cultures (0.2 ml each) were started and, as a control, a single bulk culture (10 ml volume) was also started in nutrient broth. Each culture was inoculated with approximately 10^3 cells/ml and allowed to incubate until a density of about

10^8 cells/ml was reached. The entire contents (0.2 ml) of each individual culture was spread on a nutrient agar plate previously treated with a high concentration of T1 phage. Simultaneously, ten 0.2 ml samples were similarly plated from the bulk culture as controls. After overnight incubation, the plates were scored for the number of resistant colonies. The results appear in Table 12.1. Notice that the variance/mean ratio for the controls was approximately 1.0 as expected for a random distribution of rare events. The same ratio for the individual cultures, however, was very large. Therefore, the numbers of resistant colonies varied greatly from one culture to another. These results support the mutation theory and argue strongly against the adaptive theory. Some cultures, by chance, experienced mutation(s) early in the growth period; in others the mutational event(s) occurred much later or not at all. This was the first experimental evidence substantiating the concept of *preadaptive mutations,* i.e., spontaneous changes in heredity that predispose the organism to survival in a different environment. The results of the fluctuation test indicate that bacteria do mutate to phage

TABLE 12.1. Fluctuation Test for the Origin of Resistance to Phage T1 in *Escherichia coli*

Individual Cultures		Samples from Bulk Culture	
Culture Number	Number of Resistant Colonies	Sample Number	Number of Resistant Colonies
1	1	1	14
2	0	2	15
3	3	3	13
4	0	4	21
5	0	5	15
6	5	6	14
7	0	7	26
8	5	8	16
9	0	9	20
10	6	10	13
11	107		
12	0		
13	0		
14	0		
15	1		
16	0		
17	0		
18	64		
19	0		
20	35		
Average	11.3		16.7
Variance*	714.5		16.4
Variance*/ Mean	62.9		0.9

From S. E. Luria and M. Delbrück, *Genetics,* **28**, pp. 503–504 (1943).

*Variances in the original paper differ slightly from the ones shown here because the former were corrected by the sampling error.

resistance before they contact the phage, and that these mutants are preadaptive, in that only the resistant bacteria can survive in the presence of phage. The phage act as a selective agent, allowing only the rare resistant mutants already present in the culture to grow into colonies; they are not the cause of mutation to resistance. This classical experiment broke the back of Lamarckian dogma and gave birth to the science of bacterial genetics.

A much simpler experiment to detect preadaptive mutations was devised by Joshua (Figure 12.19) and Esther Lederberg in 1952. A confluent growth of T1-sensitive *E. coli* cells were grown on a nutrient agar plate (*cir.* 10^7 cells/plate). A sterile piece of velvet was attached to a circular block the size of a Petri dish. The velvet pad was stamped onto the confluent growth of the master plate. The nap of the velvet functions as thousands of tiny sampling devices, picking up representatives of colonies from all over the plate. The pad is then stamped onto a series of nutrient agar plates (replica plates) treated with a high concentration of T1 phage. After overnight incubation, the replica plates were scored for phage-resistant colonies (Figure 12.20). The numbers of colonies on these plates were usually identical, but more importantly the spatial arrangements of the colonies on the replica plates were also identical. These results are only explicable if we assume that phage-resistant mutants were already present on the master plate. Representatives from each resistant colony were transferred to the replica plates by the velvet pad and, hence, formed the same geometrical pattern on each replica plate. Again the phage merely acts as a selective agent to prohibit growth of

Figure 12.19. Joshua Lederberg (born 1925); Nobel Prize laureate 1958.

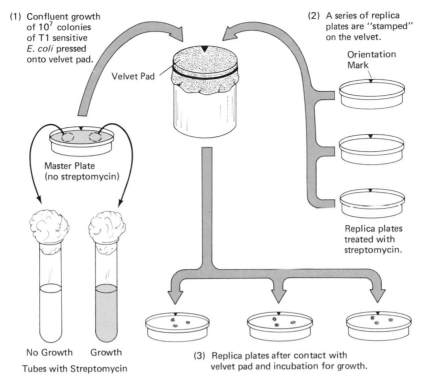

(1) Confluent growth of 10^7 colonies of T1 sensitive *E. coli* pressed onto velvet pad.

Velvet Pad

Master Plate (no streptomycin)

No Growth   Growth

Tubes with Streptomycin

(2) A series of replica plates are "stamped" on the velvet.

Orientation Mark

Replica plates treated with streptomycin.

(3) Replica plates after contact with velvet pad and incubation for growth.

(4) Pickings from area on master plate corresponding to colonies on replica plates produce growth in streptomycin–treated broth; pickings from any other area produce no growth.

Figure 12.20. Diagram of the "replica plate" technique used to confirm the theory of preadaptive mutation.

the sensitive cells also transferred from the master plate. Suppose that one returns to the master plate and picks some bacteria from a position corresponding to a colony on the replica plates, transfers them to a nutrient broth, and incubates them for several generations. Upon plating samples from this "enriched" broth on T1-treated agar, we find a large number of colonies. Pickings from positions on the master plate not corresponding to colonies on the replica plates yield few if any resistant colonies when similarly treated. This confirms the earlier inference that mutations to T1 resistance had occurred on the master plate before any contact with the phage had been made. The *replica-plate technique* has been widely used to isolate auxotrophic bacterial mutants (i.e., those that are unable to synthesize one or more essential organic compounds and therefore cannot grow in unsupplemented broth). It has also been used to isolate mutants resistant to various antibiotics (e.g., streptomycin and penicillin) and bacteriostatic drugs (such as sodium azide).

The beneficial nature of a preadaptive mutation is only realized when the mutant organism is exposed to an environment suitable for the expression of its adaptive qualities. For example, bacteria that have mutated from streptomycin sensitivity to streptomycin resistance do not enjoy a selective advantage in the absence of streptomycin from the environment. As a matter of fact, in some cases, drug-resistant mutants do not grow as well as drug-sensitive bacteria. In other cases, a drug-resistant mutation may be adaptively neutral when no drug is present. Plant and animal breeders watch continuously for potentially beneficial spontaneous mutations and have even resorted to applying mutagens to increase the mutation rate in hopes of finding useful hereditary modifications.

> Kwashiorkor (Figure 12.21) is a severe nutritional disease resulting from too little protein in the diet. In populations that subsist largely on a diet of corn, this condition commonly develops in children after weaning. Corn is low in lysine, tryptophan, and isoleucine (three of eight essential amino acids that must be supplied in the diet because they cannot be synthesized in the human body). Merely increasing protein consumption may not correct the problem; attention must also be given to the quality of the protein (i.e., its content of essential amino acids). Lysine is the most limiting amino acid in most cereals. Recently a mutant gene called opaque-2 was discovered which could increase the lysine content of corn by as much as 69 per cent. Another gene called floury-2 has a similar effect. Plant breeders hope to use these kinds of mutations to increase the nutritional value of corn and help alleviate both general and specific protein deficiency diseases.

Mutations that are deleterious in one environment may be beneficial in some other environment. This is the well-known phenomenon of *genotype-environment interaction*. There is no one best genotype for all conditions. Each population adapts to local conditions through selection of variants specialized to exploit particular niches of the local environment. Agronomists have found barley mutations that strengthen the stems. In areas of high rainfall, strong stems are desirable to prevent the moisture-laden heads of barley from bending over and lying on the ground (lodging). In dry climates, stiff brittle stems might be detrimental, because hot, dry winds would tend to shatter the heads before they could be harvested. The value of a mutation must always be judged *relative* to the environment in which it will be expressed.

> Perhaps one would think that monstrous or deforming mutations could not possibly be beneficial under any set of conditions. Take the vestigial wing condition in *Drosophila* as

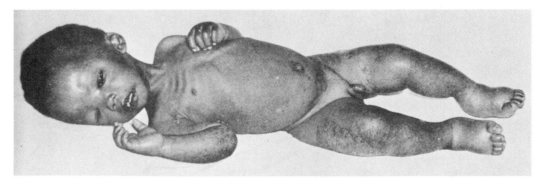

Figure 12.21. Marasmic kwashiorkor (skin lesions and oedema; age twenty-one months). (From Brock, J. F., *Recent Advances in Human Nutrition*, Little, Brown & Company, 1961)

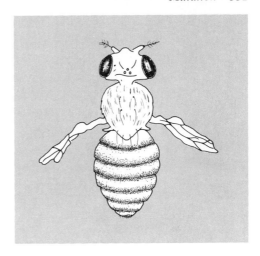

Figure 12.22. Vestigial wings in *Drosophila melanogaster.*

an example (Figure 12.22). This recessive autosomal mutation reduces the single functional pair of wings of the fly into shriveled nonfunctional rudiments. How could such a fly find food or a mate, and how would it escape predators? These problems might appear to be insurmountable. However, many wind-swept oceanic islands are characterized by numerous species of flightless insects whose closest relatives on the continents retain functional wings. Apparently, the ancestors of these island populations could fly, but ofttimes were swept out to sea to die. Vestigial winged mutants would stay on the ground, and thereby potentially become part of the breeding population.

Many insects that pass through a larval stage have segmented larvae with relatively little differentiation between the segments. The obvious serial homology of the larvae tends to become obscured by the metamorphosis that produces the adult. Nonetheless, the antennae, mouth parts, and legs of the adult are still considered to be serially homologous structures. In *Drosophila,* mutants are known that transform one organ or part of the body from its usual form into one characteristic of some other body segment. These are called *homoeotic* (*homeotic*) *mutations.* The "tetraltera" mutation converts the wings into haltere-like structures characteristic of an adjacent body segment. A mutant called "aristapedia" converts the antenna into a leg-like structure (Figure 12.23); "proboscipedia" similarly transforms the proboscis. Now, just because we cannot visualize how such mutations could contribute to fitness does not mean that they are necessarily worthless variations under all conditions. Perhaps the precursors of functional wings in insects began as a "freak" mutation that allowed the organism to exploit a new niche. After all, the air above the ground was a completely unoccupied habitat until the insects took wing. Richard Goldschmidt (Figure 12.24) championed the hypothesis that higher taxonomic groups such as orders and classes could have evolved in this way.

A rapid evolutionary shift to a new adaptive zone is termed *quantum evolution* (also known as *megaevolution* and *macroevolution*). These *macromutations* or *systemic mutations,* responsible for a major adaptive shift, must be extremely rare (if indeed they exist at all). The series of interrelated developmental pathways through which the organism must pass to reach the adult form is termed the *epigenotype.* A systemic mutation could represent a *switch gene,*

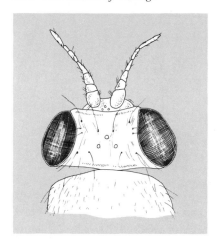

Figure 12.23. Homozygous spineless *aristapedia*.

Figure 12.24. Richard Benedict Gold-schmidt (1878–1958). [From *Genetics*, 45: Frontispiece (1960)]

i.e., a gene that causes the epigenotype to switch from one developmental pathway to another so that the adult structure might be greatly modified. Goldschmidt thought that chromosomal rearrangements could possibly play a major role in this respect. One line of support for the "hopeful monster theory" is the fact that new orders and classes of organisms seem to appear suddenly in the fossil record without any trace of intermediate transitional forms. According to the Darwinian concept, the traits that now characterize taxonomic groupings above the species level were accumulated gradually over a long span of geological time through the action of natural selection on newly arisen variations. If this really happened, we would expect to find evidence of transitional forms connecting different orders in the fossil record. Failure to find these postulated intermediate types is a critical deficiency in the Darwinian theory. Negative evidence, however, should not be construed as positive proof against this aspect of Darwinian theory. The highly biased, nonrandom record of prehistoric life in the earth's rocks gives an extremely sketchy view of organic changes. Furthermore, during the geological revolutions that separate major time zones, profound environmental changes were occurring quite rapidly. During such times, intense selection pressures may have been operative to move certain populations rapidly from one adaptive zone to another by the accumulation of numerous small variations. This is the conventional view of most modern evolutionists, but it does not absolutely exclude the haunting possibility of a "hopeful monster." There are at least two reasons for this modern view. First, taxonomic orders differ in many characters. Adaptation to a completely new way of life generally requires numerous coordinate modifications. It is difficult to imagine how a single macromutation could produce all the harmoniously integrated traits necessary for a shift to a completely new adaptive zone. Second, if these quantum jumps are extremely rare, a singular mutant forming a new species would find itself without a mate. Unless it was capable of self-fertilization, the mutation would disappear in the same generation in which it arose. If the mutation could backcross to one or both of its parents, reproductive isolation would not have been attained and therefore it could hardly represent a new taxon.

Point Mutational Effects on Protein Structure

Now that we understand the nature of the genetic code, at least three major classes of point mutations can be identified. Included in the first class are the "silent" *samesense mutations* that cause no phenotypic change. Degeneracy of the code allows many codons to be changed in the third position without altering the amino acid thereby specified. For example, changing the mRNA codon from UUU to UUC would not change the codon meaning; phenylalanine would be inserted into the polypeptide chain by either codon. Mutations in the first position of a codon need not always result in an amino acid substitution. For example, mutation from UUG to CUG still codes for leucine.

A second class of point mutations are those that cause amino acid substitutions; these are *missense mutations*. For example, the mutation from AUG to AUA would cause isoleucine to replace methionine in the polypeptide chain. The phenotypic ramifications of such substitutions can be quite variable. There may be little or no damage produced, if the lesion occurs in

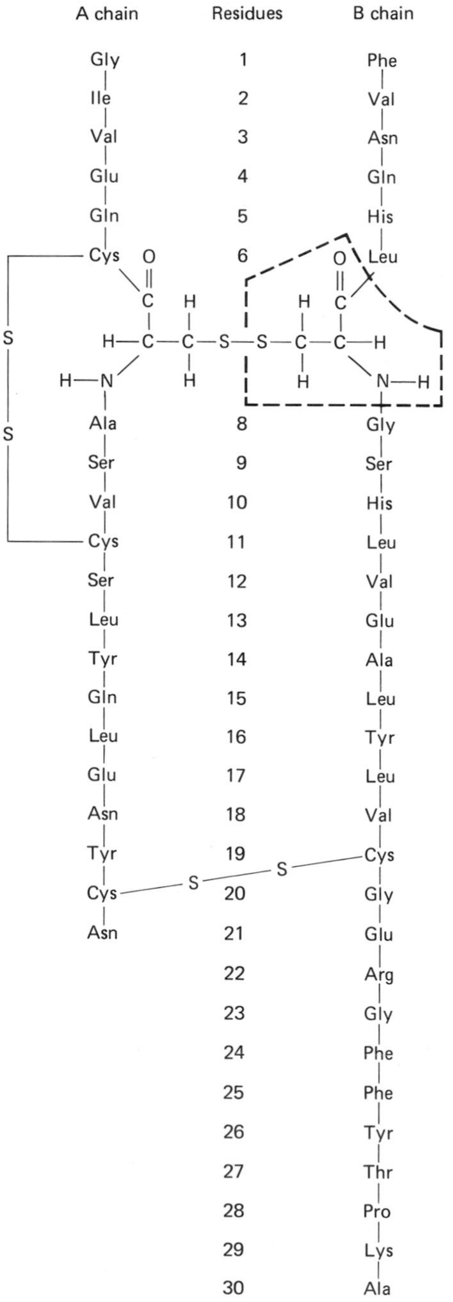

A chain	Residues	B chain
Gly	1	Phe
Ile	2	Val
Val	3	Asn
Glu	4	Gln
Gln	5	His
Cys	6	Leu
	8	Gly
Ala	8	Gly
Ser	9	Ser
Val	10	His
Cys	11	Leu
Ser	12	Val
Leu	13	Glu
Tyr	14	Ala
Gln	15	Leu
Leu	16	Tyr
Glu	17	Leu
Asn	18	Val
Tyr	19	Cys
Cys	20	Gly
Asn	21	Glu
	22	Arg
	23	Gly
	24	Phe
	25	Phe
	26	Tyr
	27	Thr
	28	Pro
	29	Lys
	30	Ala

Figure 12.25. Beef insulin consists of two polypeptide chains (A chain = 21 amino acids; B chain = 30 amino acids). Two interchain disulfide (S-S) bridges and one intrachain disulfide bridge are shown. The structure of one of the two cysteine residues at position 7 is enclosed by a broken line.

a region of the protein that is not critical in formation of the secondary (weak) bonds that stabilize the amino acid chain in its biologically active form. Mutation from CUU to CCU causes proline to replace leucine. Proline is the only amino acid whose α-amino group is a secondary rather than a primary amine. For this reason it cannot participate in alpha helix formation. Proline residues are commonly found in nonhelical "corners" of polypeptide chains. This type of mutation could seriously impair the biological activity of the protein (perhaps destroying the catalytic activity of an enzyme) and thereby produce a distinctive phenotypic variation. Cysteine is one of the sulfur-bearing amino acids. Stability of the tertiary protein configuration is enhanced by formation of disulfide bridges between cysteine residues (Figure 12.25). A point mutation from UGU to UAU would cause cysteine to be replaced by tyrosine. The minimal effect of this mutation would be to reduce the thermodynamic stability of the protein, perhaps rendering it easily denatured by a slight elevation in temperature. Mutations such as these provide opportunities for the environment to modify the phenotype and thereby allow phenotypic plasticity (nongenetic variations) to be expressed.

Nonsense mutations constitute the third class. If the tryptophan codon UGG is changed to UGA, a nonsense codon is created that prematurely interrupts protein synthesis. Shortened proteins almost invariably have lost their biological functions and therefore usually contribute to a reduction in fitness.

An additional class of mutations might be considered, although they are not simply nucleotide substitutions. These are *frameshift mutations* that result from single base additions or deletions. A single base addition (or deletion) near the 5′ end of a mRNA molecule would undoubtedly be highly deleterious, because it would shift the reading frame out of register and create mostly missense amino acid substitutions through the rest of the message. Frameshifts can also generate nonsense codons that would terminate protein synthesis. A frameshift mutation near the end of the message (i.e., near the 3′ end) may or may not be deleterious depending largely on how critical the carboxyl end of the polypeptide chain is to the normal three-dimensional structure of the protein. Thus, single nucleotide changes (point mutations) may have phenotypic consequences ranging from lethality through semilethality and subvitality to adaptive neutrality. To the extent that a new mutation can create proteins with new functional capabilities, it is also possible that the mutant heterozygous individual would be adaptively superior to other genotypes in the population. This would be a case of overdominance with respect to fitness.

Hemoglobin Evolution

One of the most enlightening examples of how mutations have contributed to the evolution of proteins is found in a study of hemoglobin, the molecule that transports oxygen in the red blood cells of most vertebrates. Hemoglobin is a tetramer, i.e., four polypeptide chain monomers form its most functional quarternary structure. A pair of identical polypeptide chains forms a *homodimer*. Adult hemoglobin consists of two types of homodimers; e.g., two α chains constitute one homodimer and two β chains form the other homodimer. The major fraction of adult human hemoglobin is designated hemoglobin A (Hb-A or HbA$_1$) represented by the tetrameric formula $\alpha_2\beta_2$. The α chain contains 141 amino acids, the β chain has 146. Both chains are similar in three-dimensional conformation despite the fact that they differ considerably in amino acid composition. There are sixty-four sites occupied by identical residues and seventy-seven sites by different amino acids (Figure 12.26). The genes responsible for the α and β chains appear to assort independently. Nonetheless, a common ancestry

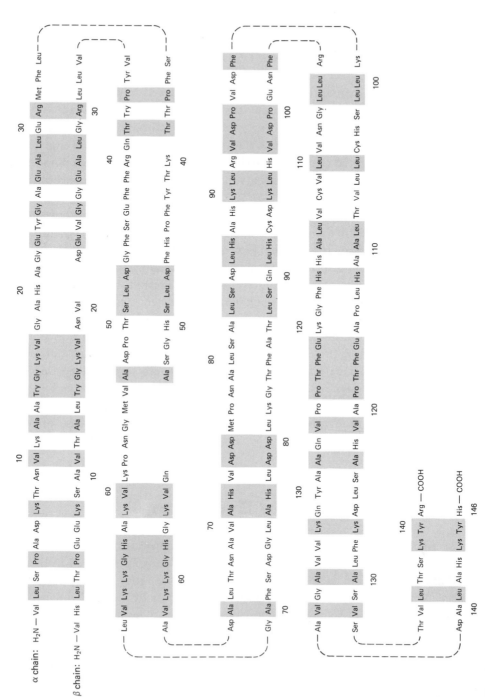

Figure 12.26. Comparison of primary structures of human hemoglobin chains alpha and beta, aligned to maximize the number of identical residues (shown in boxes).

for both chains is postulated, because it is highly unlikely that unrelated polypeptide chains could evolve independently to resemble each other at so many corresponding sites, with the same conformation and function. The relatively high number of amino acid differences, however, indicates that they diverged by gene duplication from a common ancestor long ago. Numerous independent point mutations (including base additions and deletions) have since brought these two genes to their present degree of divergence.

During fetal development, three other kinds of hemoglobin chains appear (Figure 12.27). The epsilon (ϵ) chain is synthesized for only a brief period early in development. Gamma (γ) chains also appear early in development, but they continue to be synthesized until shortly before birth. In contrast, β-chain synthesis does not peak until shortly after birth. Delta chains (δ) are made at low levels during the embryological period and on into adult life. Alpha chains are produced early and stay at high levels throughout development and into adult life. The β, δ, and γ chains contain 146 residues each and closely resemble each other in amino acid sequence. There are thirty-nine site differences between β and γ chains, and ten between β and δ chains. The sequence of amino acids in ϵ chains has not yet been deciphered. Therefore β and δ chains probably represent recent evolutionary departures. Further, there is evidence that the β and δ genes are linked. The β and γ chains probably diverged at a more remote time, but more recently than the α chain. Fetal hemoglobin (Hb F), with the structural formula ϵ_4 (a tetramer of epsilon chains called Gower I), is found in embryos 1 to 2 cm long. In embryos one to three months of age the predominant form is $\alpha_2\epsilon_2$ (Gower II). During most of development, however, Hb F has the formula $\alpha_2\gamma_2$. A minor fraction (about $2\frac{1}{2}$ per cent) of adult hemoglobin (Hb A$_2$) has the formula $\alpha_2\delta_2$. Within a given hemoglobin molecule, the two chains forming a homodimer are identical even in heterozygotes for point mutations.

Myoglobin is a single polypeptide chain of about 155 amino acid residues, with a conformation similar to that of hemoglobin chains. Myoglobin does not circulate in the blood, but is localized in muscles where it functions as an oxygen repository. The complete structure of human myoglobin has not yet been analyzed, but indications are that it will be more nearly like whale myoglobin than human hemoglobin. Whale myoglobin shares only thirty-seven

Figure 12.27. Differential synthesis of various human hemoglobin chains during prenatal and early postnatal periods.

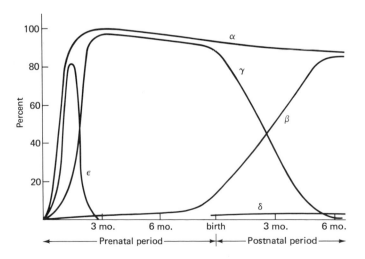

residues in common with the α chain of human hemoglobin and thirty-five with the β chain. Therefore, it is inferred that myoglobin and the human hemoglobins diverged from a common ancestor at a very remote time in the past. Figure 12.28 depicts the relationship between myoglobin evolution and that of various hemoglobin chains. Homologous segments in polypeptide chains of unequal length are maximized by postulating appropriate additions or deletions in their controlling genes. The horizontal scale of Figure 12.28 is used to plot chain differences in terms of the maximum permissible difference (100 per cent when no homologous regions can be matched by the aforementioned method). The origin of each branch in this molecular "phylogenetic tree" represents a duplication presumed to initiate two evolutionary lines leading to different contemporary polypeptide chains.

In general, the number of differences separating homologous polypeptide chains of different species is roughly proportional to the degree of genetic relationship between the species. It is quite likely that differences at a given position in closely related polypeptide chains result from single mutational events. It is expected that at least some of the differences separating more distantly related chains have incurred multiple mutations at a given site. For example, a mutation might change A to G, and the G to a C, and the C back to an A.

Linus Pauling has made allowance for this process in estimation of the ages of ancestral proteins shown in Figure 12.28. If we assume that the α and β chains have evolved at the same rate, we can pool the number of differences and then compare their mean with a mean similarly derived from α and β chains in four other animals (horse, pig, cattle, rabbit). The mean difference between humans and these animals is twenty-two, or eleven changes per chain. It is estimated from fossil evidence that the common ancestor of humans and these animals lived about eighty million years ago. Thus, on average, about seven million years is required to establish a successful amino acid substitution (point mutation) in any of these

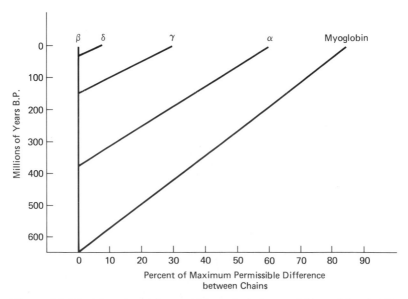

Figure 12.28. Inferred evolutionary history of whale myoglobin and four human hemoglobin chains.

species. The β and δ chains differ at only ten sites, because of their recent common ancestry. Therefore only five changes occurred in each. If seven million years is required to establish each amino acid substitution, we estimate that a duplication occurred about thirty-five million years ago from which the β and δ chains evolved through independent mutations to their present state of divergence. By applying Pauling's correction factor, it is estimated that β and γ chains (differing at thirty-seven sites) arose by gene duplication about 150 million years ago. Similar calculations for the β and α chains (differing at seventy-six sites) places the point of departure at 380 million years ago. If this method is valid as a rough approximation, the various human hemoglobin chains existing today have probably evolved from a common ancestor in early amphibians of the Devonian period. This example gives us some insight into the methodology used by the new discipline of *chemical paleogenetics.*

Phylogenic trees have been constructed from analyses of mutational changes inferred from differences in homologous proteins of various species. Computers are used to construct the tree that best fits the data. One such tree, represented in Figure 5.23, is derived from cytochrome c data. A computer determines the most probable number of mutations along each branch of the tree. Thus, the length of a branch is proportional to its *mutational distance* (indicated by numbers on each branch). Only a few phylogenetic trees have been constructed in this manner, but the topology (order of branching) of these chemical trees matches those constructed from conventional morphological considerations.

The adaptive value of the various hemoglobin chains is not at all clear. Around the time of birth, the human γ gene begins to shut down its synthesis of γ chains concomitant with activation of the β gene into full capacity synthesis of β chains. This is an example of the well-known phenomenon of differential gene action presumably responsible for embryological development. Perhaps the γ chain in fetal hemoglobin is better adapted to *in-utero* conditions than the β chain. However, a few adults have been discovered with γ chains rather than β chains, and they appear to be functioning normally.

If mutations are being produced spontaneously all the time, we would expect to find many hemoglobin mutants being tested for survival value in present human populations. As a matter of fact, over two hundred mutant hemoglobins have already been detected. Many harmful hemoglobin mutations probably are lethal early in embryological development, and we therefore cannot obtain the corresponding protein for amino acid sequencing analysis. Most of the mutant hemoglobins that have been analyzed appear to function normally as expected. A few abnormal hemoglobins are known to be harmful in homozygotes, but of little consequence in heterozygotes.

Normal Hb A has a lysine residue at the sixteenth position from the amino end of the α chain. A point mutation in the corresponding codon substitutes asparagine for lysine at site 16, producing the abnormal hemoglobin I (Hb I) with structural formula $\alpha_2^I \beta_2^A$. Many of these abnormal hemoglobins migrate at different rates than normal hemoglobin in an electrical field (electrophoresis; Figure 12.29).

Normal hemoglobin (Hb A) has glutamic acid at site 6 in the β^A chain. A point mutation in hemoglobin S (Hb S) has substituted valine for that glutamic acid (Figure 12.30). Heterozygotes for Hb S have some hemoglobin molecules as $\alpha_2^A \beta_2^S$ and some as $\alpha_2^A \beta_2^A$. These heterozygotes usually appear to function normally, but under severe oxygen stress (as in athletics or rapid ascent to high altitudes) these individuals may not be able to supply their tissues with sufficient oxygen. They go into what is called a

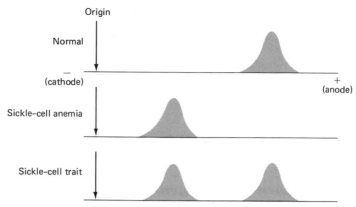

Figure 12.29. Electrophoretic mobility of solutions of human hemoglobins from normal red blood cells ($\alpha_2{}^A \beta_2{}^A$), those with sickle-cell anemia ($\alpha_2{}^A \beta_2{}^S$), and those with sickle-cell trait ($\alpha_2{}^A \beta_2{}^A + \alpha_2{}^A \beta_2{}^S$). The substitution of valine (net charge $= 0$) for glutamic acid (net charge -1) causes the Hb S molecules to move less rapidly toward the anode.

"sickling crisis." The red blood cells are deformed into a sickle shape; they cannot move easily through the capillaries and circulation is impaired. It can be very dangerous. Homozygotes for Hb S are severely anemic and must live their lives with moderation. Many homozygotes die before adulthood is reached. In certain African tribes, the frequency of the sickling allele is unexpectedly high (Figure 12.31). A. C. Allison discovered that heterozygotes for the sickling allele are more resistant to malaria than individuals homozygous for the normal allele (Hb A). Apparently the malarial parasite cannot survive inside red blood cells containing this abnormal protein. This over-dominant gene action keeps the sickling allele in the population despite its lethal effect in homozygotes. Under these conditions, the population bears a segregational (balanced) load. If the mosquito vector could be eradicated from the environment of these populations, heterozygotes would no longer enjoy a selective advantage. The sickling gene would then become part of the mutational load of deleterious variants. Natural selection would then begin to reduce the frequency of the sickling gene to a low equilibrium frequency dependent on the selection intensity and the back mutation rate.

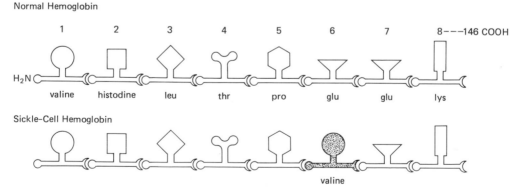

Figure 12.30. The beta chain of normal human Hb A (top) and that of Hb S (bottom) are identical at all 146 amino acid positions with the exception of position 6.

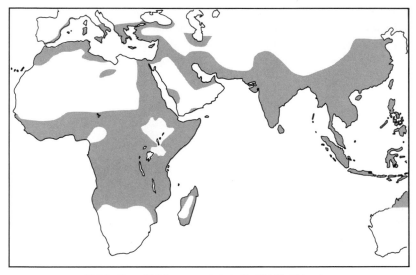

Distribution of falciparum malaria in the Old World tropics.

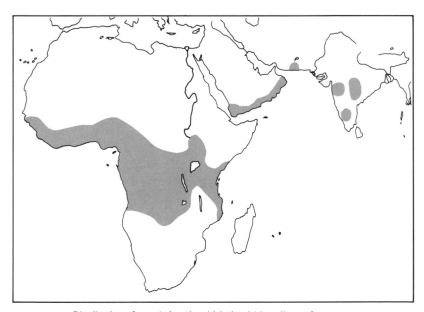

Distribution of populations in which the sickle–cell gene frequency
exceeds five percent.

Figure 12.31.

Formal Aspects of Mutation

For reasons stated earlier in this chapter, forward mutation rates are usually much greater
than back mutation rates (as determined by phenotypic changes). Let normal gene *A* forward
mutate to an alternative allelic form *a* with frequency *u* per generation; let the back mutation

rate be v. Also let the frequency of A and a alleles in the gene pool of the population be represented by p and q respectively. After mutation has occurred, the frequency of A becomes $p - up + vq$. This, of course, assumes that all conditions of the Hardy-Weinberg law are in force save that of mutation. At equilibrium, there is no change in gene frequencies, so

$$up - vq = 0,$$

and therefore

$$\widehat{p} = v/(u + v) \text{ and } \widehat{q} = (1 - \widehat{p}) = u/(u + v).$$

This is a *stable* equilibrium. That is, if the value of p is displaced from $\widehat{p}$, it will return to the equilibrium value determined by the magnitude of forward and back mutation rates. This equilibrium would be established even in a population that initially contained only A or a alleles, because forward or back mutation, respectively, would restore the missing allele to the gene pool. Opposing mutation pressures alone are usually so weak that the cause(s) of persistent equilibria should be sought elsewhere. One exception to this generalization is found in lethals. Here, mutations at many different loci can produce the same phenotype (death). The cumulative mutation pressure in this case could be substantial and could account for marked changes in chromosomal frequencies.

Since two breaks in a chromosome are required to produce an inversion, each inversion is virtually an unique variant with a vanishingly small chance of recurrence. In order to predict the fate of this singular aberration, let us consider the following model. All members of a finite population are initially homozygous for the standard chromosome structure (AA). An inversion occurs in a single individual (Aa) of this population. Assume that the variant is adaptively neutral, i.e., selection is inoperative. The probability that a is lost is a function of family size. The mutant heterozygote must mate with AA homozygotes. If Aa leaves no progeny, a will disappear from the population. If Aa leaves one offspring, there is $\frac{1}{2}$ chance that a will not exist in the next generation. In a family of size two, the probability that a is lost becomes $\frac{1}{4}$, etc.

Family size (n)	0	1	2	3 ... n
Probability of loss of a (m)	1	$\frac{1}{2}$	$\frac{1}{4}$	$\frac{1}{8} \ldots (\frac{1}{2})^n$

If the population size is stable from one generation to the next, the average family size would be two. The frequency of families of different sizes tends to approximate a Poisson distribution in which the mean and variance is approximately two.

Family size (n)	0	1	2	3	... n
Family Frequency (f)*	e^{-2}	$2e^{-2}$	$\frac{2^2}{2!}e^{-2}$	$\frac{2^3}{3!}e^{-2}$	... $\frac{2^n}{n!}e^{-2}$

The probability of extinction for a is found by summing the products ($m \cdot f$) over all family classes from 0 to n. This total is 0.3679; thus, the probability that a will not be lost is $1 - 0.3679 = 0.6321$. As an example, if one hundred unique mutations occurred, only about

*e is the base of the Naperian (natural) logarithm = 2.7183818. . . .

sixty-three of them would survive one generation; only about fifty-three would survive a second generation, etc. Eventually all the mutants would be lost. The only way such a unique mutant is likely to be retained in the population depends on its possessing a superior adaptive value at its inception.

Other kinds of mutations are not necessarily unique, but are, by nature, essentially irreversible (e.g., gene deletions). A few simple calculations will reveal that, in the absence of selection, these one-way mutations will eventually replace all the original alleles. Let p_0 represent the initial frequency of gene A; let u be the rate of one-way mutation from A to its allele a; and let p_1 symbolize the frequency of A after one generation of mutation. The change in frequency of A (Δp) from generation 0 to 1 is $\Delta p = up_0$. Therefore,

$$p_1 = p_0 - \Delta p = p_0 - up_0 = p_0 (1 - u).$$

To find the frequency of A in the next generation, we replace p_1 in the above formula by p_2 and p_0 by p_1:

$$p_2 = p_1 (1 - u).$$

Substituting $p_0 (1 - u)$ for p_1, we have

$$p_2 = [p_0 (1 - u)] (1 - u) = p_0 (1 - u)^2.$$

It is obvious that, after t generations of this replacement process, the frequency of A allele will be

$$p_t = p_0 (1 - u)^t.$$

Eventually, A would disappear from the population and all remaining individuals would be *aa*. In the above formula, the mutation rate u is very small compared to the value 1; therefore, $(1 - u)^t$ is approximated by e^{-ut} and $p_t = p_0 e^{-ut}$. To demonstrate how weak an evolutionary force mutation pressure is, we can calculate how long it would take to reduce the initial gene frequency (p_0) by one half.

$$(\tfrac{1}{2})p_0 = p_0 e^{-ut}$$
$$e^{-ut} = 0.5$$
$$ut = -\log 0.5 = 0.7$$
$$t = \frac{0.7}{u}$$

If an average gene mutates at rate 10^{-5} (i.e., 1 in 100,000 alleles of a particular kind mutates to a given alternative state in one generation), then we can substitute $u = 10^{-5}$ in the above formula; the answer is 70,000 generations. If p_0 was 0.5, then 70,000 generations later gene A would have a frequency of 0.25; after another 70,000 generations, it would have a value of 0.125, etc. Perhaps now we can see clearly that mutation pressure alone cannot begin to account for observed rates of evolutionary change in natural populations. The role of mutations in evolution is in the production of variations; the role of natural selection is to reject the harmful ones and sculpture the beneficial ones into the most adaptive combinations.

Essay Questions

1. Give two reasons why "systemic mutations" cannot be used to support the hypothesis of "quantum evolution."

2. With respect to the Luria-Delbrück "fluctuation test," outline is (a) purpose, (b) experimental design, (c) results, and (d) conclusions.

3. There are two major classes of forward mutations: (1) loss of function mutations and (2) change of function mutations. Which class generally has the higher mutation rate? Explain fully.

4. How can a mutation in one cistron compensate for the damage inflicted by a mutation in a different cistron?

5. Either defend or dispute the following statement: "Population fitness is maximized when all genotypes have equal fitness values."

SELECTION MODELS

Fitness

Charles Darwin's concept of evolution through natural selection continues to be regarded by most biologists as the preeminent force in natural populations for the production of adaptive changes. "Survival of the fittest" is an expression coined by British philosopher Herbert Spencer, and this phrase is often used to summarize the essence of natural selection. However, Darwin realized that the survival of individuals is not directly equated with fitness as much as is their reproductive contributions to the next generation. We can never know what the fittest of all possible genotypes might be, because the number of possible genotypes is an astronomically large number, only a tiny fraction of which has been produced and tested against a given environment. The *fitness* or *adaptive value* of a given genotype must always be understood to be a value relative to the fitter genotype, i.e., the one with the highest average reproductive ability currently in the population. Therefore, it would be more correct if we modified Spencer's phrase to the "survival and reproduction of the fitter." For an individual to have fitness, it is not sufficient merely to survive to a "ripe old age"; unless that individual leaves offspring, its fitness in the Darwinian sense is zero. Fitness is a complex of attributes specified by the genotype, chief among which are the following:

1. The individual must survive at least to sexual maturity (reproductive age) if it is to have any chance of contributing to the next generation. All other factors being equal, the longer the reproductive life, the greater the reproductive potential.
2. Meiosis must proceed normally in the gonads to produce viable, functional gametes.
3. The *fertility* or *fucundity* of an individual is measured by the number of viable and fertile offspring it produces.
4. In many animals, the gap between the physiological potential for leaving offspring and the realization of that potential involves numerous behavioral characteristics summarized as "sexual activity."

Each quantitative attribute outlined above is obviously influenced, at least partly, by numerous genetic loci. Mathematical models that attempt to define the effects of selection on a single genetic locus are gross oversimplifications of a process that, by necessity, must involve most, if not all, the total genotype.

Alleles that do not contribute differentially to fitness are referred to as "adaptively neutral." Adaptive neutrality of genetic variants at a given locus should not be equated with lack of importance of the locus itself to survival (and/or other aspects of fitness). Among isoalleles that make different forms of the same functional enzyme (isozymes), one genotype may be just as adaptive as another in this respect. However, the locus itself may be absolutely essential for the production of an enzyme critical to the developmental process. Deletion of this locus would then be lethal. Whether any characteristic can be truly adaptively neutral is still a moot question among evolutionists.

Natural selection does not act at the level of a single gene locus, but rather at the level of the total phenotype. Because of genetic interactions and/or genotype-environment interactions, the adaptive value of a given allele may vary with genetic background and/or environment. Therefore, selection models use the statistical average adaptive value of a given allele in a population.

The easiest models to construct are *deterministic,* i.e., ones in which the outcome of a series of events can be predicted, because all the initial parameters are invariate. Most selection models have invariate fitness values for the genotypes under consideration. These deterministic models are useful pedantic tools for introducing general selection concepts. However, the student should bear in mind that models seldom (if ever) correspond to the real world. For example, the Hardy-Weinberg equilibrium is a well-accepted model, but it is understood that it is constructed from a set of invariate parameters that do not exist in nature (e.g., no mutations, infinitely large population size, complete random mating, etc.). So it is with selection models. Fitness values may not be invariate even under the assumption of uniform abiotic environmental conditions, because as the composition of the gene pool changes under selection, the fitness values themselves may be altered (e.g., see frequency-dependent selection in Chapter 14). Because of these changing parameters, the evolutionary process is *indeterminate* in nature. As long as one remembers this, a study of the deterministic selection models in this chapter should not mislead the student into believing that life processes are so simple that they can be reduced to mathematical formulas.

Bruce Wallace has coined the terms "hard" and "soft" selection for cases involving invariate and variable fitness scales, respectively. Under "hard" selection, a fixed cut-off point exists for all populations of a species beyond which variability in fitness values is rejected. "Soft" selection allows flexibility in fitness values that results in a relatively uniform proportion of survivors among the various populations of a species (Figure 13.1). Note that hard selection need not imply that it always is more rigorous (in terms of proportion of the population rejected) than soft selection.

Genetic lethals that appear to cause death in all environments are classical examples of hard selection. Calculations of genetic loads also has been primarily concerned with hard selection. The concept of lethal equivalents usually connotes that less-rigorous causes of inviability are also invariably determined. These fitness values can be treated as invariables only if soft selection is relatively negligible. Wallace defines soft selection as being both density- and frequency-dependent, whereas hard selection is both density- and frequency-independent.

Territorial animals offer good examples of density-dependent selection. A given area can only support a limited number of territories. The death of individuals that cannot establish

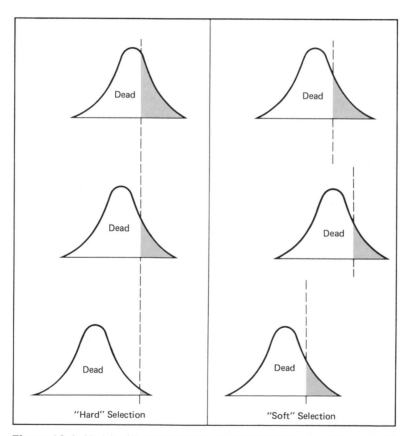

Figure 13.1. Models of three populations under "hard" selection (left) vs. "soft" selection (right). Note the inflexible limit of environmental rejection (dashed line) under "hard" selection and the relatively constant proportion of survivors under "soft" selection.

territories is a density-dependent phenomenon. If fewer individuals were in the population they might all have established territories (no differential mortality). On the other hand, deaths due to a catastrophic flood would be considered a density-independent event because the population would be decimated independently of the number of individuals per unit space (density).

Drosophila melanogaster prefers to pupate on the walls of a culture bottle at a certain distance above the food pad. There is only room for a limited number of pupae at the preferred pupation site. Larvae arriving late at the pupation site tend to dislodge the early arrivals (that are in the process of pupation). The dislodged pupae drown in the liquefied medium. When various strains of *D. melanogaster* were tested in this fashion, it was found that although the total number of pupae formed varied considerably, the number of survivors was relatively constant. A culture bottle of mutant flies would thus be expected to yield approximately the same number of progeny as a culture of wild type flies. From these observations, there is no indication of the low fitness value which mutant flies typically exhibit in competition with wild type flies grown in the same culture bottle.

If a fly culture contained genotypes of varying fitnesses (with respect to successful competition for limited pupation sites) in constant proportions, then the greater the number of young larvae the better the chance that all survivors would be of the fitter genotypes. In this case, the frequencies of genotypes are being held constant, but the density of the population is allowed to vary. Thus, these situations in which the number of survivors is limited by the number of occupyable sites or territories demonstrates that soft selection is density-dependent.

Now suppose that in a culture containing genotypes of varying fitnesses there is a constant number of larvae competing for a limited number of pupation sites (e.g., three times the number of pupation sites). In this case, density is being held constant and the relative frequencies of different genotypes is allowed to vary. Obviously, the number of survivors of the less-fit genotypes will depend on the number of fitter genotypes with which they have to compete. This illustrates that soft selection is dependent on the frequencies of the different genotypes (frequency-dependent). Therefore, soft selection is both density-dependent and frequency-dependent.

Wallace believes that soft selection is far more common in nature than hard selection. The invariate fitness values that population geneticists use in simplified models are based on hard selection. By themselves, these models are inadequate for describing the evolution of populations.

Selection is the differential (nonrandom) reproduction of genotypes, the intensity of which is expressed as a coefficient of selection. The *selection coefficient* (*s*) may be defined as the proportionate reduction in the gametic contribution of one genotype compared with the most fit genotype presently in the population. Fitness of a given genotype then becomes $1 - s$. Both selection coefficient and fitness may take any value from zero to one.

> In the single locus case in which the recessive phenotype, on the average, produces ninety-five zygotes for every one hundred zygotes produced by the dominant phenotype, the intensity of selection acting against the recessive allele is expressed as $s = 0.05$. The fitness of the phenotype that proportionately produces the greatest number of zygotes is by convention assigned a fitness of 1.0. In the present example, the dominant phenotypes (homozygous dominant and heterozygous genotypes) have fitness = 1.0, and the

recessive type has fitness $= 1.0 - 0.05 = 0.95$; i.e., the recessive type has only 95 per cent of the reproductive capacity of the dominant types.

Selection Against the Recessive Allele

When selection acts against the recessive phenotype, the frequency of the recessive allele (q) is changed by the increment

$$\Delta q = -\frac{sq^2(1-q)}{1-sq^2},$$

where s is the selection coefficient against the recessive allele. From this formula it is clear that the rate of evolution is a function both of selection intensity and gene frequency.

A population in equilibrium should exhibit the zygotic ratio $p^2\, AA : 2\, pq\, Aa : q^2\, aa$. After one generation of selection against the recessive phenotype, the zygotic ratio should be $p^2\, AA : 2\, pq\, Aa : q^2\,(1-s)\, aa$, because the proportionate contribution of a given genotype is found by multiplying the genotypic frequency by its fitness value. The gametic ratio $A : a$ is $(p^2 + pq) : [pq + q^2(1-s)]$, respectively, or

$$p(p+q):pq+q^2-q^2s = p:(pq+q^2-q^2s) = p:[q(p+q)-q^2s] = p:(q-q^2s).$$

The total population after one generation of selection is then expressed as

$$p + q - q^2s = 1 - q^2s.$$

Therefore the allelic frequencies in the next generation are

$$p_1 = p(1 - sq^2) \quad \text{and} \quad q_1 = (q - sq^2)/(1 - sq^2).$$

The change in frequency of the recessive allele due to one generation of selection against the recessive allele is

$$\Delta q = q_1 - q = \frac{q - sq^2}{1 - sq^2} - q = -\frac{sq^2(1-q)}{1-sq^2}.$$

Natural selection will be most influential in changing the frequency of a recessive allele when the recessive genotype is lethal (i.e., when the recessive individual dies before sexual maturity or is completely sterile). In either case, $s = 1$ and the formula for change in gene frequency of the recessive allele becomes

$$\Delta q = -\frac{q^2(1-q)}{1-q^2} = -\frac{q^2(1-q)}{(1+q)(1-q)} = -\frac{q^2}{1+q}.$$

Eugenics is the application of genetic principles to the improvement of human heredity. *Negative eugenics* proposes to improve human heredity by removing certain genotypes from the population either by sterilization or by segregating them into institutions so that they cannot reproduce. The Nazi regime under Adolf Hitler slaughtered millions of Jews in the name of eugenics; only madness would dictate killing as a eugenic measure. *Positive eugenics* proposes to improve human heredity by encouraging certain genotypes to contribute proportionately more than their share to subsequent generations. The efficiency of negative eugenics (apart from humanitarian considerations) is so low for rare recessive defects as to be impractical.

> In order to predict how many generations of selection against a recessive allele would be required to achieve a certain reduction in the incidence of a given trait, the following calculations need to be made. In the previous section it was shown that one generation of complete selection ($s = 1$) against a recessive allele caused its frequency to decrease by the amount $q^2/(1 + q)$. The frequency of this allele in the first generation following selection (q_1) can be expressed in terms of the initial gene frequency (q) as
>
> $$q_1 = q - \Delta q = q - \left(\frac{q^2}{1 + q}\right) = \frac{q}{1 + q}.$$
>
> Similarly, the gene frequency of the second generation following selection (q_2) can be expressed in terms of the allelic frequency of the previous generation as $q_2 = q_1/(1 + q_1)$; in terms of the original gene frequency this becomes $q_2 = q/(1 + 2q)$. Generalizing for any number of generations (t) of complete selection against the recessive allele, we have $q_t = q/(1 + tq)$. Solving for t, we obtain a formula expressing the number of generations of complete selection against the recessive allele required to lower the gene frequency from its initial value (q) to any specified value (q_t). Thus,
>
> $$t = \frac{1}{q_t} - \frac{1}{q}.$$
>
> Now for a specific example, consider the lack of pigmentation in humans called albinism. This trait is produced by a rare recessive autosomal gene. If albinos occur in 1 out of 20,000 individuals of a population, and we let the frequency of the albino gene be q, then $q = \sqrt{1/20,000} = 1/141$. Suppose that it is planned to reduce by half the incidence of this genetic condition through compulsory sterilization of all albinos. How long would this take? The goal in this case is to reduce the incidence of albinism to 1/40,000; then $q_t = \sqrt{1/40,000} = 1/200$. This will require $t = 1/(1/200) - 1/(1/141) = 59$ generations. If the average generation time is 25 years, it will take about 1,500 years for such a plan to accomplish its objective.

The efficiency of complete selection against a recessive allele decreases markedly as the frequency of that allele diminishes. When the recessive allele becomes rare in the gene pool, q in the above formula becomes so small that the denominator approximates unity; then $\Delta q = -q^2$. That is, the change in gene frequency per generation of a rare allele is roughly equivalent to the frequency of homozygous recessive lethal zygotes. Selection against rare recessives in general becomes very ineffective at low gene frequencies, and at frequencies less than 0.1 the efficiency decreases almost logarithmically.

A recessive allele, called ebony, produces a dark body color in *Drosophila melanogaster*. A large number of ebony heterozygotes were released into an area of England normally devoid of this species. The fate of the ebony allele in this population was followed for several generations. The rate of elimination of the ebony allele was more rapid than expected even for a recessive lethal gene. From this it was inferred that the ebony allele also reduced the fitness of heterozygotes; i.e., the ebony allele (although recessive with respect to phenotype) was dominant with respect to fitness. Let s represent the harmful effect of a recessive allele (a) when in homozygous condition; let h represent in heterozygotes the fraction of the harmful effect that *a* has in homozygotes. Thus, *h* is a measure of dominance of *a* with respect to fitness.

	Genotypes		
	AA	Aa	aa
Initial frequencies	p^2	$2pq$	q^2
Fitness values	1	$1 - hs$	$1 - s$

The average fitness of the population after selection is

$$p^2 + 2pq(1 - hs) + q^2(1 - s) = 1 - 2hspq - sq^2.$$

If we assume that this harmful allele is rare in the population, the homozygote (*aa*) would be extremely rare and may be eliminated from consideration. Under this simplifying assumption, the average fitness of the population is approximately $1 - 2hspq$. The frequency of *a* after selection is

$$\frac{pq(1 - hs) - q^2(1 - s)}{1 - 2hspq}.$$

Again, when *q* is very near zero and *p* is very near 1.0, the frequency of *a* after selection simplifies to $q(1 - hs)$. The change in gene frequency (Δq) is $q(1 - hs) - q$, or $-hsq$.

These calculations show that the loss of a rare deleterious allele through selection acting primarily against heterozygotes is directly proportional to the frequency of the harmful allele. If the allele is only harmful in homozygotes, its loss is proportional to the square of its frequency in the population.

Partial selection is said to exist when the selection coefficient (*s*) has a value greater than zero but less than one ($1 > s > 0$). When partial selection is acting on a recessive allele, the effectiveness of that selection in causing genetic change is greater at intermediate gene frequencies than near the extremes.

Selection Against the Dominant Allele

Visualize a population being shifted to an environment in which the dominant phenotype specified by a given locus suddenly becomes lethal. In one generation of complete selection against the dominant allele, all dominant phenotypes disappear leaving only the viable recessive phenotypes. A dominant lethal mutation occurring in a gamete would be eliminated from the population by the death of the individual formed therefrom. In one generation of selection against the dominant allele, the change in frequency of the recessive allele becomes

$$+\Delta q = \frac{sq^2(1 - q)}{1 - s(1 - q^2)}.$$

Consider a locus in equilibrium for a pair of alleles, A and a, with frequencies p and q, respectively.

Genotypes	AA	Aa	aa	Total
Initial frequencies	p^2	$2pq$	q^2	1
Relative fitness	$1 - s$	$1 - s$	1	
Gametic contribution	$p^2(1 - s)$	$2pq(1 - s)$	$q^2(1)$	$1 - s(1 - q^2)$

The frequency of the recessive allele after one generation of selection against the dominant gene (q_1) is

$$[pq(1 - s) + q^2]/[1 - s(1 - q^2)];$$
$$\Delta q = q_1 - q = + \frac{sq^2(1 - q)}{1 - s(1 - q^2)}.$$

Note that when selection is against the dominant gene, the recessive allele will increase, hence, the positive Δq value. If the dominant gene is lethal ($s = 1$), then $\Delta q = p$; i.e., the change in frequency of the recessive allele is equivalent to the frequency of the dominant allele.

As was the case for partial selection against a recessive allele, the magnitude of evolution that occurs when partial selection acts against the dominant allele is greater at intermediate than at extreme allelic frequencies. It might seem logical that selection against a gene dominant with respect to fitness would always produce greater genetic changes than the same intensity of selection directed against a gene recessive with respect to fitness. This assumption follows from the fact that dominant genes are not hidden from selection in heterozygotes as is the case with recessives. Although this is true at lower frequencies of the recessive allele, it

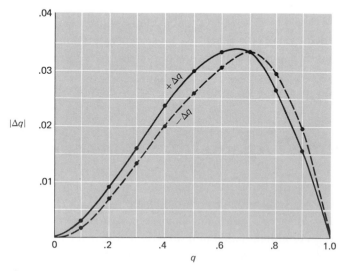

Figure 13.2. Magnitude of absolute change in the frequency of a recessive allele $|\Delta q|$ at different frequencies of the recessive allele (q) where selection ($s = 0.2$) favors the recessive phenotype ($+\Delta q$) and where selection favors the dominant phenotype ($-\Delta q$).

does not hold true at higher frequencies because the majority of genotypes are homozygous recessive at high frequencies of the recessive allele.

> The curves in Figure 13.2 show the absolute change in frequency of the recessive allele $|\Delta q|$ for a selection coefficient of 0.2 at various frequencies of the recessive allele (q). The curve labeled $+\Delta q$ represents selection against the dominant allele; $-\Delta q$ represents selection against the recessive allele. Notice that at certain gene frequencies (e.g., $q = 0.8$) the absolute genetic change $|\Delta q|$ is actually greater when selection acts against the recessive than when it acts against the dominant allele. As the value of s decreases, the difference between the effectiveness of selection against dominants and recessives is diminished.

Selection in the Absence of Dominance

When the fitness of heterozygotes is exactly intermediate between the two homozygous genotypes, maximal genetic change for a given selection intensity occurs closer to the intermediate gene frequency of 0.5 (though not exactly at 0.5) than in the previous cases where selection acted against dominants or recessives. The magnitude of $|\Delta q|$ produced by a given selection coefficient in the absence of dominance is not always greater than in the case of that same amount of selection against dominant or recessive alleles. If s is small, the $|\Delta q|$ is essentially equivalent to selection directly against the gametes.

> Let the frequencies of codominant alleles A^1 and A^2 be represented by p and q, respectively.

Genotypes	A^1A^1	A^1A^2	A^2A^2	Total
Initial frequencies	p^2	$2pq$	q^2	1
Relative fitness	1	$1-s$	$1-2s$	
Gametic contribution	p^2	$2pq(1-s)$	$q^2(1-2s)$	$1-2sq$

The new total after one generation of selection is found as usual:

$$
\begin{aligned}
p^2 + 2pq(1-s) + q^2(1-2s) &= p^2 + 2pq - 2pqs + q^2 - 2q^2s \\
&= (p^2 + 2pq + q^2) - 2pqs - 2q^2s \\
&= 1 - 2sq(p+q) \\
&= 1 - 2sq
\end{aligned}
$$

The frequency of the A^2 allele after one generation of selection is

$$
q_1 = \frac{pq(1-s) + q^2 - 2q^2s}{1 - 2sq} = \frac{q - qs - q^2s}{1 - 2sq}.
$$

The amount of evolution due to one generation of selection is

$$
\Delta q = q_1 - q = -\frac{qs(1-q)}{1-2sq}.
$$

Figure 13.3 shows the change in gene frequency when selection acts against the heterozygote with an intensity of 10 per cent (i.e., $s = 0.1$) and against the A^2A^2 homozygote with twice this intensity ($2s = 20$ per cent).

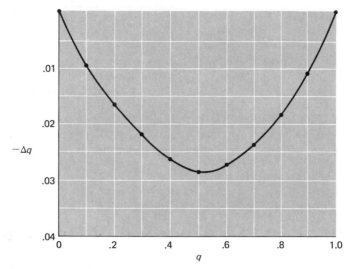

Figure 13.3. Change in gene frequency of A^2 allele ($-\Delta q$), given the fitness values of genotypes A^1A^1, A^1A^2, and A^2A^2 to be 1.0, 0.9, and 0.8, respectively.

Note in Figure 13.3 that maximum evolution occurs quite near, but not at $q = 0.5$; i.e., the curve is not quite symmetrical. Comparison of this curve with the previous ones (all $s = 0.2$) in this chapter reveals that $|\Delta q|$ at certain gene frequencies is less than when selection acts against dominants or recessives.

Selection Against Haploids

Selection may occur in the haploid or gametic stage rather than in the diploid zygotic phase of the life cycle. This is more likely to occur in plants than in animals because gene activity is usually greater in plant gametophytes than in animal gametes. This is especially true in the lower plant forms such as mosses and liverworts in which the sporophytic (diploid) stage is dependent on the gametophyte for sustenance. Gametic selection implies that gametes bearing different alleles fail to function equally in the fertilization process. However, a form of gametic selection may also occur during the formation of gametes when abnormal meiotic segregation produces ratios other than 50:50 for a pair of alleles. A term commonly applied to this latter process is *meiotic drive.*

> An apparent example of meiotic drive is known for the *t*-locus in certain strains of the mouse (*Mus musculus*). Homozygotes for the recessive *t* allele are either lethal or sterile. Heterozygotes (*Tt*) segregate these alleles abnormally in the average ratio of 95 per cent *t*:5 per cent *T*. Meiotic drive can maintain the frequency of the *t* allele at a relatively high value in the gene pool despite complete selection against the homozygous recessive genotype. A stable equilibrium could exist under such conditions if the loss of *t* alleles in the zygotic stage is compensated by the gain made in the gametic stage.

Selection against sex-linked genes is equivalent to gametic selection in hemizygous males

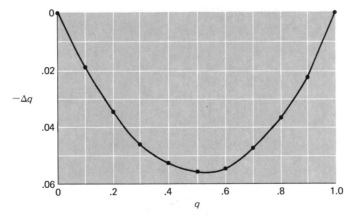

Figure 13.4. Change in gene frequency of an allele $(-\Delta q)$ when $s = 0.2$ and selection acts in the gametic stage.

(XY) and to zygotic selection in homogametic females (XX). Thus, evolution is expected to proceed faster at a sex-linked locus than at an autosomal locus with the same intensity of selection.

Gametic selection causes greater genetic change at a given selection intensity than any form of zygotic selection.

Consider a case in which selection acts in the gametic stage.

Alleles	A^1	A^2	Total
Initial frequencies	p	q	1
Relative fitness	1	$1-s$	
After selection	p	$q(1-s)$	$1-qs$

The frequency of the A^2 allele after selection is

$$q_1 = \frac{q(1-s)}{1-qs}.$$

The genetic change wrought by one generation of selection is

$$\Delta q = q_1 - q = \frac{q(1-s)}{1-qs} - q = -\frac{qs(1-q)}{1-qs}.$$

The asymmetrical curve in Figure 13.4 shows the genetic change caused by a selection intensity of 20 per cent against the A^2 allele. Note that the curve reaches maximum $-\Delta q$ at gene frequency very near, but not exactly at, 0.5. Comparison of this curve with any of the preceeding ones in this chapter (all with $s = 0.2$) shows that selection in the gametic stage is more efficient (i.e., produces a larger $|\Delta q|$) at all gene frequencies than in the zygotic stage.

Selection Favoring Heterozygotes

Adherence of a population to Hardy-Weinberg conditions is not the only way to establish genetic equilibrium. When the heterozygous genotype is adaptively superior to either homozygous genotype, the population will evolve toward a stable equilibrium determined entirely by the relative selection coefficients acting against the less fit homozygotes. *Overdominance* and *euheterosis* are terms commonly applied to allelic interactions manifested as heterozygote superiority with respect to fitness. When considering aspects of the phenotype other than fitness characters (such as size of flowers, fruits, leaves, and general vigor), heterozygote superiority (heterosis) is termed *luxuriance.*

Consider a genetic locus with two alleles, A^1 and A^2, with freqiencies p and q, respectively. Selection acts against the A^1A^1 homozygote with intensity s_1 and against the A^2A^2 homozygote with intensity s_2.

Genotypes	A^1A^1	A^1A^2	A^2A^2	Total
Initial frequencies	p^2	$2pq$	q^2	1
Relative fitness	$1 - s_1$	1	$1 - s_2$	
Gametic contribution	$p^2(1 - s_1)$	$2pq$	$q^2(1 - s_2)$	$1 - p^2s_1 - q^2s_2$

The frequency of the A^2 allele following one generation of selection favoring heterozygotes is

$$q_1 = \frac{pq + q^2(1 - s_2)}{1 - p^2s_1 - q^2s_2}.$$

The change in gene frequency due to one generation of selection is

$$\Delta q = q_1 - q = \frac{pq + q^2(1 - s_2)}{1 - p^2s_1 - q^2s_2} - q = +\frac{pq(ps_1 - qs_2)}{1 - p^2s_1 - q^2s_2}.$$

At equilibrium $\Delta q = 0$, and in the numerator of the above equation $ps_1 = qs_2$. The equilibrium frequency of the A^1 allele is $\hat{p} = s_2/(s_1 + s_2)$, and the equilibrium frequency of the A^2 allele is $\hat{q} = s_1/(s_1 + s_2)$. The curve in Figure 13.5 gives Δq values over all gene frequencies when $s_1 = 0.2$ and $s_2 = 0.5$; equilibrium exists at $\hat{q} = 0.2/(0.2 + 0.5) = 0.286$. Note that when q is greater than its equilibrium value, the negative Δq indicates that q is decreasing. When q is less than its equilibrium value, the positive Δq indicates that q is increasing. The curve is obviously not symmetrical on either side of the equilibrium value unless $s_1 = s_2$, in which case $\hat{q} = 0.5$.

In the special case in which the selection coefficients against both homozygotes are $s = 1.0$, the equilibrium frequencies of both alleles are maintained at 0.5. Only the heterozygotes survive in this *balanced lethal system* (recall the curly-plum example in *Drosophila*; Chapter 9). A population is said to be *polymorphic* when it consists of two or more Mendelian traits, the rarest of which is still so common that recurrent mutation cannot be the only force maintaining the responsible alleles in the gene pool. When a polymorphism is attributed to overdominant allelic interaction with respect to fitness, it is termed a *balanced polymorphism.* At least two alleles must stay in the gene pool because of the adaptive superiority of the heterozygous genotype. Maintenance of more than one allele at a locus in the gene pool is equivalent to storage of genetic variability. The evolutionary flexibility of a population is equivalent to its store of genetic variability. In comparison to a genetically uniform population, a highly flexible one is more likely to recombine this store of genetic material into complexes adapted to meet the challenges of a shifting environment. With overdominance, the population pays the price of continually segregating less-fit homozygotes from the fitter heterozygotes; however the heterozygotes maintain the flexibility that contributes to long term evolutionary potential.

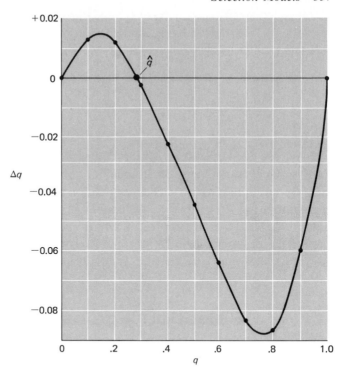

Figure 13.5. Change in frequency of A^2 allele (Δq) when selection favors heterozygotes (A^1A^2); relative fitness of $A^1A^1 = 0.8$ and of $A^2A^2 = 0.5$.

A human polymorphism involving a variant form of hemoglobin exists in some parts of Africa. An unusual form of hemoglobin called Hb S and the normal hemoglobin Hb A are produced by the codominant alleles S^1 and S^2, respectively. Homozygotes for the S^1 allele produce only Hb S and suffer from a disease called "sickle cell anemia" that warps red blood cells. Many of these individuals die before sexual maturity. Heterozygotes have both Hb A and Hb S, but may not have any noticeable debility. Homozygotes for the S^2 allele have only normal Hb A. In some African Negroid populations, the incidence of individuals with the abnormal hemoglobin is as high as 40 per cent. Approximately 3 per cent of those with Hb S suffer from sickle cell anemia. The cause of the relatively high frequency of the S^1 allele, even though it is deleterious when homozygous, has been attributed to the superior fitness of heterozygous carriers. Apparently certain malarial parasites cannot complete a normal life cycle in red blood cells with the abnormal hemoglobin. Heterozygous individuals then leave more progeny on the average than either homozygous genotype. The fitness of S^1S^1 homozygotes is estimated to be 25 per cent of the fitness of heterozygotes. An estimate of the selective advantage that heterozygotes have over normal S^2S^2 homozygotes can be obtained as follows. Let p and q represent the frequencies of the S^1 and S^2 alleles, respectively. Under the assumption that the population has reached a stable equilibrium due to overdominance, the frequency of individuals with Hb S is

$$p^2 S^1S^1 + 2pq\, S^1S^2 = 0.4.$$

Because 3 per cent of those with Hb S are homozygous, $p^2 = 3$ per cent of 0.4 or 0.012; thus, $2pq = 0.4 - 0.012 = 0.388$. The frequency of S^1 allele = frequency of S^1S^1 homozygotes + $\frac{1}{2}$ frequency of heterozygotes; thus, $p = 0.012 + \frac{1}{2}(0.388) = 0.206$.

Genotypes	S^1S^1	S^1S^2	S^2S^2
Relative fitness	$1 - s_1$	1	$1 - s_2$

If the fitness of S^1S^1 is $\frac{1}{4}$ that of S^1S^2, then $s_1 = 0.75$. Earlier in this chapter it was shown that the equilibrium value for S^1 should be $\hat{p} = s_2/(s_1 + s_2)$. Again under the assumption that this locus has reached its equilibrium values, we have $0.206 = s_2/(0.75 + s_2)$, from which $s_2 = 0.195$. We can now express the selective advantage of heterozygotes relative to normal homozygotes as $1/(1 - s_2) = 1/(1 - 0.195) = 1.24$; i.e., the heterozygotes have a selective advantage 24 per cent greater than that of normal homozygotes due to their resistance against malaria.

A given genotype has no intrinsic fitness of its own apart from the environment in which it exists. One cannot speak of a "good" or "poor" genotype without specifying the environmental conditions in which such an evaluation is made. In the above example of the abnormal hemoglobin Hb S, we saw that allelic interaction in heterozygotes produced overdominance when such genotypes live in areas in which malaria is endemic. These same genotypes living in other environments devoid of the malarial organisms do not enjoy any detectable adaptive advantage and may even be slightly less fit than those with normal hemoglobin.

Selection Against Heterozygotes

When homozygous genotypes are adaptively neutral with respect to one another and heterozygous genotypes are less fit, an unstable equilibrium will exist only at gene frequency 0.5. An allele with frequency greater than 0.5 will move toward *fixation* (allelic frequency = 1.0) in the gene pool; below 0.5 it will move toward *loss* (allelic frequency = 0) from the population.

Let p and q represent the frequencies of alleles A^1 and A^2, respectively.

Genotypes	A^1A^1	A^1A^2	A^2A^2	Total
Initial frequencies	p^2	$2pq$	q^2	1
Relative fitness	1	$1 - s$	1	
Gametic contribution	p^2	$2pq(1 - s)$	q^2	$1 - 2pqs$

The frequency of the A^2 allele after one generation of selection against the heterozygote is

$$q_1 = \frac{pq(1 - s) + q^2}{1 - 2pqs} = \frac{q - spq}{1 - 2spq}.$$

The change in gene frequency per generation which accrues by this selection is

$$\Delta q = q_1 - q = \frac{q - spq}{1 - 2spq} - q = \frac{spq(2q - 1)}{1 - 2spq}.$$

When s is small, the denominator approximates unity and $\Delta q = 2spq(q - \frac{1}{2})$. It is obvious from this formula that when $q < \frac{1}{2}$, Δq is negative and hence q will decrease; when q is $> \frac{1}{2}$, Δq is positive and q will increase. An unstable equilibrium exists at $\hat{q} = 0.5$. Figure 13.6 illustrates the genetic change that occurs when $s = 0.2$ against heterozygotes. Unlike the curve for selection favoring heterozygotes, this curve is symmetrical on either side of the equilibrium value. Maximum $|\Delta q|$ values occur close to, but not at, $q = 0.2$ and $q = 0.8$.

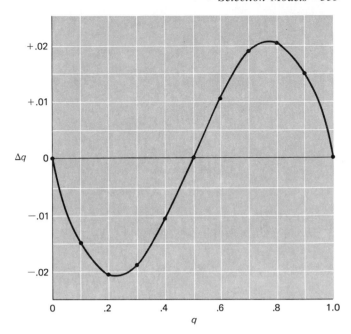

Figure 13.6. Change in frequency of the recessive allele (Δq) when the selection coefficient against heterozygotes is $s = 0.2$.

It is possible for a stable equilibrium to be established by opposing selection forces acting on different components of fitness (e.g., the t locus in mice). Selection may act against a given genotype during the embryonic stage, but favor that same genotype in the adult stage (or *vice versa*).

The human blood group antigen called Rh factor is produced by a dominant gene (R). People who have this gene are called Rh positive; those homozygous for the recessive allele (rr) do not have this antigen and are called Rh negative. Heterozygous babies born to recessive mothers may stimulate (sensitize) the mother to produce antibodies against the Rh factor. Such antibodies may cross the placenta and cause destruction of the baby's red blood cells. In severe cases of this disease (called hemolytic anemia of the newborn or erythroblastosis fetalis), the infant may die (Figure 13.7). Selection in this case acts not against all heterozygotes, but only those born to recessive mothers. If $(1 - s)$ is the fitness of heterozygotes born to Rh negative mothers, it can be shown that $\Delta q = spq^2(q - \frac{1}{2})/(1 - spq^2)$, where p = frequency of R allele and q = frequency of r allele. When s is small, Δq is approximately $spq^2(q - \frac{1}{2})$. An unstable equilibrium exists at $q = 0.5$. Thus, the frequency of the recessive allele should have fallen to a very low frequency if initially less than 0.5 and risen to a high value if initially greater than 0.5, if this condition has existed for a very long time, as it probably has. In actual fact, the frequency of r in the United States is approximately 0.38. Several possible explanations for the high incidence of the gene have been proposed. One explanation suggests that the viable heterozygotes (whether born to recessive mothers or not) may compensate for the loss of heterozygotes born to recessive mothers by enjoying a superior survival rate. The equilibrium frequency of heterozygotes is given as $2pq = (2pq - spq^2)(1 + h)$, where the fitness of viable heterozygotes is $(1 + h)$. Solving for h, we have $h = sq/(2 - sq)$. When s is small, the denominator is approximately 2; thus $h = sq/2$. If $s = 0.035$ (small) and $q = 0.38$, $h = 0.00165$; notice what a small heterozygote advantage is required to maintain this balance.

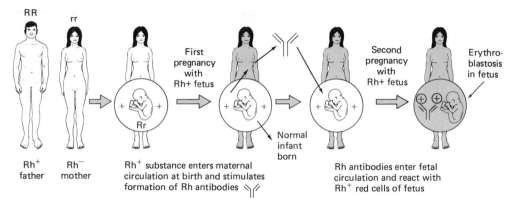

Figure 13.7. Rh disease as an example of a disadvantaged heterozygote. An Rh positive baby in a sensitized Rh negative mother can suffer immune attack by maternal antibodies.

Essay Questions

1. *Is it possible for a population to make gains in immediate fitness (i.e., adaptation to present environmental conditions) and lose evolutionary flexibility in the process?*

2. *Does overspecialization predispose a population to extinction?*

3. *How is it possible for selection at many loci (often with high intensities) to act without any noticeable reduction in size of the population?*

4. *Can a genotype be extremely fruitful in terms of the relative numbers of zygotes it spawns and yet be at a selective disadvantage?*

5. *Can selection over many generations push one allele to fixation without being responsible for physical deaths?*

NATURAL SELECTION

The term *adaptation* is widely used and commonly misused. In one sense, it may be used for the evolutionary process whereby organisms become modified for survival and reproduction in specific environments. In another sense, it refers to any morphological (anatomical), physiological (biochemical), developmental (embryological), or behavioral (ethological) characteristic of an organism that enhances its probability for survival and leaving descendants. The genotype that, on the average, leaves the largest proportion of fertile progeny is considered to be the most fit. Because of the astronomical number of possible genotypes, it is unlikely that the one genotype with the absolute maximal theoretical average fitness has ever been produced in any finite population. Fitness, therefore, is a *relative* attribute in at least two respects: (1) it is relative between individuals of a population, and (2) it is relative to a particular set of environmental conditions. No one genotype can have maximum fitness in all environments. The longevity and the phenotypic vigor associated with a particular genotype do not, by themselves, constitute fitness (although they may be important components of fitness to the extent that they contribute to the production of viable, fertile offspring). Therefore, we should not speak of "survival of the fittest," but rather survival and reproduction of the fitter genotypes. Because many genes have pleiotropic effects on fitness, dominance

**TABLE 14.1. Model of a Genetic Locus
with Two Alleles in Which Phenotypic
Dominance Is Distinct from Dominance
with Respect to Fitness**

Genotype	Phenotype	Reproductive Ability
A^1A^1	tall	sterile*
A^1A^2	tall	fertile
A^2A^2	short	fertile

* Genotypes that are inviable (lethal) or result in sterility have no fitness. They do not make any contribution to the gene pool of subsequent generations.

with respect to fitness need not be correlated with phenotypic dominance. In Table 14.1, gene A^1 exhibits phenotypic dominance, while its allele A^2 exhibits reproductive dominance (fitness).

Natural selection operates most fundamentally at the level of the individual's total phenotype and thereby only indirectly on the individual's total genotype. Single-locus selection models are gross oversimplifications of biological reality. High fitness involves the cooperation of a large number of genes (virtually the total genotype). A population does not benefit from selection for viability unless, at the same time, there is selection for reproductive traits, and vice versa. What good is an individual that produces large numbers of gametes, if gametogenesis is so abnormal that none of them have the haploid complement of chromosomes? Individuals do not evolve; this is a function of the population gene pool. Selection, therefore, tends to adjust gene frequencies in the gene pool to maximize the probability of generating coadapted gene complexes (i.e., integrated combinations of genes that contribute positively to all aspects of fitness). What is "good" for the individual is not necessarily "good" for the population, and *vice versa*.

> Genotypes that predispose an individual to sacrifice itself for the good of the population would probably have a low survival value, but a population harboring such genotypes could have higher reproductive values as a consequence. For example, think how the gene pool benefits when parents preferentially feed their young rather than themselves in times of food scarcity. The undernourished parents are likely to succumb to disease or predators, but the genes that foster this kind of behavior tend to perpetuate the species and strengthen the probability of such beneficial altruistic behavior in subsequent generations.
>
> Alternatively, some genotypes may have high individual reproductive values, yet depress the average fitness of the population. Consider the tailless gene in mice (discussed in Chapter 9). Meiotic drive favors production of gametes bearing the tailless allele, but the population fitness is lowered by the production of lethal tailless homozygotes.

Failure of a population to realize its reproductive potential should not be ascribed totally to selection. Many factors contributing to environmental resistance may treat all genotypes alike. For example, neither seeds with "good" genotypes nor "poor" genotypes can germinate if they happen to settle on rocks or other barren substrate. A volcanic eruption destroys all genotypes within the path of its lava flow. An oil leak in a marine environment indiscriminately wipes out many life forms in the affected area. Flash floods can drown many animals and uproot plants regardless of their genotype. When a prolonged drought dries up ponds, it generally kills organisms irrespective of their genotypes. These kinds of nonselective factors can thwart the reproductive potential. Natural selection, however, may then be expressed among the survivors by certain genotypes leaving more progeny than others.

A single gene or a set of polygenes may be selected for its beneficial effects on the phenotype. Probably all genes have pleiotropic effects, but not all pleiotropic effects are necessarily beneficial. If the original genes have harmful pleiotropic effects, any other genes that would tend to compensate for these disadvantages would have a high selective value. These compensating genes would produce distinctive phenotypes of their own. Therefore, two groups of characters (one for the original adaptation and another specified by the compensating genes) would tend to evolve as an adaptive unit. This is the principle of *adaptive compensation;* the two groups of characters are kept together by *selective correlation*. The traits produced by the compensating genes may have no selective value by themselves. Many of these "nonessential" character differences between different populations could be generated by selective correlation acting on different adaptive compensation systems.

Mimicry

When two widely divergent groups come to occupy similar habitats, selection for common functions or adaptations may produce superficially similar phenotypes. This is known as convergent evolution. The wings of birds and insects bear superficial resemblances, but are only analogous structures that have convergently evolved to exploit an aerial niche. An example of convergent evolution on a much smaller scale is found in the phenomenon of *mimicry.* Here we have two (or more) species resembling each other, because at least one of these species thereby gains a selective advantage. Three elements are involved in classical mimicry: (1) a model, (2) a mimic, and (3) a signal receiver. The model is a distasteful or otherwise obnoxious species that transmits a signal advertising itself as such. The mimic is a signal-transmitting species to which the receiver directs a response that is not of advantage to itself (but is of advantage to the mimic). The signal receiver is a species (commonly a predator) that fails to distinguish between the model and mimic and therefore responds to both with similar behavior. The signals (visual, acoustical, or olfactory) emitted by the mimic resemble those of the model.

> Visual signals may be color or behavioral patterns. Color patterns that resemble large eyes ("eye spots") on the hind wings of a moth may be suddenly revealed, startling predators, or confusing the predator's attack away from the head end (Figure 14.1). Some beetles perform mock stinging motions resembling those of bees. Acoustical signals may also be displayed by these beetles (e.g., humming like a bee). Some hole-nesting birds (e.g., tits and chickadees) hiss like a snake when disturbed on their nests. Olfactory signals may be emitted by some "stinky" flowers, serving to attract carrion insects for pollination.

Figure 14.1. Polyphemus moth with "eye spots" on wings. (Courtesy Carolina Biological Supply Company)

The original (classical) observations of mimicry were made by English naturalist Henry W. Bates while in Brazil (from 1849 to 1860) collecting, among other things, butterflies. In 1862, he published his "mimetic theory" in the *Transactions of the Linnean Society, London*. Bates observed that a butterfly of the family Pieridae differed sharply from other species in this family and closely resembled certain common butterflies of the family Heliconiidae. These latter butterflies were extremely abundant, conspicuously colored, and slow fliers (therefore easy to catch). He noted little predation on Heliconiids by birds and assumed that these butterflies were simply unpalatable. If the Pierids were edible, any variation in color pattern that would tend to resemble the pattern of Heliconiids might also tend to confuse the bird predators. Having learned by previous encounters with Heliconiids to associate their warning coloration with their noxious qualities, the predator would tend to ignore those Pierids that imitated the Heliconiid pattern. Natural selection would be expected to perfect the initial resemblance in each generation so that the color pattern of the Pierid would more nearly match that of the Heliconnid. Bates called the noxious Heliconiid a *model* and the palatable Pierid a *mimic*. This type of mimicry is now termed *Batesian mimicry* (Figure 14.2) and has the following assumptions:

1. The model is obnoxious in some way to the signal receiver (e.g., a predator); the mimic is benign.
2. The model emits conspicuous signals perceived by some other species in the community; the mimic bears a close resemblance in one or more ways to the signals of the model.
3. Models are much more frequent in numbers than mimics, and both occur in the same geographical location as the signal receiver.

Figure 14.2. An example of Batesian mimicry in butterflies. The North American monarch (*Danaus plexippus*) at top is the model, and the viceroy (*Basilarchia archippus*) at bottom is the mimic. The monarch belongs to the Danaidae, a family that the world over contains butterflies characterized by a toughness of body, strong odor, and presumably evil taste. Repeated experiments with birds and monkeys indicate that danaids are almost always rejected as food and appear to be eaten, if at all, only under stress of extreme hunger. The viceroy, on the other hand, is a member of the Nymphalidae, the species of which are, in most cases, readily eaten and are therefore classified as palatable. It is important to note that the viceroy has departed widely in pattern and color from other species of the genus *Basilarchia*. (Courtesy of Ward's Natural Science Establishment, Inc.)

4. The mimic gains a selective advantage by resembling the model; the model is harmed in proportion to the relative abundance of the mimic.

Obviously, when the mimic becomes very numerous, it will take a predator longer to learn to associate the signal (e.g., warning coloration) with the noxious quality of the model. The smaller the mimic population relative to the model, the more accurate the mimetic features are likely to become. This is anticipated because as the phenotype of an abundant edible species moves toward that of an obnoxious species, natural selection would tend to favor mutations in the obnoxious species that would make it distinctive from the edible species. Mimicries do not appear suddenly; they must be evolved and perfected over a long period of time. When a mutation appears within an unprotected species that happens to produce a phenotype resembling that of a protected species, natural selection would tend to spread this favorable variant throughout the population. During this process, the control the gene exerts over morphological (developmental) processes gradually undergoes modifications by many modifier genes in the residual genotype so that the mimetic similarity it promotes is improved.

Some weeds have become useful to humans through the process of mimicry. A classic example of a converted weed is rye (*Secale cereale*). Wild rye grows as a weed in wheat (genus *Triticum*) fields. Separation of weeds from wheat must be attempted after the seed is harvested (on the basis of weight and size differences). Since wheat has been selected for large seed size and rigid panicle (the end of the stalk bearing the seeds) spindle, rye also was unconsciously selected by the same process. Wild rye is a perennial, but occasionally an annual form is produced. Since wheat is harvested annually, rye also comes under selection for its annual variants. In this way, the annual, hard-spindled rye weed evolved from the perennial, weak-spindled wild rye without attracting human attention. Since rye is less demanding of climate and soil requirements than wheat, it could outgrow wheat in marginal areas; recognizing this, people began to cultivate it as a separate crop. The same evolutionary scheme has produced the cultivated oat (*Avena sativa*) from a wild grass species (*Avena fatua*).

The terms *unpalatable, defensive, protected,* and the like, are only valid within certain limits; there is no all-or-none effect. For example, rabbits can eat the poisonous toadstool (*Aminita phalloides*); the red-backed shrike (*Lanias collurio*) is skillful at removing the stingers from bees and wasps; the mongoose feeds on poisonous snakes, etc. Protection is not absolute in any "noxious" species.

Mimetic anatomical characters usually evolve on a background of more ancient behavioral patterns. It is entirely possible that a character presently important to a mimetic relationship may have once had a quite different function, and only recently become incorporated into a mimetic pattern. For example, leaf-like fossil insects (Figure 14.3) have been found in rocks of Upper Jurassic age before deciduous trees had evolved. In mimicry, the ultimate stage in the evolution of a signal is already present in the model. This is a unique advantage in evolutionary research.

A second type of mimicry was proposed in 1878 by German zoologist Fritz Müller, hence termed *Müllerian mimicry*. His theory was proposed to explain observations wherein two inedible and unrelated species of butterflies were very similar in appearance (Figure 14.4). It seems logical to assume that if two (or more) noxious species could converge toward a common recognition pattern, each would benefit thereby. As a matter of fact, the greater the number of noxious species employing the same warning signals, the lower the losses to each species individually; i.e., this type of mimicry increases in efficiency with the number of noxious species involved in emitting a common signal. The trend here is toward reducing the number of different color patterns (or other signals) predators (or other signal receivers) must learn to associate with an unpleasant experience. This phenotypic convergence does not place the more common species at a disadvantage (as is the case in Batesian mimicry), because all the signal transmitters are noxious. The rarer species, however, is expected to evolve faster than the common species because a larger proportion of the rarer species will be sacrificed to educate predators.

Genuine Müllerian mimicry probably should not be classified as mimicry at all, because there is no division of the system into model and mimic. Nonetheless, the term persists. The species in a Müllerian mimicry may be related or may have adopted similar protective signal patterns by convergent evolution from unrelated ancestors. It is possible for one of the related species in a Müllerian mimicry to lose its noxious qualities and become completely dependent for protection on the distasteful species. In the evolutionary process of losing its repellent

Figure 14.3. Two oriental dead leaf butterflies (*Kallima inachis*) are shown resting on a branch among the leaves. The insects' wings are leaflike in shape, the apices of the fore wings being drawn out like a leaf tip, and the anal angle produced to form a "petiole." Even an apparent "midrib" and "veins" are shown, and to make the resemblance even more convincing, "worm holes" and "fungus spots" are added. It is to be noted that this butterfly does not always rest in the "leaf position" shown here. Apparently a large enough percentage does and hence this particular adaptation has survival value. Adaptations of this kind involve not only structural modification but special behavior patterns as well. The insect must not only look like a leaf but must also act like one. (Courtesy of Ward's Natural Science Establishment, Inc.)

taste, there would be a time when one species was much more distasteful than the other. At this point, a Müllerian mimicry exists to the extent that both species are distasteful, but at the same time a Batesian mimicry exists because the less noxious member is dependent to some degree on the protection afforded by the more unpalatable member. Therefore, Batesian and Müllerian mimicries are not mutually exclusive. Both types of mimicry must have representatives living in the same area; mimetic patterns would be extremely unlikely to develop in geographically isolated populations unless the predators (or signal receivers) have a wide home range.

A third mimetic concept was introduced by E. G. Peckham (now called *Peckhammian mimicry*). This is an *aggressive mimicry* in which the predator itself is the pretender (*mimic*).

Cleaner symbionts are an example of protocooperation. A wrasse-type fish (*Labroides dimidiatus*) scours the external surface of their larger "customers" for exoparasites. The customers learn to distinguish *Labroides* by size, shape, color, and swimming patterns. They allow the cleaners to approach them and probe their bodies, often lapsing into a

Figure 14.4. An example of Müllerian mimicry in butterflies. Common warning colors are shown by the tropical American *Heliconius xenoclea* (Heliconiidae) and *Actinote diceus callinira* (Acraeidae), both species belonging to families that are distasteful to birds. The family Acraeidae, while very limited in extent in the New World, is a characteristic feature of the African butterfly fauna, and there the species of this family are much mimicked by insects of other groups. (Courtesy of Ward's Natural Science Establishment, Inc.)

trance-like state while being cleaned. Fish of the species *Aspidontus taeniatus* resemble the cleaners in size, coloration, and swimming behavior (Figure 14.5). Customer fish, accustomed to being cleaned by *Labroides,* will usually position themselves for cleaning by *Aspidontus.* When the mimic gets close enough, it bites a piece from its victim's fin and eats it. The victim immediately jerks around, but the mock cleaner stays put as though it was innocent. The victim sees nothing but "a cleaner." After many such unpleasant experiences, fish become wary of cleaners and eventually may learn to

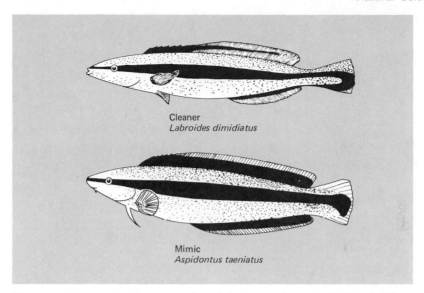

Figure 14.5. Example of an aggressive mimic. A true cleaner *Labroides dimidiatus* (top) and its mimic *Aspidontus taeniatus* (bottom).

distinguish the mimic from a true cleaner. The mimic thus becomes largely dependent on young fish who have not yet learned to make this distinction.

Some orchids of the genus *Orphrys* have flowers that not only mimic the shape of insects, but also give off an odor similar to (and in some cases, stronger than) an odor emitted by the female of the insect species (Figure 14.6). Male insects are thereby attracted to these flowers and, mistaking them for females, attempt to mate with them (*pseudocopulation*). In the process, pollen sacs become attached to the insect's body and are thus transported to another flower. The insects derive no benefit from this, because

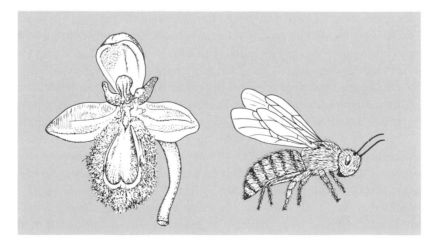

Figure 14.6. Flower of *Orphrys speculum* is pollinated by a male bee (*Scolia ciliata*) because of its resemblance to a female bee.

Figure 14.7. A true coral snake (*Micrurus fulvius*). Repeating sequence of color bands are yellow, red, yellow, black. (Courtesy Carolina Biological Supply Company)

the flowers have no nectar. This symbiotic association could best be classified as a commensalism; the orchid benefits, but the insect is essentially unaffected (other than useless waste of its energy in pseudocopulation).

In 1956, an herpetologist by the name of Mertens proposed yet another form of mimicry as a result of his observations on coral snakes of the New World tropics and subtropics. True coral snakes of the genus *Micrurus* are highly poisonous. They usually advertise their noxious nature by a conspicuous red, black, and yellow banding pattern (Figure 14.7). Some species of the genus *Erythrolamprus* and *Pseudoboa* are only moderately poisonous and at least one species in the genus *Simphis* is nonpoisonous. These "false coral snakes" tend to use the same banding pattern as the true coral snakes. Extremely poisonous snakes usually kill the enemy after a single encounter, thus providing no opportunity for the enemy to "learn" from its experience. This learning experience often can be gained from the mildly poisonous false coral snakes, which are thus considered to be the models, whereas the fatally poisonous species are regarded as mimics. This type of mimicry, in which one species is too poisonous to allow the signal receiver to learn, is called *Mertensian mimicry*. Members of the completely harmless species, using the same color pattern, are Batesian mimics dependent on the intermediate (in terms of their noxious qualities) forms. It might be tempting to consider the mildly poisonous and the fatally poisonous species as Müllerian mimics, but since enemies cannot learn from fatally poisonous snakes, all these snakes do not mutually "teach"; hence, they should not be regarded as Müllerian mimics.

Intraspecific mimicry is a form of mimicry in which both model and mimic are found in

Figure 14.8. Secured threat behavior. Center animal threatens the one at the left and at the same time presents his backside as an appeasement to the higher ranking male at the right.

the same species. The terms "self-imitation" and "automimesis" are sometimes used synonomously.

> The monarch butterfly is sometimes polymorphic in terms of its palatability to predators. If most of the larvae (caterpillars) subsist on plants rich in substances toxic to birds, then the few larvae that live on nontoxic plants are perfect mimics of the more abundant unpalatable members of the same species.
>
> Male hamadryas baboons have conspicuous red buttocks that mimic the estrous swellings of female baboons. The latter present their backsides with tail raised as an invitation to copulation. This display inhibits aggression by males of high social rank (dominant). Males of lower social rank also may present their backsides to a more dominant male as a greeting that acknowledges their inferior social status. It is hypothesized that the dummy estrous swellings of males evolved to mimic those of females as a means of recognition in these highly social groups. A phenomenon called "secured threat" (Figure 14.8) demonstrates the strong protection these appeasement displays offer. One baboon threatens another with angry squeals. This commotion immediately attracts the highest ranking male to the scene. The initiator of the squabble presents his backside to the dominant male who then bypasses the initiator and chases off the threatened animal.

Camouflage

Mimicry is usually thought of in terms of protective patterns resembling some other biotic form. Camouflage, on the other hand, is directed at the substrate (which may be biotic or abiotic). Some animals, such as the octopus, are notorious for the speed with which they can change color patterns to match their substrate. An animal incapable of color change may exhibit *cryptic behavior* by seeking out a suitable background before coming to rest. This is actually the most common type of camouflage. Camouflage is most effective when coupled with cryptic behavior.

Probably the most widely known example of camouflage is that in the peppered moth (*Biston betularia;* Figure 14.9) of England. Before the advent of the industrial revolution, the woods of Britain were clean and virtually the only form of peppered moth found there was

Figure 14.9. Left: *Biston betularia,* the Peppered Moth, and its melanic form *carbonaria,* at rest on soot covered oak trunk near Birmingham, England. Right: *Biston betularia,* the Peppered Moth, and its black form *carbonaria,* at rest on lichened tree trunk in unpolluted countryside. (From the experiments of Dr. H. B. D. Kettlewell, University of Oxford)

light in color. The first report of a dark (melanic) variant appeared in 1849. By 1886 the melanic form had increased in some areas such that it had become far more common than the light form. This period of rapid evolution paralleled the explosive growth of industrialization in the British Isles and the concomitant human population growth. Coal was used extensively to fire the furnaces of industries and homes. Smoke belched from chimneys everywhere and blanketed nearby trees with layers of soot. Because air pollution was the source of the environmental change that sparked the rapid evolution of dark moths, thc phenomonon has been called *industrial melanism* (Figure 14.10).

In the 1950s, H. B. D. Kettlewell captured, marked, released, and recaptured moths in both rural unpolluted and urban polluted woods of England. In the nonindustrial woods of Dorset he recaptured 14.6 per cent of the light moths released, but only 4.7 per cent of the melanic forms. In the woods near industrial Birmingham, he recaptured 13 per cent of the light forms and 27.5 per cent of the melanic moths. Kettlewell found that birds preyed more intensely on the more conspicuous forms. Light moths are more easily seen on the soot-darkened tree trunks that are their normal daytime resting places; dark moths are more easily seen on the lichen-covered bark of nonpolluted trees. Natural selection has radically changed the composition of these moth populations within a span of a mere fifty years. Moreover, it was not only the peppered moth that responded to this environmental change. More than seventy other species of night-flying moths in Britain have gone through similar changes, as have moths of industrial regions in continental Europe, Canada, and the United States. Other

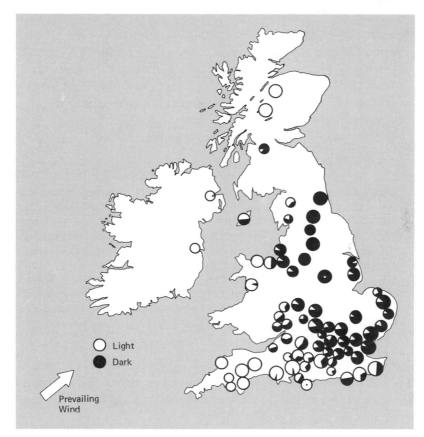

Figure 14.10. Composition of various populations of the peppered moth (*Biston betularia*) in the British Isles.

insects (e.g., ladybird beetles) and other arthropods (e.g., spiders) have also experienced industrial melanism.

Since the enactment of antipollution laws following World War II, the foliage of Great Britain is beginning to emerge from its layers of soot. The percentage of light moths is already significantly higher in some polluted woods than it was in the 1950s when Kettlewell made his initial observations. In one study, light moths increased from 5.2 per cent in 1961 to 8.9 per cent in 1964; by 1974 it had increased to 10.5 per cent. One woodland near Manchester had no observed light forms during the early 1960s, but now they constitute about 2.5 per cent of the samples.

Microevolutionary changes (i.e., changes in gene frequencies at one or a few loci), as exemplified by industrial melanism, appear to be easily reversible phenomena. Probably the most remarkable feature of industrial melanism is the speed with which these populations adapt to altered environments through natural selection.

More recently, detailed studies of the peppered moth and several other species of moths have revealed that the story of industrial melanism may not yet be complete. For example, in numerous localities the light and dark forms exist together in frequencies considerably higher than can be accounted for by recurrent mutation, i.e., they exist in a *polymorphism* (more than one form in the population). Melanic forms remain common in areas where theoretically the birds should have exterminated them. What maintains these polymorphic states? Why are the frequencies of melanic forms in different moth species living in the same area so different? The answers to these questions are not yet known. One theory proposes that heterozygotes for the dominant melanic allele are fitter for some nonvisual trait than homozygotes. The degree of this overdominant gene action could vary widely in different species of moth. Another theory suggests that polymorphisms may be maintained at both extremes by migration between polluted and unpolluted localities.

> The scalloped hazel moth (*Gonodontis bidentata*) exhibits greater local differentiation of their populations than is seen in the peppered moth. In some areas, the scalloped hazel moth forms extremely dense populations (perhaps 50,000 to 100,000 moths per square kilometer). Peppered moth populations are not nearly as dense (perhaps only 10 per square kilometer). The peppered moth must therefore travel considerably further to find a mate than the scalloped hazel moth. Studies have shown that peppered moths fly much further from their release sites than scalloped hazel moths during the same time period. The latter do not disperse their genes as far as the peppered moth and consequently local populations of the scalloped hazel moth tend to be more sharply differentiated.

Moths vary greatly in their preferred resting sites. A limited number of these sites provide a good match to the moth's wing patterns. Perhaps density-dependent selection operating at various resting sites may be at least partially responsible for the difference in gene frequencies in various moth species.

Many other examples could be cited for the adaptive value of camouflage. An ecological principle called *Gloger's rule* states that among mammals and birds there is a tendency for those populations living in the colder, drier parts of the range to be lighter in color than those living in warmer areas of higher rainfall. This principle undoubtedly involves natural selection having preserved light-colored variants in populations where the substrate is also light colored and vegetation is sparse (as it tends to be in cold, dry areas). Conversely, in warmer regions that receive more rainfall, dark green vegetation is more abundant and the substrate tends to be darker in color. Under these ecological conditions, natural selection would tend to preserve variants that were darker in color and would blend in with their surroundings. Prey species would thus be camouflaged and difficult for predators to locate. It is also advantageous for predators to become similarly camouflaged so that they could stalk their prey without being detected.

Though not related to camouflage, several other ecological rules are attributed to natural selection. It is common knowledge that the largest game animals are found in northern climates such as Canada and Alaska. Populations of the same species (or related species) living further south tend to be smaller on average than their northern counterparts. This is *Bergmann's rule,* and it applies only to homoiothermic vertebrates (Figure 14.11). The rationale behind this rule relates to physiological problems in maintaining a constant body temperature. All other things being equal, the larger the animal, the less surface area is exposed per unit volume. This means that larger animals tend to lose proportionately less heat to their environment by radiation from their surfaces. Despite the scant vegetation of the far

Figure 14.11. Illustration of Bergmann's rule in various species of penguins. (A) Humboldt penguin (*Spheniscus humboldti*) found on the west coast of South America from Corral, Chile, north to Paita, Peru (body length about 20 inches). (B) Magellan penguin (*Spheniscus magellanicus*) found as far as 52° south latitude; breeds in Tierra del Fuego (body length about 28 inches). (C) King penguin (*Aptenodytes patigonica*) found in southern oceans to 55° south latitude from Tierra del Fuego to Macquarie Island (body length about 40 inches).

north and the attendant problems of obtaining enough food to maintain a large body, natural selection has apparently placed a great deal of pressure on the volume-surface area ratio in the evolution of these northern species. The polar bear, for example, is the largest land carnivore in the world; the arctic hare is almost the size of a small dog; the wolf is the largest undomesticated member of the dog family, etc. According to *Allen's rule,* the extremities (e.g., wings, legs, noses, ears, tails, etc.) tend to be shorter in colder climates than in warmer ones (Figure 14.12). According to *Rensch's rule,* bird populations with relatively narrower and more

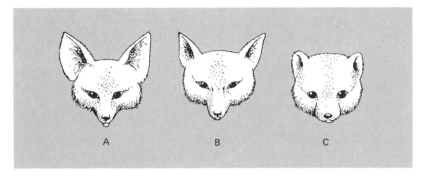

Figure 14.12. Allen's rule demonstrated in ears of (A) desert fox *Fennecus zerda* (North Africa), (B) red fox *Vulpes fulva* (Europe), and (C) arctic fox *Alopex lagopus*.

acuminate wings tend to occur in colder climates; those with broader wings tend to occur in warmer climates. Here again, selection has undoubtedly favored a reduction in size of projecting parts in order to conserve body heat.

Sexual Selection

Charles Darwin devoted only a couple of pages in the *Origin of Species* (1859) to the subject of sexual selection. It was much more fully discussed in his *The Descent of Man and Selection In Relation To Sex* (1871). "This form of selection depends, not on a struggle for existence in relation to other organic beings or to external conditions, but on a struggle between the individuals of one sex, generally the males, for the possession of the other sex. The result is not death to the unsuccessful competitor, but few or no offspring. Sexual selection is, there-fore, less rigorous than natural selection." Apparently Darwin viewed sexual selection in a different light from all other forms of "natural" selection. Today, however, sexual selection (if such really does exist) is considered to be merely a special case of natural selection. Darwin was convinced that females of certain species (especially birds) actually choose their mating partners from a group of males displaying their charms via elaborate courtship dances, brilliant plumage, etc. He also acknowledged that in some species the females have no choice in mating partners. The male who could successfully defeat other males in battles over reproductive rights would win the females in its territory. The winners usually were endowed with special advantages of size, weapons (such as tusks, antlers, or spurs), or defensive adaptations (as the mane of the lion defends against neck attacks). According to Darwin, ". . . the shield may be as important for victory, as the sword or spear." Thus, males tend to evolve special traits not selected for in females; this leads to a *sexual dimorphism* (i.e., conspicuously different phenotypes in males and females of the same species; Figure 14.13). These sexual dimorphisms occasionally are so elaborate that some taxonomists have mistak-enly classified males and females of the same species in different taxonomic groups. In order for selection by female choice to operate within a species, males should either be more frequent than females, or polygamous matings should be the rule. Not much evidence has accrued to date in support of this aspect of sexual selection. In most species the number of males and females is approximately the same. There is little doubt that in some cases the secondary sexual characteristics of males does serve to arouse sexual excitement in females, but there is little indication that the females actually prefer to mate with any particular male (or as Darwin puts it ". . . choose the most attractive partner."). For the other aspect of sexual selection, we do know of many polygamous species in which one male wins reproductive rights for its "harem" of females through battles with rival males. Fur seals, which breed on the Pribilof Islands near Alaska, are a case in point. The males arrive at the rookery before the females and do battle with each other for breeding territories. Losing males are driven off; the relatively few winners gather large harems of females as they arrive. Among birds it is common for males to be adorned more colorfully than females. In the phalarope (a type of sandpiper), however, the female has the brilliant plumage and the male is of dull coloration. Here, the female does the courting, and the male builds the nest and incubates the eggs. This

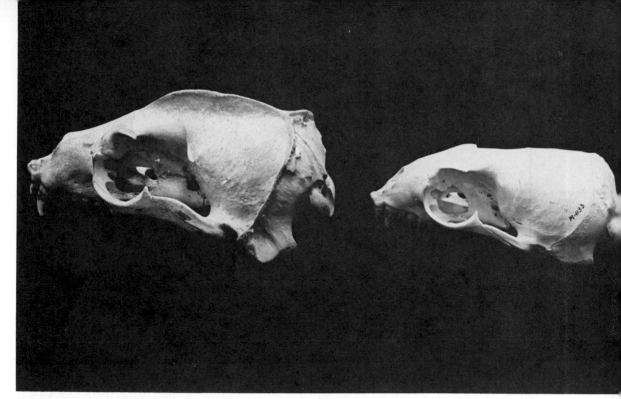

Figure 14.13. Sexual dimorphism in the skulls of adult male (left) and adult female (right) California sea lions (*Zalophus californianus*).

would seem to support the hypothesis of sexual selection. Perhaps too much stress has been placed on the evolution of bright colors and not enough attention directed toward the selective advantage of dull colors. It seems logical that much higher selection pressures would be placed on the evolution and maintenance of dull colors in the sex that cares for the young. Birds sitting on nests, for example, would be less visible to predators if their plumage matched that of their background (camouflage).

Traits that serve an adaptive function only via the nervous systems of other organisms are called *allesthetic traits*. Any trait that serves to attract or stimulate members of the opposite sex during courtship or in other ways contributes to the successful union of gametes is termed *epigamic*. Some allesthetic traits perform an epigamic function. Female moths emit a *pheromone* (a chemical that influences the behavior of members of the same species), which attracts males from perhaps more than a mile away. The enlarged antennae of male moths (Figure 14.14) is a sexual dimorphism adapted to detection of a few molecules of the female sex attractant. By following the concentration gradient of the pheromone, males arrive at the female source. The calls of male frogs, toads, and birds serve to attract females of the same species via the acoustical senses. Visual (and in some cases tactile) signals are often important components of courtship rituals, which attract and stimulate members of the opposite sex, certify that the courting members are of the same species, and synchronize the release of gametes.

Allesthetic traits may perform many functions other than epigamic. For instance, when the female killdeer (a bird) is disturbed from her nest, she flops along the ground dragging a wing as though it was broken. This would appear to be an easy catch to a predator, but the

Figure 14.14. Sexual dimorphism in antennae of the giant silkworm moth; Family Saturnidae. Male left, female right.

killdeer manages to stay ahead of her pursuer until she has led the predator far from the nest. At this point she flys back to brood her eggs leaving the predator to marvel at her "powers of healing." In some cases, the same colors used in courtship display are also used as threat displays or warning signals against other males or members of other species. Cryptic coloration, cryptic behavior, and mimicry are additional examples of allesthetic traits that become adaptive through the nervous systems of other organisms.

Selection Modes

Natural selection may operate in any one of three basic ways: (1) stabilizing (normalizing) selection, (2) directional (progressive) selection, and (3) disruptive selection. Under *stabilizing (normalizing) selection,* the intermediate phenotype is favored and extreme types are rejected. This is probably the form of selection that predominates in populations that are already well adapted in an environment that is relatively stable from one generation to the next. Using a single locus model with codominant alleles, we have three phenotypes produced by genotypes A^1A^1, A^1A^2, and A^2A^2, respectively. If A^1A^2 is of intermediate phenotype and is more fit than either homozygote, both alleles tend to stay in the population and gene frequencies remain stabilized. The sickle cell example of balanced polymorphism in the previous chapter fits this single locus model of overdominant gene action. It is possible for strong selection pressure to be operative in a population without a corresponding change in gene frequency. A balanced lethal system can maintain a pair of alleles at 0.5, while fully half of the zygotes die in each generation. For a metric trait, selection would tend to prune back the extreme variants (produced by recombination and mutation) in each generation so that the mean phenotypic value would remain unchanged (Figure 14.15A). Thus, when a population has reached an adaptive peak for a given trait, natural selection becomes a "conservative" force in the sense that it tends to stabilize the gene pool at allelic frequencies that maximize the fitness values of genotypes constructed therefrom. If a population has not yet reached its equilibrium values,

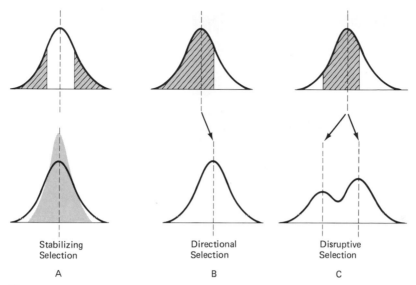

Figure 14.15. Three fundamental ways by which selection may cause change in genetic variability of populations.

there may be some reduction in variability from one generation to the next because phenotypes at one extreme are not allowed to reproduce with one another.

> In 1899, H. C. Bumpus (Figure 14.16) measured the wing size of a flock of sparrows killed by a severe snowstorm. He found that a significantly higher percentage of birds with abnormally long or short wings were killed by the storm relative to the average of the population. The same trend was obvious in other morphological traits measured on these birds. This is one way that stabilizing selection could act to maintain the average for quantitative morphological traits from one generation to the next.
>
> Studies on the number of eggs in the nests of birds (clutch size) have revealed that each species has an optimum. Obviously the more eggs laid, the greater the reproductive potential. However, the more chicks in a clutch, the less food each receives from its parents; the birds that do survive and are fledged from the nest are poorly nourished and more easily succumb to disease, predators, and hardships. Therefore, stabilizing selection strikes a balance between two factors contributing to fitness (viz. clutch size and fledgling survival) to maximize the fitness of the population.

Humans commonly select for one extreme in attempts to improve production in crops and livestock. This is termed *directional (progressive) selection.* In a similar manner, natural selection may tend to favor variants at one extreme for a particular trait. For the single locus model, the A^1A^1 phenotype might be superior in fitness to A^1A^2 and A^2A^2. In this case, the frequency of A^2 would become lower in each generation until it was lost entirely from the population. In reality, an equilibrium would be established at very low frequencies involving selection against the A^2 allele and backmutation from A^1 to A^2. For the multilocus model (see Figure 14.15B), the phenotypic mean would be progressively increasing (or decreasing, depending on the extreme favored) in each generation. The magnitude of this change in the mean of each generation is a function of the selection intensity and the heritability of the trait.

Figure 14.16. Hermon Carey Bumpus (1862–1943). Biologist, educator; director, American Museum of Natural History; president, Tufts College. (From the *Dictionary of American Portraits*, Dover Publications, Inc. 1967. Courtesy Tufts University)

Recall from Chapter 8 that if the trait is inherited entirely by additive genes the heritability of the trait approaches zero as selection progressively accumulates the favorable alleles in homozygous condition. Thus, the same selection pressure operating after many generations of progressive selection is expected to yield less genetic gain than was possible in earlier generations. Directional selection is expected to operate when the environment is changing progressively in a given direction.

> The environmental resistance may progressively increase, as when climates become colder, dryer, or in other ways more inhospitable to life. Abiotic factors need not change, however, in order for directional selection to be operative. Consider the predator-prey biotic relationship. As predators become more efficient in locating and capturing prey species, directional selection also is at work in the prey species to modify its morphology, physiology, and behavior for greater efficiency in detection and escape from its predators. Furthermore, the physical environment of a region may be stable, but if a population is expanding from one biome into another, the immigrants will find themselves in new environmental conditions. Directional selection would be operating strongly on the pioneers at the boundary of the expanding population, forming a gradient of selection pressures to almost zero at the center of the original range in which adaptation is optimal.

The more drastic the environmental change, the greater the selection pressure; under

these conditions, fewer individuals may survive to reproduce the next generation. A hereditary lineage can survive as long as it possesses an *adaptive norm* (i.e., an array of genotypes in
a given population compatible with the demands of the environment). If the environment is
shifting too rapidly for the population to generate an adaptive norm, the lineage is headed for
extinction. This is presumably the explanation for the disappearance of many species from the
fossil record at the end of geological eras. It also explains why new fossil species appear rather
suddenly during and following geological revolutions, because these were times of widespread
and rapid environmental changes that placed strong selection pressures on many traits in
many different organisms. The shifts in mean phenotypic values wrought by directional
selection are the kinds of microevolutionary steps that form the core of Darwin's mechanistic
theory.

In progressive selection and stabilizing selection, we see the revolutionary and conservative roles, respectively, played by natural selection. Progressive selection creates adaptive
genotypes that would be extremely unlikely to be formed in its absence. Stabilizing selection
tends to hold a population at its adaptive peak by eliminating poorly adapted extreme
segregants and mutants.

Disruptive selection favors more than one phenotypic optimum. This mode of selection
can become operative, if a formerly homogeneous environment is disturbed in some way,
creating new subniches for occupancy by different phenotypic classes. It is also conceivable
that a migrating population could move into a heterogeneous environment in which these
various subniches were unoccupied. In the single locus model, genotypes A^1A^1 and A^2A^2
would be adapted to different subniches, but the intermediate phenotype produced by A^1A^2
would not be well adapted to either niche. In one niche, most of the matings would be
between A^1A^1 individuals, which would tend to fix the A^1 allele in that breeding group;
similarly the A^2 allele would tend to become fixed in the other breeding group; heterozygotes
would tend to disappear. A polymorphism is thereby produced in which each morph is
adapted to different niches of the same habitat.

> Populations of bentgrass found growing in the tailings from lead and copper mines have
> adapted to the high concentration of these metal ions that are toxic to most plants.
> Apparently the resistant plants have evolved by disruptive selection from nonresistant
> populations nearby. Bentgrass is normally cross pollinated, and there is ample oppor-
> tunity for genetic exchange between the resistant and nonresistant populations via
> wind-borne pollen. Despite this fact, these two populations persist side by side. Appar-
> ently, strong selection for resistance operates on seedlings that germinate on metalic
> soils; selection against resistant plants probably operates on unpolluted soils because
> nonresistant plants grow faster there.

In the polygenic model, different nonmodal phenotypes for a given trait become adaptive
to different niches. Disruptive selection, working by directional processes in opposite directions on different parts of the phenotypic distribution, begin to tear the Gaussian curve apart
(Figure 14.15C). A multimodal curve soon appears with increased phenotypic variability. If
the process continues long enough and there is sufficient genetic variability, eventually two (or
more) nonoverlapping curves of phenotypic values may be produced. At this stage, the
subpopulations may appear so different that they could be classified as different races or
subspecies.

Most populations of the African swallow tail butterfly (*Papilio dardanus*) contain two or three mimics of different noxious species of butterflies common to the same area. Recall that Batesian mimicry is most efficient when the numbers of mimics relative to models is small. By copying different models, *P. dardanus* has been able to increase the total number of mimics without an attendant loss of protection from predators. The evolution of these mimics presumably began with mutations that vaguely resembled different models; disruptive selection then began to collect those modifier genes that tended to perfect the mimetic patterns and create a genetic background that enhanced the dominance of the corresponding genes. Only the females evolved mimicry; apparently the normal color pattern and wing shape was retained by males because of their adaptive value in triggering sexual receptiveness in the females. In areas in which models are abundant, imperfect mimics are rare; in other areas in which models are rare, imperfect mimics are common. It seems likely that disruptive selection is too weak to maintain the proper modifiers for the major (switch) alleles in areas where models are rare.

Genetic Assimilation

Much of the genetic heterogeneity of a population is hidden beneath a relatively uniform phenotype (wild type) in the usual range of environments. Unusual environments, however, can reveal some of this genetic variability through the production of phenocopies (see Chapter 8). Not all members of a species can produce phenocopies. The ability of the organism to produce a phenocopy in an unusual environment depends on the individual's genotype. Those genes that contribute to the production of the phenocopy can be accumulated by selection in the abnormal environment until they cross a genetic "threshold" after which the new trait appears even without the help of the unusual environment. C. H. Waddington called this process "the *genetic assimilation* of an acquired character." It is also termed the *"Baldwin effect."*

Waddington subjected wild-type *Drosophila melanogaster* to high temperature shocks during the pupal stage. A few of the imagos (adults) from these treated pupae were found with a break in one of the supporting structures of the wing (the "crossveinless" phenotype; Figure 14.17). He bred only from the crossveinless phenocopies in each generation, using heat shocks to reveal those genotypes that contributed positively to the

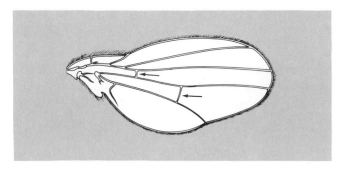

Figure 14.17. Wild type wing of *Drosophila melanogaster*. Arrows identify crossveins missing in crossveinless phenotype.

crossveinless condition. After about fifteen generations of such selection, more than 90 per cent of imagos from treated pupae possessed the crossveinless phenotype; moreover, some imagos from untreated pupae also had crossveinless wings. From then on, heat shocks were no longer necessary. Selection under standard rearing temperatures could be applied to increase the percentage of individuals with the crossveinless condition.

The explanation for this genetic assimilation is as follows. Many genes contribute to the crossveinless phenotype. Some genotypes in the original population had enough of these polygenes to allow production of crossveinless wings when aided by high temperatures at a critical stage of development. Selection under this unusual environment increased the frequency of these genes in the population until the genotype was sufficiently "strong" to canalize development of crossveinless wings without the "help" of heat shocks. Selection is then said to have lowered the threshold for the appearance of crossveinless phenotypes.

At first glance, genetic assimilation might appear to be a Lamarckian phenomenon. It seems that a new phenotype has appeared as a consequence of exposure to a particular environment, and that the new phenotype has subsequently become heritable (the inheritance of an acquired characteristic). What has actually happened is that the unusual environment has lowered the threshold for producing the character, and selection can then operate on the genetic variability thereby revealed. By acquisition of numerous genes that contribute to the development of the trait, it eventually is capable of becoming expressed even in those environments in which it formerly failed to appear.

Genetic assimilation provides us with an explanation for the origin of new adaptive phenotypic variants during times of environmental stress. These changes may then persist in the population even after the environment regains stability. Under fluctuating environmental conditions, selection might favor evolution of developmental pathways easily modified by environmental factors and, as a consequence, phenotypic plasticity. Where the demands of the environment are relatively stable generation after generation, selection would probably favor evolution of highly canalized developmental pathways, which lead to a highly adaptive uniform phenotype (Figure 14.18).

Another example of genetic assimilation will be given to demonstrate that the altered conditions need not be of exogenous origin. Wild mice have an almost invariate number (19) of vibrissae (nose "whiskers"). Selection cannot be applied on vibrissae number if there is no phenotypic variability. A semidominant, sex-linked mutant gene, called Tabby, causes considerable variation to be exhibited in vibrissae number (average for Tabby males is 8.7; for females is 15.1). In one experiment, selection was applied among heterozygous Tabby females ($Ta/+$) for low vibrissae number in one line and for high vibrissae number in another line. Every other generation, matings were made between Tabby males and heterozygous Tabby females; this produced wild-type males in which vibrissae number developed in the absence of the Tabby gene. After six generations of such selection, wild-type males in the "high" line had extra vibrissae; those in the "low" line were missing some vibrissae. Under the unusual *genetic* conditions provided by the Tabby gene, selection was able to be applied to the genetic component of variability in vibrissae number. Accumulation of polygenes for high or low vibrissae numbers eventually created genotypes that allowed vibrissae development to escape the canalization that formerly produced an invariate number.

Whenever an unusual environment (either exogenous or endogenous) reveals previously

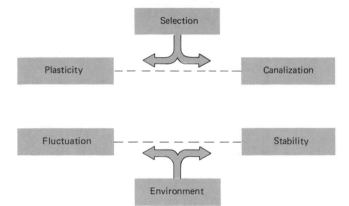

Figure 14.18. Diagram depicting the influence the environment has upon selection for different types of developmental pathways. Broken lines represent a continuum.

hidden genetic variability in the population, the selection process of genetic assimilation can begin to operate (Figure 14.19). During genetic assimilation, selection acts at two levels: (1) it changes gene frequencies so that a new phenotype becomes common, and (2) it broadens the range of genotypes capable of producing the new phenotype. The first level merely selects extreme phenotypes, but the second level is a more complex process of shifting the zone of canalization of development of one trait to that of another. Thus, *canalizing selection* not only produces a new phenotype but also tends to reduce the phenotypic variability of the genotypes expressed within the range of the new phenotype.

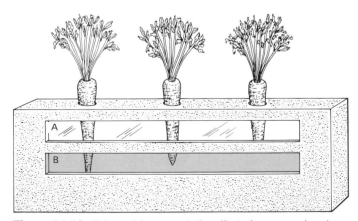

Figure 14.19. This model represents the effect of an unusual environment (window B) on revealing phenotypic variability in taproot length of carrots. Under normal environments (window A), selection cannot be applied to root length for lack of phenotypic variability in this trait.

The Fundamental Theorem of Natural Selection

One of the most influential books in the field of population genetics is R. A. Fisher's *Genetical Theory of Natural Selection* (1930). In this work, Fisher (Figure 14.20) states in mathematical terms that the increase in population fitness at any point in time is proportional to the genetic

Figure 14.20. Sir Ronald A. Fisher (1890–1962). [From *Genetics,* **61**: Frontispiece (1969)]

variance in fitness at that time. This principle has come to be known as the *Fundamental Theorem of Natural Selection.* In other words, the amount of improvement in population fitness from one generation to the next is directly proportional to the amount of genetic variability upon which selection for fitness may act.

In order to present a simplified proof of Fisher's Theorem, we will make two simplifying assumptions for our model: (1) at all loci, dominance with respect to fitness is lacking (e.g., A^1A^2 is exactly intermediate in fitness between A^1A^1 and A^2A^2), and (2) there are no epistatic interactions between genes at any of the loci contributing to fitness. Under these conditions, each gene has an effect on fitness that is independent of other alleles at the same locus or of genes at other loci. Therefore, sexual reproduction need not be considered, and we may treat the population as though it consisted of a number of asexual clones.

Initially our model population has a genotypic array $(g_1, g_2, g_3, \ldots, g_n)$, each genotype having a specific frequency $(f_1, f_2, f_3, \ldots, f_n)$ and a specific fitness value $(w_1, w_2, w_3, \ldots, w_n)$, respectively. The average fitness of the population $(\bar{w})$ equals $f_1 w_1 + f_2 w_2 + f_3 w_3 + \cdots + f_n w_n$. The variance among genotypes of the population with respect to fitness is calculated as follows: $f_1(w_1 - \bar{w})^2 + f_2(w_2 - \bar{w})^2 + f_3(w_3 - \bar{w})^2 + \cdots + f_n(w_n - \bar{w})^2$. If for each term we square the deviation (in parentheses) and multiply by the frequency we obtain an expansion that can be regrouped into three series:

(a) $f_1w_1^2 + f_2w_2^2 + f_3w_3^2 + \cdots + f_nw_n^2$

(b) $-2\overline{w}(f_1w_1 + f_2w_2 + f_3w_3 + \cdots + f_nw_n)$

(c) $+ \overline{w}^2(f_1 + f_2 + f_3 + \cdots + f_n)$.

Since the terms in parenthesis of series (b) equal $\overline{w}$ and those in series (c) equal 1.00, all three series can be reduced to

$$(f_1w_1^2 + f_2w_2^2 + f_3w_3^2 + \cdots + f_nw_n^2) - \overline{w}^2$$

After one generation of selection, the new frequencies for each of the genotypes $g_1, g_2, g_3, \ldots, g_n$ will be proportional to $f_1w_1, f_2w_2, f_3w_3, \ldots, f_nw_n$ (because the fitness of each genotype determines its own frequency in the next generation). The average fitness in this new population is represented within the parentheses of series (b). In order for the sum of frequencies to equal 1.00, we can apply the following correction:

$$(f_1w_1^2 + f_2w_2^2 + f_3w_3^2 + \cdots + f_nw_n^2)/\overline{w}.$$

The change in average fitness of the population ($\Delta\overline{w}$) is the new average minus the old average or

$$(f_1w_1^2 + f_2w_2^2 + f_3w_3^2 + \cdots + f_nw_n^2 - \overline{w}^2)/\overline{w}.$$

In all these calculations, the w values are relative fitnesses. Now, if we choose the original w values so that $\overline{w} = 1.00$, we see that $\Delta\overline{w}$ is the same as the genetic variance in fitness. That is to say, the amount by which a population is expected to increase in average fitness ($\Delta\overline{w}$) through selection in each generation is equivalent to the genetic variance in fitness at that time.

The fundamental theorem of natural selection makes it quite clear that a population should maintain a liberal supply of genetic variability as "insurance" against extinction. No one genotype is best adapted to exploit all the niches in a given habitat. Selection therefore does not always favor one genotypic optimum, but often favors an array of genotypes that can most fully use the available resources of the heterogeneous environment (the adaptive norm). In this way, genetic variability is maintained in the gene pool. This, in turn, provides the opportunity for continued improvement in fitness even if a changed environment requires a shift in the present adaptive norm.

> Does a high degree of phenotypic specialization predispose a population to extinction? For example, some arctic fish cannot survive outside a very narrow range of temperatures. Because they are highly specialized in this regard, does this mean that they will probably become extinct long before other fish (with a broader temperature optimum)? The answer to this question depends largely on the genetic variability in fitness of these populations. If the stenothermic (narrow temperature optimum) species has lost most of its genetic variability in adapting to its niche, the prognosis for surviving relatively minor temperature changes is not good.

Some of the mechanisms that function to maintain genetic heterogeneity in the gene pool are presented in the next two topics.

Frequency-Dependent Selection

Some alleles may have a selective advantage when they are comparatively rare in the population. For example, many plant species have a self-incompatibility locus with three or more alleles $(a_1, a_2, a_3, \ldots, a_n)$. A pollen grain carrying any one of these alleles cannot germinate on the stigma or grow a pollen tube down the style of a plant containing that same allele. A pollen grain bearing the a_1 allele is incompatible with a plant of genotype a_1a_2, but can fertilize a plant of genotype a_2a_3. Selfing leads to homozygosity and typically to inbreeding depression (loss of vigor and fertility). Hence, a mechanism that ensures self-incompatibility would probably have some selective advantage. Consider a population containing three self-incompatibility alleles (a_1, a_2, a_3) in equal frequencies. All plants in such a population are heterozygous $(a_1a_2, a_1a_3,$ or $a_2a_3)$. Pollen bearing the a_1 allele can only fertilize one third of the plants (*viz.* a_2a_3), a_2 pollen another third (*viz.* a_1a_3), and a_3 pollen another third (*viz.* a_1a_2). If a new allele (a_4) appears by mutation, it would immediately enjoy a large selective advantage because pollen bearing a_4 could cross with any plant in the population.

An equilibrium is attained in a self-incompatibility system when all of the alleles reach the same frequency. For example, when a_1 is rare, pollen bearing a_1 will most likely encounter and fertilize the more common a_2a_3 heterozygotes, thereby increasing the numbers of a_1a_2 and a_1a_3 types in the next generation. As the frequency of a_1 increases, its initial advantage diminishes. When all three alleles exist in equal frequencies, none of them will have a selective advantage and the population will be in equilibrium. Does each self-incompatibility allele newly introduced into a population (where the n old alleles are at their equilibrium values of $1/n$ each) have the same selective advantage? To answer this question we need to develop several formulas. First we need to determine how many different kinds of heterozygotes can be formed with n alleles. Consider a gametic table with three alleles.

♂ \ ♀	a_1	a_2	a_3
a_1		a_1a_2	a_1a_3
a_2	—		a_2a_3
a_3	—	—	

Homozygotes on the diagonal are not possible with self-incompatibility alleles. The table is symmetrical above and below the diagonal. We want only those heterozygous genotypes on one side of the diagonal. There are n^2 theoretical gametic combinations; $n^2/2$ represents all of the genotypes on one side of the diagonal plus half of the diagonal itself; $n/2$ represents half of the diagonal. Therefore, the total number of different kinds of heterozygotes formed by n alleles is

$$\frac{n^2}{2} - \frac{n}{2} = \frac{n^2 - n}{2} = \frac{n(n-1)}{2}.$$

Next, we need to find how many of the heterozygous genotypes in a population can be fertilized by a pollen grain carrying a given self-incompatibility allele. A pollen grain carrying a_1 allele cannot fertilize any of the genotypes on the same line as a_1 pollen in the above table. There are $n - 1$ heterozygotes on that line which a_1 pollen cannot fertilize. The same is true for any of the other self-incompatibility alleles. Therefore, the number of heterozygous genotypes that can be fertilized by a given kind of pollen grain equals the total number of heterozygotes minus the ones with which that pollen grain is incompatible:

$$\frac{n^2 - n}{2} - (n - 1) = \frac{n^2 - n - 2(n - 1)}{2} =$$

$$\frac{n^2 - n - 2n + 2}{2} = \frac{n^2 - 3n + 1}{2} = \frac{(n - 1)(n - 2)}{2}.$$

For any given kind of pollen grain, the proportion of all heterozygous plants with which it is compatible is expressed as

$$\left[\frac{(n - 1)(n - 2)}{2}\right] \div \left[\frac{n(n - 1)}{2}\right] = \frac{n - 2}{n}.$$

When a new allele (a_{n+1}) arises by mutation (or migration), it has a fitness value of 1.0 because it can fertilize any other plant in the population. All other (old) alleles can fertilize only the fraction $(n - 2)/n$ of all plants; hence their fitness is $(n - 2)/n$. Fitness of the old alleles is $1 - s$, where s is the selection coefficient acting against the old alleles. Thus $(n - 2)/n = 1 - s$. Solving for s, we have

$$n - 2 = n - sn; \quad sn = n - n + 2; \quad s = \frac{2}{n}.$$

Now we are ready to answer the original question. The selection coefficient acting against the old alleles decreases with the addition of each new allele; therefore, the selective advantage of each new self-incompatibility allele diminishes correspondingly.

Polymorphisms

A *genetic polymorphism* exists when two or more alleles occur at the same locus in the same population with appreciable frequency (conventionally taken at a minimum of about 1 per cent). When a favorable mutant phenotype appears, selection can increase its frequency. During a considerable portion of the evolutionary march from low initial gene frequency to high frequency under the guidance of natural selection, the population exhibits two (or more) alternative forms (*phenotypic polymorphism*). This type of polymorphism is governed by only one force, namely selection favoring one allele. Consequently this type of polymorphism lacks stability and is referred to as a *transient polymorphism*.

When at least two opposing forces are active, a *balanced polymorphism* can be produced. The classical explanation for balanced polymorphism is heterozygote advantage (overdominant gene action). There are at least six other ways by which a balanced polymorphism can be maintained.

(1) Opposing mutation pressures. Let p and q represent the frequencies of alleles A and a, respectively; let A mutate to a at rate u, and let the backmutation rate be v. An equilibrium will be established when $pu = qv$. If $u = v$, a theoretical equilibrium will form with $p = q = 0.5$. According to the above definition, a polymorphism exists when the ratio u/v lies between .01 and .99. It seems highly unlikely that a balanced polymorphism due to opposing mutation pressures could be established at intermediate gene frequencies, because back-mutation rates are usually much smaller than forward mutation rates (as discussed in Chapter 12). Furthermore, this type of polymorphism is highly unlikely, because evolutionary pressures other than mutation are commonly at work on all loci.

(2) Frequency dependent selection. As discussed earlier in this chapter, some genes increase in fitness when they become rare in the gene pool. Frequency-dependent selection will thus tend to maintain at least two alleles in the population at appreciable frequencies. Perhaps sexual selection could also be involved if rare phenotypes become preferred mates (consider the hypothetical mating preference for rare or unusual human phenotypes such as blue eyes or blond hair). The genes responsible for Batesian mimicry have high selective values as long as the ratio of mimics to models remains small. There is no genetic load on a population when frequency-dependent alleles are in equilibrium, because all genotypes have equal fitnesses. This mode of selection is therefore more likely to explain the high incidence of balanced polymorphisms in nature than overdominance. The latter mechanism, when operative at more than 100 loci, is likely to induce intolerable loads on the population.

(3) Selection acting in opposite directions in males and females. It can be mathematically shown that a balanced polymorphism can occur (for sex-linked and autosomal traits) when selection favors the female heterozygote or when selection favors a gene in one sex and disfavors the same gene in the other sex. In the same way, differential fertility acting in a direction opposite to differential viability in the sexes can also lead to a stable polymorphism. That is to say, a viability deficiency of a genotype in one sex may be compensated by high fertility when mated to a different genotype in the opposite sex.

(4) Selection intensity varying in time and space. Although selection may not favor the heterozygote in every generation, a polymorphism could still exist if the heterozygote was favored (on average) over all generations. The same is true even though the heterozygote might be mostly at a disadvantage, but at least occasionally gains a strong advantage (as during periodic epidemics). The same could be true for a migrant population shuttling seasonally between different feeding ranges, where one allele is favored in say the winter range and the other allele is favored in the summer range. Although the polymorphism may not be especially stable under this scheme, it is very likely to aid in maintaining both alleles in the population for considerable periods of time. Probably no polymorphism is absolutely stable on an evolutionary time scale anyway.

(5) *Selection acting in opposite directions in gametes and zygotes.* If gametes bearing a particular allele are favored with respect to fertilization over those with an alternative allele in heterozygotes, a stable polymorphism is possible. It may be possible for a gene to express itself in the gamete by enhancing its chemoattraction to opposite gametes, its ability to fuse with opposite gametes, etc. This is probably not an important mechanism, however, because few genes appear to be active in gametes such as spermatozoa. When heterozygotes fail to form equal numbers of gametes bearing alternative alleles, *segregation distortion* is said to exist. The t-locus in the mouse was cited in Chapter 9 as an example of meiotic drive in which many more t-bearing gametes were produced than those bearing normal alleles. These t-alleles are favored during the gametic stage, but they are lethal to zygotes when in double dose.

(6) *Disruptive selection.* A polymorphism can be produced when both extremes of a phenotypic distribution are favored in different subniches within a habitat, provided that mating is not entirely at random over all subniches. Disruptive selection is generally thought to be more active in the central regions of the species range, where environmental conditions tend to be less hostile and offer a variety of subniches for colonization. At the margins of a species range, conditions are often much less favorable and the populations that exist there tend to be adapted more narrowly because fewer subniches are available. Thus, a species tends to be more diversified (polymorphic) both morphologically and chromosomally in the center of an old established range than at its margin. This is sometimes referred to as the *Ludwig effect* after W. Ludwig who described it in 1950. Ludwig reasoned that a genotype that could successfully exploit an unoccupied subniche might survive, even though it was at a selective disadvantage in parts of the niche used by other genotypes. Disruptive selection applied to populations in different geographical locations can produce diversifications that are the basis of race formation. Polymorphisms produced by the Ludwig effect need not create a balanced load on the population as long as each genotype is well adapted to its own subniche.

Some polymorphisms may be maintained by combinations of two or more of the above mechanisms. For example, overdominance and the Ludwig effect are not mutually exclusive of one another. The mechanisms that maintain balanced polymorphisms are only part of the explanation for the maintenance of genetic variability in natural populations. In the next section we will investigate the importance of genetic variability to the phenomenon of genetic homeostasis.

Until the late 1950s it was commonly believed that each species was genetically highly uniform, and that almost all of the intraspecific genetic variation existed as differences between its highly homozygous populations. This concept was shattered in the 1960s following several reports on the amount of heterozygosity in human and *Drosophila* populations as indicated by their allozyme variations. An *allozyme* is a unique electrophoretic form of an enzyme or protein determined by a specific allele. The term *isozyme* is used more generally for different forms of a protein irrespective of cause (either genetic or nongenetic). Some authors proposed that one quarter or more of all gene loci may be segregating in small Mendelian populations, and within an individual perhaps as many as one locus in five or six might be heterozygous for dissimilar alleles. Since this revelation, geneticists and evolutionists have broken into two major camps in attempting to account for all of this protein variation, *viz.*, neutralists and selectionists. The neutralists championed the idea that most of the observed variation is generated by selectively neutral isoalleles. If this is true, then most of these molecular polymorphisms are largely irrelevant to adaptive evolution. Selectionists, on the other hand, tended to favor the notion that balancing selection was responsible for these polymorphisms. The main argument supporting balancing selection has been the purported discrepancies between observed patterns of gene frequencies (in populations of a widespread species) from that predicted by random processes.

In 1965, R. Levins predicted that genetic variation in fitness should increase with environmental variability. If patterns of geographic variation in heterozygosites exist for statistically correlated ensembles of loci, they could be interpreted to represent adaptive shifts in response to changes in the variability of specific environmental factors. However, it has been shown mathematically that random events, coupled with gene flow between populations, could also account for these patterns.

> A study was made of the distribution of electrophoretic variation over chromosomes in species of the *repleta* group of *Drosophila*. It was found that there was no correlation in the amount and distribution over the genome between cytological (inversion) and electrophoretic variation. This suggests that allelozymes (allozymes) are not members of the coadapted complexes of genes characterizing inversions, and therefore argues against the maintenance of allelozymic variation by balancing selection.

Not enough data are available as yet to resolve the dispute. Neutralists have charged that evolutionary theory has so many alternative pathways that almost any phenomenon can be explained. For example, at least one model (based on gene control rather than on the structural gene) has been proposed to allow natural selection to favor heterozygotes even though the corresponding allozymes are selectively neutral. Selectionists claim that the neutral allele hypothesis is also excessively permissive and cannot be subjected to a "set of potential falsifiers." The problem is an extremely important one, and it should be very interesting to observe the progress toward its solution.

Genetic Homeostasis

Just as the individual's genotype canalizes its embryological development toward a relatively uniform phenotypic expression over a wide range of environments (developmental homeostasis), so also does the population gene pool tend to canalize the range of all possible genotypes into that of the relatively uniform adaptive norm. Gene frequencies, at all loci, are continually selected for their integrated ability to maximize the average fitness of the entire population. Selection molds the composition of the gene pool so that reproduction therefrom generates the maximum number of coadapted gene complexes (i.e., groups of genes that operate harmoniously with each other to yield combinations of characters with high fitness values). Any major shifts from these established gene frequencies are likely to reduce the average fitness of the population. Stabilizing selection, favoring intermediate phenotypes rather than extremes, thus tends to operate against deviant individuals that would disturb the average population fitness. This resistance to change that characterizes a population gene pool has been dubbed *genetic homeostasis* by I. M. Lerner.

Matings between closely related individuals (inbreeding) tends to increase homozygosity at the expense of heterozygosity. Continued inbreeding in many laboratory populations fails to increase homozygosity beyond certain limits (often far short of the maximum 100 per cent theoretically possible). Similarly, directional selection experiments sometimes reach plateaus beyond which the population fails to respond to selection in the same direction. Yet reverse selection or relaxation of artificial selection meets with immediate success, indicating that considerable heterozygosity remained in the gene pool at the plateau stage. The gene pool tends to resist changes (especially rapid ones) in its gene frequencies beyond certain limits (genetic homeostasis). Lerner suggests that certain levels of heterozygosity *per se* must be maintained in the gene pool in order to ensure normal development of individuals. As an example, an organism may fail to develop normally when heterozygosity (in the total genotype) drops below 20 per cent. One explanation for this is that some genes have deleterious pleiotropic effects on fitness. For example, a recessive autosomal gene in the mouse (obese) causes much fat to be deposited. This layer of adipose might appear to be potentially advantageous to mice in cold climates. However, sterility is one of the pleiotropic effects of this gene. Therefore, homozygosity for certain allelic combinations can reduce fitness. Overdominant gene action at the selected loci is another reason. Population fitness will be reduced to the extent that heterozygosity is reduced at these overdominant loci. Lerner believes that the totality of traits contributing to individual fitness exhibits overdominance.

> Directional selection for small body size in *Drosophila* was shown to be initially effective, but eventually it became ineffective (Figure 14.21). However, genetic variability had not been exhausted in this line (i.e., homozygosity at all loci contributing to small body size had not been attained). Reverse selection (for large body size) in this same line was immediately effective proving that much genetic variability remained in the unresponsive line. Eventually a similar plateau was reached in selection for large

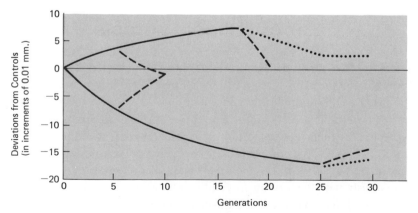

Figure 14.21. Idealized results of a selection experiment for thorax length in *Drosophila melanogaster*. Selection in the "high" line favored flies with a long thorax; selection in the "low" line favored flies with a short thorax. Broken lines indicate reverse selection; dotted lines indicate relaxed selection.

body size. When heterozygosity was reduced through selection to a critical level, the adverse pleiotropic (and possibly linkage) effects on reproductive traits (fitness) counteracted artificial selection on polygenes for body size. The population then became refractory to further selection in the same direction. This resistance of a population to disturbance of its equilibrated gene frequencies is termed genetic or population homeostasis.

Lerner presents a simple model to explain how heterozygosity *per se* can lead to the production of normal phenotypes (Figure 14.22). Assume that genes at many loci are interchangeable in their developmental effects. The organism can tolerate a certain amount of homozygosity at these loci provided that it does not exceed certain limits. In this model, it is immaterial which of the loci are heterozygous and which are homozygous. With three loci, a normal phenotype could develop with, say, two or three loci heterozygous; abnormal development occurs when fewer than two loci are heterozygous.

Figure 14.22. A three locus model relating the degree of heterozygosity to normal development. Arrows indicate the developmental pathways mediated by each genotype. (After Lerner)

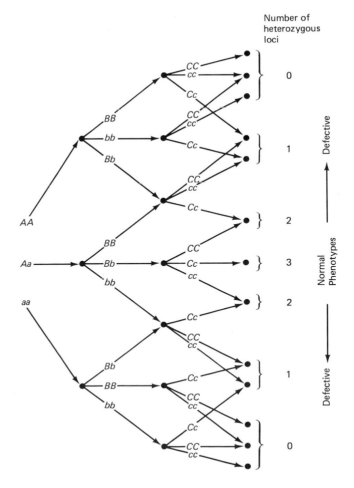

Joint Effects of Selection and Mutation

Selection Against Recessive Mutants

Up to this point, the formal aspects of mutation and selection have been considered separately. In this section we begin to take a tiny step toward the realities of the living world by considering the joint effects of these two evolutionary forces.

Mutation of a common gene to an uncommon allele is most effective in increasing the uncommon allele when that allele is rare, because there are more of the common genes to mutate. Selection, however, becomes increasingly less effective in removing a gene from the gene pool as it becomes rare. The magnitude of change in gene frequency wrought by either of these forces is dependent upon the gene frequency itself. Eventually, these two forces working against each other will establish a stable equilibrium.

Consider a pair of alleles A and a, with initial frequencies p and q, respectively. If the mutation rate of A to a is u, then the change in frequency of A allele (Δp) equals $-up$. If allele a is rare, p is nearly 1.0, and $\Delta p = -u$ (therefore $\Delta q = u$). If selection acts against the a allele, a stable equilibrium will eventually be established between the forces of selection and mutation. Two types of equilibrium can be attained depending on whether or not the a allele is completely recessive with respect to fitness or at least partially dominant with respect to fitness.

Case 1. If allele a is completely recessive with respect to fitness,

$$\Delta q = \frac{-sq^2(1 - q)}{1 - sq}.$$

When q is nearly 0, Δq is approximated by $-sq^2$. At equilibrium, the gain of a by mutation is balanced by its loss due to selection.

Thus, $\Delta q = 0$ when $u - sq^2 = 0$.
$$u = s\widehat{q}^2 \text{ (Note that } sq^2 \text{ is the mutational load)}$$
$$\therefore \widehat{q} = \sqrt{u/s}$$
If a is a lethal mutation ($s = 1.0$), $\widehat{q} = \sqrt{u}$.

At equilibrium, the frequency of aa homozygotes (lethal) is roughly equal to the mutation rate of A to a (i.e., $\widehat{q}^2 = u$; the mutational load is q^2). If a lethal arises by mutation once in a million gametes ($u = 10^{-6}$), the equilibrium frequency $\widehat{q}$ would be $\sqrt{10^{-6}} = 10^{-3}$ (i.e., one gamete in a thousand would carry the mutation).

Case 2. If allele a is dominant (to any degree) with respect to fitness, $\Delta q = -hsq$ when q is near zero (under these conditions, the frequency of aa homozygotes is negligible). Recall from Chapter 13 that hs represents the reduction in fitness of heterozygotes attributed to the a allele. At equilibrium, $\Delta q = 0$ when $u = -hsq$.

$$u = hs\widehat{q} \text{ (Note that } hsq \text{ is the mutational load)}$$
$$\therefore \widehat{q} = u/hs$$

For a gene that is lethal in homozygotes ($s = 1$), but that only reduces the fitness of heterozygous by 1 per cent ($h = .01$), the equilibrium frequency $\hat{q}$ that balances the mutation rate ($u = 10^{-6}$) is 10^{-4} (i.e., only one gamete in 10,000 would carry the mutation). Note that this is tenfold less than for completely recessive alleles (Case 1).

Most mutation rates are approximately 10^{-5} in magnitude. Very mild selection suffices to hold these mutant genes at very low equilibrium frequencies. For example, if a recessive gene arises by mutation at the rate of 10^{-5}, a selective disadvantage of 10 per cent will hold the frequency of the recessive homozygote at 1/10,000; a 50 per cent selective disadvantage will hold it at 1/50,000. These figures reveal that it is selection, not mutation, that determines whether a gene spreads through a population or not.

A recessive sex-linked trait governed by a rare gene will occur almost exclusively in males. An equilibrium will be established between mutation and selection acting only on the heterogametic sex.

> Hemophilia is a disease involving a deficiency in the blood-clotting mechanism governed by a rare sex-linked recessive gene. It is semilethal in males, but practically unknown in women (because of extreme rarity and/or lethal manifestations). Let us assume that a population is in equilibrium at this locus. If q is the frequency of the recessive hemophilic gene, the frequency of hemophilic males is also q. If the survival value of hemophilic males is $1 - s$, then sq hemophilic genes will be lost in each generation from the males. To be in equilibrium, an equivalent amount of hemophilic genes must be introduced by mutation from normal alleles. Females have two X chromosomes; males have only one. Assuming a 1:1 sex ratio, only $\frac{1}{3}$ of the X chromosomes of the population reside in males. Thus, at equilibrium, the mutation rate u is balanced by $\frac{1}{3}sq$.

A single neutral mutation will ultimately be lost by chance from a large population. The vast majority of mutations that confer a selective advantage on individuals who carry it (heterozygotes) will also be lost a few generations after their occurrence. Their chance of survival is approximately $2s$ (when s is small). For example, when $s = 0.01$, the new mutation must occur about thirty-five times in order for the chance mechanism of reproduction to favor its establishment in the population. If the mutation rate is 10^{-5}, it would require a population of 3.5 million individuals to allow this mutation to occur thirty-five times in a single generation. Therefore, even if a new mutation does have a slight advantage over other established alleles, it would have practically no chance of survival in a small population. For a mutation that is neutral in heterozygotes, but favorable in homozygotes, its chance of survival is much smaller than if it was favorable in heterozygotes. This is due to the fact that selection cannot begin to favor its retention in the gene pool until after two such recessive mutations occur together in the same individual. The frequency of a mutation would have to be fairly high in order to generate the recessive homozygote. Most likely the mutation would be lost by chance and never reach this required frequency. Therefore, the initial increase in frequency of a recessive allele depends on its mutation rate and chance. Selection cannot begin to act (either positively or negatively), until the recessive gene reaches a sufficiently high frequency to produce a homozygous recessive genotype. Mutation is an inconsequential evolutionary force in small populations; a major factor in the evolution of small populations is chance fixation of genes. The evolutionary consequences of small population size will be discussed in the next chapter.

Selection Against Dominant Mutants

A stable equilibrium is also established by the opposing forces of mutation and selection operating against a homozygous dominant genotype. If v is the mutation rate of the recessive allele a to the dominant allele A, then the equilibrium frequency of A is $\hat{p} = \sqrt{v/s}$.

> In humans, short fingers (major brachydactyly) are produced by a dominant mutation that is essentially lethal in homozygous condition. Thus, brachydactylous people are heterozygotes. The equilibrium frequency of this dominant gene is approximately $\hat{p} = \sqrt{v}$. The mutation rate (v) can be calculated directly from the incidence of brachydactyly in the population (H). $H = 2pq$, but since q is nearly 1.0 in this case, $H \approx 2\hat{p} = 2\sqrt{v}$; and $v = \frac{1}{4}H^2$. The mutational load is $\frac{1}{4}H^2$.
>
> If heterozygous dominants are also inviable ($s = 1$), as is the case for human retinoblastoma (a cancer of the eye), the mutation rate is $v = \frac{1}{2}H = p$, where H is now the incidence of retinoblastoma in the population. In other words, the frequency of a dominant gene, which kills before sexual maturity, equals the mutation rate of that gene or half the frequency of the trait (H) in the population. The mutational load is p.

Now let us consider a case in which the homozygous dominant genotype is lethal and the heterozygote has fitness $W = 1 - s$.

Genotypes:	AA	Aa	aa
Frequency before selection:	0	H	$1 - H$
Relative fitness:	0	$1 - s$	1
Frequency after selection:	0	$H - sH$	$1 - H$

When $(1 - H)$ is nearly 1.0, almost all mutations of $a \rightarrow A$ will occur in aa genotypes. If the mutation rate per a allele is v, then new A mutants will appear with a frequency approximated by $2v(1 - H)$, because there are two a alleles in each recessive homozygote. At equilibrium, the elimination of dominant genes in heterozygotes is balanced by new dominant mutants arising almost exclusively in recessive homozygotes; thus,

$$2v(1 - H) = sH \quad \text{and} \quad v = \frac{sH}{2(1 - H)}.$$

But when H is small, the mutation rate is approximated by

$$v = \frac{sH}{2}, \quad \text{or} \quad v = \frac{H(1 - W)}{2}.$$

A knowledge of relative fitness values can be used as an indirect method of estimating mutation rates.

> A dominant mutant gene in humans produces chondrodystrophic dwarfism. All newly mutated dwarfs are heterozygotes. The relative average fitness of these heterozygous dwarfs compared to their normal sibs is estimated from their family records. In one study, there were 108 dwarfs that produced 27 children and there were 457 normal sibs

that produced 582 children. The relative average fitness of these dwarfs is $(27/108)/(582/457) = 0.196\ (= 1 - hs)$, where hs is the amount by which the mutant allele reduces fitness of heterozygotes; thus, $hs = 0.804$. The frequency of dwarfs is estimated from records on 94,075 births of which 10 were dwarfs; thus $2pq = 0.000106$. Because the frequency of the normal allele is almost 1.0, q is approximately 0.000053. The mutation rate u (from normal gene to dwarfing allele) is hsq, or $(0.804)(0.000053) = 0.000043$ or 4.3×10^{-5}. This indirect method yields an estimate of the mutation rate in good agreement with the direct method. Out of 94,075 children born in one hospital in Copenhagen, 8 were dwarfs born to normal parents. This furnishes a direct estimate of the mutation rate $u = 8/94{,}075$ or approximately 4×10^{-5}.

Selection Against Heterozygotes

Recall from Chapter 13 that when selection acts only on heterozygotes (homozygotes have equal fitness values) an unstable equilibrium exists at $\hat{p} = \hat{q} = 0.5$. If q is less than 0.5, the frequency of the recessive allele will progressively diminish (in the absence of mutations). An equilibrium will be established when mutations from $A \rightarrow a$ (rate u) balance the loss of the recessive allele due to selection against heterozygotes. Again from Chapter 13, we take the frequency of the recessive allele after selection:

$$q_1 = (q - spq)/(1 - 2spq),$$
$$\Delta q = -(spq + 2spq^2)/(1 - 2spq).$$

Therefore, the frequency of the dominant allele after selection is

$$p_1 = 1 - q_1 = (p - spq)/(1 - 2spq).$$

The gain of recessive alleles through mutation is reflected in

$$\Delta q = up_1 = u(1 - q_1) = u(p - spq)/(1 - 2spq).$$

At equilibrium, the same amount of recessive alleles is lost through selection against heterozygotes. Cancelling the identical denominators on either side of the equation, we have

$$u(p - spq) = spq(1 - 2q)$$
$$u = \frac{spq(1 - 2q)}{p - spq}.$$

When both q and s are small, we can modify this formula by successively eliminating values that approximate 1.0.

$$u = \frac{sq(1 - 2q)}{1 - sq} = sq(1 - 2q) = sq, \qquad \text{or} \qquad \hat{q} \approx \frac{u}{s}.$$

Note that $\hat{q} = u/s$ is a much smaller value than $\hat{q} = \sqrt{u/s}$ as was the case when selection acts against the recessive homozygote rather than the heterozygote. This is to be expected because when q is small most recessive alleles are found in heterozygotes rather than homozogytes. In the limiting case, in which complete selection acts against the heterozygote ($s = 1$), $\hat{q} = u$

(analogous to complete selection against a dominant mutation in both heterozygous and homozygous genotypes).

Essay Questions

1. *Give three types of mimicry and explain the conditions under which they best develop and are maintained.*

2. *Present three ways by which selection can maintain a polymorphism (other than by overdominant gene action).*

3. *Why do populations tend to become refractory to directional selection, even though genetic variability is far from exhausted in the gene pool?*

4. *State Fisher's fundamental theorem of natural selection. Explain its implications for long-term evolution.*

5. *List three major modes by which natural selection operates and specify the environmental conditions under which each is most likely to be effective.*

POPULATION STRUCTURE

The evolutionary potential of a species resides in the store of genetic variability in its gene pool. The capacity to rapidly extract new adaptive combinations from this heterogeneous gene pool is dependent on two important attributes of the species; viz. its genetic architecture and its population structure. Chapter 9 dealt with genetic systems and their influences on recombination. In the present chapter, attention will be focused on the manner in which species are subdivided into local breeding groups and the evolutionary effects of finite population size, gene flow, and nonrandom mating patterns. These are all aspects of *population structure*.

As any field biologist knows, individuals belonging to the same species are seldom (if ever) distributed evenly over their geographical range. One reason for this is that essential resources (e.g., food and shelter for animals; light, moisture, and soil fertility for plants) are seldom uniformly available in all parts of the species range. Secondly, many species of animals tend to travel in groups (herds, flocks, etc.) while foraging or at least tend to congregate at certain seasons for reproductive purposes. No singular genotype is best adapted to exploit all the various subniches of the species habitat. The problem of maximizing fitness in the entire species is resolved by selection working in each local population to perfect adaptations to *local* conditions. In this way, different populations tend to acquire divergent gene pools. Genetic heterogeneity is thereby fostered in the species. This in turn, enhances its long-term evolutionary potential.

Nonrandom Mating

Panmixis is a requirement for Hardy-Weinberg equilibrium. This implies that every gamete has an equal opportunity to combine with every other gamete of opposite type, including selfing in random amount. It is usually not possible for the gametes of individuals at opposite ends of a species range to unite. Therefore, the much smaller unit within which panmixis is at least theoretically possible has been variously called a *panmictic unit,* a *local Mendelian population*, a *gamodeme*, or simply *deme*. The alternative to random mating within a deme is *assortative mating* (nonrandom). This term indicates a tendency for males of a given kind to breed with females of a particular kind. Under *negative assortative mating*, males and females tend to be of different kinds; *positive assortative mating* implies that the breeding partners tend to be similar. Assortative mating can exist at the genetic and/or phenotypic levels. Negative genetic assortative mating, involving crosses between genetically dissimilar individuals, will be discussed in Chapter 18 under the title Hybridization. Positive genetic assortative mating is discussed in the next section under the heading "Inbreeding." A discussion of phenotypic assortative mating follows that of inbreeding.

Inbreeding

It is common to find individuals of the same genetic family living fairly close together in the same geographic area. Even though mating may be essentially at random, under these conditions, mates will often be genetically related. *Inbreeding* is the term for matings between closely related individuals.

The most extreme form of inbreeding is *self-fertilization* (*selfing*). If we start with a hypothetical population in which all members are heterozygous at a given locus (*Aa*), we find that after each generation of selfing, the percentage of heterozygotes is reduced by half that expressed in the previous generation (Figure 15.1). Correspondingly, the frequency of homozygotes increases. The population rapidly approaches 100 per cent homozygosity (50 per cent homozygous for allele *A* and 50 per cent homozygous for allele *a*). Gene frequencies in the population, however, have not changed; they remain at the initial frequencies 0.5 *A* and 0.5 *a*. Under Hardy-Weinberg equilibrium, we expect 25 per cent *AA*, 50 per cent *Aa*, and 25 per cent *aa*. When inbreeding is complete, we observe 50 per cent *AA*, 0 per cent *Aa*, and 50 per cent *aa*. Thus, inbreeding is an evolutionary force that changes zygotic frequencies but does not disturb gene frequencies as long as both homozygotes are equally viable.

For example, if a population consists initially of 54 per cent A^1A^1, 32 per cent A^1A^2, and 14 per cent A^2A^2, the initial gene frequencies are $A^1 = 0.7$ and $A^2 = 0.3$ (the starting population is obviously not in Hardy-Weinberg equilibrium). We can predict that when selfing is complete, there will be 70 per cent A^1A^1 and 30 per cent A^2A^2, because the attainment of homozygosity is directly proportional to initial gene frequencies.

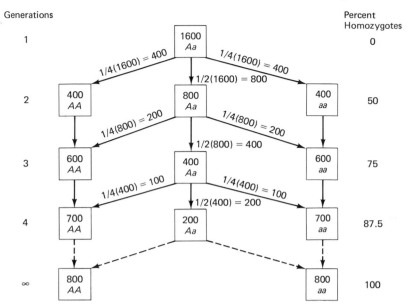

Figure 15.1. The effects of selfing (extreme inbreeding) in an initial population of 1,600 individuals.

The same is true for multiple alleles. Given an initial population with gene frequencies $q_1 = A^1$ allele, $q_2 = A^2$ allele, $q_3 = A^3$ allele, then heterozygosity will be reduced by $\frac{1}{2}$ in each generation. When inbreeding is complete, we expect zygotes in the ratio $q_1(A^1A^1) : q_2(A^2A^2) : q_3(A^3A^3)$.

If at any point during the inbreeding process, random mating is resumed, zygotic frequencies will be restored in one generation to the Hardy-Weinberg expectations of p^2, $2pg$, and q^2.

In the case of a selfing autotetraploid, the rate of approach to homozygosity is much slower than for diploids. Inbreeding usually results in harmful effects correlated with the degree of increase in homozygous recessive types. For this reason polyploidy may be beneficial in retarding the appearance of recessive homozygotes that attends mating of close relatives, as often happens in small populations. Continued selfing is expected to ultimately produce homozygotes in the same proportions as the initial gene frequencies, viz. $p = A$ allele $= AAAA$ homozygotes and $q = a$ allele $= aaaa$ homozygotes.

Given an initial population of autotetraploids in the proportions 20 per cent *AAAA* (quadruplex), 30 per cent *AAAa* (triplex), 30 per cent *AAaa* (duplex), 10 per cent *Aaaa* (simplex), and 10 per cent *aaaa* (nulliplex), the initial gene frequency of the *A* allele is

$$p = [4(20) + 3(30) + 2(30) + 10]/400 = 0.6.$$

After inbreeding is complete, we expect to find 60 per cent *AAAA* and 40 per cent *aaaa*.

In a population of duplex autotetraploids (*AAaa*), the total amount of heterozygosity (*H*) produced by selfing through *n* generations is given by the recurrence formula:

$$H_n = \frac{7}{5}\left(\frac{5}{6}\right)^{n+1} - \left(\frac{1}{6}\right)^{n+1}$$

The following table compares the reduction of heterozygosity by selfing a duplex autotetraploid population vs. selfing in a diploid heterozygous population (*Aa*).

	Percentage Total Heterozygosity	
n	Autotetraploid	Diploid
0	100.00	100.00
1	94.44	50.00
2	80.55	25.00
3	67.43	12.50
4	56.25	6.25
⋮	⋮	⋮
∞	0	0

Notice how much quicker the loss of heterozygosity occurs at the diploid than at the polyploid level.

Selfing reduces heterozygosity by one half in each generation. This rule is true for each heterozygous locus in the population and, therefore, for all heterozygous loci jointly. In reality, we never know how many loci are heterozygous in a natural population. However, we may predict that after each generation of self-fertilization, approximately half the hetero-zygosity existing in the previous generation will have been lost; homozygosity will score corresponding gains. If a group of identical individuals is heterozygous at k loci, the average individual in the next generation will have $\frac{1}{2} k$ loci heterozygous. Eventually, all loci become homozygous, and the population consists of 2^k different homozygous "lines." This sounds quite straightforward, but in reality this is unlikely to be attained because if we start with just 100 loci heterozygous in a natural population (a gross underestimate), then there simply would not be room on the earth for 2^{100} "lines" theory predicts should develop. Some of the lines must die out. In theory, this can be attributed to chance, but in reality many of the lines are expected to become homozygous for deleterious alleles (usually recessives), and these are the lines most likely to disappear during the inbreeding process. Selection will undoubtedly be at work to preserve only those lines that happen to become homozygous for the most adaptive gene combinations. Therefore, the residue of genotypes remaining when inbreeding is complete is anything but a random sample of the multitude of genetic lines theoretically possible. This helps to explain the origin of many species of flowering plants that are almost complete selfers, despite the fact that the inbreeding process is generally considered to be harmful.

The term *inbreeding depression* is used to express the reduction of the mean phenotypic value (displayed by traits connected with reproductive capacity or physiological efficiency) accompanying inbreeding. Much of this loss in fitness is thought to be due to the phenotypic expression of deleterious recessive alleles brought to homozygosity by the inbreeding process. As inbreeding proceeds, the genetic variance within lines decreases, while that between lines increases. Heritability of metric traits decreases within each line as inbreeding increases, because more and more of the phenotypic variability becomes environmentally induced (because the genetic variability is approaching zero). Homozygotes are generally less well buffered physiologically against environmental changes and tend to loose developmental canalization. The phenotype becomes highly plastic. In the limiting case in which all loci have become homozygous, selection is no longer effective *within a line,* because there is no genetic variability on which it can act. The species is then dependent, for its evolutionary potential, on the genetic diversity *between* its inbred populations (lines). If relatively few lines have become established, the prognosis for survival of the species is not good.

Self-incompatibility is thought to be a primitive condition in most plant taxa and therefore self-compatibility is always a derived state in flowering plants. Four main lines of evidence support this contention:

1. Self-fertilizing species tend to be more specialized in morphological traits than most of their self-incompatible, cross-pollinating relatives.
2. Numerous selfers still retain relict traits that could only have been useful to their more primitive outcrossing ancestors (e.g., colorful flowers that formerly attracted insects for cross-pollination).
3. Genetic studies have revealed that closely related genera or families have the same multigenic basis for self-incompatibility and hence are inferred to have an ancient common origin.
4. Within historical times, self-fertilization has been observed to evolve in formerly self-incompatible species. For example, the cultivated snapdragon (*Antirrhinum majus*) is the only self-fertilizing species in this genus. The change from multigenic self-incompatibility to self-compatibility requires relatively few genetic alterations, whereas the evolution of a multigenic self-incompatibility system requires so many alterations that it probably arose only a few times during the history of the angiosperms (or perhaps even preceded the divergence of this group).

Self-fertilization has several advantages that may compensate for the generally harmful effects mentioned earlier. One obvious advantage is in long-range species dispersal. It would be possible for a single plant that happened to colonize a new area to perpetuate the species if it could self-pollinate. For a self-incompatible species, two pioneer plants would need to be within pollinating distance if the species was to survive in the new locality. The chance establishment of outcrossing species through long-range dispersal is much less than for self-compatible species. This is the central idea of *Baker's rule* (named after H. G. Baker, 1955). G. L. Stebbins suggests that the success of many weedy species is partly due to their self-compatible systems, which allow them to colonize temporary habitats (especially those disturbed by human activities). Wind pollination requires that much pollen be shed in order for this random mechanical process to affect fertilization. Much less pollen needs to be produced by selfers. Self-fertilization is a highly efficient process that greatly reduces biological wastage of gametes. In cold climates, insects for cross-pollination might not become available until far into the growing season. Dense foliage in tropical regions may act as an impediment to movement of pollen by wind. Deserts have too little water to aid in carrying gametes. In all these ecological situations, self-compatibility could be advantageous.

In addition to the physiological changes that accompany the transition from self-incompatibility to self-compatibility, structural adaptations commonly evolve to enhance the effectiveness of the system. For example, flowers that fail to open (cleistogamous) contribute to maximal selfing. In flowers that do open, reduction in length of the style to (or below) the level of the anthers is also an aid in this respect. Rotation of the anthers so that they open (dehisce) toward the style instead of outward would also serve the selfing mechanism.

One of the most fundamental principles in population genetics is *Wright's Equilibrium Law*. The symbol *F* is used to denote the *coefficient of inbreeding*, first defined (1922) by Sewall Wright (Figure 15.2) as the genetic correlation between uniting gametes. The only practical way to use *F* is to define some base population in which all genes are presumed to be independent (i.e., not identical by descent). This is obviously merely a convenience for making calculations because all genes are related to preexisting genes. The base population is considered to have an inbreeding coefficient of zero ($F = 0$). The value of *F* denotes the additional increase in homozygosity (above that already present in the base population) as a result of the inbreeding process. When inbreeding is complete ($F = 1.0$), all loci are homozygous. An inbred population may be considered to have two components: (1) a panmictic component ($1 - F$), representing loci that segregate, and (2) a fixed (homozygous) component (*F*) that cannot segregate. Wright's Equilibrium Law expresses the zygotic proportions of an inbred population thusly:

Figure 15.2.   Sewall Wright in 1954. (Photograph by The Llewellyn Studio)

Genotypes	A^1A^1	A^1A^2	A^2A^2
Original frequencies	p^2	$2pq$	q^2
Change due to inbreeding	$+Fpq$	$-F2pq$	$+Fpq$

Consider a population in which $p = q = \frac{1}{2}$. Before inbreeding, the population was in Hardy-Weinberg equilibrium with 25 per cent A^1A^1 : 50 per cent A^1A^2 : 25 per cent A^2A^2 zygotes. After one generation of selfing ($F = \frac{1}{2}$) the frequency of heterozygotes is

$$2pq - F2pq = 2pq(1 - F) = 2(\tfrac{1}{2})^2(1 - \tfrac{1}{2}) = \tfrac{1}{4}$$

or half its value in the previous generation. The frequency of A^1A^1 zygotes has increased from $p^2 = 25$ per cent to

$$p^2 + Fpq = (\tfrac{1}{2})^2 + \tfrac{1}{2}(\tfrac{1}{2})(\tfrac{1}{2}) = \tfrac{3}{8}$$

or 37.5 per cent.

For less severe forms of inbreeding, the rate of increase in homozygosity (i.e., the rate of loss of genetic heterogeneity) is correspondingly reduced (Figures 15.3 and 15.4). For example, $F = \frac{1}{4}$ for an offspring whose parents are full sibs (brother and sister having both parents common). That is to say, an average of about 25 per cent of all loci should become homozygous for identical alleles in the progeny from full sib parents. Similarly, the inbreeding coefficient for offspring from half sibs (only one parent in common) is $\frac{1}{8}$; for first cousins it is $\frac{1}{16}$; etc. In general, heterozygosity decreases by $1/2N$ per generation, assuming complete random union of gametes within a group of N individuals. As an example, when $N = 4$, heterozygosity is expected to be reduced by $1/(2 \cdot 4) = \frac{1}{8} = 12\frac{1}{2}$ per cent per generation if individuals mate completely at random (including selfing in random amount). Though the calculations are too complex to be presented here, the corresponding average rate of reduction in heterozygosity when all matings are confined to double first cousins (N also equals 4) is about 8 per cent. Note that this formula also applies to the extreme case of self-fertilization where $N = 1$.

Almost half our economically important plants are self-compatible, including wheat, barley, rice, soybeans, cotton, sorghum, peas, and beans. Even among these so-called "selfers," there is commonly a small amount of cross-fertilization. Evolution in natural plant populations tends to strike a balance between the advantages and disadvantages of self- vs. cross-pollination.

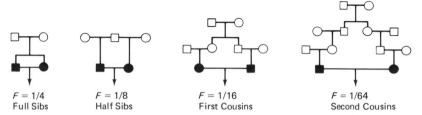

$F = 1/4$	$F = 1/8$	$F = 1/16$	$F = 1/64$
Full Sibs	Half Sibs	First Cousins	Second Cousins

Figure 15.3. Family pedigrees illustrating the principle that the inbreeding coefficient (*F*) of offspring decreases with decreasing genetic relationship of parents (solid symbols).

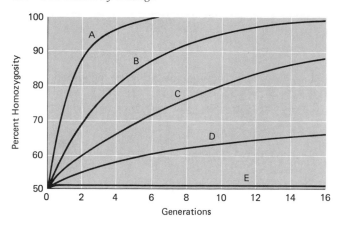

Figure 15.4. Increase in percentage of homozygosity under various systems of inbreeding. (A) self-fertilization, (B) full sibs, (C) double first cousins, (D) single first cousins, (E) second cousins.

Phenotypic Assortative Mating

When males and females of the same phenotype tend to mate in nature, this is called *positive phenotypic assortative mating*. If dominant phenotypes only mate with their kind and recessive phenotypes only with their kind (complete positive phenotypic assortative mating), the population will gradually become more homozygous. However, the approach to homozygosity is expected to be much slower than under close inbreeding. Gene frequencies are not expected to change under positive phenotypic assortative mating. However, natural populations rarely breed completely in this way. With imperfect positive phenotypic assortative matings, the population should gradually reach an equilibrium characterized by a constant proportion of heterozygotes. Although both inbreeding and positive phenotypic assortative mating reduces genetic heterozygosity, the end results are strikingly different. As an example, consider a metric trait governed by two loci each with a pair of alleles both additive and equal in effect. Inbreeding among the five phenotypes would ultimately fix four homozygous lines: *AABB*, *AAbb*, *aaBB*, and *aabb;* whereas positive phenotypic assortative mating would only fix two lines: *AABB* and *aabb*. The rate at which genes can be fixed in a population can be greatly accelerated by combining a system of close inbreeding with the additional restriction of positive phenotypic assortative mating.

Sexual dimorphism is an example of complete *negative phenotypic assortative mating*, because males mate only with females in dioecious populations. As a further example, consider a population with *AA*, *Aa*, and *aa* genotypes. If breeding is only between different phenotypes, there will only be two kinds of matings, *viz.* (*aa* × *AA*) and (*aa* × *Aa*). No *AA* genotypes will appear in the next generation; hence, all subsequent matings will be (*aa* × *Aa*). The equilibrium is immediately reached, because in every generation there should be $\frac{1}{2}$ *aa* and $\frac{1}{2}$ *Aa* offspring. This scheme can be used to explain *distylic heteromorphism* (*heterostyly*) in many angiosperms. Suppose we hypothesize that short styles are dominant to long styles, and that homozygosity for the dominant allele is lethal. Therefore short-styled plants are *Aa* and long-styled plants are *aa*. If short-styled plants receive only pollen from long-styled plants, there should be 50 per cent short- and 50 per cent long-styled offspring in each generation. At equilibrium, dominants and recessives should occur in equal proportions even when negative phenotypic assortative mating is far from complete.

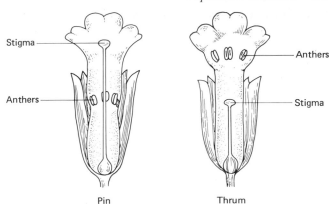

Stigma — Anthers

Anthers — Stigma

Pin   Thrum

Figure 15.5. Heterostyly in flowers of the primrose *Primula officinalis.* The "pin" phenotype has its stigma located in the same general region as the anthers in the "thrum" phenotype (and *vice versa*).

Heterostyly (Figure 15.5) has been well studied in primroses (genus *Primula*). Long-styled flowers, called "pin" phenotype, have their stigmas near the top of the corolla (the whorl of petals in a flower) tubes; anthers are located near the middle of the corolla tubes. The other type of flower, called "thrum" phenotype, has a short style with a stigma located approximately at the level of the anthers in pin flowers. Likewise, the anthers of thrum flowers are located near the top of the corolla tube (where the stigma is located in pin flowers). Thus, when an insect visits one of these types of flowers, it tends to receive pollen on that part of its body which corresponds to the location of the stigma in the other type of flower. This greatly enhances cross-pollination in these species. In nature, they seldom self-fertilize, but it can be done artificially. When pin is selfed, it breeds true (as expected for a homozygous recessive genotype). When thrum is selfed, the offspring approximate 3 thrum : 1 pin (as expected for a heterozygous genotype).

Pin and thrum flowers differ in at least three respects other than position of flower parts: (1) they have different sized pollen grains, (2) the morphology of their stigmas are different, and (3) their pollen grains have different kinds of papillae on their surfaces. Furthermore, the surface of a pollen grain from a pin flower fits the stigma of a thrum flower better than thrum pollen itself. Pin stigmas tend to inhibit growth of pin pollen, but actually enhance the growth of pollen tubes of thrum origin. Now all of these functionally related physiological and anatomical characteristics can hardly be attributed to the pleiotropic effects of a single gene. A much more likely hypothesis suggests that there is a cluster of several closely linked genes involved in this character complex. This hypothesis is supported by findings of rare recombinants in this "supergene." It is highly unlikely that these genes originated by mutation *in situ*. They probably originated by mutations in different regions of the genome and subsequently were brought together by fortunate structural rearrangements such as translocations and inversions and stabilized by natural selection because of the advantage of transmitting these genes as a unit.

Genetic Drift

If one tosses a coin ten times, it would not be surprising to obtain eight heads and two tails. However, in a sample ten times larger (100 tosses), obtaining the same percentage deviation (eighty heads and twenty tails) would be considered highly unusual. This serves to illustrate that large deviations from expected values are common in small samples, but chance deviations should become proportionately less as the sample size increases. The random (chance) changes in gene frequencies that commonly occur in small populations are called *genetic drift* (or the *Sewall Wright Effect*). Drift is viewed as a *stochastic process,* i.e., a series of steps, at each of which the movement made is random in direction.

Because of geographical or ecological discontinuities, the large base populations of most species may be conceptually (if not actually) subdivided into numerous lines or demes in which random mating is at least theoretically possible. An *idealized population structure* is one in which

1. The base population is subdivided into a very large number of demes.
2. Within each deme, all the Hardy-Weinberg restrictions are met excepting deme size (i.e., mating is at random, including selfing in random amount; no selection; no mutation; each deme is totally isolated from other demes).
3. The number of breeding individuals in each line is the same for all lines and in all generations (numbers of males equals numbers of females).
4. Each generation is distinct; there is no overlap.

Under these simplifying assumptions of the idealized population structure, it is possible to predict mathematically the evolutionary effects of genetic drift. We cannot predict, for any one deme, in which direction the gene frequency change will occur, but we can predict the variance in change of gene frequencies among all the lines following one generation of sampling errors in small populations of stated size N. Consider an idealized base population in which all demes have initial gene frequencies p and q for alleles A^1 and A^2, respectively. The variance in change of frequency of the A^2 allele after one generation of drift is

$$\sigma_{\Delta q}{}^2 = \frac{pq}{2N}.$$

Obviously, the variance increases as deme size (N) becomes smaller. That is to say, drift assumes greater importance as an evolutionary force in proportion to reduction in population size. For any value of N, $\sigma_{\Delta q}{}^2$ is maximized when $p = q = 0.5$.

Visualize an idealized population consisting of a large number of isolated panmictic units, each of size 50. Two alleles, A^1 and A^2, are initially at frequencies $p = q = 0.5$. After one generation of genetic drift, the A^2 allelic frequencies in all lines is expected to form a normal distribution around a mean of 0.5. Therefore, the standard deviation of change in frequency of A^2 allele (hence a corresponding change of A^1 allele) is

$$\sigma_{\Delta q} = \sqrt{\frac{0.5 \times 0.5}{2 \times 50}} = \pm 0.05.$$

The mean gene frequencies of A^1 and A^2 alleles over all lines does not change from one generation to the next, though it is changing in at least some of the lines individually. We can predict the 95 per cent confidence limits within which the A^2 allelic frequency should be found. This is approximated by $\mu \pm 2\sigma$ or $0.5 \pm 2(0.05)$. Thus, in about 95 per cent of the lines, drift should have moved the A^2 allelic frequency no further than 0.4–0.6. A correlative prediction can be made with 95 per cent accuracy that no one line will have drifted beyond the range $q = 0.4$–0.6.

The ultimate effect of dispersion of gene frequencies by genetic drift is fixation (or loss) of one of the alleles in each line. In this and several other ways, the random sampling process in small populations is akin to the effects of inbreeding: (1) homozygosity increases within lines at the expense of heterozygosity, (2) subpopulations become progressively genetically differentiated, (3) genetic heterogeneity becomes distributed *between* lines rather than within lines, (4) variance of gene frequencies is expected to increase in each generation until all lines have become homozygous, and (5) average gene frequencies over all lines remains constant.

The loss of heterozygosity that occurs in finite breeding groups is not simply due to occasional close inbreeding by chance in panmictic units. A population still looses heterozygosity even with maximum avoidance of inbreeding in each generation (albeit at a slower rate than under random mating). The decrease in heterozygosity is a function of limited size of the panmictic unit *per se*.

In a large number of demes of small size (all with initial gene frequencies p and q), the percentage of these lines in which an allele is either fixed or lost is directly proportional to the starting allelic frequency. For example, if $p = q = 0.5$ in all demes we expect 50 per cent of the lines to fix one allele, and 50 per cent to loose the same allele (or fix the alternative allele). If $p = 0.7$ and $q = 0.3$, we expect 70 per cent of the lines to fix A^1 allele and 30 per cent to fix A^2 allele. The time required to fix (or loose) an allele through drift in a deme depends on the initial gene frequency q and deme size N. When $p = q = 0.5$ the average fixation time is roughly equivalent to $2N$ generations. For example, in demes of ten breeding individuals with $p = q = 0.5$, it would take about $2(10) = 20$ generations to fix half of them. Obviously, the mean fixation time becomes shorter as q approaches 0 or 1, because there is less distance remaining to the nearest boundary.

S. Wahlund (1928) proved mathematically that the effect of subdividing a population into numerous panmictic units is to increase the proportion of homozygotes by an amount equivalent to the variance of gene frequencies (σ_q^2) at the expense of heterozygotes. The following tabulation reveals that subdivision of a population into panmictic units is equivalent to the practice of inbreeding within the total population:

Genotype	Wahlund's Principle	Inbreeding
AA	$p^2 + \sigma_q^2$	$p^2 + Fpq$
Aa	$2pq - 2\sigma_q^2$	$2pq - 2Fpq$
aa	$q^2 + \sigma_q^2$	$q^2 + Fpq$

The previous discussions on inbreeding and genetic drift have related deviations in subpopulations from Hardy-Weinberg conditions to variance in gene frequencies, variance to deme size, deme size to inbreeding, and inbreeding to variance.

In reality, natural populations do not conform to the idealized breeding structure on which our previous models were built. For example, in polygamous species there are many fewer breeding males than breeding females. This obviously is equivalent to an inbreeding situation. To treat mathematically these deviations from the idealized structure, we need to entertain the concept of *effective population size* (or the effective number of breeding individuals). This is considered to be that number of individuals in an idealized population required to produce the same sampling variance (dispersion of gene frequencies) as that of the nonidealized population under consideration. By converting the actual population size (N) to the effective number (N_e), the rate of inbreeding can be expressed as

$$F = \frac{1}{2N_e},$$

corresponding to the rate of inbreeding as previously calculated for an idealized population ($\Delta F = 1/2N$). For a population consisting of unequal numbers of breeding males and females the formula that estimates the effective population size is

$$N_e = \frac{4N_f N_m}{N_f + N_m},$$

where N_f and N_m are the actual numbers of breeding females and males, respectively. The rate of inbreeding is approximated by

$$\Delta F = \frac{1}{2N_e} = \frac{N_f + N_m}{8N_f N_m}.$$

Elephant seals return to breed each year at the rookery on Ano Nuevo Island off the coast of California. According to one report, out of about 150 males, the top 4 per cent routinely get 90 per cent of the females, or about 40 apiece. The other 10 per cent of the females are perhaps divided among six other males. For that segment of the female population served by the six males (.04 × 150) the effective population size is estimated to be

$$N_e = \frac{4(6)(240)}{6 + 240} \approx 23.$$

Thus, even though the total population is almost 250 breeding individuals, the gross inequality of the sexes causes the population to respond genetically in the same fashion as an idealized population consisting of less than twenty-five breeding individuals. An estimate of the loss of heterozygosity under these conditions is obtained by

$$\Delta F = \frac{240 + 6}{(8)(6)(240)} = 0.021.$$

Therefore, this elephant seal population is expected to lose approximately 2 per cent of its heterozygosity each generation. This seems to be a very high rate of erosion of genetic variability. Despite these grim calculations, elephant seal populations are flourishing today. At one time they were nearly exterminated by whale hunters for their blubber. Elephant seals were rediscovered in 1892 on Guadalupe Island, about 150 miles off the Baja California coast. They were put under protective law and were finally removed from the endangered species list in 1974. During the period when the entire population was very small, genetic drift had the potential to cause considerable alterations in gene frequencies.

An abnormal sex ratio may sometimes be beneficial to the species. Plants seldom use a sex chromosome mechanism of sex determination. An exception is found in the dioecious campion *Melandrium album,* in which sex is determined by an XY chromosome mechanism. An excess of female plants is commonly produced, because pollen grains bearing an X chromosome grow faster down the style than those bearing a Y chromosome. When female plants far outnumber male plants, there will be fewer pollen grains available to fertilize the larger number of ovules. Under these conditions, the slower growing pollen tubes bearing the Y chromosome will not be at a disadvantage in fertilizing their share of ovules. A balance is struck between these opposing forces such that the largest number of female plants (and therefore the largest number of seeds) is produced in every generation. Male plants produce more flowers (hence more pollen) than female plants, and this further accentuates the abnormal sex ratio. Dioecious plants in which the staminate (male) plant is the heterogametic sex (XY chromosomal composition) has the opportunity to adjust its sex ratio to maximize population fitness as *Melandrium* has done. Plants in which the female is heterogametic are deprived of this opportunity. This seems to explain why staminate plants are heterogametic in most species of dioecious plants, despite the fact that pistillate and staminate heterogamy are equally attainable evolutionary possibilities.

Population size is seldom invariate over many generations. Genetic drift is expected to be most influential in changing gene frequencies in those generations with the smallest numbers of breeding individuals. Therefore, the effective number over t generations is approximated by the harmonic mean of actual numbers ($N_1, N_2, \ldots, N_t$) in each generation:

$$\frac{1}{N_e} = \frac{1}{t}\left[\frac{1}{N_1} + \frac{1}{N_2} + \cdots + \frac{1}{N_t}\right].$$

Given four generations of actual population sizes 100, 800, 20, and 5000, the average effective size is

$$\frac{1}{N_e} = \frac{1}{4}\left[\frac{1}{100} + \frac{1}{800} + \frac{1}{20} + \frac{1}{5000}\right] = 0.0153625$$

$$\therefore N_e = 65 \text{ (approx.)}.$$

Notice that the average effective size is much closer to the minimum number than to the maximum number over any range of generations. Obviously, the generations with the smallest numbers have the greatest evolutionary effect on the population.

Large fluctuations in population size are probably common seasonal or annual occurrences in many species. For example, huge populations of insects that appear in the warm spring and summer months usually experience great mortality during the cold winter months. Relatively few insects (their eggs, pupae, or other dormant stages) survive to become part of the breeding population when conditions become more favorable the following spring. The few survivors represent a much smaller gene pool than existed the previous summer. This is called a "bottleneck effect," in which large gene pools are periodically funneled through intermediate populations of small size to once again become large breeding units. These bottlenecks of small size offer opportunities for the random sampling process of genetic drift to alter substantially gene frequencies.

A bottleneck also exists when a small group of individuals from some larger population invades a new geographical region and becomes founders of a noncontiguous subpopulation. These founders carry only a small sample of the genetic variability in the gene pool from which they came. Different founding colonies will somewhat fortuitously possess distinctive (and diminished) genetic endowments at their inception. This is the basic idea behind what

Ernst Mayr has called the *founder principle.* Drift can be an evolutionary force of considerable importance during the early stages of a founding colony when its size is still relatively small.

Mayr also postulates that these isolated marginal populations of small initial size are particularly subject to what he calls *genetic revolutions,* i.e., drastic genetic reconstructions and rapid evolution. The "marginal isolates" or "founder populations" are not likely to be well adjusted to the new habitat and therefore will come under strong selection for adaptations to the new conditions. Perhaps it is in these marginal isolates that conditions are optimal for the evolution of the kind of novelties characteristic of new genera (or higher taxonomic categories). Rapid evolution in ancestral founder populations might account for the sudden appearance of new forms in the fossil record. Since the adaptive value of any gene is influenced to some extent by the residual composition of the gene pool, it is quite likely that the entire genetic system of a founder population will need readjustment, because the gene pool of the isolate is no longer the same as that of its parental population. Thus, the average fitness values of two alleles need not be identical in two founder populations, even though the fitnesses might have been identical in the parental population. In all but very small populations (e.g., fewer than ten or twenty breeding individuals), even small selection pressures are usually sufficient to counteract the relatively weak effects of genetic drift. Nonetheless drift may supplement selection favoring homozygotes that are narrowly adapted to the more restrictive conditions of the colonized region. H. L. Carson uses the term *homoselection* to describe this process in marginal isolates. The large parent population has more genetic variability; heterozygosity tends to buffer each individual for a generalized ability to exploit all the subniches in the habitat. Carson proposes that *heteroselection* typifies these larger breeding units because of the advantages that polymorphisms and buffering systems offer in this respect. Structural polymorphisms would tend to be disadvantageous in marginal isolates, because they restrict recombination, and liberal recombination would almost be essential to a small, rapidly evolving system. Homoselection in small populations would therefore tend to favor structural homozygotes (e.g., inversion homozygotes) in which recombination is relatively unrestricted (i.e., recombinationally "open"). Heteroselection in large populations favors structural heterozygosity for their "supergene effect" and therefore restricts recombination (i.e., recombinationally "closed").

Some evolutionists favor the notion that polymorphisms for certain characters that appear selectively neutral could have become established by genetic drift during times when the population was small. After the population becomes large these loci could remain in Hardy-Weinberg equilibrium at the frequencies attained earlier by random effects.

> Several traits in the "Dunker" religious community of Franklin County, Pennsylvania (formally known as the German Baptist Brethren) seem to be polymorphic without any apparent selective value. This community was founded by about fifty families immigrating from centers in Western Germany during the years 1719–1729. The community now numbers about 300 individuals, but has been less than 100 for several generations. They have tended to marry only within their own religious group, forming a semi-isolated founder population. It is estimated that about 12 per cent of the parents in the community came in from the surrounding population in each generation. This "gene flow" from outside would be expected to render the Dunker gene pool more nearly like that of their neighbors. Despite this equalizing tendency, Dunkers are neither like the West Germans nor like the surrounding Americans, nor anything in between. For

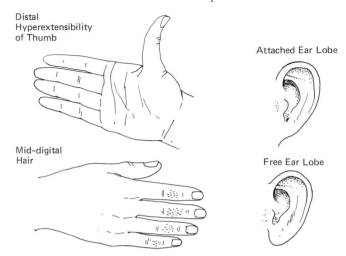

Distal
Hyperextensibility
of Thumb

Attached Ear Lobe

Mid-digital
Hair

Free Ear Lobe

Figure 15.6. Some characters in the "Dunker" population which are of lower frequency than in the surrounding populations. These differences have been hypothesized to be the effects of genetic drift.

example, the frequency of blood group A is 45 per cent in West Germany and 40 per cent in the United States; but it is nearly 60 per cent in Dunkers. Blood groups B and AB together are about 15 per cent in both major populations, but are only about 5 per cent in Dunkers. These differences are statistically significant. No race of West European origin has anything resembling the Dunker's frequencies; the nearest frequencies are to be found in American Indians, Polynesians, or Eskimos. Perhaps the original Dunker founders did not represent a random sample of the West German parental population from which they were derived.

In another study, the Dunkers were divided into three age groups roughly corresponding to three successive generations: (1) 3–27 years, (2) 28–55 years, and (3) 56 years and older. Gene frequencies for the M-N blood group system were found to differ markedly among these age groups. In the oldest group, the frequency of *M* allele was about the same as the surrounding population (30%). The frequency of *M* in the intermediate age group was 66 per cent, and in the youngest age group it was 74 per cent. It seems as though these genes have been detected in the act of drifting. Additional evidence for the drift hypothesis was found in other traits. Dunkers had lower frequencies of middigital hair, distal hyperextensibility of the thumb ("hitchhiker's thumb"), and attached ear lobes than other United States communities (Figure 15.6). H. B. Glass suggests that most inherited racial differences involve adaptively unimportant traits of this kind. However, this hypothesis is not universally accepted by all anthropologists and evolutionists.

Gene Flow

Most Mendelian populations are not closed as assumed by the Hardy-Weinberg equation. Local breeding units tend to communicate genetically with each other by sending out emigrants to other subpopulations and receiving immigrants from same. This *migration* or *gene flow* links all the demes in the common gene pool of the species. Gene flow tends to counteract the loss of genetic variability due to drift in small populations. Thus, small amounts of migration can be beneficial to each local breeding population. However, if the amount of gene flow is heavy, it tends to swamp any distinctiveness that might otherwise be built up in semi-isolation. In this sense, gene flow may act as a retarding element in speciation. Migration is the only process other than mutation by which a population can receive new kinds of genetic information. Perhaps the easiest way to summarize the evolutionary significance of population structure is in a form such as Table 15.1. For convenience, we may assume three arbitrarily defined populations: (1) large, panmictic (probably over 1,000 breeding individuals), (2) intermediate, partly subdivided (perhaps 50–1,000 individuals in each deme), and (3) small, isolated (under 50 breeding individuals). Migration has its most striking effects in subdivided populations of intermediate size (Figure 15.7). This kind of population structure offers a species more evolutionary flexibility than almost any other status. The species as a whole tends to maintain heterogeneity when subdivided into local populations despite inbreeding or genetic drift within demes. Each population constitutes a separate adaptive experiment. Those groups that are particularly successful can transmit this genetic information via gene flow to other semi-isolated groups in the species, or they may spread into adjacent areas and replace less successful groups by interpopulation competition (group selection). The origin of evolutionary novelties is more likely to occur in a species that is constantly experimenting with new gene complexes in subgroups semi-isolated geographically from each other. Adaptations to local conditions by these demes allows the species to extend its range over a diversified environment. When the environment itself is unstable, the genetic variability residing in gene pools of local populations generally offers the species a greater survival opportunity than if its members formed a singular confluent breeding pattern.

TABLE 15.1. Relationships Between Population Structure and Evolutionary Flexibility

Factors	Large, Panmictic Population	Intermediate-Sized, Partly Subdivided Population	Small, Isolated Population
No. of mutations	Many	Variable	Few or none
Rates of response to selection	High at intermediate gene frequencies; otherwise low	Fluctuating; depends on amount of gene flow	Negligible except in early stages of isolation
Genetic drift	Negligible	Occurs locally or periodically	Important
Gene flow	None	Important	None
Rate of Evolution Under: Static environment	Slow or static	Fluctuating	Rapid approach to equilibrium; then slow or zero
Slowly changing environment	Slow and progressive adaptation	Fluctuating, but generally progressive adaptation	Probable extinction depending on degree of individual homeostasis
Rapidly changing environment	Extinction or reduction to small isolated population or intermediate, partly subdivided population	Rapid, with maximal opportunity for origin of new forms	Certain extinction

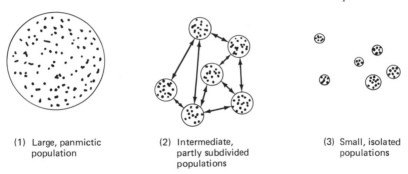

(1) Large, panmictic population

(2) Intermediate, partly subdivided populations

(3) Small, isolated populations

Figure 15.7. Three models of population structure.

Migration can be thought of in two ways: (1) gene flow between demes and (2) origin of founder populations. This latter function can be a single migratory event by which the range of the species is extended. Dispersion of a species is essentially a migratory phenomenon. The margins of a species range should be conceptualized as a highly dynamic region in which the species is constantly testing localized gene pools for survival value in environments beyond the limits of the adaptive norm at the center of the species range.

There are at least three models to represent migratory schemes (Figure 15.8). First, the "island model" considers a group of semi-isolated subpopulations exchanging genes in various quantities and directions through periodic or regular migrations. Second, the "river model" visualizes unidimensional gene flow along a linear array of subpopulations. Third is the concept of "isolation by distance." In a group of organisms continuously distributed over a large geographic area, individuals at opposite ends of the distribution have virtually no opportunity to mate. Although we cannot see any discontinuities in the distribution of individuals, there do exist localized neighborhoods in which gametic interchange is possible. The amount of local differentiation in such a population largely depends on the size of the potentially panmictic neighborhoods of which the species is composed and the amount of gene flow between them. Gene pools of adjacent neighborhoods are more likely to be similar than those of distant neighborhoods. And so it is not uncommon to find, both in plants and animals, species that exhibit distinctive characteristics in their local populations distributed

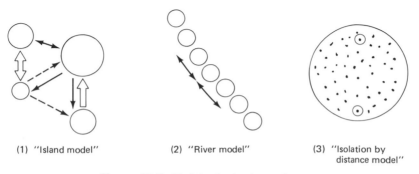

(1) "Island model"

(2) "River model"

(3) "Isolation by distance model"

Figure 15.8. Models of migratory schemes.

over wide geographical regions. Some of these migration concepts will be resurrected in discussions concerning the origin of races in the next chapter.

Formal Aspects of Migration

Let us consider the effect of gene flow on a single isolated deme (Figure 15.9). Alleles A and a are at initial frequencies p_1 and q_1, respectively. The deme then receives a number of immigrants that represents the fraction m of the deme size (there being $1 - m$ natives remaining after the immigration event). Immigrants come from a population with allelic frequencies p_2 and q_2, respectively. After migration, the average frequency of a is

$$\bar{q} = q_1(1 - m) + mq_2$$
$$= q_1 + m(q_2 - q_1)$$

The absolute change in gene frequency due to migration is

$$\Delta q = |m(q_2 - q_1)|$$

The actual change will be positive if $q_2 > q_1$; it will be negative if $q_2 < q_1$. Thus, it is clear that the evolutionary effect of gene flow depends on (1) the proportion of immigrants and (2) the difference in allelic frequencies of the two populations.

If we know (or can estimate) q_1, q_2, and $\bar{q}$, the immigration "pressure" (m) can be determined by the formula

$$m = \frac{\bar{q} - q_1}{q_2 - q_1}.$$

The infusion of "White" (Caucasian) genes into the "Black" (African Negro) population, which began after they were brought to America as slaves in the seventeenth century, has been partly responsible for the evolution of the American Negro population (race). Because Negro-"White" hybrids are socially considered Negroes in the United States, interbreeding between these two populations has resulted in an essen-

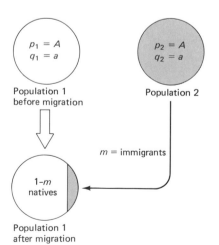

Figure 15.9. Diagram of one-way gene flow (magnitude m) from Population 2 to Population 1.

tially "one-way" gene flow from the "White" to the Negro population. How much total gene flow has occurred in this racial intermixture? In attempting to answer this question we need to choose a locus (1) for which selection is essentially neutral (or for which the adaptive values are essentially the same in both populations), and (2) with markedly different allelic frequencies in the two populations. All other things being equal, the greater the absolute difference $|q_2 - q_1|$, the more accurate will be the estimate of gene flow (m). The locus governing the ability to taste the chemical PTC appears to qualify in these respects. The inability to taste PTC is controlled by an autosomal recessive gene (t). The frequency of this gene obviously is unknown in the ancestral African population. The best we can do in this respect is to estimate it from an average of several African tribes at the present time (assuming that there has been essentially no change at this locus in the interim). In one study, this estimate from the African population was $q_1 = 0.18$; for American Negroes it was $\bar{q} = 0.30$, and for "Whites" it was $q_2 = 0.55$.

$$m = \frac{0.30 - 0.18}{0.55 - 0.18} = 0.324.$$

Thus, it is estimated that approximately 32 per cent of the American Negro gene pool consists of genes derived from "Whites" by almost two centuries of gene flow. It is estimated that, on the average, approximately 3–4 per cent of the genes in the American Negro population have been derived each generation from the neighboring "White" population. Similar estimates have also been obtained using other suitable loci such as those for the ABO and Rh blood group systems.

Now let us consider some of the joint effects of migration and other evolutionary pressures. For simplicity, assume that heterozygotes are of intermediate fitness between the two homozygotes (i.e., fitness of $AA = 1$, $Aa = 1 - s$, and $aa = 1 - 2s$); further assume that the frequency of the recessive allele in the deme is q and that of immigrants is q', and that the migration pressure is m. Then, the joint effect of selection and gene flow on changing gene frequency is expressed as

$$\Delta q = \overbrace{-sq(1 - q)}^{\substack{\text{effect of} \\ \text{selection}}} - \overbrace{m(q - q')}^{\substack{\text{effect of} \\ \text{migration}}}.$$
$$= sq^2 - (s + m)q + q'm$$

At equilibrium, $\Delta q = 0$, and applying the general formula for solving a quadratic equation we have

$$\hat{q} = \frac{(m + s) \pm \sqrt{[(m + s)^2 - 4msq']}}{2s}.$$

A positive value of s indicates that selection acts against the recessive allele, whereas a negative value of s denotes that selection favors it. In order to evaluate the effects of m and s on equilibrium gene frequencies we can derive the approximate values of $\hat{q}$ for three cases: (1) $m = |s|$, (2) $m < |s|$, and (3) $m > |s|$.

Case 1. Migration pressure is the same order of magnitude as the absolute coefficient of selection.

In this instance, if s is positive, then $m + s$ can be expressed as $2s$; if s is negative

$m + s = 0$. Substituting these values into the above equation, we obtain

$$\hat{q} = \sqrt{q'} \text{ when } a \text{ is favored } (s \text{ negative});$$
$$\hat{q} = 1 - \sqrt{(1 - q')} \text{ when } a \text{ is unfavored } (s \text{ positive}).$$

For example, given $q' = 0.4$, $\hat{q} = \sqrt{0.4} = 0.6325$ if selection favors the recessive allele, and $\hat{q} = 1 - \sqrt{0.6} = 0.2254$ if selection acts against it. It is clear then, that local populations can be significantly differentiated under different selective forces when $m = |s|$.

Case 2. Absolute selection coefficient is greater than migration pressure.

If local populations can differentiate under Case 1 ($m = |s|$), then they are certainly going to be even more strikingly differentiated when $m < |s|$. As an illustration, take $q' = 0.4$, $m = 0.01, s = \pm 0.15$. Substituting these values into our equation, we find that when selection favors the recessive allele $\hat{q} = 0.9611$, but when selection acts against it $\hat{q} = 0.0256$. Notice that migration in this case exerts a very weak force that barely prevents genes from becoming fixed or lost in the gene pool. In this sense, migration performs a function similar to that of new mutations.

Case 3. Migration pressure exceeds the absolute selection coefficient.

Any tendency for different selective forces to differentiate local populations can be swamped by relatively higher levels of immigration. The equilibrium values are expected to vary little from that of the total population. With $q' = 0.40$, $m = 0.15$, and $s = \pm 0.01$, $\hat{q} = 0.416$ if selection favors the recessive allele, and $\hat{q} = 0.384$ if it is disfavored. Both equilibrium values are close to the average frequency of the entire population (0.40). The larger the mutation pressure relative to the selection coefficient, the more closely we expect gene frequencies in local breeding units to cluster around the mean value for the total population. Relatively small adaptive advantages will not exert enough pressure to differentiate markedly local populations. This may actually be the most common situation in nature.

What about the joint effects of migration and inbreeding? Recall that a reduction in population size (or inbreeding) results in increased homozygosity within demes at the rate of about $1/2N$ per generation (with random mating). The kinds of genes fixed (or lost) will undoubtedly be different in the various local breeding groups. Therefore, migration between demes would tend to counteract the loss of heterozygosity within demes. Although the calculations are too complex to present here, it can be proven that an equilibrium between these two forces exists when (negating selection)

$$\hat{F} = \frac{(1 - m)^2}{2N - (2N - 1)(1 - m)^2}.$$

The larger the migration pressure (m), the smaller the equilibrium inbreeding coefficient ($\hat{F}$). To illustrate this, assume that all demes are size $N = 50$, mating at random, with migration pressure $m = 0.05$ between demes (i.e., 5 per cent of natives are replaced by immigrants in each generation). At equilibrium, the inbreeding coefficient is $\hat{F} = .0847$ (i.e., about $8\frac{1}{2}$ per cent of loci heterozygous when the deme was founded will have become homozygous at equilibrium). Under the same conditions, a migration pressure $m = 0.10$ yields $\hat{F} = .0409$. The smaller the deme size, the greater the equilibrium inbreeding coefficient (e.g., when $m = 0.05$ and $N = 25$, $\hat{F} = 0.1562$).

For a large population with a continuous distribution over a wide area, local panmictic groups (neighborhoods) exist because of isolation by distance. A convenient model is to think of a circular neighborhood of N panmictic individuals, uniformly distributed over a localized area so that N becomes directly proportional to the area of the neighborhood. Since the area of a circle is proportional to the square of its radius, increasing the area of a neighborhood nine times only lengthens the radius three times. It can be shown mathematically that there is likely to be a significant amount of local differentiation if the random breeding neighborhood size is as small as $N = 10$, even within a total population area of one hundred times the neighborhood size (the radius of the population is then only ten times that of the local neighborhood). As the neighborhood size increases, differentiation becomes important only at greater distances. For neighborhoods exceeding 1000 individuals, the population behaves essentially as one giant panmictic unit. It is clear then, that the size of the local panmictic neighborhood is a critical determiner of genetic differentiation within a population conforming to the "isolation by distance" model of migration.

When a large population is distributed continuously along a narrow strip (e.g., along a riverbank), the size of the local panmictic unit is roughly proportional to distance, rather than to the square of the distance (as is the case for area continuity). Much greater genetic differentiation of neighborhoods per unit distance is expected in this "river model" than in the model of area continuity.

Although it may be true that large populations have identical *degrees* of differentiation as indicated by the same F value, the *pattern* of differentiation is expected to be quite different in various models of migration systems. In the "island model," adjacent groups may be well differentiated and uncorrelated. In models of area continuity, panmictic neighborhoods are expected to be genetically similar with gradual differentiation increasing radially. Simplified models of population structure are essential for mathematical considerations, but in nature the exact correlatives are probably nonexistent. For example, a uniform distribution of individuals is a rarity in any natural population. Density differences in various regions of the species range is an important complicating factor. Highly irregular distributions may lead to much greater degrees of local differentiation than predicted by these models.

Essay Questions

1. *Discuss the evolutionary effects inbreeding is expected to produce in* (*a*) *a deme and* (*b*) *the total population.*

2. *What are the expected consequences of dividing a species into a number of isolated panmictic subpopulations.*

3. *Explain Mayr's concept of genetic revolutions. Outline the conditions he views as optimal for this phenomenon.*

4. *What kind of population structure offers the greatest evolutionary flexibility. Why?*

5. *Expound upon the interactions of migration and* (*a*) *selection and* (*b*) *population size.*

part

IV

THE GENERATION OF TAXA

RACES

Artificial Groupings

This chapter deals with infraspecific variation (below the species level) and the attempts of biologists to organize this variation into groups as an aid to understanding the evolutionary process. Much of Darwin's thinking concerning natural selection was influenced by his knowledge of agricultural breeding practices. Humans control the breeding of their cultivated plants and domesticated animals. For crops, those members of the same species that are usually distinguished by several common characteristics belong to a group called a *variety*. This term can also be applied to a genetically uniform group produced by artificial propagation in the laboratory or in agriculture. For livestock (including laboratory animals such as rats, mice, guinea pigs, and rabbits), a *breed* is an artificial mating group for genetic study or domestication, sharing many traits in common and derived from a common ancestor. A *strain* (or line) is an artificial breeding group (either plant or animal) that usually differs from others in the same variety or breed by only one (or a few) traits. The terms are not themselves rigorously defined, although the criteria established for a specific variety, breed, or strain may be highly restrictive. The amount of gene exchange among strains of a breed might be fairly frequent, but is rarer among breeds, and usually absent among species. These artificial infraspecific groups with distinctive characteristics and varying amounts of reproductive isolation serve as models for corresponding populations in nature.

Natural Groupings

A *race* is a genetically, and as a rule, geographically distinct interbreeding natural subdivision of a species. The semi-isolated local breeding populations discussed in Chapter 15 have the potential to become races. The basic problem is to agree on how much genetic differentiation must exist before two populations warrant being called separate races. The problem is further compounded when the species consists essentially of one continuous population over a relatively large area. Even here it is often obvious that groups of individuals in different parts of the species range are markedly different from one another; however, there can be all degrees of gradation in between so that any racial designations would necessitate an arbitrary boundary between them. Scientists do not like to be arbitrary, but sometimes the nature of the data requires that such decisions must be made. Some biologists simply refuse to make racial designations even though they recognize that samples taken from a gene pool in different localities may have significantly different properties. Assigning a group of individuals to a "race" fosters the idea of discontinuity even though it may not exist in a species. This is probably an important part of the argument against assigning individuals to racial categories. Another important aspect is the rhetorical question "How different is different?"

In the final analysis, racial designations need only be made when they serve some purpose in biological investigation of a species. The number of such races one recognizes often depends on the purposes of the investigation. For example, an anthropologist studying major patterns of human variation all over the world may only wish to consider five races based mainly on skin color (black, brown, yellow, red, white), whereas an anthropologist studying human variation in Northern Europeans may find it convenient to subdivide the "White" group in that region into perhaps a dozen or more minor races. Therefore race, of necessity, must receive an *operational definition,* i.e., a definition in terms of properties significant to a given experimental situation, without consideration of whether there may be more fundamental characteristics of that which is defined.

Most evolutionists today maintain that racial variability must be described, not in terms of abstract average phenotypic values, but rather in terms of gene frequencies in groups of individuals occupying particular habitats or living in different geographical regions. In the most common situation, races are not distinguished by the presence or absence of any particular gene. Rather, they are recognizably different in gene frequencies.

The following excerpt from a 1972 newspaper article, illustrates how commonly the concept of "race" can be misconstrued.

Sickle Anemia Not Racial

Dear Dr. T: I am curious what might happen. Black people have a disease, sickle cell anemia. Now when there is a biracial child, what could come of this?

> Answer: Well, sickle cell anemia is by far more prevalent in black people, but it is not confined just to them. Furthermore, not all blacks carry the sickle cell trait. . . . Sickle cell anemia is not totally dependent on color, you see, even though, for whatever historical reasons, it is primarily a disease of blacks.

In his reply, Dr. T. attempts to explain that one of the distinguishing characteristics of the "Black" race (a very large and genetically diverse group of people) is a much higher incidence of sickle cell trait than in the "non-Black" races. The title of the article however, is misleading. Certainly the sickle cell gene is not the exclusive property of the "Black" gene pool. No individual can be assigned to any one human race based on whether or not he/she has the sickle cell gene. However, since sickle cell anemia occurs much more frequently in Blacks than non-Blacks, it is one of a number of traits that can be collectively used to define and differentiate major gene pools of mankind.

Taxonomic Problems

Taxonomists work primarily with morphological characteristics in making distinctions between infraspecific categories. Of course, if biochemical, chromosomal, behavioral, or physiological traits are also investigated, this broadens the base of possible character differences by which races may be defined. Some single gene mutations can produce gross modifications of the phenotype. For example, the gene *apterous* in *Drosophila* prevents the development of wings. Now if different populations of winged and wingless flies, respectively, were found in nature (nothing being known concerning the genetic origin of these forms), a taxonomist might be tempted to place them in different species or perhaps in different subgroups of the same species. If the taxonomist wished to formally recognize these subgroups for taxonomic purposes, he or she could assign each population to a different *subspecies*. In order for the subspecies to be officially recognized by the scientific community, a complete valid description of the two populations would need to be published in a journal or other organ intended primarily for consumption by biologists. A subspecies receives a formal name (usually in Latin, Greek, or Latinized, preferably descriptive, uncapitalized, and italicized) following the binomial species name.

Botanists also can use the subspecies category, but more commonly use a category below that of the subspecies, viz. *varietas* (called "variety" in common usage). Perhaps the preference for "variety" stems partly from the fact that it requires less formal changing of names than subspecies. There is a possibility of confusing botanical varieties with horticultural varieties (which are only minor variants of commercial or aesthetic significance). Scientific names are followed by the author's name (i.e., the person who described the species or subspecies); author's names are not italicized.

> There are two recognized subspecies of the dogtail dragonfly. The northern subspecies is *Tetragoneùria cynosùra cynosùra* (Say); the southern and western subspecies is *T. c. simulans* Muttkowski. When the author's name is in parenthesis, it indicates that he described the species (or subspecies, if a trinomial) in some genus other than the one in which it is now placed. When the author's name is not in parenthesis, this indicates that he described the species (or subspecies) in the genus indicated. Species originally named

by Linnaeus may be followed by the name Linnaeus or by some abbreviation (e.g., Linn. or simply L.). Other well-known individuals may also have their names abbreviated (e.g., Nutt. for Nuttall, etc.). The rules of nomenclature are quite involved. Zoologists have different rules from botanists. Botanists sometimes recognize one or more taxonomic levels below the subspecies (or varietas) as represented in the name *Saxifraga Aizoon* var. *typica* subvar. *brevifolia* forma *multicaulis* subforma *surculosa*. In such names we see the extent to which the emphasis on relatively minor characteristics can be carried. The value both to taxonomy and phylogeny of recognizing groups below the subspecies (or varietas) level is questionable.

Races may be considered genetically differentiated groups that have not attained formal taxonomic status through the procedure described above. It is not considered good practice to separate subspecies on the basis of single gene differences. This would be tantamount to assigning people with different ABO blood groups to different races. On the other hand, a trait governed by numerous polygenes might require relatively extensive genetic modifications in order to differentiate races at the phenotypic level. Therefore, the magnitude of morphological distinction between two subgroups of a population does not always accurately indicate the extent of genetic differentiation (evolutionary distance). The species on which genetic studies have been made to date constitute an extremely trivial percentage of all known species. Only an extremely small percentage of those species on which these kinds of investigations have been made are relatively well defined genetically. Thus, it becomes obvious that taxonomists are essentially "in the dark" when they attempt to define subpopulations in nature on the basis of known genetic differences. The best they can usually do for the present (and the forseeable future) is to use the extent of morphological differentiation as their diagnostic criterion.

It should be self-evident that populations belonging to the same subspecies are more nearly alike than are those belonging to different species; species belonging to the same genus are more nearly alike than those belonging to different genera; etc. It is impossible to evaluate the degrees of similarity (or of difference) among three or more populations belonging to the same *taxon* (general name for any taxonomic group irrespective of rank) from the taxonomic rank iself. For example, given three subspecies of the same species (A, B, and C), we do not know if A is more similar to B than to C or if B is more similar to C than to A, etc. Some taxonomists have attempted to quantify the degree of similarity by assigning numerical values to the traits used in making taxonomic distinctions. A single numerical value constructed for each population from the quantified characters serves as an estimate of "taxonomic distance." This methodology is called *numerical taxonomy* (*taxometrics* or *morphometrics*). It is defined as "the numerical evaluation of the affinity or similarity between taxonomic units and the ordering of these units into taxa on the basis of their affinities."

Numerical taxonomy is a *phenetic* scheme, distinguishing groups solely on the basis of their phenotypic similarity; no phylogenetic (evolutionary) relationships are implied. The word root "phene" refers to a phenotypic character controlled by genes. Most of the work using numerical taxonomy is done with higher taxa, but the principles are essentially the same even for subspecies.

As a simple example of the distinction between the phenetic and phylogenic approaches, consider three taxa (A, B, and C) and five characteristics (of the "all or none" type) numbered 1 through 5. The phenetic approach would group taxa A and C because they agree in $\frac{3}{5}$ of the total observations, whereas A and B agree in only $\frac{2}{5}$ of the observations.

	A	B	C
1	+	+	0
2	0	+	0
3	0	+	0
4	0	+	0
5	+	+	0

However, a phylogenist wants to distinguish *primitive* characters (i.e., those that are older; more like the ancestral condition) from *derived* or *advanced* characters (i.e., newer; less like the ancestral condition).

The terms *primitive* and *derived* do not imply that one is more beneficial or more complex than the other. They simply refer to relative position in time. They should not be confused with the relativistic terms generalized and specialized. A *generalized* condition is one that is broadly adapted to a variety of habitats and niches. A *specialized* condition is one that is narrowly adapted to a few (or a single) habitats and niches. Generalized characteristics tend to be more primitive, and specialized ones tend to be more advanced, but this is not an absolute rule. Systematists want to define the most parsimonious, internally consistent scheme of evolutionary development. They also may want to "weight" certain characters more heavily than others according to the degree of genetic differentiation represented (a highly subjective interpretation). Perhaps there is evidence from the fossil record that characters 1 and 5 are primitive (therefore 2, 3, and 4 are inferred to be derived). The phylogenetic approach would group A and B together, because they share in common two primitive characteristics. The advanced traits shared by A and C would then be viewed as evolutionary convergence.

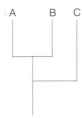

The *practice* of taxonomy is concerned with the preparation and use of dichotomous keys that tend to emphasize the differences between organisms. The *theory* of taxonomy is concerned primarily with evolutionary relationships between groups of organisms. This schism between theory and practice is both ancient and artificial. In the fifth century B.C., Pythagoras first attempted to describe the shapes of organisms mathematically. In the fourth century B.C.,

Aristotle introduced a dichotomous classification based on empirical syllogistic logic (deductive reasoning). Linnaeus adopted the syllogistic classifications of Aristotle. The Linnaean scheme has failed to satisfy the desire of taxonomists to understand, as well as to describe, evolutionary relationships. Conventional taxonomy uses single traits to distinguish groups of organisms into hierarchical subgroups. The characters used for this purpose are those that, in the judgment of the individual taxonomist, reflect the generality of the contrast being made. Not all taxonomists agree on which traits to use because of differing views on the degree of "primitiveness" of characters used in dichotomies. Assigning different weights to these characters is also a highly subjective task. In 1763, Adanson proposed that all traits are of equal potential worth as taxonomic discriminants. In order to avoid the subjective elements of classical taxonomy, morphometrics follows the Adansonian principle of using large numbers of characters from all parts of the organism and initially assigns all of them equal weight. The complexity of calculations involving many traits is too great to be done in practice without the help of a computer. Methods are now available for objectively weighting characters in each contrast by a value that maximizes the separation between the groups. The weighting procedure of morphometrics is automatic and *a posteriori,* based on all available data rather than on *a priori* or character-by-character weighting as is done in conventional phylogenetic procedures. All the weighted characters become discriminant functions that can be summarized in terms of a "generalized distance," which is a measure of the degree of resemblance between groups.

Humans can only perceive in three dimensions at a time. However, the relationships existing simultaneously between many characters can be handled mathematically in "hyperspace" and ultimately reduced to a relatively few major vectors representing major patterns of development. Natural selection acts on these interlocking patterns of growth and development rather than on their constituent characters.

The value of numerical taxonomy in helping biologists to understand evolutionary relationships in organic diversity is far from being universally accepted at the present time. However, it goes a long way in the direction of answering the question "How different is different?" by an objective methodology capable of independent verification.

The Origin and Distribution of Races

As discussed in the previous chapter, local Mendelian populations may conform to the island model, the uniform continuous distribution model, or (more commonly) all degrees of intermediate structure. Geographic isolation is obviously important in allowing local gene pools to differentiate from one another. Generally, the greater the distance between local breeding populations, the less gene flow occurs between them. Selection within each isolate will be modifying the gene pool for maximal adaptation to local conditions. The size of these isolates will vary with time; during periods of small population size, genetic drift may become an important process for changing gene frequencies. All these factors can function to genetically differentiate isolated populations. They also function to differentiate various geographically defined populations of individuals that form a continuous distribution. Racial groups may be relatively easy to define if the species is broken up into genetically distinctive isolated populations spatially separated from one another. Even though just as much diversity might exist locally in various regions of a large continuously distributed species, the problem of defining racial boundaries becomes more difficult. The incidence of a given trait can gradually change along one or more transects of a continuous distribution. Such a character gradient is referred to as a *cline* (Figure 16.1). Clinal variation at the phenotypic level is indicative of progressive gene frequency changes in local populations along the transects. Presumably the local gene pools would respond to selection for one or more environmental factors that gradually change along the transect (e.g., annual temperatures or rainfall).

One of the most famous clines is that of the distribution of the human ABO blood groups. Four blood groups (A, B, AB, O) are controlled by three alleles (I^A, I^B, i). In Figure 16.2, we see a well-defined cline for the frequency of I^B allele, which is highest in central Asia and progressively diminishes westward across Europe; another cline of lesser magnitude exists to the northeast and southeast. Gene I^A is common in western

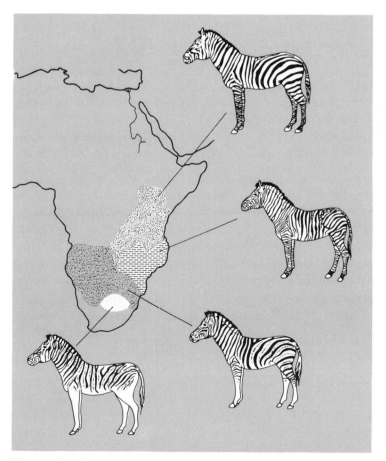

Figure 16.1. Example of a cline in the amount of striping on the legs of a common species of zebra (*Equus burchellii* or *quagga*) as seen in various parts of its range in Africa.

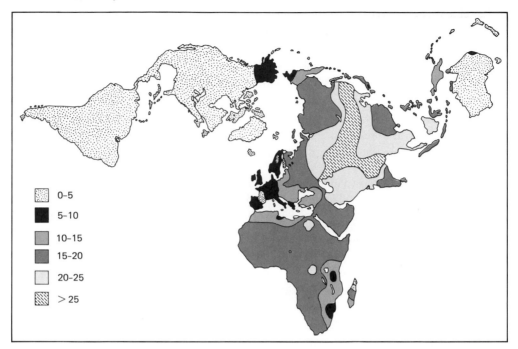

Figure 16.2. A cline formed by the incidence of the I^B allele of the human ABO blood group system. Gene frequencies shown for the Americas and Australia are based on indigenous populations.

Legend:
- 0–5
- 5–10
- 10–15
- 15–20
- 20–25
- > 25

Europe and parts of Africa and Australia. The recessive allele is common everywhere, especially among American Indians. There is really no convincing evidence to date that the ABO blood groups are associated with adaptations to particular environments. The cline in this case probably reflects to a large extent the influence of gene flow (migrations) between human populations. The low frequency for I^B in the New World and Australia might have its origin in a founder effect and/or drift.

Drosophila has giant chromosomes in its salivary gland cells in which chromosomal rearrangements such as inversions are easily detected. Many natural populations of these flies are characterized by different inversions and are referred to as *inversion races*. It is possible to define relationships between these races based upon overlapping inversions. Most investigations have been conducted in the ubiquitous species *D. pseudoobscura* and *D. persimilis*. The methodology is illustrated by the following example. Consider three races (1, 2, 3) with segments of the third chromosome labeled A through K.

Race

1	A B C D E F G H I J K
2	A B F E D C G H I J K
3	A B F E I H G C D J K

Two breaks are required for an inversion. Four breaks would be extremely unlikely. Therefore the most plausible evolutionary scheme is as follows. The segmental order in Race 1 can be inverted (by the breaks shown as verticle lines) to yield the segmental order in Race 2; another inversion in Race 2 can produce the segmental order in Race 3. However, since we cannot determine the ancestral arrangement, three schemes are possible:

$$1 \rightarrow 2 \rightarrow 3 \quad \text{or} \quad 3 \rightarrow 2 \rightarrow 1 \quad \text{or} \quad 1 \leftarrow 2 \rightarrow 3.$$

By postulating the fewest number of breaks separating one race from another, 1 could not give rise directly to 3 (or the reverse).

Figure 16.3 depicts a phylogenetic sequence inferred from overlapping inversions in *Drosophila* of the western United States. No one inversion type occurs over the entire species range. Some populations harbor as many as eight different inversions (Figure 16.4). The adaptive value of specific inversions or inversion complexes is indicated by Figure 16.5. Populations of *D. pseudoobscura* at the base of the Sierra Nevada Mountains in California are rich in the *Standard* (ST) chromosome arrangement; as we progress to higher elevations, the *Arrowhead* (AR) arrangement increases largely at the expense of *Standard* chromosomes. The frequency of the *Chiricahua* (CH) inversion varies little among these populations.

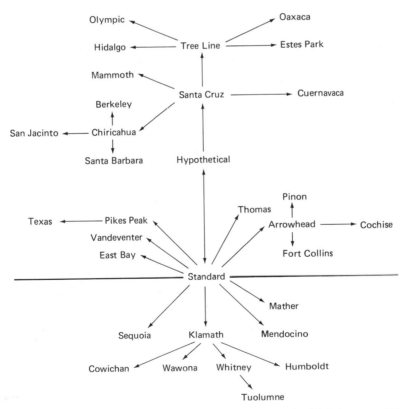

Figure 16.3. Family tree of segmental arrangements in the third chromosomes of *Drosophila pseudoobscura* and *D. persimilis*. The Standard arrangement occurs in both species. Those below the line are inversion types found only in *D. persimilis;* those above the line are found only in *D. pseudoobscura*. Arrangements connected by an arrow differ by a single inversion.

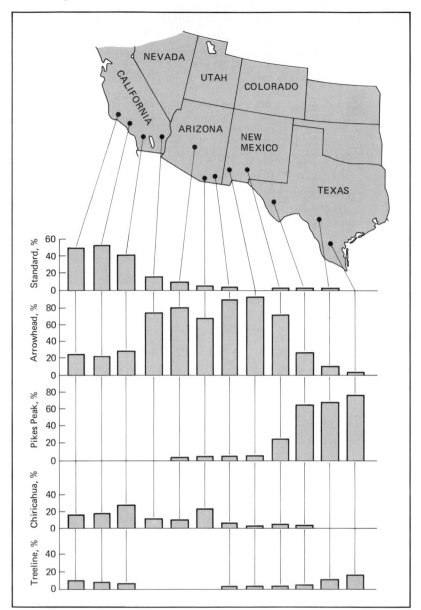

Figure 16.4. Variation in frequencies of five arrangements of the third chromosome in *Drosophila pseudoobscura* in twelve populations along an east-west transect through the southwestern United States.

A slightly different situation exists at Mount San Jacinto in California. Here ST is also more common at lower than at higher altitudes, but CH increases with altitude at the expense of ST; AR varies little. If the Mount San Jacinto cline is produced by a temperature gradient, we expect ST to be most frequent during hot summer months and CH to be most frequent during the cooler months of the year. This is actually the way the population appears to respond (Figure 16.6). Further confirmation of the adaptive value of these inversions is

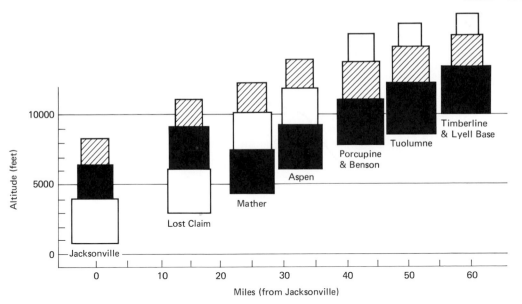

Figure 16.5. Variation in the frequencies of three structural arrangements (Arrowhead = black, Chirica-hua = hatched, Standard = white) of the third chromosome in *Drosophila pseudoobscura* associated with altitude in the Sierra Nevada Mountains. (After Dobzhansky, 1955)

obtained by rearing them in the laboratory. A population of *D. pseudoobscura* was started with 11 per cent ST and 89 per cent CH chromosomes. After about nine months at a constant temperature of 25°C, the composition of the gene pool had shifted and established an equilibrium at about 70 per cent ST and 30 per cent CH (Figure 16.7). When the experiment was repeated at 16°C, no marked change was noted. The seasonal changes in selective values

Figure 16.6. Annual variations in frequencies of four inversion types in the third chromosomes of *Drosophila pseudoobscura* at Pinon Flats, California (combined data of six years observations). Names of arrangements: ST = Standard, CH = Chiri-cahua, AR = Arrowhead, TL = Tree Line.

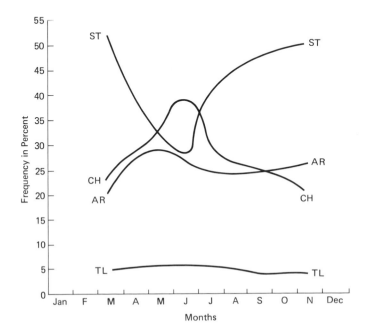

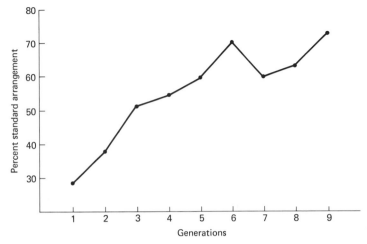

Figure 16.7. Results of an experiment involving competition between two arrangements (ST and CH) of the third chromosome in *Drosophila pseudo-obscura* at a constant temperature of 25°C. From an initial frequency of about 10 per cent, the standard arrangement tended to increase and equilibrate around 70 per cent.

for various inversion types tends to maintain structural polymorphism in the population; these inversion races are in a constant state of flux. Furthermore, structural heterozygotes (e.g., ST/CH) have higher fitness values than homozygotes (ST/ST or CH/CH). Recall that inversion heterozygotes tend to prevent linked gene combinations from being recombined through single crossovers. This protection does not necessarily extend to multiple crossovers, however. In general, the shorter the inversion, the less opportunity for multiple crossovers and consequently the greater the integrity of the inverted regions. But the shorter the inversion, the fewer genes can be protected from recombination. Therefore, an optimal size for inversions does exist. Different mutations may accrue in the undisturbed segments, causing them to diverge from each other. Because most mutants are harmful, inversion homozygosity tends to be disfavored; selection will favor the structural heterozygotes and thereby establish a balanced polymorphism. The superiority of structural heterozygotes is a function of not only the genes within the inverted segment, but also their interactions with other chromosomal arrangements in the population.

Three different arrangements that are closely related to each other through single step inversions are seldom found in the same population. As an illustration, we find ST and AR together or AR and PP (Pikes Peak inversion) together as coadapted systems, but ST, AR, and PP do not occur together in frequent proportions in the same population. It has been suggested that a third arrangement would allow recombination to break up the adaptive complexes and therefore would be selected against. Through loss of the *hypothetical* arrangement (not known in any population), a population could use CH together with ST and AR without endangering the integrity of these complexes.

Both experimental clines (using laboratory populations of *Drosophila*) and computer-simulated clines have revealed that the effect of gene flow will be very effective in preventing local differentiation if there are only a few demes in a cline (using the stepping-stone model). The maximum effect of gene flow is seen in the two-deme model. However, in clines consisting of fifty demes located on a smooth environmental gradient, the attenuating effects of gene flow are suppressed; the tendency to elevate gene frequencies by demes higher on the gradient will be counteracted by gene flow from demes lower on the gradient. In one laboratory experiment, fifteen demes (separated by increments of selection coefficients of 0.04 between demes) differentiated under 40 per cent gene flow to essentially the same extent as identical clines without gene flow. Computer simulations indicate that marked local differentiation can also occur along a relatively weak environmental gradient (increments of selection coefficient = 0.008 between demes). These weak gradients would almost certainly escape detection in the field. Theoretical conclusions derived from the simplistic two-deme model cannot be applied to species spatially distributed among many demes connected by gene flow. Local populations situated along irregular environmental gradients become more susceptible to the effects of gene flow in proportion to the increase in gene frequency difference between adjacent populations. However, even large amounts of gene flow are unable to prevent the formation of steep clines if a large difference in selective effects exists between two environments. Asymmetry of gene flow does not impede differentiation as long as the environmental gradient is smooth; it only shifts the position of the transitional zone between the differentiated areas away from that expected on the assumption of symmetry. Therefore, abrupt changes in gene or morph frequencies should not be interpreted as evidence for environmental changes in the immediate vicinity of the steepest part of the cline.

We do not perceive any adaptive significance in the cline of the ABO blood groups; but the cline of inversion types in *Drosophila* appears to be highly correlated with at least one environmental factor. Turesson coined the term *ecotype* to represent a race that is genetically adapted to a particular environment. Many variations on this theme have been proposed. A *climatype* and *topotype* (geoecotype) represent populations that are differentiated mainly in response to climatic and topological (geographical) factors respectively. Since broadly different geographical regions commonly have different climates, a distinction between climatic and geographic ecotypes is usually impossible. An *edaphic ecotype* is a race that has adapted primarily to particular soil properties in the range of the species. A *biotic ecotype* (synecotype, agroecotype) is a special kind of edaphic ecotype adapted to cultivated soils.

How can one determine whether racial characteristics in different geographical regions are really genetic adaptations to different environments or merely the phenotypic plasticity of similar or identical genotypes? Parts of the same plant can be removed and grown into new individuals, all of which are of identical genotype (cloning). Representatives from different geographical races can be cloned in environments normally occupied by other races. By comparing the survival and reproductive qualities of clones grown in their own environment with those in different environments, an appraisal of genotype-environment interactions can be made.

> The yarrow plant (*Achillea lanulosa*) forms a cline in California (Figure 16.8). A race of yarrow on the humid Pacific Coast near San Francisco is characterized by compact, thick-stemmed plants that grow continuously throughout the year. Another race at intermediate altitude in the forest zone of the Sierra Nevada Mountains consists of relatively tall plants with slender, erect stems, which become dormant during the winter.

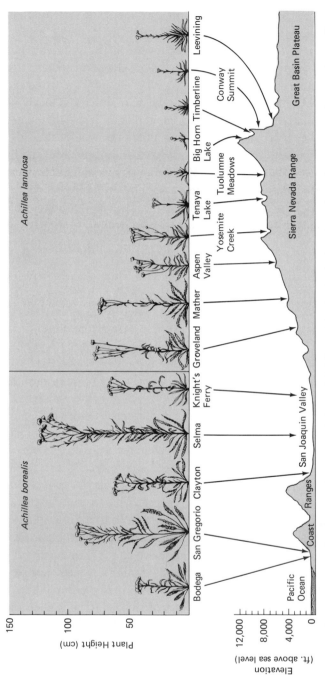

Figure 16.8. Mean plant heights of different populations of *Achillea* originating in the locations shown by arrows and grown in a uniform garden on the coast at Stanford, California. (After Clausen, Keck, and Hiesey)

A third race in the high alpine zone of the Sierra Nevada has dwarf plants that can mature and set seed quickly during warmer weather, but remain dormant during most of the year. A team of botanists from the Carnegie Institute (Clausen, Keck, and Hiesey) subdivided each of the three races and grew clones at each of the three locations (Figure 16.9). Each race grew best in its native location. The coastal race transplanted to the intermediate forest zone became dormant during winter and flowered much later than the native race. The genotype of the coastal race proved to be lethal in the alpine zone. The forest zone race grew quite tall on the coast, but mostly became dormant during winter. When transplanted to the alpine zone, the forest race was highly variable; some died, some grew slowly, but not fast enough to ripen seed. The alpine race was able to grow at intermediate altitude and on the coast, but more weakly than the native races. The alpine race was the only one able to grow successfully in the harsh alpine habitat. This experiment demonstrates conclusively that yarrow races have different genotypes specifically adapted to the conditions of their native habitat. It also shows that the environent itself causes considerable modification of the phenotype (plasticity) in races transplanted to foreign habitats. A yarrow race (ecotype) is clearly the product of a developmental pattern produced through interaction of a distinctive heredity with a certain environment.

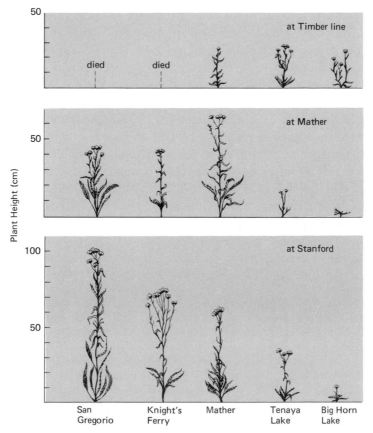

Figure 16.9. Typical growth responses of *Achillea* clones native to five localities in California when grown at each of three altitudes; sea level (Stanford), 4,600 feet (Mather), and 10,000 feet (Timber line).

What then is the difference between subspecies and ecotypes? Subspecies are usually geographically defined on the basis of easily recognized morphological differences for taxonomic purposes. Ecotypes, on the other hand, are distinguished primarily by their adaptive responses to particular environments whether or not they exhibit easily recognized morphological differences. Clausen, Keck, and Hiesey consider a morphologically distinguishable ecotype the basis of a subspecies. In a similar vein, clines and ecotypes need not be mutually exclusive concepts. If clinal variation exists for adaptive traits, this could form the basis for recognition of ecotypes. Alternatively, ecotypic variation may consist of a series of parallel and/or antiparallel clines; furthermore, ecotypic variation can also follow a "patchwork" of suitable habitats devoid of clinal gradients. The ecotype concept is almost entirely restricted to experimental verification in plants at the present time. As cloning techniques become perfected in animals, the concept may become more universal. Fortunately, the environmentally induced modifications that so often mask racial differences in plants are much less prevalent in animals, so that zoologists do not need a purely genetic-ecological term in addition to one that is primarily taxonomic.

Although racial groups may sometimes be recognized in a clinal pattern of variation, they often fail to exhibit a regular pattern of variation over their geographic range of distribution. Clines and/or racial groups are sometimes distributed more circularly than linearly. The German word *Rassenkreis* is used to denote a "ring of races." The evolutionary significance of a Rassenkreis lies in the extreme differentiation that can sometimes be seen in adjacent populations at the ends of the ring. The differences in phenotypes may be striking enough, but if the differentiation extends to the physiological levels that control reproductive compatibility, the result is often spectacular. We shall see in the next chapter that reproductive isolation is the hallmark of the biological species definition. Adjacent populations in a ring of races would be expected to exchange genes freely with one another. Occasionally we find that terminal populations in a ring fail to do so. So many fundamental differences have accumulated in these terminal populations that they have become reproductively isolated from one another. Hence, they behave as separate species even though they are connected to one another through the intermediate races of the ring. From such a Rassenkreis we gain a most valuable insight into the process of speciation. For if the intermediate races were to disappear, the terminal populations would behave as different species. If the intermediate forms were not preserved in the fossil record, we would have no direct evidence of how these species originated.

> The distribution of a Pacific coast salamander *Ensatina eschscholtzii* forms a classical Rassenkreis in the mountains of California (Figure 16.10). The hot, dry central valley of California acts as a barrier to the dispersal of this species. Taxonomists recognize seven subspecies usually connected by intergrading populations wherever the ranges of two subspecies overlap. It is probable that the species originated in the northwest and spread downward into California following the cooler, moister mountain ranges (one on the coast and the Sierra Nevada range inland). Four coastal subspecies are brownish dorsally; the three interior subspecies form a cline with orange or yellow spotting increasing from north to south. The ends of the ring meet in southern California, where *Ensatina eschscholtzii eschscholtzii* becomes adjacent to *E.e. croceator* in one locality and *E.e. klauberi* in another. The phenotypic differences at the ends of this Rassenkreis are

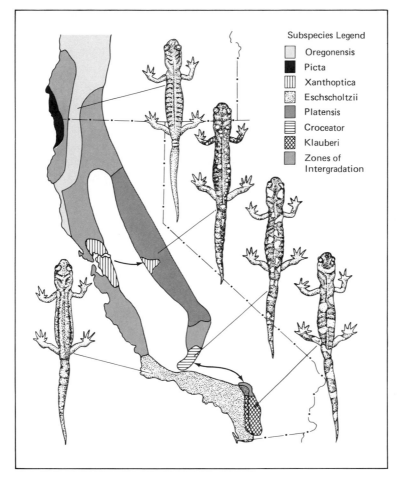

Figure 16.10. Rassenkreis formed by subspecies of the salamander *Ensatina eschscholtzii* in California. (After Stebbins, 1949, University of California Publications in Zoology 48)

striking. Since no intermediate types have been found between these adjacent terminal subspecies, R. C. Stebbins believes that they have become reproductively isolated from one another.

Adaptive Topographies

Within each local population, selection functions to adjust allelic frequencies in the gene pool to maximize population fitness. In order to gain some insight into how various gene frequencies affect average population fitness values, we can construct what are called *adaptive surfaces* (landscapes, topographies). The simplest adaptive surface is a two-dimensional map for a single locus with a pair of alleles.

Consider the case of a pair of alleles (A, a) wherein the heterozygote is adaptively superior to either homozygote.

Genotypes	AA	Aa	aa
Frquency (before selection)	p^2	$2pq$	q^2
Fitness	$1 - s_1$	1	$1 - s_2$

Recall from Chapter 13 that the average fitness of a population ($\bar{W}$) is calculated by multiplying frequency times fitness for each genotype and summing over all genotypes. In this example, $\bar{W} = 1 - s_1 p^2 - s_2 q^2$. An equilibrium is established when $\hat{q} = s_1/(s_1 + s_2)$. It can be shown by the calculus that $\hat{q}$ is the frequency that maximizes $\bar{W}$. As selection moves allelic frequencies toward their equilibrium values, it simultaneously maximizes the average fitness of the population. Given $s_1 = 0.2$ and $s_2 = 0.4$, we can construct an adaptive map that reveals how gene frequencies are related to average fitness values in the population. When q is $< \hat{q} = \frac{1}{3}$, selection will tend to increase the frequency of a; when q is $> \hat{q}$, selection will tend to decrease a. At equilibrium ($\hat{q}$), the population is sitting on its "adaptive peak" (Figure 16.11).

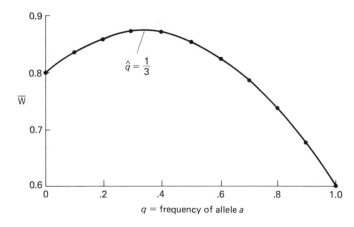

$\hat{q} = \frac{1}{3}$

$\bar{W}$

q = frequency of allele a

Figure 16.11. Two-dimensional landscape for a single locus with a pair of alleles exhibiting overdominant interaction ($s_1 = 0.2$ for AA and $s_2 = 0.4$ for aa). $\bar{W}$ is the average fitness of the population.

Now let us consider the two locus case. Suppose that we have two independently assorting pairs of alleles (A, a and B, b). There are nine possible genotypes. Let the fitness of each genotype (W_i) be as shown in the following table.

	BB	**Bb**	**bb**
AA	1.00	0.65	0.65
Aa	1.00	0.40	0.65
aa	0.70	1.00	0.50

If we let the frequency of allele a be q_1 and b be q_2, then at equilibrium the average population fitness values for all possible combinations of q_1 and q_2 form an adaptive surface (three dimensional). In order to calculate one point on this surface, we would do the following. Given $q_1 = 0.4$ and $q_2 = 0.8$, the fitness value for genotype *AABB* is

$$p_1^2 p_2^2 W_{AABB} = (0.6)^2(0.2)^2(1.00) = 0.0144.$$

Likewise, the fitness value for genotype *AaBB* is

$$2p_1 q_1 p_2^2 W_{AaBB} = 2(0.6)(0.4)(.2^2)(0.65) = 0.0192.$$

These calculations are summarized for all nine genotypes in the following table.

(*i*)	Genotype	(*f_i*) Expected Frequency	(*W_i*) Fitness	(*f_i W_i*) Product
1	*AABB*	$(.6^2)(.2^2)$	1.00	0.01440
2	*AaBB*	$2(.6)(.4)(.2^2)$	1.00	0.01920
3	*aaBB*	$(.4^2)(.2^2)$	0.70	0.00448
4	*AABb*	$(.6^2)2(.2)(.8)$	0.65	0.07488
5	*AaBb*	$2(.6)(.4)2(.2)(.8)$	0.40	0.06144
6	*aaBb*	$(.4^2)2(.2)(.8)$	1.00	0.05120
7	*AAbb*	$(.6^2)(.8^2)$	0.65	0.14976
8	*Aabb*	$2(.6)(.4)(.8^2)$	0.65	0.19968
9	*aabb*	$(.4^2)(.8^2)$	0.50	0.05120

Average Population Fitness $= \Sigma f_i W_i = 0.62624$

This average population fitness establishes only one point on an adaptive surface of all possible combinations of q_1, q_2 values. Obviously, the labor involved in these calculations is too great to be done by hand for large numbers of points on the adaptive surface. A computer can be programmed to do these calculations for q_1 and q_2 from 0 to 1.0 by increments of say 0.05 in all possible combinations. This will generate $21^2 = 441$ plot points and should give a fair picture of the adaptive surface.

The adaptive surface from these calculations appears as shown in Figure 16.12. Note that the landscape of all possible population conditions consists of peaks and valleys of various heights or depths. A population at any point on the landscape would be expected to move uphill (increasing average fitness) under the direction of selection. At equilibrium, the population may be resting on top of a hill, but it may not be on the highest hill. The population cannot move to a higher hill without passing through a valley of lower adaptive values. Selection within a population (*intrademic selection*) usually tends to prevent the population from moving downhill. However if fitness of a given genotype is frequency dependent, intrademic selection may prevent a population from reaching the summit of a hill.

Consider a locus under frequency-dependent selection, with two alleles (*A, a*) at frequencies p and q, respectively.

Genotypes	*AA*	*Aa*	*aa*
Fitness	1	1	$1.5 - q$

When $p = q = 0.5$ the average fitness of the population is 1.0. When $q < 0.5$, average

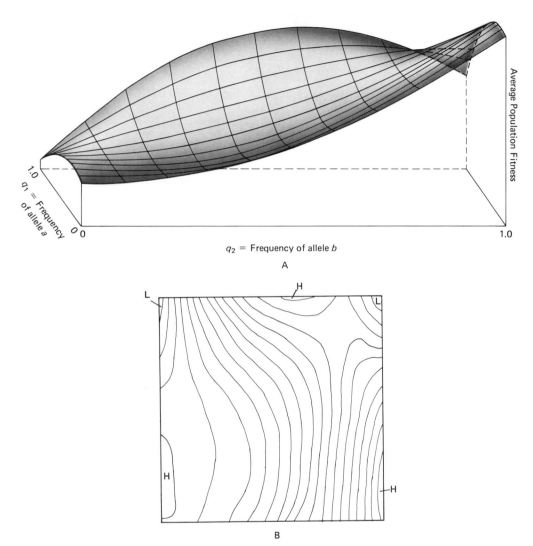

Figure 16.12. Diagram of a three-dimensional landscape formed by plotting 441 average population fitness values as explained in the text (two loci each with two alleles). (A) Cutaway model showing some of the adaptive peaks and valleys; (B) a contour map of the surface shown in part B. H = high point; L = low point.

In figure A:
- Vertical axis (right): Average Population Fitness
- Left axis: q_1 = Frequency of allele a, from 0 to 1.0
- Bottom axis: q_2 = Frequency of allele b, from 0 to 1.0

A

B

fitness is >1.0 because the fitness of *aa* genotype is >1.0. Nonetheless, intrademic selection would tend to increase q until it reached $\hat{q} = 0.5$, even though this causes the population to move part way downhill from its adaptive peak.

Two-locus models of adaptive surfaces are far from representing biological reality. Thousands of segregating loci exist in most natural populations, and the adaptive complexes that can form in hyperspace boggle the mind. Nonetheless, it is from such biological complexity that adaptive solutions can evolve in ways that the human brain could not perceive.

Perhaps *non-Darwinian evolution* (i.e., nonselective evolution, especially by genetic drift or gene flow) can ultimately benefit a population by shifting gene frequencies so that a deme becomes relocated to a new spot on the adaptive surface. If now selection can predominate over non-Darwinian processes, the population might be in a position to climb a nearby hill that is higher than the one it formerly occupied.

We should not expect the adaptive surface to remain constant. As environmental conditions change, the adaptive surface becomes modified; former valleys become elevated, and hills subside. One species' peak may be another species' valley. As a species evolves, the adaptive landscape itself may also shift.

Group Selection

Although intrademic selection may actually prevent a population from attaining the highest peak on an adaptive landscape, *interdemic* (interpopulation) *selection* (i.e., selection between local Mendelian populations) may lead to improved fitness in a *metapopulation* (a cluster of populations belonging to the same species).

The notion that animal population size is directly limited by (and balanced with) the food supply of the environment has been very popular among ecologists and evolutionists. When the numbers exceed the carrying capacity, some individuals starve; only the best adapted individuals tend to survive. An alternative theory has been extensively developed by V. C. Wynne-Edwards in his book *Animal Dispersion in Relation to Social Behavior* (1962). According to this theory, unlimited competition for food would lead inevitably to overexploitation of the resource and damage to the habitat or to mass starvation. Competition among individuals for such things as territories and social rank tends to adjust the population density at an "optimum" level that would prevent overexploitation and depletion of future yields. Social interactions between individuals in a population often generate feedback information concerning density (*epideictic* function) that serves to modulate population size relative to consumable resources. Therefore, the welfare of individuals becomes subordinate to that of the group to which they belong. Wynne-Edwards believes that the only way such mechanisms could evolve is through group selection. Populations (groups) with defective social organizations and/or homeostatic regulations tend to overexploit their habitats and thereby contribute to their own demise. Groups that regulate their densities to the available resources survive and eventually replace the unsuccessful groups. A social dominance hierarchy is a population attribute. In altruistic behavior, individual welfare appears to be sacrificed in order to benefit the group. Neither dominance hierarchies nor altruistic behavior, according to Wynne-Edwards, could have evolved apart from group selection. He views individual selection as tending to maximize individual fecundity, whereas group selection tends to "optimize" population fucundity. The two appear at cross purposes with one another and when this happens group selection becomes more important than selection at the individual level.

The theory of group selection has many opponents and the conflict is far from being resolved. Sewall Wright proposed that evolutionary opportunities are most likely to be found in a large population subdivided into many local populations (demes) of intermediate to small size, among which variable amounts of gene flow occur. The store of genetic variability in such a population is potentially greater than that expected in a panmictic population of equivalent size. Both Wright and Wynne-Edwards admit that selection acts between local populations, but Wright still views the individual as "the direct object of selection"; selection between populations is only supplementary to that between individuals. The amount of gene flow allowed by Wynne-Edwards' model appears to be very small compared to that allowed by Wright's model. Wright's model has been criticized for requiring an unrealistic degree of interdemic isolation, and the question of how much gene flow can be allowed without disrupting local divergence is still an open one. Wynne-Edwards' model has been criticized, because it does not explain how a self-regulating behavior initially arises within a group, but only accounts for its persistence and spread to other groups. Some suggest that group selection seems to be too slow a process to function effectively, because the duration of existence of a group is to group selection as the life span of an individual is to individual selection. Furthermore, since the number of groups in a population or species is far less than the number of individuals, variability between groups would be much more restricted than between individuals and hence selection would be far less effective.

Perhaps fitness has been too glibly associated with maximization of fecundity. A moment's reflection will reveal that selection does not favor a maximal reproductive rate *per se*, but rather one that is maximal relative to the environment in which the species lives. Therefore, it is better to speak of "optimization" of a reproductive rate rather than "maximization." In retrospect we see that natural selection favored those individuals who have, in the long run, left an optimal number of offspring (i.e., neither too few nor too many) in relation to their environmental resources (including long-term carrying capacity). Therefore, it is suggested that the regulation of the reproductive rate, just as for other homeostatic phenomena peculiar only to populations, can be explained by selection acting at the level of the individual without invoking a group selection process.

> Territorial behavior serves to regulate population size and ensure an adequate supply of resources (mainly food) for the participants. Theoretically, territorial behavior has an optimum level; if an individual is too aggressive, mating may not take place or proper rearing of offspring may suffer (agressive neglect). On the other hand, an excessively timid individual may fail to establish a territory and thereby fail to attract a mate. Similarly, the size and type of territory optimal for any species could be established by selection of individuals that occupy the most efficient territories relative to the function they serve in their own environment.
>
> Dominance hierarchies play an important role in regulating populations by virtue of the fact that the "lower end" of the hierarchy may be disposed of or retained according

to the available resources. A hierarchy always functions to identify the surplus individuals whenever the population density requires thinning. Traits with social meaning (such as horns, or plumage colors) may serve only to facilitate individual recognition rather than to affect social status *per se,* because individuals exhibiting dominant behavior commonly have only average development of such characters. Once a hierarchy is established, relatively little energy is wasted in useless aggressive behavior. Since age itself often confers a higher social status, it may be an advantage for a young individual to submit and live to breed later than to persist in fighting and leave few or no offspring. This type of behavior at the population level may exert a feedback on individual adaptation.

Any genetically determined individual behavior that tends to enhance the survival of offspring will be advantageous (as long as it is not overbalanced by disadvantages in other contexts). This is also true of altruistic behavior. The term *altruistic* seems to imply disadvantage to the individual that displays such traits, but this is not necessarily true. Birds who feign injury in order to lure predators from their nests (or young) are rarely caught. The progeny of altruistic parents are selectively favored by "kin selection," which is really no different from individual selection favoring their survival by behavioral patterns of their parents. The population itself is part of the environment for each individual. Therefore, whether a particular genotype is adaptive or not depends to some degree on the behavior of others in that same population.

The concept of group selection involves differential extinction of genetically different populations (or demes or families) within a species. Group selection can also operate at a level involving clusters of related species. Indeed, it is by this process that most evolutionists explain the familiar paleontological patterns of dynastic succession of major groups such as ammonites, sharks, graptolites, and dinosaurs through geological time. Extinction is most likely to occur under two conditions: (1) initially, when pioneer colonists are attempting to establish themselves in a new site (*r extinction*); and (2) later, when the population has reached (or exceeded) the carrying capacity of the site (*K extinction*) and is in danger of destroying the habitat and suffering mass starvation. Here the symbol *r* denotes the reproductive potential (intrinsic rate of increase or "Malthusian parameter") of the population, and the symbol *K* represents the population size equivalent to the carrying capacity of the environment.

When a patchwork of similar habitats is opened periodically for a relatively short period of time (e.g., new clearings in a forest, new river bars, temporary ponds, etc.), selection tends to favor colonization by species that (1) can discover the habitat quickly, (2) reproduce rapidly and use the available resources before they are forced into severe competition with other organisms, and (3) have a large dissemination potential so that there is a high probability of locating another ephemeral habitat before the old one becomes inhospitable. Species with the ability to colonize these fluctuating environments are called "*r* strategists" or "opportunistic species." The *r* strategy is to maintain numerous populations on the lower, rapidly ascending parts of the growth curve. Under these extreme conditions, genotypes with a high intrinsic rate of increase (*r* value) will be favored by what is called *r selection*.

On the other hand, a "*K* strategist" or a "stable species" tends to occupy a more stable habitat at or near its carrying capacity (*K* value). Under these conditions, it becomes more important for genotypes to confer competitive ability under crowded conditions of high density and less important to maximize *r*. This type of adaptation is fostered by "*K selection*."

Most of the classical models and theorems of natural selection have been constructed under the assumption of *r* selection rather than *K* selection. In reality, both types of selection probably are operative at various times, although one may predominate over the other. During early colonization of a new habitat, selection for maximization of *r* is paramount. After the population reaches the carrying capacity of the environment, intrinsic density-dependent controls are most likely to appear under *K* selection.

If population size fluctuates in a manner that permits it to increase most of the time, there is a tendency for *r* selection to predominate. But if these fluctuations cause population size to decline most of the time, *K* selection will generally tend to favor genotypes that defer reproduction, maximize longevity, and slow the rate of decrease (Figure 16.13). In general, higher forms of social evolution should be favored more by *K* than by *r* selection. Table 16.1 summarizes some of the most obvious features of *r* and *K* selection.

To explain the evolution of sterile casts of social insects (e.g., worker bees), Charles Darwin introduced the concept of natural selection operating at the family level rather than at the level of the individual. Theories of group selection were initially prompted by such examples of altruistic behavior. It is readily apparent how altruistic behavior can enhance an individual's fitness, when it aids in the survival of that individual's offspring. After all, that is what fitness is all about, *viz.* leaving viable, fertile descendants. It is perhaps a little more difficult to understand altruism when it benefits more distant relatives. William D. Hamilton's concept of *inclusive fitness* is helpful in this regard. Inclusive fitness is the sum of an individual's own fitness plus the sum of all the fitness effects it causes to the related parts of the fitness of all its relatives. For example, when an animal performs an altruistic act that benefits a full sib (e.g., a brother), the inclusive fitness of the altruist is its own fitness (which may have been reduced by its altruistic behavior) plus the increment of increased fitness enjoyed by that portion of the sib's genes common by descent with the altruist. Hamilton's models are unstructured, i.e., they are not integrated with other parts of evolution theory such as group size, migrations, mutations, allelic frequencies, and epistasis. Therefore, Hamiltonian concepts

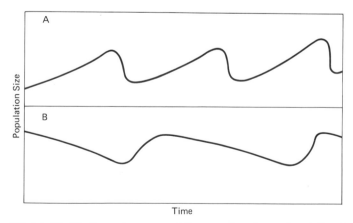

Figure 16.13. Two extreme population growth patterns. Population A is usually increasing (*r* selection tends to predominate); population B is usually decreasing (*K* selection tends to predominate).

TABLE 16.1. Characteristics of *K* Selection and *r* Selection (after Wilson, 1975).

Factor	*r* selection	*K* selection
Habitat	Ephemeral and/or variable	Relatively constant
Extinction	Often catastrophic; density independent	Tendency to be density dependent
Survivorship	Usually great initial loss; few survivors of advanced age	Usually linear loss with age or very little loss until advanced age
Population Size	Well below carrying capacity; variable	Equilibrium at or near carrying capacity; fairly constant
Recolonization of Habitat	Frequent (e.g., each year)	No recolonization necessary; continuous occupation.
Competition (intra- and inter-specific)	Variable; often weak	Usually strong
Population life span	Short (mostly less than 1 yr)	Long (generally more than 1 yr)
Emphasis in energy utilization	Productivity	Efficiency
Colonizing ability	High	Low
Social Behavior	Weak	Often well developed
Attributes favored by selection	1. rapid development 2. high *r* value 3. early reproduction 4. small body size 5. single reproduction (semelparity)	1. slower development 2. competitive ability under high density 3. delayed reproduction 4. larger body size 5. repeated reproductions (iteroparity)

can only be loosely coupled with evolution theory, and the number of predictions that can be made therefrom are highly restricted.

That fraction of the genes of two individuals identical by descent from common ancestors is represented by the coefficient of relationship (r; not to be confused with the same symbol used earlier for the Malthusian parameter). In the absence of inbreeding, full sibs share about half their genes in common ($r = \frac{1}{2}$). Suppose that the altruist leaves no offspring, but that its altruistic act more than doubles the fitness of its brother. Since half the genes of the altruist are in its brother, the altruist has its fitness increased by half the gain in fitness enjoyed by its brother. That is, the altruist actually gains genetic representation in the next generation by virtue of the genes it shares with its brother. The increased inclusive fitness of the altruist then causes the spread of altruistic genes through the population and thereby promotes the evolution of altruistic behavior in the entire group. This model can also be extended to other relatives. Again, if the altruist leaves no offspring, and only uncles ($r = \frac{1}{4}$) are benefited, the fitness of uncles would have to be increased more than fourfold in order for the inclusive fitness of the altruist to exceed unity. If combinations of relatives are benefited, the genetic effect of altruism is weighted by the number of relatives of each kind whose fitness is thereby increased and by their coefficients of relationship.

In general, k (the ratio of gain in fitness to loss in fitness) must exceed the reciprocal of the average coefficient of relationship ($\bar{r}$) to all the relatives. That is,

$$k > \frac{1}{\bar{r}}.$$

In the case of brothers, $1/\bar{r} = 1/(\frac{1}{2}) = 2$. If the altruistic act prevents the altruist from reproducing, its loss in fitness is 1.0. In order for the altruistic genes that are identical by descent in its brother to increase in the next generation, k (the gain to loss ratio) must exceed 2; i.e., the brother's fitness must be more than doubled by the altruistic act.

Selfishness can be treated by the same model. Selfishness is compatible with individual fitness as long as the fitness of relatives is unaffected by the selfish act, because when close relatives are harmed, genes shared in common by descent will be lost in the relatives' contribution to the next generation. Inclusive fitness must again exceed 1.0 in order for "selfish genes" to increase in the population.

Deductions derived from at least two different mathematical models agree that evolution of an altruistic gene entirely by interdemic selection is an improbable event. Therefore, the wide array of "social conventions" postulated by Wynne-Edwards (and others) is probably incorrect. Furthermore, these analyses indicate that altruism is least likely in the larger, more stable populations in which social behavior is generally most highly developed. A majority of biologists currently believe that altruistic behavior can better be explained by natural selection acting at the individual level.

If selection operates only on groups of closely related individuals (say by at least the degree of third cousins) as a unit, or operates on an individual in any way that affects the frequencies of genes shared by common descent in relatives, the process is referred to as *kin selection*. If the entire breeding population (consisting of many families) is the unit of selection, so that genetically different demes are extinguished differentially, or disseminate different numbers of colonists, the process is called interdemic (interpopulation) selection. Both "kin selection" and "interdemic selection" are included in the term "group selection." Selection operates on a gradient between pure kin selection at one extreme and pure interdemic selection at the other extreme. They are sufficiently different to require different mathematical models. A transition zone between the two extremes is between approximately 10 and 100 individuals for most organisms. At the upper limits, family size becomes so large that groups of families (demes) are likely to be formed. This range of population size is the same order of magnitude for the effective population numbers (N_e) in many vertebrate demes.

Asexual reproduction is the most efficient way for an organism to infuse its genes into its descendants. Sexual reproduction involves the union of gametes; meiosis and gamete formation actually reduces the genetic representation of an individual in its own offspring. Any genetic success that offspring achieve is shared equally by the mother and father. But a partheogenic female shares her reproductive success with no one. It has long been conceded that sexuality functions to facilitate long-term evolutionary adaptation, even though it is detrimental to the reproductive interests of the individual (relative to asexual reproduction). Sexuality is a nearly universal phenomenon of paramount importance to the biological world, and yet it is exactly the kind of phenomenon for which evolution by group selection must be postulated. Paradoxically, awareness of this concept has not yet reached consciousness in many biologists. Perhaps someday this will be known to historians as one of the most bizarre aspects of biological thinking in the twentieth century.

Essay Questions

1. *How do the methodologies of pheneticists differ from those of pylogenists? On what philosophical bases do these differences rest?*

2. *Can interdemic selection improve the average fitness of the entire population if each deme has already reached its adaptive peak through intrademic selection? Explain.*

3. *Are races artificial groupings or do they actually exist in nature? Explain. What factors determine how many races are recognized within a species?*

4. *How can kin selection foster the evolution of altruistic behavior?*

5. *What light can be shed on the problem of speciation through a study of a Rassenkreis?*

SPECIES

Species Definitions

Through the ages *species* has meant different things to different people. Even today, a museum curator, a taxonomist, a paleontologist, a geneticist, a physiologist, and an ethologist may all have quite different conceptions of a species. When we compare grossly different groups such as rats with ducks or ferns with oak trees, there is no disagreement among biologists that these are distinctive species. However, when we compare two rather similar groups of organisms, such as the eastern and western meadowlark, or the red oak and the scarlet oak, the disagreement begins. There are three major ways in which scientists view a species: (1) morphological, (2) biological, and (3) evolutionary.

Morphological (Typological) Species

Probably the earliest concept of a species was that of the *morphospecies;* i.e., placing organisms that look alike into the same taxonomic group (species). Pre-Lamarckian taxonomists used the morphospecies concept almost exclusively for two main reasons. First, the virtually universal belief in special creation predisposed the naturalist to think in terms of static (immutable) morphospecies, essentially unchanged since creation, breeding essentially "true to type" since that event. Second, in that frame of mind, there was no point in collecting more than one specimen (the *type specimen*) to represent the entire species. Minor variations from the type specimen were simply viewed as imperfections allowed by the Creator within certain limits of the ideally perfect type (Platonic Idealism) represented in the original creation. Local naturalists found the morphospecies concept very useful for their purposes. The minor variations within a morphospecies that one is likely to encounter in one's immediate neighborhood usually are not a source of confusion for purposes of "identification." At that stage in the history of taxonomy, this was really all that was required. Being able to pigeonhole a specimen as belonging to this or that species was the sole purpose of taxonomic work. Any phenotypic resemblance between species was fortuitous, and there was no need to look for more fundamental biological implications. A naturalist tended to view each local population where he or she lived as a distinctive species. Little thought was given to the variation in geographically separated populations. The morphological discontinuities present in the local populations of different species in the neighborhood provided obvious gaps to aid in identification. Today we would say that a species which consists of a single uniform population, devoid of differentiated and semi-isolated local populations (races, subspecies) is *monotypic* (i.e., all of one basic kind). Naturalists who traveled widely were more apt to be impressed with the diversity represented by different Mendelian populations of the same species in different geographic regions. As we would say today, a species is *polytypic* if it is subdivided into two or more races (or subspecies).

Individuals of all species that live in the same geographic location are said to be *sympatric;* those that live in different locations are *allopatric.* As a general rule, allopatric populations in juxtaposition (contiguous) with one another tend to be more similar than allopatric populations separated by a considerable distance (geographic isolation) or by some physical barrier such as a river or a mountain range. With little or no gene flow between disjunct allopatric populations, they each tend to go their own evolutionary way, adapting to different niches in their own locations. Under these conditions, distinctively different subgroups of a species may be easily recognized by taxonomists; they may unofficially recognize them as different races or officially name them as subspecies. Contiguous allopatric populations of a species may exhibit the same degree of phenotypic variations throughout its range as in the previous case, but the lack of geographical discontinuities makes the recognition of subgroups an arbitrary decision. Some taxonomists would avoid naming subspecies under these conditions.

Polytypic species obviously cannot be represented by a single type specimen. In fact, a single specimen cannot adequately represent any species, because it fails to display the range of variability in that population. Even if the extreme types were also included, this would still fail to tell us anything about the frequencies of various traits in the species or how these traits are distributed in various parts of the species range. Modern taxonomists realize that the concept of a type specimen is no longer tenable. They are also aware of the polytypic nature of most natural populations and the need to consider the total range of variability in grouping individuals into a morphospecies. The only criterion used for this purpose is the degree of resemblance in form. This method does not consider possible differences between genotypes and phenotypes. We know that sometimes "look-alikes" may differ considerably in physiology, behavior, chromosomal structure, etc.; quite different phenotypes may sometimes still belong to the same gene pool (e.g., sexual dimorphisms). But a typological species is relatively easy to construct, because it is not concerned with diciphering phylogenies or defining gene pools. If individuals can be grouped unambiguously on the basis of their phenotypic resemblance, they can be recognized as a morphospecies. What degree of total phenotypic resemblance is required of individuals belonging to a morphospecies is a value judgment that must be made by the taxonomist. The cliché often heard in discussions of this problem is "A species is what a taxonomist says it is." To a great extent this is true not only for morphospecies, but for other species definitions as well. The morphospecies is the most pragmatic of the species concepts. Linnaeus established the species concept on this basis, and it has been used with great success ever since. In most cases, organisms that appear different are usually also genetically different, so that different morphospecies commonly represent different gene pools (although this is not implied in the definition). Therefore, a species defined by the topological concept very often includes the same individuals grouped together by other species concepts.

Biological (Genetic) Species

Ernst Mayr has defined a biological species as "groups of actually or potentially interbreeding natural populations which are reproductively isolated from other such groups." Theodosius Dobzhansky defines it as "groups of populations the gene exchange between which is limited or prevented in nature by one, or by a combination of several, reproductive isolating mechanisms." In the biological definition of species, emphasis is placed on intrinsic impediments to free exchange of genetic information between Mendelian populations. When two Mendelian populations exist sympatrically without any gene flow, they are considered to be different biological species regardless of their degree of phenotypic differentiation. Intrinsic (physiological) barriers to gene exchange usually reflect considerable differentiation at the genetic level, and this is the ultimate concern of taxonomists who employ the biological species concept. In theory, this sounds tidy, but in practice it is very often unworkable. Consider just a few of the major difficulties with this definition of species.

1. It is a unidimensional concept in practice, because it can only be observed or experimentally investigated in living organisms. This excludes from consideration the overwhelming variety of life forms that have existed in the past.
2. Allopatric populations grouped in the same species may be geographically isolated entities (not experiencing gene flow with other populations). How is the taxonomist to know whether or not units designated as races or subspecies can freely exchange genetic information? Again, just because they appear similar does not exclude the possibility that more fundamental physiological changes have accrued to impede or prevent gene exchange. In practice, the taxonomist subjectively infers that the degree of reproductive isolation is correlated with the degree of phenotypic differentiation despite well-known examples to the contrary.
3. To test disjunct allopatric groups for reproductive compatibility, we would need to bring each population together in all pair combinations to see if they interbreed. This

itself might be an insurmountable task. Relocating a population for this purpose, be it in the field or in the laboratory, subjects it to a set of conditions different from those in its normal environment. This new set of conditions may contribute to disruption of the reproductive isolating mechanisms that function well in its usual environment. Therefore, while in theory the test of reproductive isolation is open to experimental investigation, in practice the results and their interpretations are subject to question.

4. Suppose that interbreeding can occur between two populations. This, in itself, is inconclusive evidence for the lack of reproductive isolation. Consider the hybrid mule (cross between horse and ass); mules are sterile. Therefore, genes cannot flow from the horse species to the ass species via the hybrid, and both gene pools remain intact. Even though the F_1 hybrids survive and are fertile, the F_2 and/or backcross generations may be poorly adapted so that gene flow again is repressed.

5. Two sympatric biological species that do not exchange genes are easily recognized as "good" species; those that exchange genes only rarely are "fair" species; those in which gene exchange is not uncommon are "poor" species; and those that offer little or no resistance to gene flow are not really different species at all. Representatives of all stages or reproductive isolation have been found in nature. Where, then, does the taxonomist "draw the line" on a biological species? Again the taxonomist's subjective judgment must ultimately be called on to make these kinds of distinctions, just as it is for the morphospecies concept.

6. Asexually reproducing organisms, by definition, are virtually excluded from consideration as biological species. Bacteria, viruses, and some fungi reproduce almost exclusively by vegetative means. But even here, a rare recombinational event may occur by *parasexuality* (i.e., any mechanism that achieves nonmeiotic recombination). For example, a bacterium can donate a segment of its chromosome to another bacterium and the new piece than may become integrated into the recipient cell by crossover type events. Parasexual gene exchange has been observed not only between widely different kinds of bacteria but also between bacterial cells and eukaryotic cells (including those of humans). Now even if the biological species definition could be modified to accomodate parasexual reproduction, the problem would remain as to how drastically gene exchange should be curtailed in order to qualify a group as a "species."

Evolutionary (Phylogenetic) Species

George Gaylord Simpson defines an evolutionary (phylogenetic) species as "a lineage (an ancestral-descendent sequence of populations) evolving separately from others and with its own evolutionary role and tendencies." Evolutionary lineages are often conveniently represented in a tree-like diagram called a *cladogram* (Figure 17.1). Construction of evolutionary trees is termed *cladistics*. Bifurcations represent products of the speciation processes, where one species becomes subdivided into two or more groups that go their own evolutionary ways to become new species coexisting at the same time. Certainly different branches of a cladogram reflect the idea that these populations are considered to have evolved separately with their own evolutionary roles and tendencies. A group of species derived from a single common

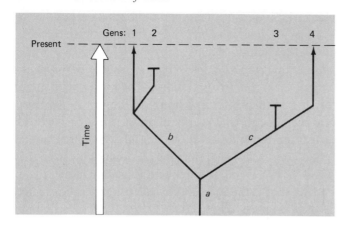

Figure 17.1. Hypothetical cladogram. Gens 1 and 2 have a monophyletic origin in ancestor b; gens 1 and 4 are polyphyletic at the level of ancestors b and c, but are monophyletic at the level of ancestor a.

ancestral population is said to have a *monophyletic* origin. A group of species classified together, some members of which are descended from different ancestral populations, are said to have a *polyphyletic* origin. Obviously these terms must be used with reference to some initial time zone, because all organisms are presumed to be related if we go back far enough in time. Each branch of a cladogram is a separate evolutionary lineage called a *gens*. The height of each branch is proportional to time; a branch that reaches the present time zone indicates the lineage exists today; if a branch terminates before the present, it indicates that the lineage became extinct. At different points along the same gens, the ancestral and descendent populations are genealogically related through heredity but isolated by time. Populations higher in a gens may have evolved so many differences that, by comparison with those lower in the same gens, they can be considered different genetic (evolutionary) species. If species are gradually transformed through time, according to the Darwinian theory, the paleotaxonomist would be expected to encounter the same kind of difficulty in subjectively recognizing evolutionary species as others have in defining extant races (or subspecies) along a uniform gradient (e.g., a cline). Fortunately for paleontological taxonomists, the fossil record is riddled with gaps that facilitate the recognition of distinctive successional species or *paleospecies* (i.e., morphospecies in different segments of a gens).

The paleotaxonomist has several serious problems not encountered in the study of living organisms. One of the most obvious problems is the highly biased nature of the fossil record itself. As discussed in Chapter 4, usually only organisms with hard parts (shells, bones, teeth) are well preserved in the rocks. Usually, little or nothing is known of vital soft organs, whose structure would often reveal clues as to how the system functioned physiologically. Also the paleontologist has to do without chromosomal information (number, structure, degrees of homology, etc.) and behavioral information (especially as it relates to reproductive potentialities). Furthermore, sampling of ancestral populations by fossilization may be far from random so that the paleontologist receives a distorted view of organic variability in times past. Paleontological gaps must be filled conceptually by hypothetical intermediate forms to complete each gens of a cladogram. These and other disadvantages are common problems of the evolutionary species concept.

It should be clear from the preceding discussions that there is no single, universally accepted species definition that satisfies all taxonomists and all kinds of organisms through all time zones. Indeed, if such a definition did exist, it would pose a major stumbling block for evolutionary theory. The gradual nature of the transmutation of species is devoid of gaps; but discontinuities are the cardinal determinants of classification. The desire of taxonomists to place unambiguously each organism into a unique species category is basically incompatible with the evolutionary nature of the biological world. Most taxonomic controversies ultimately spring from this singular fundamental principle.

Reproductive-Isolating Mechanisms

Sympatric populations that retain their genetic integrity must be reproductively isolated from each other. The nature of reproductive isolation varies greatly from one species to another. In general, we find that species are isolated not by just one, but by several mechanisms of varying importance to the overall reproductive barrier. Biological impediments to gene flow can be broadly classified into prezygotic and postzygotic mechanisms. Those that function to prevent the union of gametes from different gene pools (*prezygotic*) tend to be more efficient than those that function after fertilization (*postzygotic*) in the sense that less energy is wasted by mechanisms which prohibit zygote formation. Table 17.1 summarizes most of the important reproductive-isolating mechanisms identified to date.

TABLE 17.1 Reproductive Isolating Mechanisms

Prezygotic Mechanisms

1. *Ecological or habitat isolation.*
   Different populations are adapted to such different habitats that they seldom get together in nature.
2. *Temporal or seasonal isolation.*
   Gametes from different populations are released at such different times that they seldom form zygotes.
3. *Mechanical incompatibility.*
   Structural differences in reproductive organs preclude copulation (in animals) or pollination (in flowering plants).
4. *Gametic (physiological) incompatibility.*
   Gametes from one population fail to function in the reproductive tissues (or tracts) of another population.
5. *Ethological (behavioral, sexual) isolation.*
   Courtship signals (in animals only) of one population fails to trigger receptiveness in members of another population.

Postzygotic Mechanisms

1. *Hybrid inviability or weakness.*
   Zygotes formed from different gene pools fail to develop normally and die prematurely (before sexual maturity).
2. *Hybrid sterility.*
   (a) *Developmental (genic) hybrid sterility.*
       Gonads of hybrids fail to develop normally resulting in aberrant endocrine function and/or meiotic abortion.
   (b) *Segregational (chromosomal) hybrid sterility.*
       Failure of chromosomal synapsis and/or mispairing of chromosomes, producing irregular segregation to gametes; genetically unbalanced gametes are thereby rendered nonfunctional.
3. *Hybrid (F_2) breakdown.*
   F_1 hybrids are viable and fertile, but F_2 and/or backcross progeny contain many weak or sterile recombinants.

Prezygotic Mechanisms

Ecological (habitat) isolation is common among sedentary organisms (especially terrestrial plants). Because of specific ecological requirements (amount of sunlight, soil moisture and pH, etc.), two populations that live in the same general area would not experience much gene flow because of these habitat preferences. Even if some hybrids were formed, their genotypes would not be particularly adapted to exploit either parental niche and therefore would not be expected to survive.

> The wood frog (*Rana sylvatica*) lives in woods of the northern United States and Canada and rarely goes near water except to breed. The leopard frog (*Rana pipiens*) lives in damp meadows and by the edges of streams in most of North America and Mexico. In areas in which the ranges of the two species overlap, they remain well-isolated reproductively, largely because of their distinctive habitat preferences.
>
> Two species of oak trees are sympatric over most of the eastern United States (Figure 17.2). The scarlet oak (*Quercus coccinea*) is adapted to swamps or poorly drained bottom lands with acid soils, whereas the black oak (*Q. velutina*) lives on drier, well-drained upland soils. Hybrids between these two species are sometimes found where intermediate habitats permit their survival. This demonstrates that they are not yet strongly reproductively isolated and that their ecological specialization is the major factor keeping them apart in nature.

Probably most organisms (both plant and animal) have definite breeding seasons. In birds, the physiological control of reproductive cycles is fairly well known. Increasing hours of daylight in the spring operate through the eye and optic nerves to stimulate the pituitary gland at the base of the brain to release gonadotropic hormones; these hormones are carried by the blood stream to the gonads and there stimulate the production of gonadal hormones

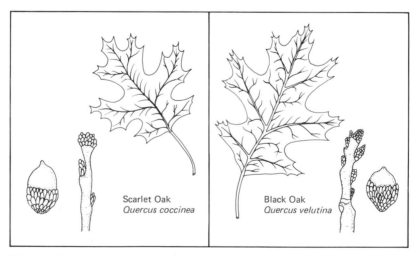

Scarlet Oak
Quercus coccinea

Black Oak
Quercus velutina

Figure 17.2. The scarlet oak (*Quercus coccinea*) and the black oak (*Q. velutina*) prefer different habitats and therefore exhibit some degree of ecological reproductive isolation. They are not completely isolated by this mechanism, however, because in some areas they form sizable hybrid populations.

(estrogens in females and androgens in males). The gonadal hormones are the last link in this chain of reactions that ultimately influences gonadal development, gametogenesis, and overt sexual behavior. Yet this sequence of reactions does not proceed at a uniform rate in all species of birds even though they may be sympatric. Some species have largely completed their reproductive cycle before other species have even begun. Under these conditions, gene flow is effectively prohibited by what is termed *temporal* or *seasonal isolation*. Many warm-blooded vertebrates have relatively long breeding seasons, providing ample opportunity for overlaps between species. This, plus the fact that great environmental fluctuations can occur from year to year, generally renders seasonal isolation relatively ineffective by itself; it functions most effectively as part of a complex of isolating mechanisms. Among vertebrates lacking endogenous control of body temperature, as well as among the invertebrates, breeding seasons may be quite restricted, thus enhancing the effectiveness of temporal isolation.

> In southern Michigan, breeding seasons for various species of frogs follow sequentially as the water temperature rises during the spring. The wood frog begins breeding early when the water is about 45°F. The leopard frog begins to breed after the water is about 55°F, and the green frog (*R. clamitans*) begins late in the season after the water reaches about 60°F. From the previous example, we have seen that the wood frog and the leopard frog are ecologically isolated in addition to being seasonally isolated. Thus, we can visualize how several isolating mechanisms (even weak ones) could reinforce each other, and in concert erect a strong barrier to gene flow between sympatric species.
>
> On the Monterey peninsula of California, the Monterey pine (*Pinus radiata*) and the bishop pine (*Pinus muricata*) form sympatric populations with little hybridization. Hybrids have been produced experimentally between these two species, but they are usually prohibited from doing so in nature because of temporal isolation. The Monterey pine commonly sheds its pollen before March. Cones of the bishop pine are not even open at this time.

At one time, it was believed that incompatibilities in the size and/or shape of genitalia of different species could act as a *mechanical barrier* to fertilization. Morphological attributes of the genitalia are often used as diagnostic traits for many insect species. It was proposed that the male and female genitalia had complementary structures so that the male organs fit the female organs as a key fits a lock. Males of one species would not have the proper "key" to fit female "locks" of a different species. This may sound very plausible, but there is little supporting evidence. Many observations and experiments have shown that in most animals mechanical factors are of little or no consequence in isolating species. Even extreme differences in size of genitalia may not be an effective deterrent to copulation as evidenced by the fact that the giant dog breed called St. Bernard has been successfully mated to the much smaller Dachshund breed. Among the higher plants, however, the size and the structure of the flower are frequently correlated with its ability to attract specific kinds of pollinators and, in this way, contribute to interruption of gene flow between species. Observations on individual bees have revealed that they nearly always visit flowers of a single species on any one flight from the hive. In his book, *Forms of Flowers* (1862), Charles Darwin concluded that most of the differences in floral structures of related species of plants have probably evolved by natural selection favoring different variations that reduce or prevent interspecific pollinations.

One of the most striking examples of the correspondence between floral structure and specific pollinators is found in the beard tongues (genus *Penstemon*) of the western United States (Figure 17.3). The mountain *Penstemon* (*P. Grinnellii*) has bright blue flowers with large, two-lipped, widely gaping corollas. On the other hand, the scarlet bugler (*P. centranthifolius*) has tubular, bright red corollas. It is well known that bees can perceive wavelengths in the violet and ultraviolet range of the spectrum better than those near the red end; hummingbirds see colors best in the red and infrared range of the spectrum. Both color and flower structure of the mountain *Penstemon* are adaptations for pollination by large carpenter bees (*Xylocopa*); those attributes of the scarlet bugler fit them for pollination by the long, thin beaks of hummingbirds. Many *Penstemon* species have been artificially crossed to produce fertile hybrids. They appear to be separated in nature largely by ecogeographic and floral isolation.

Species of insects with fairly different genitalia have been observed to copulate without difficulty. Copulation by itself, however, does not ensure the production of viable, fertile hybrids. Starving *Drosophila* larvae produces small-bodied flies that easily mate with normal-sized members of the same species. The evidence for mechanical isolation in animals is scanty. The kind of exceptional report that keeps the theory alive is represented by an example from moths of the family Sphingidae. Males of *Chaerocampa elpenor* may mate with females of *Metopsilus porcellus,* but sometimes are unable to withdraw the copulatory organ; the reciprocal cross encounters no mechanical difficulty.

Although pollen of one species of flowering plant may be deposited on the stigma of another species, this does not guarantee that a pollen tube will germinate and successfully grow down the style to deliver the sperm nuclei into the ovule. *Gametic incompatibility* can be manifested at any stage during this critical process. In some cases, pollen tubes from foreign species (*heterospecific*) may grow, but at a slower rate than *conspecific* (i.e., same species) tubes. In the race to the ovules, conspecific pollen tubes deliver the gametes earliest and thereby negate the activities of foreign pollen tubes (Figure 17.4). Abortion of pollen tube

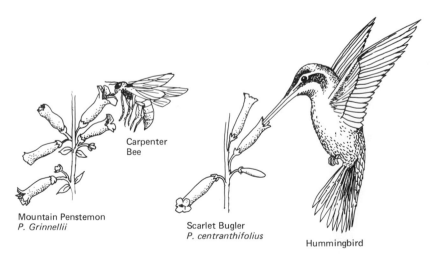

Carpenter
Bee

Mountain Penstemon
P. Grinnellii

Scarlet Bugler
P. centranthifolius

Hummingbird

Figure 17.3. Two species of *Penstemon* in California with characteristic flower structures and colors adapted to specific pollinators.

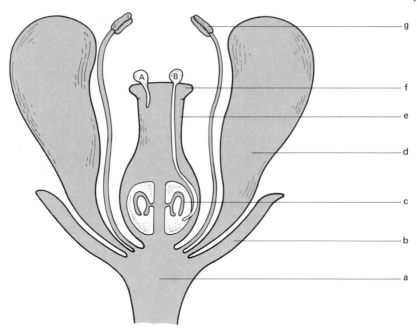

Figure 17.4. Differential growth rate of (A) heterospecific pollen tube vs. (B) conspecific pollen tube. Parts of the flower are (a) receptacle (pedicel), (b) sepal, (c) ovule, (d) petal, (e) style, (f) stigma, and (g) anther.

growth and failure of fertilization are rarely the primary cause of reproductive isolation between closely related species, but they may cooperate with other "weak" mechanisms and jointly form a strong barrier to gene flow.

In some animals with internal fertilization the environment of the female reproductive tract is hostile to spermatozoa from foreign species. When foreign sperm are killed or immobilized by *physiological incompatibility* factors such as pH, osmotic pressures, or antigen-antibody reactions, gene flow is greatly restricted between species.

> Pollen tubes are particularly likely to burst in crosses wherein the pollen parent has a larger diploid chromosome number than that of the seed parent. It seems that when the chromosome number of two species is equal, there is less physiological incompatibility, for the ratio of chromosomes in the pollen to that in the stylar tissue is the normal 1:2 (haploid to diploid). As this ratio approaches 1:1, as it may in crosses with polyploid species, physiological incompatibility tends to increase. Commercial tobacco (*Nicotiana tabacum*, $2n = 48$) is an allotetraploid derivative from *N. sylvestris* and *N. tomentosa* (both with $2n = 24$). If either of the latter species is used as pollen parent in crosses with *N. tabacum*, hybrids are easily produced because the ratio of pollen tube to stylar chromosomes is 1:4. However, when *N. tabacum* is the pollen parent, this ratio is 1:1 and most of the pollen tubes burst. In some cases, pollen tube rupture appears to be induced osmotically, not as a direct consequence of extra chromosomes in the pollen, but rather as a physiological expression of the polyploid genotype. The relatively large, thick pollen tubes produced by some polyploid species have difficulty in penetrating the smaller styles of diploids, at least partly accounting for the differential success of reciprocal crosses.

Sperm from *Drosophila virilis* rapidly loses mobility in the sperm receptacles of *D. americana;* conspecific sperm remains motile a long time. Some heterospecific matings of *Drosophila* produce what is called an *insemination reaction* in the female reproductive tract. Soon after copulation, the vagina swells; fluid is secreted into and solidifies in the vagina, thereby obstructing the release of eggs.

Ethology is the study of animal behavior as it exists in nature. Many animals display courtship behavior prior to mating. Usually it is the male that displays for the female, but in some cases it is the female who courts the male. Often there is a stimulus-response sequence of several steps that must be completed before the female becomes receptive to conspecific males. In a classical case, the male initiates a visual, tactile, auditory, or olfactory sign that stimulates a conspecific female to respond with a recognition sign of her own. The male then proceeds to the next phase of the courtship ritual, and the stimulus-response sequence eventually culminates in copulation (for animals with internal fertilization) or in simultaneous release of gametes (for those with external fertilization). Many species have their own genetically programmed stereotyped mating behavior by which conspecifics recognize each other and thereby discriminate against heterospecific matings.

Female moths specifically attract conspecific males from long distances by releasing a chemical scent. The enlarged antennae of males (Figure 14.10) detect this odor, and they fly up the concentration gradient to meet the female source. It can be shown experimentally that the attraction is a chemical one by exposing females in a screened cage and in a glass container. Males are attracted to the former, but not to the latter. The attraction is species specific. Over thirty species of moths of the same genus have been found living in a single valley without interbreeding, presumably because of distinctive olfactory signals for each species.

Some bird species that are very similar in morphology can be easily recognized by their songs. Species specific auditory stimuli are commonly part of mating behavior in birds, frogs, and some insects. For example, the "clunk, clunk" mating call of the green frog (*Rana clamitans*) is distinctively different from the "barump" of the bull frog (*R. catesbiana;* Figure 17.5).

By experimentally placing males of one species of fruit fly together with both conspecific and heterospecific females, the relative frequencies of conspecific and heterospecific matings can be ascertained. When the species are only remotely related, females almost invariably mate with conspecific males. For closely related species it is common to find some heterospecific matings, but they occur less frequently than conspecific matings. The following data from an experiment with the look-alike species *Drosophila pseudoobscura* and *D. persimilis* testifies to the preference for conspecific matings.

Females	D. pseudoobscura		D. persimilis	
Males	% Mated	% Unmated	% Mated	% Unmated
D. pseudoobscura	87	13	3	97
D. persimilis	25	75	68	32

Figure 17.5. The bullfrog (*Rana catesbiana*) from Louisiana. (Courtesy Carolina Biological Supply Company)

Postzygotic Mechanisms

The gene pool is the most basic evolutionary unit. Selection tends to adjust the gene frequencies in each pool so as to maximize the number of gene combinations adapted for the special conditions in which that population lives. Different gene pools produce different adaptive norms. Harmonious gene combinations are likely to be produced from the same gene pool but unlikely to be produced from different gene pools. Therefore it is not surprising that hybrid zygotes produced by gametes extracted from different gene pools are commonly poorly adapted, weak, partially to totally sterile, or inviable. *Hybrid inviability* is one of the strongest and most common mechanisms that reproductively isolate distantly related species. Hybrid zygotes often begin development by a series of normal mitotic cleavage divisions. However, development commonly goes awry during or following the gastrulation process that forms two of the three basic germ layers of the embryo. It was formerly believed that nuclear genes were largely dormant until the gastrulation stage. Evidence is now accumulating that (at least in some organisms) mRNA is produced by nuclear genes during cleavage, but it is blocked in some way; during gastrulation this cryptic mRNA is "unmasked" and becomes functional. In any event, gastrulation is the time when nuclear gene activity begins to be expressed. Very little is presently known of the gene-controlled biochemical events responsible for cessation of embryogenesis. Recent experiments involving nuclear transplantations have revealed that the old cliché that spoke of incompatibility between heterospecific genomes is oversimplistic and somewhat misleading. The incompatibility is now viewed as one between the genome of one species and the cytoplasm of another. The maternal parent commonly supplies almost all the

461

cytoplasm of the zygote. Specific substances in the cytoplasm function to activate (derepress) certain genes and/or inactivate (repress) other genes. If these nuclear-cytoplasmic interactions fail to function on schedule, a properly programmed sequence of gene activation/inactivation does not occur and development aborts. Hybrid inviability effectively prevents gene flow between species even if a hybrid is born/hatched/germinated, as long as the hybrid dies before reaching sexual maturity.

> Crosses between certain species of flax, wherein *Linum perenne* is the seed parent and *L. austriacum* is the pollen parent, produce hybrid seeds that do not germinate in nature. There is an incompatibility between the embryo and the endosperm that surrounds it. The embryo reaches maturity, but is unable to sprout from the seed. If the embryo is dissected out, it will grow into a vigorous, fertile hybrid sporophyte. The reciprocal cross produces only a diminutive embryo, which nonetheless can be removed by dissection, cultured in artificial medium, and eventually transplanted to soil where it completes its growth.
>
> It is relatively easy to make interspecific crosses experimentally, using many of the marine animals with external fertilization. Zygotes can be formed between members of different phyla (e.g., echinoderms $\times$ mollusks), between different classes of the same phylum (e.g. sea urchin eggs fertilized by sea lily sperm), etc. These remote hybridizations never produce viable embryos. In some cases the paternal chromosomes are extruded during cleavage; in other cases the embryos soon develop abnormalities and die. A similar fate almost invariably awaits hybrids between different genera of the same family (e.g. goats $\times$ sheep or hares $\times$ rabbits). Newspaper reports of putative interorder hybrids of rabbits $\times$ guinea pigs, rabbits $\times$ cats, etc. appear sporadically, but have not been substantiated by scientific investigation.

Occasionally hybrid inviability is manifested only in one sex, while the viability of the other sex may be little affected. J. B. S. Haldane (Figure 17.6) observed that when one sex is absent, rare, or sterile in F_1 hybrids, that is the heterogametic sex. This generalization is referred to as *Haldane's rule*. Among birds, moths, and butterflies, female hybrids are defective more frequently than males because these organisms use the ZW (or ZO) chromosome mechanism of sex determination in which the female is heterogametic. Most other animals (including mammals, amphibians, and most insects) have the XY (or XO) chromosome mechanism of sex determination in which the male is the heterogametic sex and, therefore, usually less viable than the females. Recessive sex-linked genes fixed in either parental species should be as strongly expressed in the heterogametic sex of the hybrid as it was in the species that provided the X (or Z) chromosome. In the hybrid, recessive autosomal genes from one parent are more likely to be suppressed by dominant autosomal genes from the other parent. If the expression of the recessive sex-linked genes are incompatible with the rest of the hybrid genotype, then the heterogametic sex will suffer greater mortality than the homogametic sex.

> *Drosophila pseudoobscura* and *D. miranda* differ in gene arrangement; some genes that are sex-linked in one species are autosomal in the other, and vice versa. Crosses between *D. miranda* ♀ $\times$ *D. pseudoobscura* ♂ produce normal female and abnormal male hybrids. The reciprocal cross also produces normal females, but the males die. In order to explain these results, let us assume that a group of autosomal genes (*A*) in *D. miranda* are sex linked in *D. pseudoobscura;* a group of autosomal genes (*B*) in *D. pseudoobscura*

Figure 17.6. John Burdon Sanderson Haldane (1892–1964). (Courtesy of The Genetics Society of America)

are sex linked in *D. miranda.* Females of both species and the hybrids are alike with respect to these genes (*viz. AABB*). Males of *D. pseudoobscura* are *ABB*, as are males from the cross *D. miranda* ♀ × *D. pseudoobscura* ♂. Males of *D. miranda* are *AAB*, as are males from the cross *D. pseudoobscura* ♀ × *D. miranda* ♂. The residual genotype of *D. pseudoobscura* is well adjusted to the *ABB* complex; likewise that of *D. miranda* is compatible with the *AAB* complex so that normal males are produced in each species. However, the *ABB* complex is incompatible with the genome from *D. miranda,* and hybrid males fail to develop normally. Likewise, the *AAB* complex is incompatible with the genome from *D. pseudoobscura,* causing death of male hybrids.

Even if an interspecific hybrid reaches the age of sexual maturity, gene flow can still be interrupted via the mechanism of *hybrid sterility*. In animals, genic disharmony in hybrid sterility is most commonly expressed as a failure of the gonads to develop normally. The gonads of vertebrates have a dual function: (1) production of gametes and (2) secretion of sex hormones. Imbalance of gonadal endocrines in hybrids can be responsible for failure of the normal secondary sex characteristics to develop. These traits are sometimes very important in triggering receptiveness in conspecific members of the opposite sex during courtship rituals. Occasionally ethological isolation breaks down, and one or a few hybrids are produced. If the hybrid lacks the distinctive hormonal-induced coloration, morphology, or behavior of one or the other parental species, backcrossing of the hybrid to most members of the parental species is usually prohibited. Therefore, even if the hybrid gonad retained capacity to produce functional gemetes, gene flow would be restricted or prevented because the hybrid would not be recognized as conspecific by either parental species. However, in most instances, hormonal disfunction would be of only secondary importance to abortive gametogenesis. Hybrid testicular germ cells are more commonly affected than ovarian germ cells, although in some hybrids the female gonad and in others both sexes may be adversely affected. A low rate of mitosis in the hybrid seminiferous tubules commonly produces few primary spermatocytes. Even if primary spermatocytes are produced, genic control of the precise sequence of events in meiosis is often abnormal and the process seldom yields viable, functional gametes. Sometimes genic hybrid sterility is manifested only in one of the reciprocal crosses between the same pair of species. This usually indicates plasmagenes (extrachromosomal genetic elements) or maternal effects. This is because a hybrid zygote receives nuclear genes from both parents, but inherits the bulk of its cytoplasm only from the maternal parent. If the incompatibility stems from intrinsic properties of the egg, independent of its chromosomes, *cytoplasmic inheritance* is involved. On the other hand, if the incompatibility can be attributed to a predetermination of the egg cytoplasm by chromosomes present in the egg before meiosis and fertilization, *maternal effects* are implicated. Thus, *genic* or *developmental hybrid sterility* may be expressed at any of several stages, from the initial development of the gonad to the production and release of the mature gametes.

One of the most bizarre examples of genic sterility is that of the *intersex,* in which gametogenesis fails because of a breakdown in the sex determination mechanism. Crosses between geographical strains of the gypsy moth (*Lymantria dispar*) sometimes produce females with a considerable number of male characteristics (or *vice versa*), called intersexes. Goldschmidt explained the intersex by postulating female (*F*) determiners on the W sex chromosome and male (*M*) determiners on the Z sex chromosome. Within a geographic strain, the strength of these determiners is balanced so that one dose of *F* factors is sufficient to channel development toward femaleness in the presence of one dose of *M* factors. Different strains, however, have different potencies of *F* and *M* factors. The Korean strain has relatively weak sex-determining factors (*Fw, Mw*), whereas the Japanese strain has relatively strong factors (*Fs, Ms*). When Korean strain females are crossed to Japanese strain males, the F_1 consists of males (*MwMs*) and intersexes (*MsFw*). The intersex (chromosomally female, ZW) results from the inability of the weak female determiners (*Fw*) to overcome the effects of the strong male factors (*Ms*). Sometimes the F_1 males are transformed into females, indicating that female determiners are also present in the autosomes and/or in the cytoplasm.

Since the pollen parent species contributes essentially none of the cytoplasm to its offspring, it is possible to experimentally construct a plant that contains only the cytoplasm of one species and the nuclear genes of another. This is done by repeatedly backcrossing a fertile hybrid to the pollen parent for several generations (Figure 17.7). Each backcross is expected to reduce the maternal species' genetic contribution by half. After ten or more backcrosses, the progeny's nuclear genes are expected to be almost entirely those of the recurrent pollen parent species, but the cytoplasm is that of the original seed parent. Peter Michaelis performed this kind of experiment wih two species of the willow herb. He found that nuclear genes from *Epilobium hirsutum* in the *plasmon* (i.e., the total extrachromosomal hereditary complement of a cell) of *E. luteum* caused the advanced backcross generations to be male sterile. The reciprocal cross

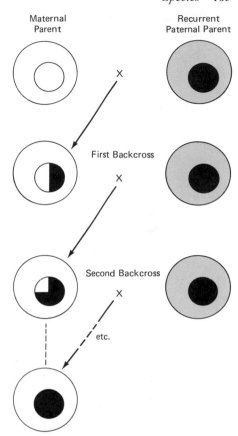

Maternal Parent
Recurrent Paternal Parent
First Backcross
Second Backcross
etc.

Figure 17.7. Diagram of repetitive backcrossing of one species (maternal line) to another species (paternal line). Within a few generations virtually all nuclear genes are of paternal origin; the cytoplasm is of maternal origin in each generation.

(*E. hirsutum* ♀ × *E. luteum* ♂) failed because the third backcross generation was either lethal or sterile (depending on the strain of *hirsutum* used). Thus, even a partial substitution of the *luteum* genome interacted lethally with the *hirsutum* plasmon.

If the chromosomes of different species bear so many structural differences that they fail to pair normally in meiotic cells of the hybrid, *segregational (chromosomal) sterility* results. Unless all the chromosomes synapse in homologous pairs, segregation of balanced sets of chromosomes to the gametes will be irregular. In plants, genetically unbalanced gametes are not expected to be functional, because they are unable to direct normal development of the haploid gametophytes (pollen grains or embryo sacs). Since the effects of segregational sterility are expressed in haploid gametes or in monoploid gametophytes where heterozygosity is absent, they must be due to epistatic interactions. Genic imbalance in animal gametes seldom renders them incapable of fertilization. However, genic imbalance in the animal zygote is likely to be expressed as embryogenic disfunctions (hybrid inviability). Translocation heterozygotes in both plants and animals are commonly semisterile (the degree of sterility being a function of adjacent disjunction from the metaphase ring of chromosomes). Even when crosses between closely related species exhibit only bivalents at first meiotic prophase, there may be numerous minor architectural differences (*cryptic structural hybridity*) that interfere with perfect pairing; the consequence is abnormal segregation of chromosomes to the gametes, genic disharmony, and gametic incapacitation.

Segregational hybrid sterility occurs about equally in both plants and animals. However, genic sterility is far more common in animals than in plants. Furthermore, genic sterility functions at an early stage of development and therefore is more likely to be the limiting factor in animals, whereas segregational hybrid sterility is more commonly limiting in plants. Recall from Chapter 9 that a sterile hybrid diploid can be converted into a fertile allotetraploid (amphidiploid) by doubling the chromosomes. Segregational hybrid sterility can be corrected by doubling the chromosome number, but developmental hybrid sterility cannot. The criterion by which developmental sterility is distinguished from segregational sterility in plants (*viz*. doubling the chromosome number restores fertility in the amphidiploid) is virtually useless in animals. Plants appear to be much less delicately balanced around a diploid chromosome number and can often survive polyploidy. Polyploid animals are rare, and those that do survive are usually sterile, presumably because of imbalance of the sex chromosome determination mechanism (intersexes). Many plants are monoecious, and polyploidy is seldom a problem in this respect.

> Crosses between the primrose species *Primula verticillata* × *P. floribunda* produce viable hybrids that are completely sterile. Both species have nine pairs of chromosomes. The hybrid forms nine loosely associated bivalents at meiosis. Doubling the chromosome number of the hybrid produces the fully fertile allotetraploid species *P. kewensis,* which forms mostly bivalents at meiosis. This is obviously a case of segregational hybrid sterility. Bivalent formation in the hybrid can be explained by assuming that the chromosomes of *P. floribunda* and *P. verticillata* are homologous in certain regions (e.g., ends and centromeres). Some nonhomologous regions, however, reduce pairing and chiasmata formation. The hybrid gametes are inviable, not because they contain abnormal numbers of chromosomes, but because they contain genetically unbalanced, disharmonious segments of chromosomal material. Doubling the chromosome number produces an exact conspecific homologue for every chromosome with which preferential pairing is exhibited in the allotetraploid. Furthermore the 2*n* gametes of the allotetraploid are functional, because they contain the same balanced genetic content as the somatic cells of the viable hybrid.
>
> The sturdy mule is a sterile hybrid from the interspecific cross of a mare (female) horse *Equus caballus* × a stallion (male) donkey *Equus asinus.* The reciprocal cross of a jenny (female) donkey × a horse stallion produces a weaker, domestically unimportant sterile hybrid known as a hinny (Figure 17.8). The diploid chromosome number of the horse is 64 and that of the donkey is 62. Both the mule and the hinny have 63 chromosomes. The difference in phenotypic vigor between the mule and the hinny can therefore be attributed to cytoplasmic factors and/or maternal effects (hinny develops in a smaller uterus than the mule and perhaps receives less milk from its mother). The haploid chromosome set of the horse includes 18 acrocentric autosomes and is so different from that of the donkey (with only 11 acrocentric autosomes) that meiosis in hybrid germ cells usually fails during the synaptic phase of the first division.
>
> Nonetheless some reports of fertile mare mules do appear in the literature. One chromosomally studied case of a putative fertile mare mule revealed that she was really a donkey that happened to look like a mule. In general, mules tend to look more like donkeys, having short manes and small hooves. Hinnies appear more like the horse, with long manes and husky tails. A thoroughly investigated male hinny revealed foci of spermatogenesis in its testis. The testes of male mules investigated previously have

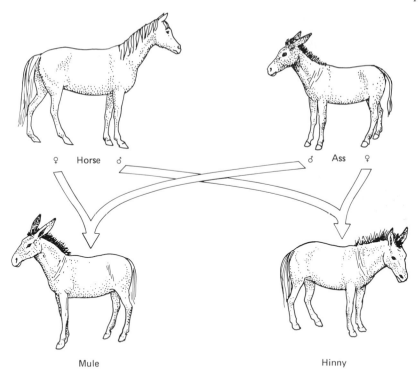

Figure 17.8. Crossing a female horse to a male ass produces a hybrid mule. The reciprocal cross produces a hybrid hinny.

normal or even excessive numbers of androgen-producing interstitial cells, but the seminiferous tubules are lined with only a single layer of cells. Therefore, the primary cause of sterility in mules is genic (developmental) rather than segregational (chromosomal). This hinny, however, had very large spermatogonia, indicating tetraploidy, and each secondary spermatocyte appeared to contain four nuclei. The size of the hinny's spermatozoa appeared to be the same as that of the donkey and the horse, favoring the view that these cells were haploid rather than diploid. Most semen samples from this hinny revealed poor counts of normal-appearing sperm, and they lacked motility. Several copulations with mares failed to produce a pregnancy, and the hinny was most likely sterile. There may also exist biochemical differences between the seminal fluids of mules, horses, and donkeys that could play a role in the lack of fertility of these hybrids.

Finally, reproductive isolation can still be manifested even though interspecific crosses produce a viable, fertile F_1 hybrid. Certain gene combinations prove to be concordant in the hybrid, allowing it to survive and form gametes. However, the genic harmony of the F_1 hybrid does not guarantee the same for its F_2 or backcross recombinant progenies. When the fitness of the F_2 or backcross progenies is noticeably reduced, the mechanism of *hybrid* (F_2) *breakdown* is operative. In many cases the backcross progenies are poorly adapted and too weak to survive, especially in competition with the more highly adapted parental types. The co-adapted gene complex of the F_1 hybrid becomes fragmented by the sexual process and recombined into new complexes, many or most of which prove to be disharmonious and nonadaptive.

The tetraploid cotton species *Gossypium hirsutum* var. *punctatum* × *G. tomentosum* intercross freely to produce vigorous, fertile F_1 hybrids. The following data on 110 F_2 seeds indicates the extent of hybrid breakdown among the recombinants.

(1) Seven seeds with small embryos failed to germinate.
(2) Thirty-six seeds with normal appearing embryos failed to germinate.
(3) Nine seedlings failed to expand their cotyledons.
(4) Twenty-two seedlings died within three weeks.
(5) Sixteen weak seedlings at three weeks.
(6) Twenty strong seedlings at three weeks.

No hybrids are produced in nature between the look-alike species *Drosophila pseudoobscura* and *D. persimilis,* but they can be crossed in the laboratory. F_1 hybrids appear fully as vigorous as the parental species, but the males are sterile. Hybrid females produce many eggs and can be backcrossed to either parental species. However, the general viability of backcross progeny is relatively low. It seems that mutant genes that do not greatly impair the viability of pure laboratory strains of these species or their F_1 hybrids nonetheless behave as semilethals in individuals developed from the eggs of these hybrid females.

The Speciation Process

While it is true that new species can arise very quickly through allopolyploidy, this is not considered to be the general process by which the bulk of species present and past have evolved. Darwin's conception of the origin of species by gradual accumulation of inheritable variations is still considered to be the most fundamental process by the overwhelming majority of biologists. Natural selection is the creative principle that molds the random variations generated by recombination and mutation into the coadapted gene complexes that typify population gene pools. Because of environmental instability, gene pools are not static entities, but highly dynamic fluxes under continual adaptive restructuring by natural selection.

New species arise (*speciation*) by three major modes: (1) phyletic evolution, (2) primary speciation, and (3) secondary speciation. These three modes are diagrammed in Figure 17.9.

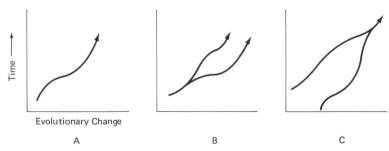

Figure 17.9. Three major modes of speciation. (A) Phyletic evolution, (B) primary speciation, and (C) secondary speciation.

Phyletic Evolution

If environmental changes have similar ramifications throughout the species range, the entire species is expected to evolve as a unit. Through time, genetic changes within a single lineage (gens) are reflected in altered phenotypes that, if sufficiently different, may be recognized by the paleontologist as different evolutionary species. This mode of speciation is termed *phyletic evolution*. It is obvious from Figure 17.9A that phyletic evolution does not increase the number of species at any one time. If the fossil record was complete, the paleotaxonomist would have to decide arbitrarily which sections of a continuous gens would be considered different species. Fortunately for the taxonomist, the fossil record is far from complete, and the gaps conveniently serve as discontinuities by which paleospecies can easily be defined. In a sense, this method of recognizing paleospecies is by default rather than by application of biological principles. We allow the fortuitous gaps in the data to determine the time transect of a gens that we shall call a paleospecies.

Primary Speciation

Race formation is a crucial step for *primary speciation*, i.e., the splitting of a single lineage into two (or more) separate evolutionary lineages (Figure 17.9B). As was pointed out in the previous chapter, geographically isolated populations of a species have the opportunity to evolve as independent or semi-independent units (depending on the amount of gene flow between isolates). Each local population tends to adapt by natural selection to the peculiar conditions in its own environment. Other evolutionary forces (e.g., mutation, drift, inbreeding, etc.) may also operate differentially among these demes to enhance the differentiation process. They thereby begin to diverge genetically from one another. If kept in isolation long enough, two or more races (or subspecies) of what was formerly a single species might accumulate so many differences that they are judged worthy of classification as separate species. Races or subspecies may thus be considered *incipient species* or *semispecies*.

In most instances, the degree of phenotypic divergence of populations in isolation gives some valuable clues as to the underlying degree of genetic difference. The correlation, however, is far from perfect and this method should not be applied without giving some consideration to this fact.

Crosses between the American bison or "buffalo" (*Bison bison*) and domestic cattle (*Bos taurus*) produce hybrids called cattalos, with most females fertile and the males usually sterile. The phenotypic differences between cattle and bison are so great that taxonomists have always placed them in separate genera. Yet the degree of genetic differentiation must be much less than indicated by the phenotypes, because genomes from both species work harmoniously during embryogenesis of the hybrid and even function normally in control of hybrid female gametogenesis. They are definitely different biological species, because some reproductive barriers are present in the form of hybrid male sterility and a preference for conspecific matings. However, genes can flow from bison to cattle via fertile female cattalos and this fact suggests that these two species are not as different genetically as their taxonomic names imply. In 1971, a California rancher was able to breach the male-sterility barrier in cattalos and has developed a new breed called Beefalo ($\frac{3}{8}$ bison, $\frac{3}{8}$ Charolais, $\frac{2}{8}$ Hereford; the last two being breeds of domestic cattle). Both males and females of the Beefalo breed are fertile. In 1974, one Beefalo bull sold for $2.5 million, a record price for a bull of any breed.

Two species of sycamore trees, *Platanus orientalis* and *P. occidentalis,* exhibit striking morphological differences. They also probably have many physiological differences, because they are native to different climatic regions; the former species is found in the Middle East, the latter in North America. They have been geographically separated for over a million years. Artificial crosses between these species produce the vigorous and highly fertile hybrid species *P. acerifolia,* widely planted as an ornamental tree. Therefore, keeping two populations geographically isolated for prolonged periods of time gives no assurance that they will evolve the kinds of differences required for biological isolation. Furthermore, the phenotypic differences between these species provide no index to the kinds of genetic changes that contribute to reproductive isolation and the formation of biological species.

Striking examples of the imperfect correlation between genetic and phenotypic differences are also found in some species that outwardly appear almost identical to the eye and yet are highly reproductively isolated in nature. These are called *sibling species*. Populations that live in the same locality without interbreeding must be accorded species recognition regardless of the degree of morphological distinctions they exhibit. Such populations are valid biological species because they "pass the test of sympatry," i.e., they live together without fusion of their gene pools. The existence of sibling species indicates that the genetic differences that contribute to reproductive isolation can be accumulated independently of those that produce the gross morphological characters by which taxonomists classically define species.

On July 6, 1962 a 100 kiloton nuclear device was detonated at a Nevada test site, creating the giant Sedan Crater. In 1968, a team of researchers returned to the area and sampled the kangaroo rats living near the crater for possible genetic damage. They found that the DNA content of cells from rats living near the rim of the crater differed markedly from those living near the limits of the throw-out zone (some 2,500 yards from the crater center). These findings were incongruous with knowledge of radiation effects in other organisms. After a long, careful analysis, it was discovered that what was thought to be a single species of kangaroo rat actually were two, so similar in outward appearance as to have escaped detection by zoological taxonomists (sibling species).

A group of inconspicuously flowered annual plant species called the *Gilia inconspicua* complex consists of at least twenty-five sibling species. They are so similar in morphology that at one time they were thought to be a single species. Detailed investigations later revealed that the Gilias were a complex of tetraploid species, completely genetically isolated from one another in nature. Experimental attempts to produce hybrids have been highly unsuccessful. The *Gilias* are self-compatible and the progeny resulting from selfing are fully vigorous generation after generation. They are widely distributed throughout arid regions of western North America and southern South America. Here again, the degree of genetic isolation cannot be ascertained on the basis of morphological distinctions.

Although special kinds of changes are involved in the development of genetic isolation mechanisms, it is not necessary to propose special kinds of processes in order to explain their origin. They are often easily explained as a consequence of natural selection for adaptations to specific ecological conditions.

In the case of the California pines mentioned earlier in this chapter, the species that sheds pollen earliest has allopatric populations found further south, lower in altitude, or nearer the coast than those of the late-pollinating species. At these locations, conditions for reproduction become favorable in early spring. Allopatric populations of the late-pollinating species are found further north or at higher altitudes where the climate does not become favorable for reproduction until late spring. It seems likely that the timing of pollination in these two species evolved as specific adaptations to local conditions in allopatric populations. Later, when their range of distribution overlapped, they passed the test of sympatry, being genetically programmed to pollinate at different seasons.

The reproductive isolating mechanisms of hybrid inviability and developmental hybrid sterility are thought to evolve mainly as a by-product of selection for adaptively different metabolic rates and temperature optima affecting developmental processes.

Populations of the leopard frog (*Rana pipiens*) form a cline (Figure 17.10) along the eastern United States with respect to several physiological attributes, including rate of development and embryological temperature tolerance. Northern species of frogs, including the northern races of *R. pipiens,* experimentally develop more rapidly than southern species (or races) at low temperatures, but at high temperatures the rates of development are more nearly equal or even inverted. Races of *R. pipiens* from Vermont can be interbred successfully with those from New Jersey, those from New Jersey can hybridize with races from the Carolinas, the Carolinas with Georgia, and Georgia with Florida races. That is, neighboring geographical races have relatively little trouble in producing viable, fertile hybrids. However, hybrids of Vermont × Florida races are commonly abnormal and inviable. The gene pool of Vermont races has been selected for a rate of embryological development concordant with colder environment of its native habitat; the gene pool of Florida races has corresponding physiological adaptations to its warmer climate. However, the mixture of genomes from these northern and southern gene pools in the hybrid is so discordant that it fails to regulate normal rates of development in all parts of the embryo and the hybrid dies. If the populations intermediate between the ends of this clinal distribution of frog races were to disappear, the northern and southern races would be able to pass the test of sympatry and be

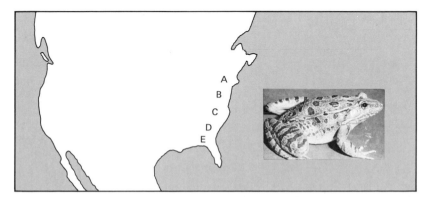

Figure 17.10. Developmental cline of races (A through E) of *Rana pipiens* on east coast of United States.

recognized as highly reproductively isolated biological species. But because they presently are connected by a series of intermediate interbreeding populations, all the *R. pipiens* races form a single gene pool and hence are classified as members of the same species. In the process of adapting rates of embryological development to peculiar climatic conditions in their own localities, geographically isolated populations of the same species have acquired, as a by-product, the reproductive isolating mechanism of hybrid inviability.

Although it is possible to develop some degree of reproductive isolation from single gene differences, the overwhelming majority of facts support the contention that most barriers to gene flow between species are highly multigenic. Progeny from experimental hybrids between species (separated in nature by habitat preferences, reproductive seasons, mating behavior, or flower structure) never show clear-cut Mendelian ratios, as would be expected if these traits were governed 'by only one or a very few genes. Instead, the progeny of species hybrids displays a spectrum of intermediate types as would be expected if these traits were under polygenic control.

Most species (including closely related ones) are generally reproductively isolated by several mechanisms. In some cases one mechanism is paramount and the others are subsidiary; in other instances, several weak mechanisms interact *synergistically* (the total effect being greater than the sum of the individual effects).

Hybrid males from *Drosophila pseudoobscura* × *D. persimilis* are sterile, but the females are partly fertile. Among the backcross males of hybrid females to either parental species, there is much variability in testis sizes (Figure 17.11). If the backcross male received its singular X chromosome from one species, then its testis size is reduced in proportion to the number of autosomes derived from the other species. This is evidence

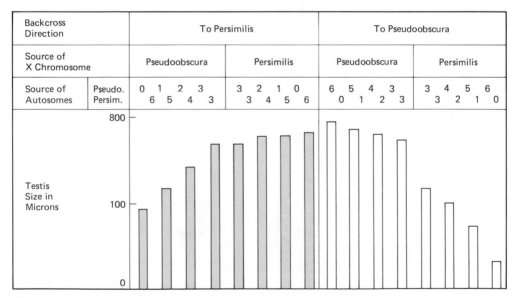

Figure 17.11. Histograms relating testis length to chromosomal content in backcross males from the hybrid (*Drosophila pseudoobscura* × *D. persimilis*) to each of the parental species.

that genes contributing to developmental hybrid sterility are located on all chromosomes of the parental species. In addition to developmental hybrid sterility, these two sibling species of flies are kept apart in nature by several other mechanisms. *D. pseudoobscura* generally lives in warmer, drier habitats and at lower elevations than *D. persimilis*. The former species tends to be more active later in the day than the latter. Females of both species exhibit a preference for mating with conspecific males. The backcross progeny commonly are deficient in vigor and vitality, unlikely to survive outside the special environments provided them in the laboratory. Thus, at least four reproductive isolating mechanisms cooperate in nature to maintain the integrity of their respective gene pools: habitat and sexual preferences, ethological isolation, genic hybrid sterility, and hybrid breakdown. None of these barriers are strong enough by themselves to prevent completely gene flow. Together in nature, however, they function to isolate completely the gene pools of these sibling species.

A wide range of fertilities has been observed in large F_2 progenies from some plant hybrids suffering partial sterility of the chromosomal type. This variability indicates that the F_1 hybrids are segregating for numerous structural differences, each with a relatively small effect on fertility.

Reproductive isolating mechanisms, developed while two populations of a species are geographically isolated, only become functional when they again come into a common region (*secondary sympatry*). Weak barriers to gene flow can then become strengthened by the direct action of natural selection. Suppose individuals that mate with members of another species tend to produce fewer viable and fertile offspring than those that only mate with conspecifics. Any genetic variations that enhance conspecific matings would then have a positive selective value and would begin to accumulate in the population. Over many generations, the direct action of natural selection tends to convert weak isolating barriers into strong mechanisms and thereby increases the average fitness of each species by eliminating the production of less fit hybrids.

Character Displacement

Closely related species tend to look very similar. Sexual isolation may be quite weak in sympatric populations of closely related animal species, because of failure of conspecifics to recognize their own kind for breeding purposes. Those genetic variations that aid conspecifics in recognizing each other would come under positive selection pressure. Species markers (visual clues, scents, mating calls, courtship rituals, etc.) tend to be more exaggerated in sympatric populations than in allopatric populations of related species. This phenomenon is termed *character displacement* and is attributed to the direct effects of natural selection intensifying allesthetic traits useful for species discrimination.

At low temperatures (16°C), sexual isolation is relatively weak between *Drosophila pseudoobscura* and *D. persimilis*. In one experiment, equal numbers of males and females of both species were placed in population cages maintained at 16°C. All hybrids were removed each generation, thereby penalizing those genotypes that participated in heterospecific matings. Promiscuous individuals wasted their gametes in the production of useless hybrids. Genotypes that fostered intraspecific matings left more progeny. Three replicate experiments produced, in the first generation, 36, 22, and 49 per cent hybrids, respectively. In the second generation, there were 24, 6, and 18 per cent hybrids,

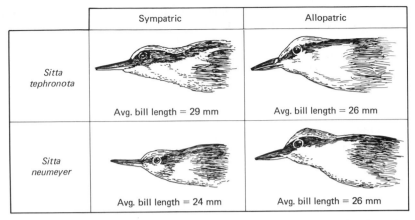

	Sympatric	Allopatric
Sitta tephronota	Avg. bill length = 29 mm	Avg. bill length = 26 mm
Sitta neumeyer	Avg. bill length = 24 mm	Avg. bill length = 26 mm

Figure 17.12. An example of character displacement in two closely related species of Asiatic nuthatches. Allopatric populations appear very similar; sympatric populations are clearly differentiated by several characters (including plumage pattern and bill length).

respectively. After the fifth generation, most experiments had less than 5 per cent hybrids. This shows the rapidity with which sexual isolation can become strengthened by the direct effects of selection.

An excellent example of character displacement is provided by two closely related species of Asiatic nuthatches: *Sitta tephronota* and *S. neumayer* (Figure 17.12). These two species are clearly differentiated by plumage pattern and bill length where their ranges overlap; outside the area of sympatry, they are very difficult to distinguish.

The phenomenon of character displacement is by no means restricted to traits important in species recognition. When two populations (or species) are attempting to use a common niche, it is possible that each will initially be somewhat better able to use different parts of the niche more efficiently than the other. If these populations can be sustained by their unique parts of the niche long enough for natural selection to be effective, each population should become more restricted to using its unique parts of the niche (and using them more effectively) and at the same time become less dependent on the shared parts of the niche. Through this process, preexisting differences (anatomical, physiological, or behavioral) tend to become exaggerated or "displaced."

Sympatric Speciation

Unless a species becomes fragmented into two or more geographically isolated subpopulations, it is difficult to understand how genetic differentiation leading to the production of new species can arise. A species commonly becomes fragmented into two or more subgroups isolated not too far apart by some barrier such as a river or a deep valley. Each subgroup goes its own evolutionary way and in time they each may become separate morphospecies. Whether or not they may also have evolved the kinds of differences that make them biological species only becomes obvious when secondary sympatry occurs. Therefore, we commonly expect that the ranges of closely related species or subspecies are not identical, nor are they widely separated, but are adjacent and separated by some kind of barrier. This generalization is known as *Jordan's rule* (after David Starr Jordan; Figure 17.13) and emphasizes the role geographic isolation plays in speciation.

Most evolutionists agree that two conditions must be met if populations are to diverge with respect to any polygenic trait: (1) natural selection must be operating to different degrees or in different directions on them, and (2) they must be kept physically isolated so that gene flow does not dissolve the initial genetic divergence. If races are to become biological species, the geographical isolation must persist until the kinds of differences that contribute to genetic isolation have been built up. Once they are reproductively isolated, they can then become secondarily sympatric without endangering the integrity of their respective gene pools.

Conceptually, the easiest way for speciation to be accomplished is in allopatric populations (*allopatric* or *geographical speciation*). Nonetheless, there is an alternative theory of *sympatric speciation* that might account for a relatively small number of species origins. This theory proposes that strong disruptive selection within a population for different subniches of a common locality can lead to genetic differentiation and speciation without the kind of spatial isolation connoted by the allopatric or geographical theory of speciation. Admittedly, the theory has few supporters at the present time, because there is little undisputed evidence

Figure 17.13. David Starr Jordan (1851–1931). Botanist, zoologist, educator, first president, Stanford University. (From the *Dictionary of American Portraits*, Dover Publication, Inc., 1967. Courtesy Stanford University Libraries)

for it. Theoretically, sympatric speciation seems possible, given the requisite kinds of genetic variants, sufficiently intense selection coefficients, and drastic reduction in gene flow through highly restrictive habitat preferences. However, serious doubts are raised in the minds of many evolutionists as to whether such powerful disruptive forces exist in nature. There is no doubt that disruptive selection can produce and sustain polymorphisms in a population. But the crux of the problem is how much gene exchange can exist between diverging subgroups of a population without arresting or reversing the divergence. As yet, there is no ready answer to this problem.

Typical ecological isolating mechanisms are usually highly polygenic. One outstanding exception to this rule may have been found in *Drosophila pachea* of the Sonoran Desert in the southwestern United States and northern Mexico. It has evolved a highly restrictive reproductive dependence on a sterol chemical found in the stems of the senita cactus (*Lophocereus schottii*). This species cannot be maintained in the laboratory without the addition of a piece of this cactus to the medium. Several alkaloids extracted from the cactus proved to be lethal to other species of *Drosophila*, but were not harmful to *D. pachea*. Relatively few genetic changes need be postulated to account for the origin of this complete ecological isolation mechanism. Perhaps a minimum of two new proteins is all that would be required. One enzyme could change function to metabolize only the cactus sterol (thereby losing the ability to use conventional sterols). A second enzyme could detoxify the alkaloid. With so few genetic changes required to account for its ecological isolation, it is tempting to postulate that *D. pachea* could have arisen sympatrically from one genotype in a polymorphic complex.

Thoday and his colleagues performed a disruptive selection experiment with *Drosophila melanogaster* that succeeded in developing pronounced sexual isolation within twenty-one generations in a random mating population. In each generation they selected the upper and lower 10 per cent of males and females for sternopleural bristle number and allowed them the opportunity to mate at random for twenty-four hours, after which time the males were discarded. By the twelfth generation there was no overlap in sternopleural bristle number of the high and low selected lines. As early as the seventh generation, obvious mating preferences were demonstrated: of fifteen high line females, twelve mated with high males and only three with low males; of sixteen low line females, twelve mated with low males and only four with high males. All that needs to be added to extrapolate these experimental results to a natural setting for sympatric speciation is the assumption that there exists only two kinds of environments (one in which only the high line survives and another in which only the low line survives). These two environments must be distributed geographically as a fine mosaic so that all the flies have an opportunity to mate at random. Most evolutionists are of the opinion that such a patchwork of environments supporting only extreme variants and leaving no niche for intermediates to occupy must be extremely rare in nature. Several other workers have attempted experiments similar to that of Thoday (one for seventy-three generations) without achieving sexual isolation. Thus, it appears that sympatric speciation is a possibility, but a highly unlikely one.

M. J. D. White found an unusual distribution of morabine grasshoppers (family Morabinae) in southeastern Australia. This complex of unnamed species is chromosomally differentiated and occupies contiguous areas with an extremely narrow zone of overlap within which hybridization usually occurs (*parapatric distribution*). Parapatric distributions are expected to be common in sessile organisms or in those of low vagility (mobility) such as in

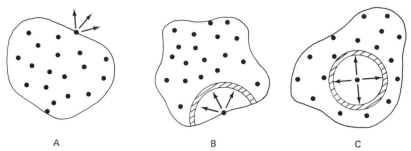

Figure 17.14. Models of geographical speciation. Black dots represent local breeding groups (demes); the border represents the geographical range of the species; narrow zones of overlap are hatched. (A) An allopatric model in which a chromosomal rearrangement becomes established in a peripheral deme and then extends the range of the species into new territory. (B) A chromosomal rearrangement may become established in a peripheral deme and then spreads along an advancing front through the existing range of the species. (C) A chromosomal rearrangement becomes established in a nonperipheral deme and spreads radially on an advancing front through the existing range of the species. Both B and C are stasipatric models.

these wingless grasshoppers. A sharp ecological discontinuity might also account for parapatrically distributed species. Because a chromosomal rearrangement, such as centric fusion and inversion, requires a minimum of two breaks, it is virtually impossible for any two rearrangements to be identical. Therefore, each chromosomally differentiated species of grasshopper is assumed to have a monophyletic origin. White proposed that a chromosomal rearrangement that first established itself near the edge of a species distribution could spread both inward throughout the range of the species (*stasipatrically*) and outward into previously unoccupied territory (allopatrically). It is statistically more likely, however, that *stasipatric speciation* (if it occurs at all) arises sympatrically within the range of the old species rather than at its margins. The new chromosomal type arising at one point in a species range begins to replace the old type until it finally encounters populations different enough genetically to be better adapted to their local conditions; this stops the spread of the new karyotype and produces the observed parapatric distribution of "species" (Figure 17.14). He also believes that the most plausible explanation for the "stasipatric" process depends on genetic isolation resulting from the lowered selective value of structural heterozygotes.

> Populations of the perennial, self-incompatible grass *Agrostis tenuis* have been found growing on abandoned mine tailings in Britain. These tailings are high in lead content and toxic to most vegetation. The population of *A. tenuis* growing on the tailings are genetically lead tolerant; those growing nearby on normal soil are genetically lead intolerant. The transition zone between these two populations is less than twenty meters wide in one locality. The lead-tolerant group must have developed recently *in situ,* i.e., within a few meters of the nontolerant forms and within pollination range of one another. Hybrids seem to be poorly adapted to normal or contaminated soils in competition with the parental populations. Hybrid inviability forms an almost absolute barrier to gene exchange. This is apparently a case of recent stasipatric speciation which theoretically could be based upon a single gene mutation (or a very small number of genetic changes).

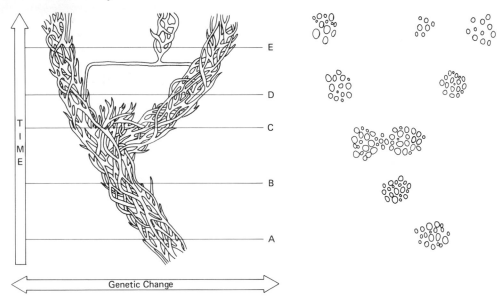

Figure 17.15. Diagrammatic summary of various kinds of speciation. At time A, a single species (diagramed as a rope) consists of numerous races (represented by strands), some of which are fusing (representing gene flow), and others are dividing into two or more subpopulations. At time B, the species has experienced much genetic change and might be considered a different species than at time A (phyletic evolution). Primary speciation is underway at time C; by time D, the primary speciation process is complete. Between times D and E, secondary speciation occurred between two populations of different morphospecies. By time E, several distinct races have developed from the initial hybridized population. Diagramed at the right are cross-sectional views at each of the time transects A through E.

Secondary Speciation

Fusion of different gens, through the process of interspecific hybridization, results in secondary speciation (Figures 17.9C and 17.15). This topic is discussed in the next chapter.

Essay Questions

1. *Give the distinguishing features of three major species definitions. Indicate their taxonomic utility and limitations.*

2. *How do reproductive isolating mechanisms arise? What advantages (if any) are gained by populations becoming genetically isolated?*

3. *What do sibling species tell us about the origin of reproductive isolating mechanisms?*

4. *How can natural selection directly intensify incipient barriers to gene flow in sympatric semispecies?*

5. *Discuss the theory of sympatric speciation, contrasting its plausibility with that of allopatric speciation.*

HYBRIDIZATION

The term *hybrid* has been used quite differently by geneticists, practical breeders, and evolutionists. Mendel used it to denote the progeny of pure (homozygous) varietal crosses of peas segregating at only one or a few loci. Geneticists still sometimes refer to simple heterozygous genotypes as monohybrids, dihybrids, etc. Practical plant and animal breeders most often use the word *hybrid* for crosses between strains, varieties, or breeds that differ at many loci. Evolutionists most commonly are referring to crosses between reproductively isolated species; less often to crosses between geographical races or subspecies. In its broadest sense, hybridization is simply *outcrossing* (*outbreeding*) or the crossing of genetically unrelated organisms (negative genetic assortative mating). To the evolutionist, however, hybridization involves crosses between populations with different adaptive gene complexes.

When genetically dissimilar individuals cross, the hybrids are recombinants. It is recombination of material already present in a gene pool that provides the most immediate source of genetic variability on which natural selection can operate to build population fitness. Geneticists make use of recombinant hybrids to decipher the nature of heredity, both in its mode of transmission and the details of its fine structure. Hybridization has great economic importance in agriculture (crops and livestock) and ornamental horticulture (landscaping, flowers, and houseplants). One of the best insights into the adaptive potential of hybridization in nature comes from our knowledge of its uses in agriculture and horticulture.

Lessons From Breeding Corn

Probably the best example of hybridization in agriculture is the development of hybrid corn. Both E. M. East (Figure 18.1) and G. H. Shull (Figure 18.2) began inbreeding maize about the same time (*circa* 1905). They observed progressive decrease in vigor with successive generations of selfing (see inbreeding depression, discussed in Chapter 15). They then crossed different inbred lines and found that the F_1 hybrids had a marked increase in vigor and reproductive potential (larger number of seeds) as well as increased phenotypic uniformity. By 1910, Shull had proposed that pure line crosses could be used to produce hybrid corn for commercial use. Shull's proposal, however, was not economically advantageous to the farmer, because the cost of the seed offset the value derived from *hybrid vigor* (or using Shull's name

Figure 18.1. Edward Murry East (1879–1938). [From *Genetics*, **24**: Frontispiece (1939)]

Figure 18.2. George Harrison Shull (1874–1954).
[From *Genetics,* **40**: Frontispiece (1955)]

for it, *heterosis*) and phenotypic uniformity. The seed producer could not offer single-cross hybrid seed to the farmer at low cost because single-cross seed is borne on the few small, unthrifty ears of one of the inbred lines. The yield of such seed is very low and consequently the price to the commercial farmer is high. This problem was circumvented in 1917 when D. F. Jones developed the *double-cross* technique (Figure 18.3). He began with four inbred lines (*A, B, C, D*). The single-cross hybrids (*A × B* and *C × D*) gave poor seed yield as expected. He then bred the two single-cross hybrids and produced a high yield of double-cross seed. The reason for the high yield of double-cross seed is that it is borne on the numerous, large cobs of the vigorous single-cross hybrid plants. Although the double-cross progeny suffers some loss of phenotypic uniformity, most of the heterosis is retained. Phenotypic uniformity is desirable from the standpoint of machine harvesting and marketability of a crop. This is much more important in the production of sweet corn for human consumption than in the production of field corn for livestock feed. The advantages of double-cross hybrid field corn were soon exploited by commercial corn growers and today virtually all commercial maize grown in the Corn Belt of the United States is of this nature.

If an open-pollinated (outcrossing) variety of corn is inbred to fixation at all loci without selection, and if the lines are then crossed at random, the entire process is equivalent to reproduction of the open-pollinated variety. Thus, in the development of inbred lines for the production of hybrid corn seed, the breeder must initially select from an open-pollinated

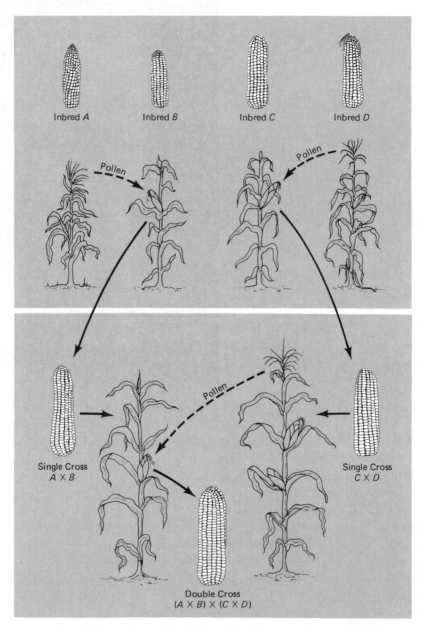

Figure 18.3. Technique for producing double-cross hybrid field corn. Highly inbred lines A,B,C,D are crossed in pairs (A♂ × B♀; C♀ × D♂), producing vigorous single-cross hybrids (A × B and C × D). Crosses between (A × B)♀ × (C × D)♂ produce the double-cross hybrid seed used for commercial plantings.

variety those plants that exhibit vigor, standing ability, disease resistance, and other desirable traits. Furthermore, rigorous selection is practiced in each generation during the inbreeding (selfing) process. As the inbreeding coefficient rises, many lines become too weak to reproduce themselves; others are discarded because they fail to meet the standards of the breeder (Figure 18.4). Only a few of the most highly productive lines survive when inbreeding is essentially complete (after about five or six generations of selfing).

Careful selection must be maintained within fully inbred lines to prevent them from accumulating changes due to occasional outcrosses and mutations. Once the selected inbred lines are available they must be screened for their general and specific combining abilities. Theoretically, the *general combining* ability of an inbred line is its average performance in crosses with all of the other lines. To determine the general combining ability of all inbred lines would require a *diallel cross*, i.e., the crossing in all possible combinations of a series of genotypes; in this case, all possible inbred lines. Even with just a few inbred lines, the number of different crosses which must be made becomes very large. For example, 190 hybrids must be made for a diallel cross involving 20 inbred lines. Thus, the development of inbred lines becomes a minor problem compared to the genetic evaluation of these lines. Fortunately, a much simpler method is available for ascertaining the general combining ability of an inbred. This is accomplished by *top crossing* inbred lines to a common pollen parent (the top cross or tester parent) of broad genetic base. Good correlation has been found between the performance of top-cross progeny and progeny from single crosses between inbred lines. No valuable material is likely to be lost by discarding the lower half of the inbred lines on the basis of their performance in general combining ability. The upper half can then be tested in single-cross combinations to identify the most productive hybridizers. The deviation in performance that a particular cross exhibits relative to that predicted on the basis of its general combining ability

Figure 18.4. Progressive loss of heterosis with inbreeding. Representative plants from two highly inbred parents (left) and their F_1 hybrid (third from left). Loss of heterosis (as evidenced by decreasing plant height) is visible in successive inbred generations (F_2 through F_8) shown to the right of the F_1.

is its *specific combining ability*. The number of combinations of inbred lines taken four at a time (in the production of double-cross hybrids) increases rapidly with the number of inbreds to be tested. A fairly accurate estimate of double-cross performance can be obtained from the mean yield of the four *nonparental* single crosses. Thus, the mean performance of the four nonparental single crosses $A \times C$, $A \times D$, $B \times C$, and $B \times D$ estimates the performance of the double-cross hybrid $(A \times B) \times (C \times D)$.

One of the factors that makes hybrid seed production so costly is the labor involved in hand cross pollination of the inbred lines. The tassels (male elements) must be removed by hand from the seed line (mother parent) so that it will only receive pollen from the other line planted (perhaps every fourth or fifth row) in the same field with it. The hand labor required for detasseling greatly inflates the price of hybrid seed. Fields used in the production of seed corn must be well isolated to prevent contamination by pollen other than the desired source. Extrachromosomal hereditary factors (plasmagenes) have been discovered in several crops, including onions and maize, that act as lethals to male gametes. Plants bearing these factors are referred to as *cytoplasmic male steriles* and therefore function only as females. Since virtually all of the cytoplasm of the zygote is derived from the maternal (seed) parent, all of the progeny of male-sterile $\times$ male-fertile crosses will also be male sterile. At least one dominant nuclear gene has been discovered in corn that restores bisexual fertility to a zygote with male-sterile cytoplasm. An ingenious method has been developed using these two systems for eliminating hand detasseling in the production of hybrid corn (Figure 18.5).

> Let S = male-sterile cytoplasm, and F = male-fertile cytoplasm; let R = nuclear gene that restores male fertility and r = nonrestorer. Inbred lines A and C, being male sterile, function only as female plants. In the AB single-cross hybrid, inbred line B provides the pollen to fertilize itself (and thus perpetuate line B) as well as to fertilize line A. Since both parents have nuclear genotype rr, the AB hybrid remains male sterile. Inbred line C is also rr, but line D is RR. Consequently the CD single-cross hybrid, being heterozygous for the restorer gene, is male fertile even though it has male-sterile cytoplasm. Thus, the CD hybrid can function as the pollen parent, and the AB hybrid can function as the seed parent in the production of the $ABCD$ double-cross hybrid. Note that half the double-cross hybrids are expected to carry the chromosomal gene R and therefore function as males, shedding sufficient pollen to fertilize all plants and thereby give the commercial corn grower a full yield of kernels on each cob. This breeding system is of such commercial value that it is the only genetic method to have a United States patent. Electron micrographs of male-sterile cytoplasm reveal what may be virus particles. One theory suggests that the virus can only survive in plants with the virus-susceptible nuclear genotype (rr). The virus apparently causes no damage in diploid cells, but kills haploid pollen cells. Chromosomal restorers of fertility are modifier genes that confer viral resistance in plants that bear them.

Plants may fail to set seed due to either incompatibility or sterility. In the case of *incompatibility*, gametes (pollen and ovule) are functional, and unfruitfulness is attributed to some physiological impediment to fertilization, usually manifested as abortion or slow growth of the pollen tube in stylar tissue. With *sterility*, the gametes are nonfunctional due to chromosomal aberrations, gene action, or cytoplasmic factors that disrupt normal development of entire flowers, stamens, pistils, pollen, embryo sac, embryo, or endosperm. Incom-

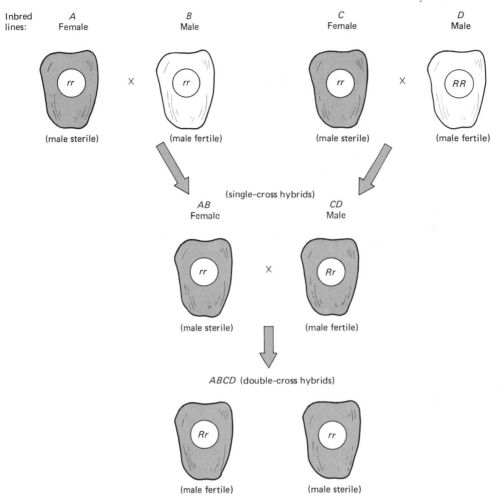

Figure 18.5. Scheme for eliminating hand detasseling in the production of double-cross hybrid field corn. Nuclear genes are shown within circles. Shaded kernels indicate that cytoplasm is male-sterile; unshaded kernels have male-fertile cytoplasm.

patibility functions as a regular mechanism for preventing hybridity in nature, because it is a common attribute of every individual in the species. Male sterile plants, however, arise sporadically in one or few members of a species, presumably by mutation at any of many loci concerned with pollen production. They are undoubtedly deleterious in nature and are thereby kept at low frequency by natural selection. The widespread agronomic use of a single source of male-sterile cytoplasm has recently had disastrous consequences for corn production in the United States. Natural populations, no less than domestic ones, also become extremely vulnerable to certain pathogens, predators, or environmental changes when their genetic base becomes depleted. Species stand the best chance of surviving when they are subdivided into many genetically diverse populations of intermediate size each adapting to its own local conditions, and yet occasionally sharing their hereditary endowments with other groups through small amounts of gene flow.

Southern corn leafblight, caused by the fungus *Helminthosporium maydis,* produced extensive damage in 1970 over most of the Corn Belt states. It is spread by air-borne spores, but these spores will not germinate unless free moisture (dew or rain) is present. Its growth is favored by wet weather and moderate temperatures (65–80°F). Because of the advantages of cytoplasmic male sterility in eliminating hand detasseling in the production of hybrid corn, most maize grown in the United States for livestock feed had been derived from a single source of cytoplasm called Texas (or T cytoplasm). Most breeders had assumed that as long as the nuclear genetic base was sufficiently diverse, the cytoplasm could be ignored. There were earlier reports of some problems with disease and insects on strains carrying T cytoplasm, but they were either not given wide publicity or simply considered to be only of local significance. After the 1970 epiphytotic, it became obvious that the use of a single cytoplasm could be just as devastating as the use of a single nuclear genotype in crop production. Maize with normal cytoplasm (or from other sources of cytoplasmic male sterility such as the J strain) is much more resistant to *H. maydis* infection than those with T cytoplasm. An effort is now being made to diversify the cytoplasmic base in the production of hybrid corn. There is potential danger to any population that lacks sufficient genetic diversity, be it nuclear or cytoplasmic.

Homozygous inbred lines and their F_1 hybrids theoretically have no genetic variance; all the phenotypic variance is environmentally induced. Yet the environmental variances for F_1 hybrid generations are usually much smaller than that of their inbred parents. This could be attributed to alternative synthetic pathways governed by different alleles in the hybrid that tend to canalize development of normal phenotypes despite vicissitudes of the environment (homeostasis). For example, in a corn hybrid one allele may function best on cold days and another on warm days. Together they buffer the hybrid's reaction to its environment, allowing it to grow well throughout the season despite daily temperature variations. This is one explanation for the fact that hybrids usually show less environmental variance than their inbred parents.

If an F_1 generation is phenotypically intermediate between its two parental populations, there is no indication of heterosis. On the other hand, a heterotic effect exists when the mean of the hybrid population exceeds the average of the two parental means (Figure 18.6). In mathematical terms, the heterotic effect in the F_1 (H_{F_1}) is expressed by

$$H_{F_1} = \bar{X}_{F_1} - \tfrac{1}{2}(\bar{X}_A - \bar{X}_B).$$

If the F_1 cross among themselves to produce an F_2, the F_2 generation usually has about the same mean phenotypic values for metric traits as the F_1, but commonly is much more variable due to segregation from loci made heterozygous in the F_1 through hybridization (Figure 18.7). The heterosis commonly manifested by an F_2 population is only half that seen in the F_1 generation due to recombination having formed less favorable gene combinations. Thus, the commercial corn grower would not save seed for planting from double-cross hybrid plants. On the other hand, F_2 progenies commonly display much more diversity than either inbred parent species or F_1 hybrids (Figure 18.8). Some of the extreme F_2 segregants can not be predicted from known characteristics in the parent, indicating that genic interactions are sometimes important factors in hybrid traits. It is at least remotely possible that some of these extreme forms could be adapted to unoccupied niches of the environment.

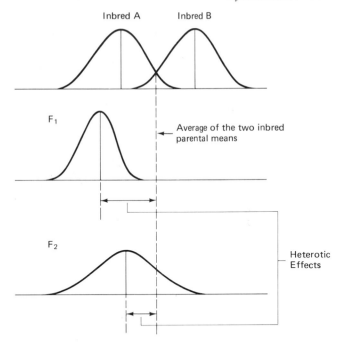

Figure 18.6. Diagram of typical heterotic effects in the F_1 and F_2 generations from crosses between two highly inbred lines. The heterosis of the F_2 is usually about half that of the F_1; the F_2 tends to be more variable than the F_1.

The term *transgressive segregation* can be applied to multigenic situations in which the phenotypic variation of the F_2 progeny exceeds that of the parents. A simple model to explain transgressive segregation assumes four loci with equipotent active ($+$) and amorphic ($-$) recessive alleles; each parent is homozygous for the active allele at two different loci.

P: *AABBccdd* × *aabbCCDD*
F_1:   *AaBbCcDd*

In this model, the F_1 has four active alleles, as do both parents, and therefore they would all have equal mean phenotypes. If each locus assorts independently of one another, we expect transgressive segregation to produce the extreme F_2 genotypes *AABBCCDD* and *aabbccdd* with frequency $(\frac{1}{4})^4 = \frac{1}{256}$ each. Obviously, if a metric trait is governed by many loci, the probability of segregating the extreme F_2 types diminishes exponentially. Many genetically different F_2 recombinant types should phenotypically resemble the F_1. Probably the most vegetatively successful segregants in nature will be those with phenotypes resembling either parental type, because the parental species are well adapted to a specific niche.

Dobzhansky makes a distinction between heterosis expressed as increased reproductive ability (*euheterosis*) and the kind of hybrid vigor that only affects vegetative development (*luxuriance*). It may be that heterotic genotypes are expressed through both increased vegetative growth and reproductive capacity, but selection would be operative only on those components of this complex that enhance fitness. We come back again to the realization that selection does not necessarily favor the largest, the strongest, or the most showy, but rather those that leave the greatest number of vigorous, fertile descendants.

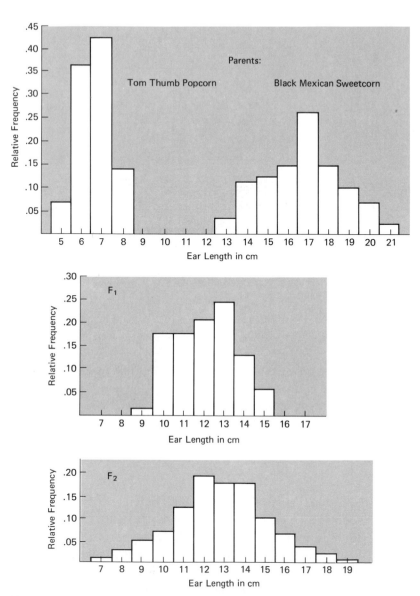

Figure 18.7. Example of mostly additive gene action in the inheritance of ear length in corn (*Zea mays*). The average ear length of the F_1 generation is intermediate between the two parental means. The average of the F_2 generation is about the same as that of the F_1, but obviously the F_2 is much more variable.

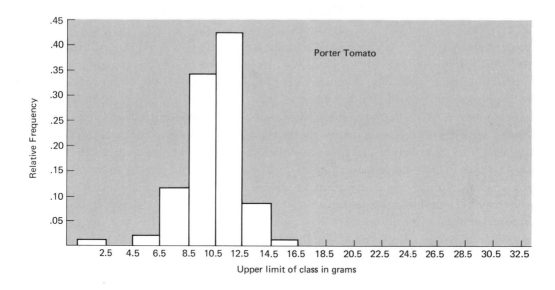

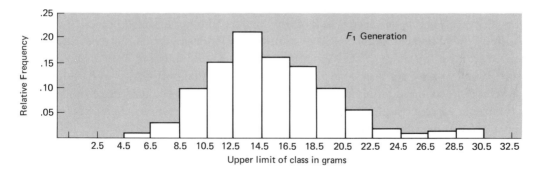

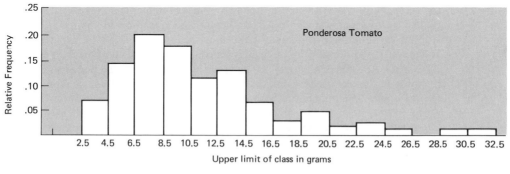

Figure 18.8. An example of heterosis in the weight per locule of F_1 hybrid tomatoes. The average weight per locule of the F_1 generation exceeds that of either parental variety (Porter × Ponderosa).

Theories of Heterosis

There are two major theories concerned with the explanation for hybrid vigor. The first was Jones' (1917) *dominance* theory of heterosis. According to his theory, each inbred line is differentially homozygous for dominant genes at many loci. These dominant genes are growth-promoting ("plus") factors, whereas their alternative recessive alleles are "null" ("minus") factors (contributing nothing to growth). Therefore, genotypes *AA* and *Aa* should have the same positive phenotypic effect on yield. If each inbred line is homozygous for equipotent growth promoters at two different loci (e.g., *AAbbCCdd* and *aaBBccDD*), the hybrid (*AaBbCcDd*) from these two lines would have twice the growth potential of either inbred parent. Under this assumption, it should be possible to find some inbreds that fortuitously became homozygous for all the plus genes during the inbreeding process and therefore are phenotypically as vigorous as the hybrids. Diligent search for such inbreds by many different workers has failed to find them. This can be explained by assuming that many more than ten loci contribute to growth and vigor. Corn only has ten pairs of chromosomes, so some of these genes must be linked. Let us make the simplifying assumption that there are thirty loci with growth factors evenly distributed throughout the genome (one at each end, and one in the middle of each chromosome) so that each locus assorts almost independently of the others. Given a plant heterozygous at all thirty loci, the number of selfed progeny required to recombine thirty pairs of factors in all possible combinations is 4^{30}. The size of the corn field required to raise this many plants is about 2,000 times the total land area of the earth. No wonder breeders have not been able to find an inbred line that performs as well as the hybrid!

According to the dominance theory of heterosis, we expect that in certain instances dominance at some loci will affect a metric trait in a positive direction, whereas dominance at other loci will affect the same trait in a negative direction. In total, the interaction could negate the expression of heterosis. Therefore, the absence of heterosis does not necessarily imply that dominance is lacking at the loci contributing to a polygenic trait.

> W. J. Robbins (1941) performed an experiment with tomatoes that sheds some light on the biochemical basis of heterosis due to nonallelic interactions. He excised roots from two varieties (Johannesfeuer and Red Currant) and grew them in liquid culture media. Adding the vitamin pyrodoxine to the media caused a greater growth response in Johannesfeuer roots; adding the vitamin nicotinamide resulted in greater response by roots of the Red Currant variety. Roots of the intervarietal hybrid, however, showed no response to the addition of either vitamin to the culture medium. It seems that the hybrid is able to manufacture both vitamins in quantities sufficient for all its growth requirements. Each of the parent varieties appears to be unable to make one of these vitamins. The situation can be diagramed as follows. Let *N* and *P* be dominant genes for the production of nicotinamide and pyrodoxine, respectively; *n* and *p* are their recessive null alleles (lacking synthetic ability).

	Parents:	Johannesfeuer	×	Red Currant
	Genotype:	*NNpp*		*nnPP*
	F_1 hybrid:		*NnPp*	

The dominant gene from one variety functions in the hybrid to mask the expression of the less-fit recessive allele derived from the other parent. Alternatively, the N and P genes need not be completely dominant; all that is required is for nn to be less efficient in vitamin synthesis than Nn and NN (similarly pp to be less efficient than Pp or PP). This is the basic idea embodied in the "dominance theory" of heterosis.

An alternative explanation for hybrid vigor was proposed by East (1936). This is the *overdominance* theory of heterosis, which attributes hybrid superiority to heterozygosity itself. In other words, the phenotype (or adaptive value) of a heterozygous genotype (A^1A^2) exceeds that of either homozygote (A^1A^1 or A^2A^2). The biochemical explanation for overdominance is presumed to reside in the biosynthetic flexibility provided by different functional alleles.

At least four mechanisms have been proposed to account for single locus overdominance (one gene heterosis), and they are summarized in Table 18.1.

TABLE 18.1. Overdominance Mechanisms

	Genotypes		
Type of Allelic Interaction	A^1A^1	A^2A^2	A^1A^2
Optimal amount	0	$2X$	X
Supplementary action	X	Y	$X + Y$
Alternative pathways	X in E_1	X in E_2	X in E_1 and E_2
Hybrid substance	X	Y	Z

X, Y, and Z are genotype products or biochemical phenotypes
E = environment.

Probably the most common type of single-locus overdominance is one in which the heterozygote (A^1A^2) produces an optimal amount of product. One homozygote (A^1A^1) makes too little product, resulting in a *hypomorphic* phenotype (a less-intense phenotype than normal or wild type); or perhaps it makes no product at all (an *amorphic* genotype). Many lethal amorphic alleles in *Drosophila,* which confer heterosis with respect to fecundity on heterozygotes, have been found to be deletions of an active allele. The other homozygote (A^2A^2) would produce more product (twice as much in the case of additive gene action) than the heterozygote. Excessive quantities of a gene product could interfere with normal metabolism and/or development. Consider the consequences of hypoglycemia (too little sugar in the blood) that could result from an overproduction of the hormone insulin. Note that overdominance with respect to fitness can result from gene action that (at the biochemical level) is essentially additive (i.e., the synthetic capacity of the heterozygote is intermediate between the two homozygotes).

Another form of single-locus heterosis is produced by supplementary allelic interaction. One gene produces product X and its allele produces product Y; the hybrid, possessing both substances, might be better buffered against environmental extremes than either homozygote.

Four alleles are known in maize at the Adh_1 locus, which specifies ten structural forms (*isozymes*) of the enzyme alcohol dehydrogenase. Three of these alleles ($Adh_1{}^s$, $Adh_1{}^f$, and $Adh_1{}^{C(t)}$) differ in net charge, and their protein products can be separated by starch gel electrophoresis. The fourth allele $Adh_1{}^{C(m)}$ produces a protein with the same electrophoretic mobility as $Adh_1{}^{C(t)}$, but with much reduced enzymatic activity. Alcohol dehydrogenase is an enzyme consisting of two polypeptide chains (a *dimer*).

There are four kinds of *homodimers* or *autodimers* (enzymes consisting of two identical polypeptide chains) and six kinds of *heterodimers* or *allodimers* (consisting of two different polypeptide chains). The $C^m C^m$ homodimer has only about $\frac{1}{30}$ of the catalytic activity of the other three homodimers but is stable at pH 10, whereas the other homodimers and heterodimers lacking the C^m protomer are unstable. Heterodimers containing the weakly active C^m protomer and an active protomer (e.g., FC^m) have only about half the catalytic capacity of a dimer consisting of two active protomers, but they remain stable at high pH. Furthermore, it was found that *FF* isozyme activity in the seedling declines much more rapidly during germination than the FC^m isozyme. For maximal functional efficiency in growth and development of an organism, an enzyme must be both active and stable. High stability and activity may be mutually exclusive properties of a single primary protein structure. By combining in one quarternary structure a relatively unstable protomer with high catalytic activity and a stable one of low activity, a hybrid enzyme (heterodimer) is made possessing both activity and stability (heterosis and homeostasis). An Adh_1 gene duplication has been found with Adh_1^F at one locus and $Adh_1^{C(m)}$ at a closely linked locus. This arrangement permits heterozygosity (and therefore heterosis) to be transmitted together in gametes with high frequency.

In *multimeric* proteins (i.e., proteins constructed of two or more polypeptide chains or *protomeric* units), alleles that make structurally defective *homopolymers* (containing identical polypeptide chains) can sometimes compensate for each other (complementation) in *heteropolymers* (containing different polypeptide chains) to produce a protein with more biological activity than homopolymers. One hypothesis for *intraallelic complementation* suggests that the protomers themselves are structurally abnormal, but able to correct each other's defects in the intact heteropolymer (Figure 18.9). An alternative theory proposes that the protomers are normal, but that irreversible misfolding occurs during maturation of the homopolymer. There

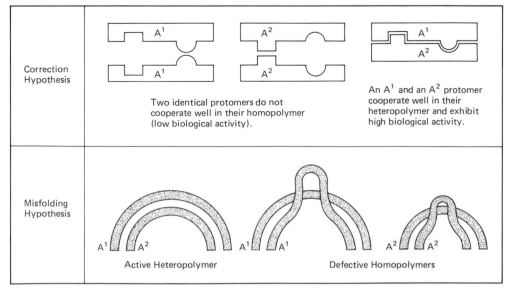

Figure 18.9. Models for two theories of intragenic complementation.

is evidence that some of the protomers of alcohol dehydrogenase mutants in maize behave in dissociation-reassociation experiments according to the latter hypothesis.

A third explanation for single-locus heterosis involves alternative synthetic pathways. This mechanism differs from that of supplementary allelic action primarily in the effect that environment has on allelic products. There are numerous examples of temperature-sensitive alleles in viruses, bacteria, plants, and animals. A simple model for this type of over-dominance has one allele producing a protein (perhaps an enzyme) with maximal activity at higher temperatures; an alternative allele produces a similar product with maximal activity at lower temperatures. If the alleles lack dominance, the hybrid can carry on normal biological activities over a broader range of temperatures than either homozygote.

Although the biochemical basis is not yet known, a fourth mechanism of overdominance has the hybrid making a unique product called *hybrid substance,* not found in either parent.

> The original (and still the most outstanding) example of hybrid substance was found by Irwin and Cole in red cell antigens of dove hybrids (family Columbidae). Serological studies of the pearlneck dove (*Streptopelia chinensis*) and the ring dove (*S. risoria*) revealed that they each possessed some distinctive antigens. The F_1 hybrid produced all the antigens of both parental species and, in addition, a uniquely hybrid antigen. Antibodies made in the rabbit against hybrid dove cells were not removed by exhaustive absorption with cells of both parents. Furthermore, antibodies to hybrid antigens were elicited by immunizing the parents with hybrid cells, whereas mixtures of cells from the parents were unable to induce antibodies with hybrid specificity in rabbits. Similar findings have been made in hybrids between the common pigeon (*Columbia livia*) and an African species (*C. guinea*), and in duck hybrids as well.

It is the general consensus of plant and animal breeders that overdominance seems to be negligible for most metric traits. Nonetheless, it is quite possible that elements of both the dominance and overdominance theories are operative in the vigor of hybrids produced either artificially or naturally.

The Role of Hybridization in Nature

Most of our knowledge concerning polygenic inheritance has come from studies on economically important traits in agricultural crops such as corn. This knowledge has also become integrated into the neo-Darwinian theory of evolution. We must understand polygenic inheritance if we are to comprehend the role hybridization can play in the improvement of species, speciation, and the evolution of higher categories. We have learned from breeding hybrid corn that selection is an important process that must accompany inbreeding if we are to obtain valuable homozygous genotypes. Furthermore, we now know that only certain matings between inbred lines will give heterosis in single or double crosses. Making crosses between genetically different lines is certainly no guarantee that the genes will work in a cooperative fashion in the hybrid, let alone produce heterosis. Interspecific crosses involving, as they generally do, widely divergent gene pools have less chance of forming harmonious gene combinations than intraspecific crosses. This is especially true in animals, because their more complex patterns of development are dependent upon equally complex and highly integrated sequences of gene action. Furthermore, higher animals (both arthropods and vertebrates) often use a mechanism to prevent hybridization that plants cannot use, namely sexual isolation. Many plants such as trees have great longevity compared with life spans of animals. The selective advantage of obtaining new adaptations through hybridization may outweigh the disadvantage of semisterility in such long-lived hybrids. Since perennial plants can tolerate lower reproductive levels per time unit than annuals in the maintenance of the species, we would expect to find more examples of hybridization in perennial groups than in annual ones, and the latter groups should therefore exhibit sharper species boundaries. Rare hybrid animals are more difficult to detect than rare plant hybrids. Relatively few animal species can be bred and hybridized in captivity to verify their hybridization potential experimentally. It is much easier to detect weak reproductive barriers in plants than in animals. For example, a hybrid animal that produces any progeny is often considered fertile; but a plant that produces only 5–10 per cent of the normal number of seeds is considered semisterile. Animal hybrids are most common in those groups with weak or no sexual isolation such as fishes, toads, and some mollusks.

In terms of their survival potential, hybrids generally have "three strikes" against them at the outset. (1) They must find a niche that they can exploit better than either of the parents. (2) Fertility must be restored in progeny of partly sterile hybrids without losing their adaptations to the new niche. (3) The attributes that make them adaptive must become stabilized so that they can be maintained through subsequent generations. This latter restriction implies "breeding true to type" due to genetic isolation from the parents. It thus becomes clear that hybridization is unlikely to be a common contributor to the evolutionary process. Nevertheless, as we have learned before, rare events in nature ofttimes have ramifications far out of proportion to their frequencies.

Since most hybrids tend to be phenotypically intermediate between the two parents, they seldom find a suitable niche that is not already occupied by other well-adapted species. Hybrids stand the best chance of securing a niche in recently disturbed habitats or in "hybridized" habitats intermediate between those of the parents. Some hybrids can be more vigorous than their parents in such a suitable niche. During the recent ice ages, glaciers advanced and retreated four times over large areas of the northern hemisphere, laying bare thousands of square miles and opening up a vast mosaic of new habitats in which many hybrids might have found a niche. Many new lakes and swamps were also formed in basins scooped out by the glaciers or dammed up by glacial debris. The level of the oceans were markedly altered during the ice ages and this caused advance and retreat of shallow seas over lowlands such as Florida. In Hawaii and elsewhere intermittent volcanic activity has provided many new habitats. In addition to these natural changes, human activities have greatly altered environments all over the world. When either nature or man alters the environment, geographical or physical barriers to gene flow between related species ofttimes break down. Species separated mainly by ecological, seasonal, or mechanical barriers can sometimes form fully fertile F_1 hybrids and in this respect are no different than hybrids between subspecies. Such species may come together and produce a *hybrid swarm* (Figure 18.10). This term is used for a continuous series of morphologically distinct hybrids resulting from interspecific crosses followed by crossing and backcrossing of subsequent generations. The commonest end result

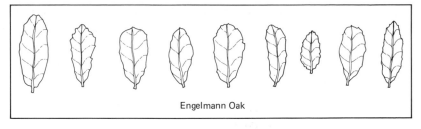

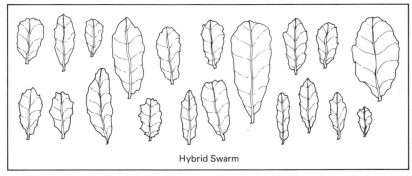

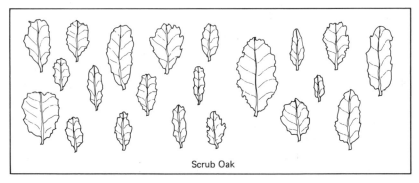

Figure 18.10. Representative leaves from two species of California oaks [*Quercus engelmannii* from Arcadia (top) and the scrub oak, *Q. dumosa,* from Pomona (bottom)], and a hybrid swarm (*Q. englemannii* × *Q. dumosa*) from the Monrovia area where these two species intergrade (middle).

of a hybrid swarm is the introduction of comparatively few genes from one species into another, because backcrosses have a much better chance of surviving than do F_2 and later generations.

When hybridization is widespread in the area of overlap, the participating species tend to lose their individual identities and fuse into a single unified gene pool. This process is termed *secondary intergradation* or *secondary speciation.* Fusion of different gene pools is not possible when strong reproductive isolating barriers exist between the species. It is at least theoretically possible that related morphospecies, having developed distinctive differences while geographically isolated from each other, might not have evolved the kinds of genetic differences that would make their genomes incompatible with one another. When such populations become sympatric, they might hybridize rather freely and become a single, more variable species. It is quite likely that this broadened base of genetic variability would result in considerable restructuring of the adaptive landscape; some adaptive peaks could become lowered or disappear into valleys, and former valleys might become hills. For the hybridized population, it could be "a whole new ball game" and many opportunities may become immediately available for the species to move to new adaptive heights through natural selection. This potential does exist in theory, but experience to date indicates that such events must have been rare in the course of evolution.

Even if interspecific hybrids do find a niche, they still face the problems of regaining fertility and stabilizing the intermediate type. Their success in resolving these problems resides in the following principle. The intermediate condition of a polygenic trait can be attained in one of two ways: (1) through heterozygosity at all relevant loci, or (2) through homozygosity at all these loci, but in combinations that differ from those in either parent. A simple model proposes four loci fixed for different alleles in each species.

$$P: \quad AABBCCDD \times aabbccdd$$
$$F_1: \quad\quad AaBbCcDd$$

In the F_2, many homozygous segregants would be of intermediate phenotype like the F_1 (*AABBccdd, AAbbCCdd, AAbbccDD, aaBBCCdd, aaBBccDD*, and *aabbCCDD*). Furthermore, upon selfing, each would breed true to type. These intermediate homozygous types would far outnumber the production of extreme types like the parents. Animal species kept apart by ethological isolation, hybrid inviability, or developmental hybrid sterility are not known to have produced through hybridization reproductively isolated intermediate combinations. This is not surprising because the genetic basis of these isolation mechanisms is complex and highly integrated. Among plants, however, examples are known of intermediate recombinants that are partly isolated from both parental species. The mechanism of segregational hybrid sterility depends at least in part on unfavorable epistatic interactions between different genes or chromosomal segments. Let us designate the parental species *AABB* and *aabb*. Gametes containing *AB* and *ab* are functional. However, the hybrid between these two species would be partially sterile, because it makes two additional gametes *Ab* and *aB*, which suffer disharmonious interactions. Now if many independently assorting loci contribute to segregational sterility, the proportion of viable gametes formed by the hybrid would be $(\frac{1}{2})^n$, where n is the number of such loci. Thus, the effectiveness of segregational sterility increases with the number of participating loci. Selection for increased fertility among the progeny of such a hybrid stands a good chance of producing an intermediate type that would be partly sterile in backcrosses to either parent. Suppose that in gametes of the hybrid *AaBbCcDd*, the only fertile interactions were *ABCD* and *abcd* (like those of the parents) and *ABcd* and *abCD*. The only homozygous and fully fertile genotypes in the F_2 would be *AABBCCDD* and *aabbccdd* (parental types) and *AABBccdd* and *aabbCCDD* (recombinant types). Hybrids between any of these four genotypes would be partly sterile. The last two genotypes could therefore be the starting points of two new species. Selection for fertility would favor those hybrids that repeatedly backcrossed to one of the parents. Therefore, stabilization of these new combinations would only be expected to develop if the parental species were normally self-fertilizing or if the hybrids became allopatrically distributed with respect to the parents.

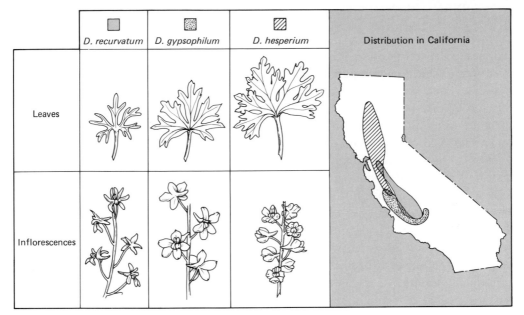

Figure 18.11. Three species of larkspur in California. *Delphinium gypsophilum* is morphologically intermediate between *D. recurvatum* and *D. hesperium pallescens* and presumably arose by hybridization of the latter two. (After Stebbins, 1966)

An example of how recombinants from an interspecific hybrid can become established as a new species is available in the larkspur genus *Delphinium*. As evident from the drawings of leaves and inflorescences in Figure 18.11, *D. gypsophilum* is morphologically intermediate between *D. recurvatum* and *D. hesperium*. All three species have eight pairs of chromosomes. *D. gypsophilum* lives in a new habitat that has probably opened only since the pluvial period of the ice age. Artificial crosses between *D. hesperium* and *D. recurvatum* produce a highly sterile F_1 hybrid (though it does produce some seed) that is morphologically and cytogentically very similar to *D. gypsophilum*. Chromosome pairing in this hybrid is nearly normal. Crosses between the F_1 hybrid and *D. gypsophilum* have produced some offspring that are actually more fertile and more regular than those produced by backcrossing the F_1 to either parent or by crossing *D. gypsophilum* with either of the other two species. It is highly probable, therefore, that *D. gypsophilum* arose by hybridization between the older more widespread *D. recurvatum* and *D. hesperium,* and became stabilized by selection for fertility among the hybrid recombinants.

The above model illustrates one way by which hybridization can produce fertile, stabilized (reproductively isolated) intermediate populations, *viz.* selection for fertility factors segregating in the offspring of partly fertile hybrids. There are at least two other ways that can attain the same result. Recall from Chapter 9 that fertile, stabilized allopolyploids can readily be produced through doubling the chromosome number in reproductive cells of a sterile interspecific hybrid. The second way is by a process called introgression (discussed in the next section).

Introgression

Interspecific hybrids are expected to be uncommon in nature. When they do occur and are at least partly fertile, there is a statistically much better chance that they will backcross to one or both of the abundant parental species than cross among themselves to produce an F_2 generation. Each successive backcross generation becomes more like the recurrent parent. Very shortly, in terms of geological time, the backcrosses become indistinguishable from the recurrent parent, and it is difficult to detect that gene flow had occurred. Some of the genes from species A have now become incorporated into the gene pool of species B (and perhaps *vice versa*). This process was named *introgressive hybridization* (or simply *introgression*) by Edgar Anderson (1938). Some of the backcross products can become stabilized by selection for favorable recombinants. Introgression is a mechanism for introducing many genes simultaneously into a species. Unlike mutations, which introduce only single genes of questionable function into the gene pool, introgression delivers blocks of genes (linked complexes in chromosomes), most of which were functionally integrated in the species of their origin. Perhaps many of the early backcross progeny would be rejected by natural selection, because of inharmoneous interactions between genomes of the two species. On the other hand, some of these introgressed genes might be neutral or even advantageous. Selection would tend to save the beneficial chromosome segments and reject the detrimental ones. Plant breeders have often made use of a similar principle in their attempts to improve crops and ornamentals.

Wheat belongs to the genus *Triticum*. *Aegilops* is a related genus possessing at least one species (*Ae. umbellulata*) with resistance to leaf rust. E. R. Sears (1956) succeeded in transferring the gene for leaf-rust resistance into the wheat genome through a sophisticated breeding scheme. *Ae. umbellulata* ($n = 7$) is a wild grass from the Mediterranean region with the genomic designation *CC*. Wheat is an allohexaploid ($n = 21$) with the genomic designation *AABBDD*. It seems that the *A* chromosome set came from either *Triticum aegilopoides* or *T. monococcum*, the *B* set probably came from the *Agropyron triticeum*, and the *D* set from *Aegilops squarrosa*. All are diploid species with $n = 7$. The origin of *Triticum* hexaploids is summarized in Figure 18.12.

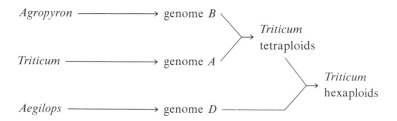

Hexaploid wheats cannot be crossed directly with diploid *Ae. umbellulata*. However, both can be crossed with the amphidiploid (allotetraploid) species *T. dicoccoides* ($2n = 14$, containing genomes *A* and *B*). Sears crossed *T. dicoccoides* with *Ae. umbellulata* and then performed two backcrosses of the F_1 hybrid to common wheat. Among the advanced backcross progeny was found a plant resistant to leaf rust that carried a single added chromosome from *Aegilops*. The addition of this single "foreign" chromosome to the wheat genome caused the plant to partially lose vigor and fertility.

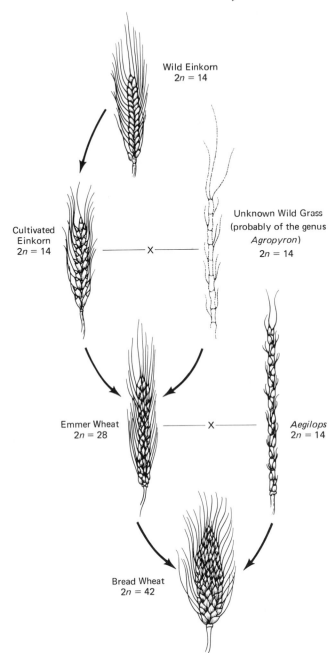

Figure 18.12. Probable origin of modern bread wheat. Cultivated Einkorn wheat (*Triticum,* 2*n* = 14) crossed with *Agropyron* (2*n* = 14) producing a hybrid that, by doubling its chromosomes, became an allotetraploid (2*n* = 28) Emmer wheat. Emmer crossed with a weed (goat grass, *Aegilops,* 2*n* = 14) to produce a hybrid that, by chromosome doubling, produced the allohexaploid (2*n* = 42) from which our modern bread wheat is descended. (After Volpe)

Some of the progeny of this plant were X-irradiated prior to meiosis, and the pollen was then used to pollinate wheat plants. The purpose of irradiation was to break the chromosomes and hopefully translocate a small segment carrying the gene for rust resistance into a wheat chromosome. There were 132 rust resistant plants among 6091 offspring produced by the irradiated pollen, including 40 that incurred a translocation of part of the *Aegilops* chromosome into a wheat chromosome. At least one translocation was of an intercalary nature that produced essentially normal pollen and was cytologically undetectable. Apparently, this translocation was a small one including the gene for rust resistance and had few other "foreign" genes to interfere with normal vigor and fertility. In essence, Sears had "introgressed" a gene from *Aegilops* into wheat. The economic value of having this gene in wheat is obvious. Introgressive hybridization in nature also has the potential to improve the adaptation of the participating populations.

There are two types of natural introgression. *Sympatric introgression* commonly increases variation of the introgressed species in its original range. *Allopatric introgression,* on the other hand, produces backcross derivatives with sufficiently altered adaptive abilities to allow the introgressed species to colonize new habitats.

An example of both sympatric and allopatric introgression can be found in two genera of shrubs of the rose family in Utah. Cliff rose (*Cowania stansburyana*) and bitterbrush (*Purshia tridentata*) are sharply differentiated morphospecies, yet they frequently form hybrids in nature. The hybrids often backcross to each of the parental species, producing two-way introgression. Cliff rose is adapted to warmer, drier conditions and is generally found in more southern states (Utah, Arizona, New Mexico, and southern California). Bitterbrush prefers less harsh conditions and is generally found further north of Utah (Idaho, Montana, northern California, Oregon, Washington, and British Columbia). The principle reproductive barrier between these two species appears to reside in different flowering periods. Bitterbrush usually completes flowering before cliff rose begins. But in Utah, both species have experienced much introgression. Here, bitterbrush growing on east-facing or north-facing slopes has its flowering delayed, and cliff rose growing on west-facing or south-facing slopes has its flowering sufficiently advanced to allow hybrid production near the summits of these slopes. Introgression of bitterbrush characters into most populations of cliff rose does not seem to have altered its original range (sympatric introgression). However, introgression of cliff rose into bitterbrush and selection for stabilized backcross segregants appears to have been responsible for the evolution of *Purshia glandulosa* (or *Purshia tridentata* variety *glandulosa*) that survives further south than *P. tridentata* (allopatric introgression). It has been hypothesized that *P. glandulosa,* having acquired genes for heat and drought resistance from *Cowania,* has been able to colonize the more xeric habitats south of the range of *P. tridentata.* Since *P. tridentata* is an important browse plant for deer and sheep, and *Cowania* is unpalatable, it has been suggested that stepwise backcrossing from one population to the next and selection for unpalatability have been responsible for finding *Cowania* characters in populations of *P. tridentata* more than 200 miles north of the *Cowania* range.

Introducing one or a few genes from a foreign source into the gene pool of another species might have no taxonomic significance and yet be much more important than independent mutations in replenishing genetic variability and enhancing adaptation in the recipient species. F_1 hybrids tend to be more uniform in phenotype than F_2 and subsequent generations. An F_2 generation segregates many different types, each requiring a slightly different habitat. Nature seldom provides the multiplicity of habitats required by the segregates. Therefore the F_2 and later generations have more problems in survival than those encountered by the F_1 hybrids. It is chiefly in areas where either nature or man has "hybridized the habitat" that any appreciable number of hybrid segregates survives. Most examples of introgressive hybridization have been detected in pastures where humans have not completely eradicated native vegetation. Many cultivated plants and weeds are largely products of introgressive hybridization. The habitat imposes severe restrictions on hybridization between well-differentiated entities. The hybrids and backcrosses most likely to survive will be those similar to one or the other parent. For independently assorting genes, the chance of recovering a parental gene combination in the F_2 is $\frac{1}{4}^n$; in a backcross it is $\frac{1}{2}^n$. Therefore the parental type is 2^n times as likely to occur in a backcross as in an F_2. F_1 hybrids tend to be intermediate between the parental populations. Backcross progeny tend to resemble the recurrent parent (i.e., the one to which they are repeatedly backcrossed). After five or six backcrosses, the mongrels are expected to become indistinguishable from the recurrent parent (Figure 18.13).

An F_2 generation is commonly much more variable than the F_1 but not infinitely so. That is to say, the range of variation seen in F_2 and later generations are not simply all possible combinations of traits from the two parental species. As a matter of fact, the range of variation is usually highly constrained, because all genes do not freely recombine. Genes are components of chromosomes. Those on different chromosomes recombine readily (independent assortment), but those in the same chromosome (linked) are recombined much less frequently through genetic crossing over. Generally, longer chromosomes incur more chiasmata than shorter chromosomes. However, chiasma formation itself is far from random in many organisms. For reasons not yet understood, certain regions of a chromosome may be highly unlikely to experience a chiasma compared with other segments of the same chromosome. The suppression of chiasma formation sometimes extends to the entire genome (e.g., crossing over is suppressed in all chromosomes of the male *Drosophila*). Structurally heterozygous rearrangements of chromosomes also tend to suppress the production of viable recombinant gametes. Certain combinations of genes function well together, but fail to do so if recombined. These three factors (linkage, structural rearrangements, and genetic interactions), and perhaps others, produce strong cohesive forces within the germplasm that prevent certain gene complexes from being broken up by recombination (Figure 18.14). These cohesive forces are more likely to be apparent in the F_3 and succeeding generations than they had been in the F_2. It is highly unlikely, therefore, that recombinations of the F_3 and later generations would produce segregates beyond the range of F_2 variability.

Because of these restrictions on recombination, it is possible to detect the effects of introgression in nature. First, if introgression has occurred, we expect a metric character to be more variable in the range of overlap of the two species. Furthermore the character should

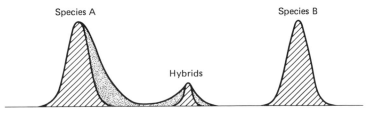

Figure 18.13. Diagram of typical effects of introgression on the variability of a metric trait. Hatched areas represent the original species and their F_1 hybrids. The stippled area represents the effects of introgression brought about by later hybrid and backcross generations.

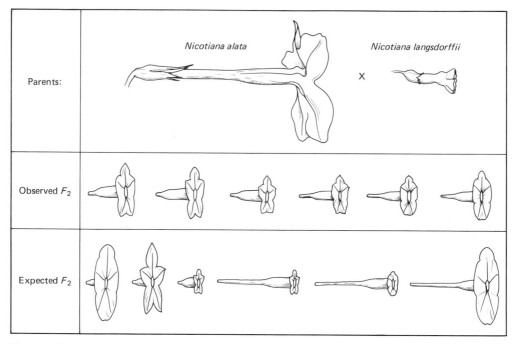

Figure 18.14. Example of constraints on free recombination. At top are typical flowers of *Nicotiana langsdorffii* and *N. alata*. The middle row shows the actual extremes of recombination found in 147F_2 plants. At bottom are the extreme F_2 recombinants expected on the basis of complete recombination of the parental characters (tube length, limb width, and lobing). (After Anderson, 1939)

tend to vary in the direction of the species with which introgression is/has been affected. Second, since introgression is characterized by correlations between characteristics that otherwise are inherited independently of one another, we can estimate the extent of hybridization through a *hybrid index* (a modification of the statistical principle of discriminant functions). In the construction of a hybrid index, we should avoid using traits that are known or suspected to be developmentally correlated. For example, corolla length, leaf length, and internode length are all expected to be positively correlated in development so that they tend to vary together in the same direction. Greater weight in the hybrid index should be given to characters that are constant in regions where the species occur alone (allopatric). Greater weight should also go to multigenic traits than to qualitative traits, because they represent differences produced by a broader genetic base. Hybrid indices are admittedly crude devices, from a statistical standpoint, but potentially very useful for initial investigations in the field. More mathematically precise methods of analysis can be effectively applied later when the nature of the problem is better understood.

As an example of hybrid index construction, let us look at H. P. Riley's (1938) work with two species of iris. Seven characters were used to distinguish between *Iris hexagona* var. *giganti-caerulea* (HGC) and *Iris fulva* (Table 18.2). All traits in *I. fulva* are given zero value. Corresponding traits in HGC receive values from 2 to 4 depending on their importance. Typical *I. fulva* plants have a total index value of zero; typical HGC plants have an index value of 17.

TABLE 18.2. Seven Characters That Serve to Distinguish *Iris fulva* from *I. hexagona* Variety *giganti-caerulea*

I. fulva	Index Value	*I. hexagona* var. *giganti-caerulea*	Index Value
1. Tube of perianth (hypanthium) yellow	0	Hypanthium green	2
2. Sepals orange-red	0	Sepals blue-violet	4
3. Sepal length 5.1–6.4 cm	0	Sepal length 8.6–11.0 cm	3
4. Petals narrowly obovate	0	Petals cuneate-spatulate	2
5. Anthers extruded beyond limbs of styles	0	Ends of anthers about 1 cm below ends of style limbs	2
6. Appendages of style branches small, barely toothed	0	Appendages of style branches large, deeply lacerate-toothed	2
7. Crest of sepals absent or very small	0	Crest of sepals present	2
Total index value	0		17

* All *I. fulva* characters have index values of zero; HGC values vary from 2 to 4 depending on their importance for taxonomic purposes. (From Riley)

On an overpastured region of a farm in the Mississippi delta, Riley found two hybrid colonies (*H*1 and *H*2; Figure 18.15). Colony *H*1 was a hybrid swarm restricted entirely to the most highly disturbed area. Colony *H*2 was much more uniform, resembling HGC very closely and yet varying somewhat in the direction of *I. fulva*. Samples of twenty-three plants from each of these four groups had index values as shown in the histograms in Figure 18.16.

The effect of introgression of some *I. fulva* genes into HGC in colony *H*2 might go unnoticed in the field, but becomes more easily discernable from the hybrid index.

Probably all multiple-factor characters of an organism are linked with each other. This genetic principle has dual ramifications for the analysis of hybridizing populations. First, if

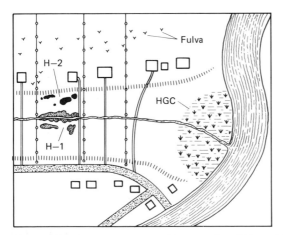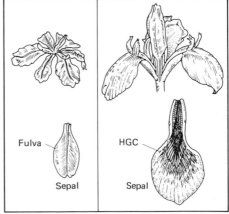

Figure 18.15. Investigations of two species of iris and their natural hybrids by H. P. Riley. (Left) Riley's research area on a farm in the Mississippi delta. H-1 and H-2 are hybrid colonies. (Right) Flowers and enlarged sepals of *Iris fulva* (left) and *I. hexagona* var. *giganti-caerulea* (right) drawn to the same scale.

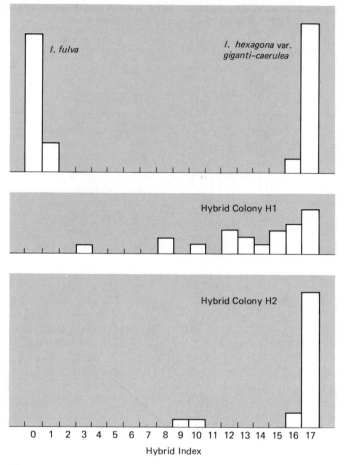

Figure 18.16. Histograms of total index values of 23 iris plants in each of four groups: two parental species and two hybrid colonies (H1 and H2). (Data from Riley)

hybrids are intermediate in one character, they will also tend to be intermediate in other traits; if they resemble either parent in one trait they will tend to resemble that parent in other traits as well. Second, the more closely the hybrids resemble a parent, the less variation is expected between individuals.

Let us now look at the kinds of relationships one might encounter in the study of just two traits (glabrous or smooth vs. pubescent or hairy leaves and light vs. dark flower color). If we score twenty-five plants for these traits and plot them on a *scatter diagram* we might see one of the four distributions shown in Figure 18.17.

In diagram A we see that plants with highly glabrous leaves are nearly always lighter in flower color than those with more pubescence. Some of the darker plants, however, are almost glabrous. This is the kind of variation one might expect to find if a light- and a dark-colored species were living sympatrically in the same region without exchanging genes.

In diagram B it is obvious that there is little or no correlation between the two sets of

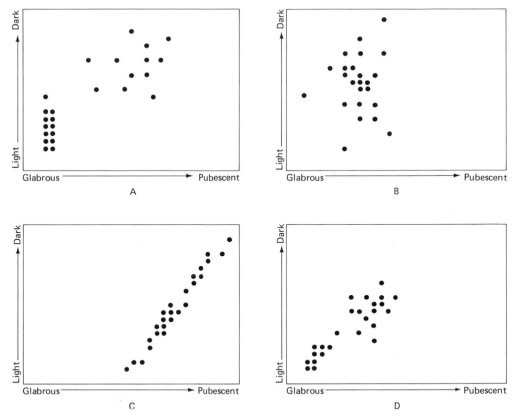

Figure 18.17. Scatter diagrams of twenty-five plants in each of four possible kinds of relationships for two traits. (After Anderson)

traits. This might simply be the normal range of variation within a species for two relatively plastic characters.

In Diagram C the traits are highly correlated; the darker the color, the more pubescent are the leaves. This might occur if the traits were both affected in the same direction by an environmental variable such as moisture; e.g., the more moisture, the darker the color and the more pubescence.

In diagram D there is evidence of introgression. Flower color and leaf pubescence definitely tend to vary together, but the relationship is not absolute. We see some plants much darker than others with no more pubescence; we also see some plants much more pubescent than others with no difference in color. If both traits have a multifactorial basis (i.e., both darkness and heavy pubescence are governed by many different genes), then the only logical explanation is introgression. To account for this correlation within a single species, we would have to postulate that gene changes affecting pubescence tend to be accompanied by gene changes affecting color intensity. Multidirectional mutation of this kind is unknown. A single instance of character associations such as this is not highly convincing evidence of introgression. However, species commonly differ from one another by many such multigenic characters. Introgression is the only mechanism known that can explain the tendency of all such traits to go together. Correlated character complexes that otherwise should be inherited independently, are the visible results of linkage systems and other cohesive forces holding introgressed blocks of genes together over many generations.

These character correlations can sometimes be used to extrapolate from a single variable population of a species to a description of another species which is introgressing into that population. This is termed the method of *extrapolated correlates*. When a species predicted by this method is actually found growing in the same area, its existence can be no mere coincidence.

As an example of the utility of extrapolated correlates, let us return to Riley's data on irises. Suppose that we had discovered *Iris hexagona* var. *giganti-caerulea* (HGC) and Colony *H*2, which looks very much like HGC but is somewhat more variable and displays some traits outside the normal range of that species. We want to attempt to predict how the source of this increased variation in *H*2 should appear. Using petal length and color of the sepal blade, we produce the scatter diagrams for HGC and *H*2 in parts A and B of Figure 18.18, respectively.

Variation in five additional traits can be accommodated by adding bars of different lengths at five different positions around the margin of each dot (see legend in Figure

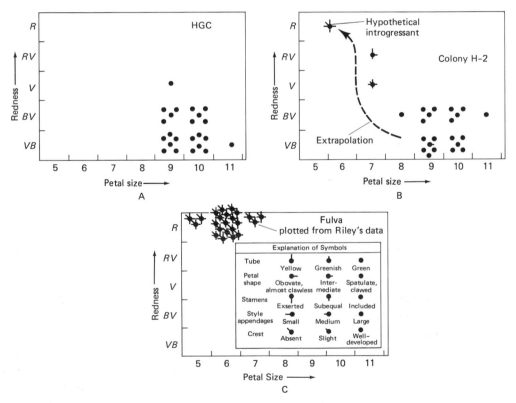

Figure 18.18. Pictorialized scatter diagram for twenty-three plants each of (A) *Iris hexagona* var. *giganti-caerulea* (HGC); (B) introgressed Colony H2; and (C) *Iris fulva;* the legend for symbols used in the pictorialized scatter diagrams is shown in the lower right-hand corner. (After Anderson)

18.18C). It is apparent in HGC and *H*2 that large petal size and violet sepals tend to go together. However, in *H*2 the few plants outside the normal range of HGC have smaller petals and redder sepals. If we assume that redness and small petal size have entered *H*2 from the same source, we predict that the other traits, tending to stay together in the two plants at the upper left, were also introduced from a common source. These seven characters are governed by polygenes, so the probability is vanishingly small that they would vary together in this way by chance. The most logical explanation is that red sepals, small petal size, yellow tubes, petal shape, stamen exsertion, small style appendages, and absence of crest all tend to go together, because they were introduced into *H*2 on chromosome segments from another species.

What kind of an iris, when crossed with HGC, would yield such variants? If it tended to cause redness in crosses with HGC, then it must be redder still; if it tended to reduce petal size in such crosses, then its own petal size must be smaller still; etc. Continuing with this line of logic, we come to the conclusion that the introgressing species would have to be an iris with very narrow, red petals, strongly exserted stamens, a yellow tube, no crest, and small styler appendages. When we examine the scatter diagram for *Iris fulva* we find that it meets all these requirements. The existence of such a species in the same area is almost indisputable evidence for the validity of the method of extrapolated correlates and confirms the hypothesis of introgression.

Introgressive hybridization is not a way of producing new characters and therefore does not contribute to evolutionary novelty (progressive evolution). It merely produces convergence between previously more distinct species. However, it probably has been important in the production of new races, ecotypes, or subspecies. It is not known to what extent introgressive hybridization has been important in evolution, but it certainly has the potential to be an important factor. Introgression is a paradox in that its biological significance increases with its degree of imperception. Hybrid swarms are interesting phenomena, but are unlikely to be of evolutionary importance except through introgression.

The Origin of Novelties by Hybridization

Hybrids usually appear phenotypically intermediate between that of the parental species. Occasionally, however, hybridization can promote the origin of entirely new characteristics, and not merely just recombinations of parental traits. One way this can be accomplished is through transgressive segregation. Genic interactions are not always of an additive nature. Sometimes unexpected characters appear in hybrid segregants from a wide cross.

> Flowers of the ornamental tobacco *Nicotiana alata* are among the largest in the genus; those of *N. langsdorffii,* are among the smallest. The hybrid between these two species is partly fertile and segregates a broad spectrum of types in the F_2 and later generations (Figure 18.19). Selection for large flower size in these hybrids produced flowers that were even larger than those of *N. alata.* Apparently genes from the small-flowered species can interact with those of the large-flowered species to produce types that exceed the range of variation exhibited in nature by *N. alata.*

Another way that hybridization can promote new characters is by providing new genetic backgrounds on which recurrent mutations, formerly rejected in the parental species, might

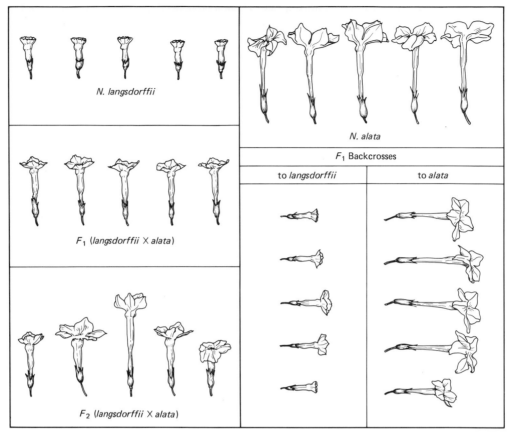

Figure 18.19. Diagrams of representative flowers of *Nicotiana langsdorffii* and *N. alata,* their F_1 and F_2 hybrids, and backcrosses of the F_1 to each parent species. Note that the F_1 is a fairly uniform, intermediate phenotype; the F_2 is much more variable. Backcross progeny tend to resemble the recurrent parent. (After Anderson)

now become adaptive. Although this is theoretically possible, it is impossible to test because one cannot distinguish between new mutations and recombinations of pre-existing genes in highly heterozygous hybrid populations.

A third possible way by which hybridization could promote novelties is by increasing the mutation rate. Gene stability is a function of the residual genotype. When this residuum is from a foreign source, some loci might become highly mutable and thereby contribute to genetic variability. Natural selection might then be able to extract from this broadened genetic base, new adaptive combinations.

Essay Questions

1. *What two major theories have been proposed to account for the phenomenon of hybrid vigor? Explain the genetic principle underlying each theory.*

2. *Using models or examples, give two mechanisms through which single locus overdominance might be manifested.*

3. *Explain introgressive hybridization and its importance to evolution of species.*

4. *How can the method of extrapolated correlates be used to predict the source from which introgressed genes are derived?*

5. *Give three ways by which hybrid derivatives can become established as stabilized, fertile types.*

HIGHER TAXA

The Role of Extinction

From the discussions in previous chapters, we have seen that disruptive selection followed by directional selection in allopatric or semi-isolated populations of a species is the way by which new races or subspecies evolve. These same processes are also considered responsible for the evolution of biological isolation mechanisms and the origin of species. Can these same principles also account for generation of taxonomic groups above the species level (so called higher categories or taxa)? Before we attempt to answer this question, perhaps we should first emphasize that higher categories (genera, families, orders, classes, phyla, and even kingdoms), though they do represent greater levels of evolutionary divergence, are still artificial groupings made by humans for convenience in taxonomic work. They are not natural groups the way that species are.

Individuals belonging to the same species freely interbreed with one another (share a common gene pool), but are reproductively isolated (at least to a marked extent) from other such groups. These biological attributes make the species a natural entity. There is no set of well-defined comparable biological attributes for any of the higher categories that make them natural entities. If higher categories are intrinsic entities that only await discovery by humans, we would expect that the more intensively a group was studied, the easier it would be to find agreement on the limits of these categories. On the other hand, if these categories are only constructs of the human mind, we would expect authorities to disagree on which traits are most important and therefore also differ in the way they define the limits of these categories. For some groups of organisms the limits of these categories are relatively easily defined; for others there is no general agreement. Groups that are declining in diversity through extinction of constituents are more sharply differentiated than groups that are still fluorishing and evolving new members. Therefore, *extinction* is the process that becomes increasingly important for the unambiguous ascertainment of boundaries as we ascend the taxonomic hierarchy.

> Among the mammals there is little disagreement on the criteria used to define orders containing strikingly different animals such as primates, whales, carnivores, and bats. Within many of these orders, the limits of families are also well recognized. However, within the rodent order and the even-toed ungulate (hooved) order, zoologists do not all agree about the number or limits of families. These are the very same groups that are flourishing today, while other mammalian orders are declining. There is even less agreement at the level of genera, the only exceptions being those families in which modern genera are relatively few.
>
> Genera of higher plants are most easily defined in the pine family and other cone-bearing trees (conifers) as well as the magnolias and their relatives. These are relatively ancient groups. Today, the conifers and magnolias are only remnants of groups that were once much more diverse and widespread. On the other hand, the grass family and the sunflower family are among those that have diversified more recently, especially in response to human activities. In these groups, genera are much more difficult to define. Thus, higher categories seem to become easier to define with the extinction of closely related intermediate groups.

Transitional Forms

Since higher categories become distinctive entities through extinction of intermediate related groups, it should be possible to find some "connecting links" in those organisms whose phylogeny is well represented in the fossil record. No one today mistakes members of the bear family for those of the dog family or vice versa. But paleontologists have great difficulty making this distinction in fossils from the Miocene and early Pliocene epochs (Figure 19.1). The lineage that gave rise to the dog family maintained the way of life of a carnivore, actively pursuing its prey. The bear's lineage evolved distinctive jaws, teeth, and faces, which were adapted to the wider range of foods typical of omnivores. Since they were no longer dependent exclusively upon a predatory way of living, there was a trend toward increased size and reduction in speed. We can easily distinguish modern dogs from modern bears, because the intermediate "kinds" have become extinct. A taxonomist studying this group when grinding-type teeth first appeared might have been inclined simply to classify those animals as somewhat aberrant members of the group. How could he or she foretell that such an adaptation would eventually be a key diagnostic trait in distinguishing members of a bear

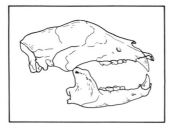 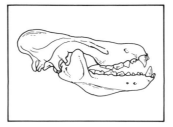 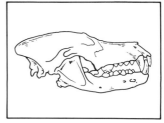

Figure 19.1. Fossil skull (middle) of an extinct form intermediate between modern bear (left) and modern wolf (right). (After Stebbins, 1966)

family from those of a dog family. Of course, numerous other changes accompanied the shift from a strictly carnivorous diet to an omnivorous one. Alteration of the structure of the digestive tract and enzymes to dissolve roughage were also adaptations in this respect. These modifications probably did not appear simultaneously, but were gradually collected and integrated into a system of increasing complexity over many generations. Soft parts are rarely preserved as fossils, so we are unlikely to ever know precisely what changes occurred or in what sequence they appeared. As the dog line separated from the bear lineage, intermediate types undoubtedly found themselves progressively less able to compete for food resources with either of these groups and eventually they became extinct. Thus, when the characteristics we now ascribe to families first appeared, they were distinctive of genera rather than of families. Similarly, when the traits we now attribute to orders first appeared, they were distinctive of families rather than of orders, etc.

Evolution of Vertebrates

From Water to Land

Even at the level of taxonomic classes there is evidence from the fossil record that evolution generally proceeds by the gradual accumulation of small genetic changes over long periods of time. Perhaps the most striking example is found in the classes of vertebrates. This is not surprising since bones and teeth are hard parts and therefore more likely to be preserved as fossils. During the middle Devonian period, a group of fishes appeared with fleshy lobe fins (*crossopterygian* fishes; Figure 19.2) supported by an internal bony skeleton. Within each fin, a single proximal bone articulated with the girdle. This was later to become the homologue of the single bone in the tetrapod (four-limbed) upper fore limb and hind limb (humerus and femur, respectively). The next two bones were to become homologous with the radius and ulna of the fore limb or the tibia and fibula of the hind limb. Beyond this was a splaying-out of bony elements that were later to become modified as bones of the wrist, ankle, hand, and foot. Unlike the related lung fishes, crossopterygian fishes did not undergo a trend toward reduction of bone in the endoskeleton. Rather, the skull and jaws remained completely bony and the pattern of these bones was comparable to that seen in other bony fishes (osteichthyes) as well as to that seen in the earliest land vertebrates (amphibians). In teeth of

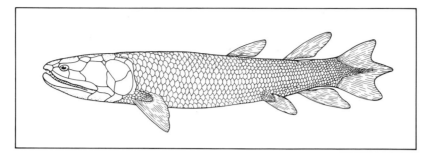

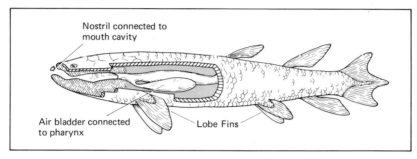

Nostril connected to mouth cavity

Air bladder connected to pharynx

Lobe Fins

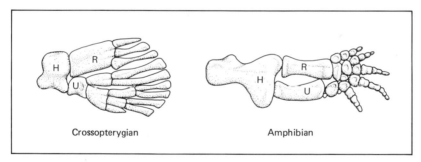

Crossopterygian

Amphibian

Figure 19.2. Crossopterygian fish. External view (top) and diagram of internal parts (middle) showing three preadaptations for terrestrial life. The bottom drawings show the homologies between the large bones at the base of appendages in a Devonion crossopterygian (left) and an early amphibian (right). H = humerus; R = radius; U = ulna.

the crossopterygians, the enamel was highly infolded, forming a complex labyrinthine pattern. This same tooth structure is seen in the early amphibians called *labyrinthodonts* (Figure 19.3). Crossopterygians had well-developed internal nares or nostrils opening into the mouth or pharynx as in modern land vertebrates. Like their cousins, the lung fish, they were capable of breathing air.

Toward the end of the Devonian period, some of the crossopterygian fishes began to come out onto land, at least for short intervals of time. Professor A. S. Romer suggests that these early land excursions were not for the purpose of exploiting unoccupied terrestrial niches, but paradoxically rather as a means for continued survival in water. Periodically, droughts caused fresh water ponds to dry up, forcing the crossopterygians to drag themselves overland in search of another pond in which they could then resume their aquatic way of life.

Eryops

Figure 19.3. A restoration of a Permian labyrinthodont amphibian (genus *Eryops*), some of which were over six feet long.

Perhaps just as an incidental byproduct of these short land journeys, the crossopterygians were able to find new sources of food. It was during the Devonian period that plants were first making a successful conquest of the land. Myriapods (centipedes and their allies) are also known from Devonian fossils. Primitive wingless insects probably evolved from myriapods by reduction in the number of legs and specialization of some body segments to produce a well-defined thorax and abdomen. The earliest insects appear in rocks of the middle Devonian period. All of these could have served as new sources of food for the early land vertebrates. An alternative theory proposes that crowding in small pools gave a selective advantage to those who could and did come out onto land to seek alternative sources of food.

> Major adaptive shifts, such as the transition from an aquatic to a terrestrial way of life, are accomplished either by (1) development of two kinds of adaptations for the same general function, or (2) use of a structure that has two functions. The evolution of the vertebrate lung serves as a good example for both (Figure 19.4). Modern bony fish (osteichthyes) have an air bladder or swim bladder that serves mainly as a hydrostatic organ, to aid the fish in maintaining position at a desired depth. In a few modern fish it also serves as a lung. Modern cartilaginous fish (sharks, rays) do not have a swim bladder. Bony fish probably arose in swamp waters, which were periodically subject to drying up and stagnation. An air bladder would be of great utility to fish under these conditions, for they could then gulp air at the surface to supplement gill respiration as some modern fish do (e.g., lung fish, garpike, bowfin). Most modern fishes belong to the group called the *teleosts*. The more primitive teleosts (such as goldfish, catfish, salmon, and trout) have a duct connecting the swim bladder with the esophagus presumably in the same manner as in their ancestors. The more advanced teleosts (such as perch, bass, tuna, and mackerel) have lost the duct and there is no connection between the esophagus and the air bladder. Modern lung fish (and probably primitive amphibians also) use both gills and lungs for respiration. Some modern amphibians (such as the frog) have gills during the larval stage and lungs as adults. Some modern slamanders retain gills as adults and can use both gills and lungs as needed. At least one group of salamanders (family Plethodontidae) has secondarily lost their lungs and obtain oxygen by pumping air in and out of their vascularized pharynx.

Lungs were adaptive for certain kinds of aquatic conditions, but at the same time they were preadaptive for terrestrial life. The evolution of amphibians from crossopterygian fishes constituted a shift from one major adaptive zone to another or from one adaptive peak to another. This kind of shift is referred to as *quantum evolution*. It did not occur overnight; it required many generations of selecting favorable variants to perfect the initial preadaptation to its present functional level. The species has always been the evolutionary unit, even during the phyletic changes that ultimately gave rise to the diversification we now (in retrospect) recognize as higher taxonomic groups. During the transition from one major adaptive zone to

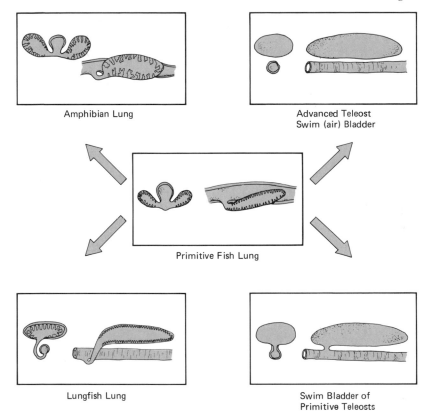

Figure 19.4. Diagrams depicting the evolution of lungs and swim (air) bladders as outpocketings from the gut. Within each frame, a cross-sectional view is at the left and a longitudinal view is at the right.

another, the species has always been well adapted. If it were not so, it would have become extinct. As the gene pools of species responded to selection tending to perfect preadaptations to terrestrial niches, the adaptive peaks themselves moved with these populations. Quantum evolution leading to a new order (for example) is most likely to occur in a single, relatively small, preadapted population of a species to which a new ecological niche becomes opened. The transition to the new adaptive zone may take place quite rapidly from the standpoint of geological time and involve relatively few individuals compared to those in the old or new orders that lived before and after this shift. Therefore it is not surprising that intermediate or transitional "connecting links" between higher taxonomic categories are seldom available in the fossil record.

The earliest amphibians, known as ichthyostegids, were a curious mixture of primitive and advanced traits. The vertebrae were primitive, and the fin rays of the fish tail were retained. In contrast, the limb girdles were strong and articulated completely with bones of the limbs. The evolution of higher groups is typified by its *mosaic* character, i.e., mixtures of primitive and advanced traits. All the characters shared in common by modern amphibians did not evolve at exactly the same time. Some traits appeared early, others much later. It is likely, however, that traits which were closely related in function were probably fairly closely synchronized in their evolution as an adaptive complex. Thus, we might suspect that the evolution of the supportive and locomotor functions of the skeletal and muscle elements were more highly synchronized than the less functionally related aspects of integumentary, respiratory, or reproductive changes.

Even modern amphibians are not completely adapted to a terrestrial mode of life, for they must return to the water for reproductive purposes. Their soft, jelly-like eggs would dry up if laid on land. Furthermore, male frogs have no intromittent organ with which to transfer sperm to the female; indeed fertilization is external just as it is in most modern bony fish, the male releasing sperm into the water as the eggs are laid by the female. Male salamanders, however, deposit their sperm in a package, called a spermatophore, attached to some object in the water. The female takes this spermatophore into her cloaca and fertilization is internal; the fertilized eggs are then laid and the embryos develop in the same environment as frog eggs. Amphibians are a declining group today, but who knows what nature will do in eons to come? We accept internal fertilization in salamanders as merely an interesting characteristic of this order. Perhaps several million years of evolution perfecting this initial pattern of pre-adaptive reproduction could eventually produce a new kind of animal completely independent of the water. It happened at least once before in the evolution of reptiles from some of the early amphibians. There is evidence that some of the early amphibians retained scales like their fish ancestors. Retention of scales would be one preadaptive way by which water loss could be retarded in animals experimenting with terrestrial environments.

Reptiles

The really important event in the conquest of the land was the evolution of the *amniote* and *cleidoic* egg, i.e., an egg that develops an amnionic sac containing fluid in which the embryo develops and that contains sufficient nutrients to carry the embryo completely through its development to hatching. This is what finally liberated the reptile from the water and set it off on an adaptive radiation that made it the dominant land animal of the Mesozoic era. Since these changes involve soft reproductive parts, they have not been preserved as fossils. Paleontologists are forced to work almost exclusively with bones. A. S. Romer confesses that the skeletons of primitive Paleozoic reptiles are so similar to some of the earliest amphibians that it is almost impossible to assign them unambiguously to either class. A taxonomist working at that time, not knowing where the reptilian lineage would eventually go, would probably be inclined to place these "aberrant amphibians" into a different family or order of the class Amphibia. Perhaps this is completely analogous to the way we now classify internally fertilizing salamanders in a separate order of the class Amphibia, because we have no foreknowledge of where this "aberrant behavior" will lead. Amphibians are a declining group today, but they were the dominant land animals of the late Paleozoic and early Mesozoic eras. Perhaps they will never again be the ancestors of another adaptive radiation, but if they ever do, it is likely that the most generalized (unspecialized) types will be the ancestors. As a rule, adaptive radiations have sprung from generalized, not highly specialized (advanced) groups. It is much easier to modify a generalized organism for particular adaptations than it is to do the same with one that is already highly specialized.

Tailless frogs and toads (class Anura) are a relatively highly advanced group with special adaptations for a saltatory mode of locomotion. They are not likely candidates for further adaptive radiation. Among the amphibians, only the tailed salamanders seem to have retained the least change from the earlier amphibian type. Indeed, some salamanders (class Urodela) such as the mudpuppy (*Necturus*) have retained gills as adults (neoteny). They swim about in an undulatory manner reminiscent of fish-type locomotion. In many, but not all respects, this is the least specialized order and hence the most likely source of further adaptive radiation. As long as mammals dominate the land and fish dominate the water, amphibians are in a relatively poor position to compete effectively in either habitat. So it is not likely that our hypothetical adaptive radiation will ever again come from the amphibians. Perhaps we can now understand why adaptive radiations generally lead to increasing levels of specialization in the various lines of descent, while the opportunities for further adaptive radiations reside with the least specialized members of the group.

Mammals

Just as reptiles sprung from the very early unspecialized labyrinthodont amphibians, mammal-like reptiles evolved from the very early generalized reptiles. These mammal-like reptiles or *therapsids* (Figure 19.5) were intermediate between typical reptiles and primitive mammals with respect to skulls and teeth. Therapsids appeared in middle Permian times. Many therapsids developed two occipital condyles on the skull (characteristic of mammals), whereas typical reptiles today have only a single median condyle. These condyles are the articulatory surfaces whereby the skull connects with the vertebral column. The legs of therapsids were well underneath the body with elbows pointing backward and knees pointing forward. This too is characteristic of modern mammals. This skeletal structure allowed much more efficient use of the limbs in locomotion and thereby set the stage for adaptive radiation of predatory forms that could compete with more typical reptiles in catching food; it also allowed them to escape being taken as prey themselves. During the Triassic and early to middle Jurassic periods, an advanced group called the *ictidosaurs* evolved from the therapsid line. Their skulls looked more mammalian than reptilian, because the temporal opening (typical of reptiles) became confluent with the orbit. The quadrate bone remained in the skull and the articular bone was retained in the lower jaw, although both were so reduced in size as to be of little use in supporting and articulating the jaw to the skull the way they did in their

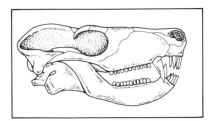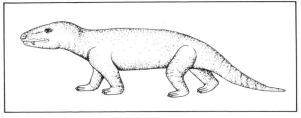

Figure 19.5. Mammal-like reptiles (therapsids) from the Triassic period. At left is a lateral view of the skull of *Bauria;* at right is a restoration of *Cynognathus* (about 7½ feet long).

remote ancestors. Perhaps for this reason only, the ictidosaurs are classified as reptiles rather than mammals. Further evolution along this line eventually converted the quadrate and the articular bones to auditory ossicles (bones of the middle ear) with a different function in conducting sounds from the tympanic membrane (ear drum) to the internal ear (cochlea). Mammals typically have three auditory ossicles, but modern reptiles have only one. Perhaps even before this transition occurred, these bones were at least partly aiding in sound conduction as well as jaw support; evolution then refined their structure to specialize only in conducting sound. This seems to be the typical way that organisms acquire new functional adaptations. Complex structures (such as the middle ear bones) do not arise *de novo,* but rather by gradual modifications from preexisting structures.

It was during the late Triassic period that the first true mammals appeared. This was before the rise of the dinosaurs. Mammals generally remained small and coexisted with the dinosaurs during the next one hundred million years. Dinosaurs mysteriously became extinct at the end of the Cretaceous period. Since that time, the Cenozoic era has been dominated by mammals (and also by birds). Today, the most primitive mammals are the *monotremes* (Figure 19.6). These are oviparous mammals that lay eggs, yet nourish their young with milk secreted by modified sweat glands that are homologous to the mammae or breasts of higher mammals. Ony two remnants of this very ancient group of mammals survive; one is the duckbilled platypus (*Ornithorhynchus*), and the other is the spiny anteater (*Echidna, Tachyglossus*) native to Australia and New Guinea. They are a mosaic of modern mammalian and primitive reptilian characters. For example, they have a primitive shoulder girdle with no true scapular spine as other mammals do. The cervical ribs are unfused. The rectum and urogenital system open to a common cloaca as in reptiles. They have no external ears or pinnae as in most other mammals. It is believed that the monotremes represent an intermediate stage of evolution between the mammal-like reptiles and the higher mammals. They persist today only because Australia became separated from the rest of Gondwana near the end of the Cretaceous period. Other members of this group apparently could not compete with the more advanced mammals that evolved in other parts of the world.

Several lines of early mammals evolved that are now defined mainly by differences in dentition. Of these, only the *pantotheres* (or *trituberculates*) have surviving descendants today. Marsupial mammals give birth to immature embryos; the embryos climb into a pouch

Duckbilled Platypus
(*Ornithorhynchus*)

Spiny Anteater
(*Echidna*)

Figure 19.6. Primitive living egg-laying mammals (Monotremes).

(marsupium), attach to nipples, and spend the rest of their embryonic development inside the pouch. Placental mammals are born at a relatively advanced stage of development. Some precocial young, such as members of the deer family, are able to follow their mothers within a few hours after birth. Both placental and marsupial mammals appear to have evolved along different lines from the pantotheres. Marsupials survive today almost exclusively in Australia, apparently for the same reasons as given above for monotremes. The only extant marsupial native to North America is the opossum; it is sometimes called a "living fossil," because it has changed very little since the Cretaceous period. An adaptive radiation of marsupials occurred in Australia comparable to the one that occurred in placentals in other parts of the world. Selection for similar niches in these two regions often produced forms that were superficially similar in appearance, although unrelated. *Convergent evolution* of this kind produced a carnivorous marsupial whose skull looked very much like the saber-toothed cats of the Pleistocene epoch in North America. Other examples of *ecological equivalents* between marsupial and placental mammals can be recognized among living forms today. Wallabies and kangaroos occupy a browsing niche similar to that of members of the deer family on other continents even though they bear no superficial resemblance to one another. Wombats can be compared with large rodents such as the woodchuck. There is a marsupial mouse that looks just like a placental mouse. Phalangers live the life of a squirrel; bandicoots take the place of rabbits; etc. (Figure 19.7).

Birds

The evolution of birds is poorly known, because of all the classes of vertebrates they are the least represented in the fossil record. Only two good skeletons of the first bird (*Archaeopteryx*) are available to tell us anything about the beginnings of this lineage. They were found in Jurassic deposits (Figure 19.8). If the imprints of feathers had not been preserved, *Archaeopteryx* would undoubtedly have been classified as a reptile, for the skeleton was typically so, including two posterior temporal openings in the skull that typified the archosaurian reptiles. Some archosaurian reptiles (*pterosaurs;* Figure 19.9) had also evolved adaptations for flight (or soaring) about the same time, but there is no indication that birds arose from them. Remains of bird skeletons in Cretaceous rocks are more extensive, but still fragmentary. By the beginning of the Cenozoic period, birds had become completely modernized in their skeletons. There appears to have been relatively little structural change in their skeletons during the last fifty to seventy million years. The demands of flight have presumably placed severe restrictions on modification of the bird skeleton.

Because of this structural uniformity, classification of Cenozoic bird fossils is usually very difficult. Modern birds are classified mainly by external characters (such as plumage shape and color) that cannot be used for identifying fossils. The distinguishing characteristics of modern birds would be considered of minor importance in other vertebrates, yet they are given ordinal rank in the classification of birds. Despite these restrictions, there have been many diversifying adaptations in other features of the body. Some birds have become proficient in snatching insects on the wing; others look for plant or animal food mainly on the ground, in trees, or in fresh or salt water. Some nest in the grass, others in trees or on bare cliff ledges. Some eat seeds or fruits; others (such as the hawk) are predators. Some migrate long

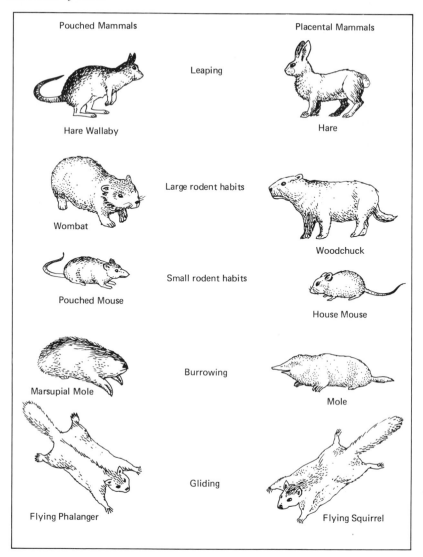

Figure 19.7. Examples of ecological equivalents and convergent evolution. The marsupials (pouched mammals; at left) on the Australian continent superficially resemble placental mammals (at right) in other parts of the world and occupy similar niches.

distances seasonally; others stay at home. Some, such as geese and ducks, are good swimmers and walkers in addition to fliers. The variation in plumage coloration and songs is well known. Some (e.g., ostriches, emus, cassowaries) have secondarily lost the ability to fly and have carved out a niche of their own on the ground in competition with the highly successful mammals. All these adaptations to particular ways of life have occurred without markedly altering the basic bird skeleton.

We should now be in a position to answer a question presented near the beginning of this chapter. When novelties (that later are used as diagnostic features of higher categories) first

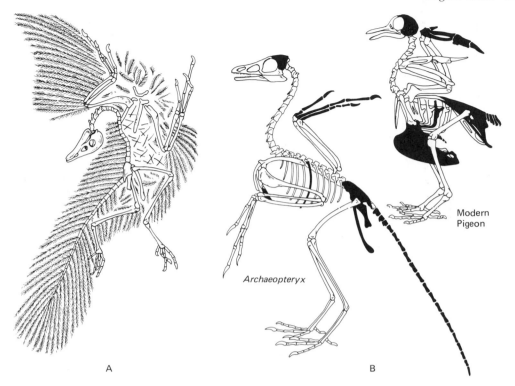

Figure 19.8. *Archaeopteryx,* a transitional form between reptiles and birds. (A) Drawing of a fossil found in Jurassic rocks showing clear impressions of feathers, claws on forelimbs, and a long reptilian tail. (B) Diagrams of skeletons of *Archaeopteryx* (left) and modern pigeon (right). Shown in black are homologous parts (brain case, forelimb, sternum, rib, pelvis, and tail).

appear, they are distinctive only of species or genera. With increasing evolutionary divergence and the extinction of intermediate types, the higher categories become easier to define. There is really nothing natural about these higher categories except the biological continuity by which they are linked through time plus extinction of groups that fail to meet the challenge attendant to environmental fluctuations. Therefore, there is no reason to invoke any new or special processes to account for the evolution of higher categories, aside from those known to function in the formation of races and species.

Figure 19.9. Reconstruction of an early Jurassic pterosaur (*Dimorphodon*).

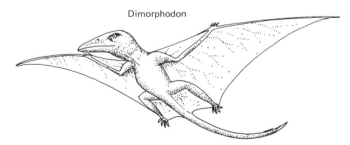

Classification vs Phylogeny

It is obvious from preceding discussions that taxonomists must often make arbitrary decisions on the limits of taxa in order to classify organisms. Their job is to establish some system (hopefully natural) by which each individual can be pigeonholed into a group and named accordingly. Difficulties inevitably arise from this attempt to superimpose discretion on the continuity inherent in the biological world through evolution. Even when we try to define what we mean by "plant" or "animal," we encounter problems, especially at the lower levels of organization. We all know that roses are plants and dogs are animals. At these higher evolutionary levels we have in mind some criteria that usually work well to distinguish plants from animals. Generally, plants are characterized by cellulose cell walls, the ability to synthesize complex organic substances from simple chemicals by photosynthesis (*holophytic nutrition*), inconstant body form, limited movement, and external organs. On the other hand, animals generally lack cellulose cell walls, have a fairly constant body form, subsist on other plants or animals (*holozoic nutrition*), are capable of relatively rapid movement, possess mostly internal organs, and exhibit pronounced irritability.

Molds and other fungi lack chlorophyll and yet in other respects are plant like. Sponges are not motile (except during the microscopic larval stage), but have other attributes that are animal like. Much greater difficulties arise in the study of the single-celled organisms called Protozoa. Zoologists claim the actively moving flagellated protozoans as their own, but are somewhat perplexed about the *euglenoids* (Figure 19.10). In this group of flagellated protozoans, many species possess chloroplasts and photosynthesize nutrients in typical plant fashion, yet have no rigid cellulose cell walls. Indeed their surfaces are so flexible that they can change shape quite readily. Many of these plant-like species can also obtain food saprophytically to supplement their growth. Botanists would also like to claim these photosynthesizing euglenoids as algae, but are unhappy that they have so many animal-like traits and have cousins that are nonphotosynthesizing. Consequently, these organisms are to be found in some zoology texts among the phylum Protozoa, and in some botany texts among the phylum (division) Euglenophyta. This mosaic of animal-like and plant-like characters is to be expected to have occurred in some forms of life during evolution of the higher groups. The fact that some of these intermediate types are alive today provides further evidence for the theory of evolution, but creates great difficulties in classification.

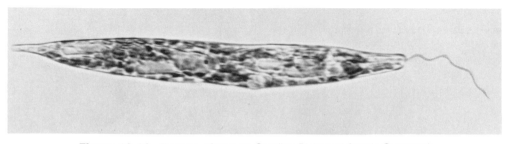

Figure 19.10. *Euglena.* (Courtesy Carolina Biological Supply Company)

One of the oldest and most widely used classification schemes has only two kingdoms: Plantae (plants) and Animalia (animals). These artificial constructions obviously do not correspond completely with the biological world. Attempts to cope with these kinds of problems have led to classification schemes with more than two kingdoms. For example, one scheme uses the kingdom Protista as kind of a dumping ground for taxonomically difficult groups including unicellular protozoans and algae (in addition to bacteria and fungi). Another scheme of classification recognizes the gulf separating unicellular organisms with nuclear membranes (eucaryotes) and those without nuclear membranes or other membrane-bound organelles (procaryotes) and places them in separate kingdoms; the procaryotic bacteria and blue-green algae belong to the kingdom *Monera* and unicellular eucaryotes belong to the kingdom *Protista* (Figure 19.11). Most bacteria (Schizomycophyta) do not manufacture their own food, but live as heterotropic saprophytes. Some bacteria are either chemosynthetic or photosynthetic autotrophs, but they lack chlorophyll *a*, which is the chief light-trapping pigment in blue-green algae and higher plants. Unlike algae, bacteria never use water as the ultimate electron donor in photosynthesis and therefore do not release molecular oxygen. Therefore, placing bacteria and blue-green algae in the same kingdom has really done nothing to clarify the distinction between "plants" and "animals"; instead it has lumped together organisms that obviously are not closely related and has created perhaps more problems than it solves. Table 19.1 lists some of the ways that organisms have been classified into kingdoms.

Very few fossils have been found from the first $1\frac{1}{2}$–$2\frac{1}{2}$ billion years of life. This is the time when chemical evolution formed protocells, and diversification took place at the unicellular and multicellular levels of organization. By Cambrian times, when fossils first became abundant, most of the major algal and animal phyla were already represented. Consequently, the most important steps in the evolution of life will probably never be revealed by the rocks. We have very little information how the bacteria are related to other living things. We know little about the relationships between major groups of algae. We do not know if fungi evolved

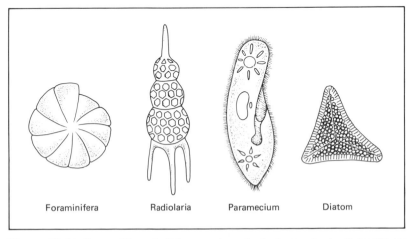

Foraminifera   Radiolaria   Paramecium   Diatom

Figure 19.11. Some of the unicellular organisms (both plant and animal) classified in the kingdom Protista.

TABLE 19.1. Some Kingdom Classification Systems.

System A	System B	System C	System D	System E
PLANTAE	MONERA	PROTISTA	PROTISTA	MONERA
Bacteria	Bacteria	Bacteria	Bacteria	Bacteria
Blue-green algae	Blue-green algae	Blue-green algae	Blue-green algae	Blue-green algae
Green algae		Protozoa	Protozoa	
Brown algae	PLANTAE	Slime molds	Green algae	PROTISTA
Red algae	Green algae		Brown algae	Protozoa
Slime molds	Brown algae	PLANTAE	Red algae	Green algae
Fungi	Red algae	Green algae	Slime molds	Brown algae
Bryophytes	Slime molds	Brown algae	Fungi	Red algae
Tracheophytes	Fungi	Red algae		Slime molds
	Bryophytes	Fungi	PLANTAE	Fungi
ANIMALIA	Tracheophytes	Bryophytes	Bryophytes	
Protozoa		Tracheophytes	Tracheophytes	PLANTAE
Multicellular	ANIMALIA			Bryophytes
animals	Protozoa	ANIMALIA	ANIMALIA	Tracheophytes
	Multicellular	Multicellular	Multicellular	
	animals	animals	animals	ANIMALIA
				Multicellular
				animals

from photosynthetic green algae or whether they arose directly from heterotropic organisms such as bacteria or some other stock. The relationships between many protozoans and multicellular plants or animals are also uncertain. Generally, the further back in time a group had its origin, the less we know about it. In many cases, early phylogenies are based on indirect evidence and/or educated guesses as to the most likely geneologies. With these reservations in mind, let us now quickly survey the major lines of diversification in higher plants and animals.

Plant Evolution

Differences of opinion as to the most natural and useful classification scheme exist among phytotaxonomists, so that no one system has yet been devised that satisfies all of them. The same is true for zootaxonomists. Therefore, the scheme presented in Table 19.2 is only one of several possible systems of plant classification. According to the Botanical Rules of Nomenclature, the plant taxon "Division" corresponds to the animal taxon "Phylum." The most primitive plants, called *thallophytes,* lack true roots, stems, or leaves and do not contain the vascular tissues of xylem and phloem. Algae are photosynthetic (autotropic; holophytic) thallophytes; fungi are nonphotosynthesizing (heterotrophic) thallophytes. Algae have probably always been the primary producers in the oceans, on which nearly all food chains in the sea are ultimately dependent. Green algae (Chlorophyta) are often flagellated, and the thallus may be unicellular, multicellular, or colonial (Figure 19.12). Early land plants probably

TABLE 19.2. A Classification of the Plant Kingdom

Formerly Algae	PHYLUM* CYANOPHYTA—blue-green algae
	PHYLUM EUGLENOPHYTA—euglenoids
	PHYLUM CHLOROPHYTA—green algae
	PHYLUM CHRYSOPHYTA—yellow-green and golden brown algae and diatoms
	PHYLUM PYRROPHYTA—cryptomonads and dinoflagellates
	PHYLUM PHAEOPHYTA—brown algae
	PHYLUM RHODOPHYTA—red algae
Formerly Fungi	PHYLUM SCHIZOMYCOPHYTA—bacteria
	PHYLUM MYXOMYCOPHYTA—slime molds
	PHYLUM EUMYCOPHYTA—true fungi
	PHYLUM BRYOPHYTA—mosses, liverworts, and hornworts
Formerly Pteridophyta	PHYLUM TRACHEOPHYTA—vascular plants
	SUBPHYLUM PSILOPSIDA
	SUBPHYLUM LYCOPSIDA—club mosses
	SUBPHYLUM SPHENOPSIDA—horsetails
	SUBPHYLUM PTEROPSIDA
	CLASS FILICINAE—ferns
Formerly Spermatophyta	CLASS GYMNOSPERMAE—conifers
	CLASS ANGIOSPERMAE—flowering plants

*The Botanical Rules of Nomenclature specify ''Divisions'' rather than ''Phyla.'' However the latter term is used here to parallel zoological usage.

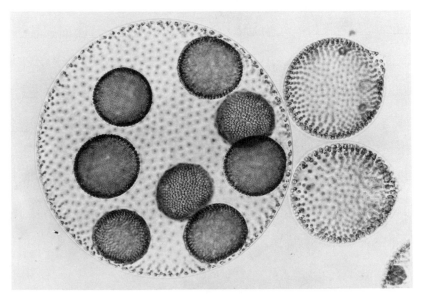

Figure 19.12. A colonial alga, *Volvox aureus.* (Courtesy Carolina Biological Supply Company)

evolved from this group of algae. Red algae (Rhodophyta) have the red pigment phyco-erythrin associated in plastids with chlorophyll. The thallus is normally multicellular. Both the blue-green algae and the red algae lack flagellated cells. The phylum Pyrrophyta includes the cryptomonads and the dinoflagellates. Most of them are unicellular with two unlike flagella, lack cell walls, and have yellow-green to golden-brown pigments in plastids. The phylum Chrysophyta also includes golden-brown algae, yellow-green algae, and diatoms (Figure 19.13). Chlorophyll is present, as always, but its green color is masked by other pigments. The cell walls are typically formed of overlapping halves, like little "pill boxes." Their walls are frequently impregnated with silica. Their fossilized remains form vast deposits of "diatoma-ceous earth" in various parts of the world, and it is often used as an abrasive or as a filtering agent. Brown algae (Phaeophyta) possess the brown pigment fucoxanthin. They are multi-cellular, ranging in size from a few cells to giant kelps over 100 feet in length. They are vegetatively the most highly specialized kinds of algae. There is also commonly an alternation of generations, a characteristic usually associated with higher plants. As expected for an extremely specialized group, they are not considered to be ancestral to any higher plant forms. All the algal phyla probably arose from autotrophic bacteria, but there is no evidence as to whether their origins were monophyletic or polyphyletic. Blue-green algae (Cyanophyta) are the earliest plants known from fossils. Whether the protistan algae evolved from the moneran blue-greens or not is unknown. The fact that both groups contain chlorophyll is suggestive, but by no means conclusive. Chlorophyll in blue-green algae is not confined to chloroplasts as it is in protistan algae. Furthermore, no sexual mechanisms have yet been found in any of the cyanophytes, whereas sex is characteristic of protistan algae.

Unlike the algae, true fungi (phylum Eumycophyta; Figure 19.14), are exclusively

Figure 19.13. Mixed diatomes (phylum Chrysophyta). (Courtesy Carolina Biological Supply Company)

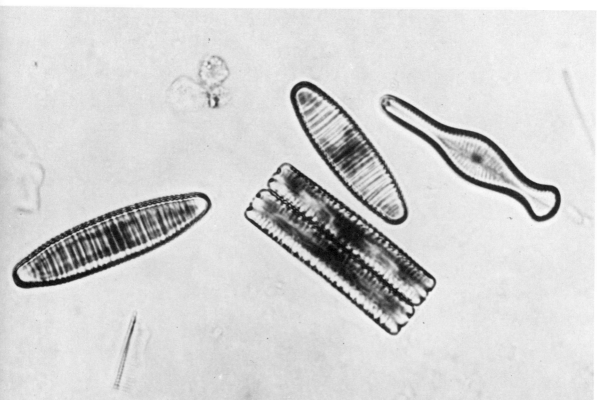

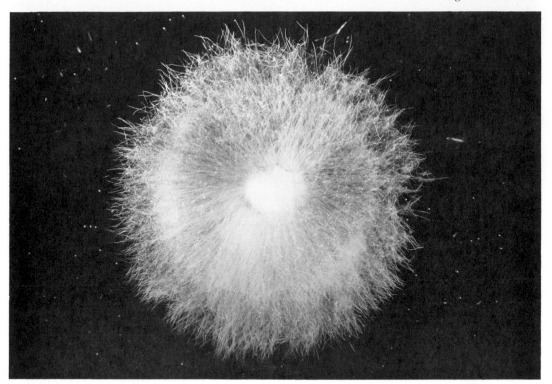

Figure 19.14. A representative of the phylum Eumycophyta (fungi); the water mold *Achlya*. (Courtesy Carolina Biological Supply Company)

heterotrophic and function as decomposers. At least three theories have been proposed for the evolution of the fungi. One theory suggests that they might have arisen from the green algae by loss of chlorophyll. Other theories variously relate them to bacteria, to slime molds (Myxomycophyta), or to Protozoa. The slime molds themselves appear to have some protozoan characteristics (Figure 19.15). For example, they consist of a naked multinucleate protoplasmic mass (plasmodium), move about in amoeboid fashion, and can ingest solid food particles. When conditions are favorable, the plasmodium becomes stationary and forms spore-bearing fruiting bodies (sporangia) in plant-like fashion.

The more advanced plants, called *tracheophytes,* possess conductive tissues. Phloem conducts food throughout the plant principally through sieve tubes, but chiefly serves to move carbohydrates from their site of manufacture in leaves (or in stems) to the rest of the plant. Xylem elements such as tracheids chiefly serve to transport water and minerals from roots in the soil upward to the rest of the plant. Unicellular plants and filamentous algae are too simple to require complex conductive tissues. Their tissues are in intimate contact with the water from which they extract their raw materials. Conquest of the terrestrial environment by plants was no less of a problem than for animals and indeed had to precede that of the animals. Perhaps plants, because of their sedentary nature, had even greater problems in coping with the transition from water to land than did animals. Adaptations for obtaining water from the soil and means for preventing its loss from tissues were undoubtedly closely

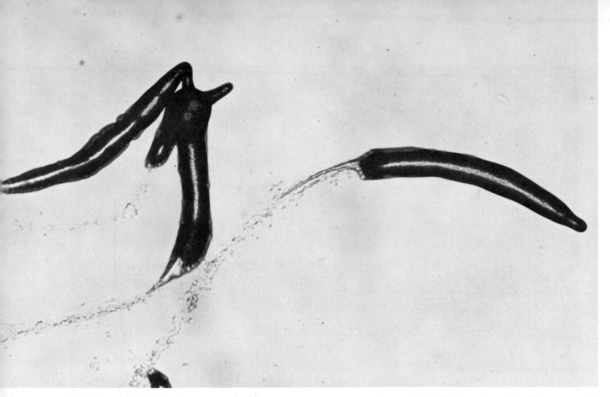

Figure 19.15. Representative of the phylum Myxomycophyta (slime molds); *Dictyostelium discoideum,* a social amoeba. The pseudoplasmodium is composed of separate cells that function as a unit. (Courtesy Carolina Biological Supply Company)

correlated traits in the evolution of terrestrial plants. There is general agreement that the first plants to colonize the land during the middle Paleozoic era were algae.

Bryophytes (liverworts [Figure 19.16] and mosses) probably arose from green algae (Chlorophyta) as did the tracheophytes. Both of these groups of so-called higher plants are termed embryophytes (subkingdom Embryophyta), because their multicellular embryos are retained in the female sex organs; furthermore, there is an alternation of a multicellular gametophyte generation with a multicellular sporophyte generation. In bryophytes, the sporophytic generation is a small, inconspicuous form, parasitic on the larger, independent gametophytic generation. In tracheophytes, however, the situation is reversed; the sporophytic generation is most conspicuous, and the gametophyte is small and ofttimes dependent on the sporophyte. Fossils of vascular plants have been found in Devonian deposits, but bryophytes do not appear until millions of years later in the Carboniferous period. Both groups are thought to have arisen independently of one another. The bryophytes were an evolutionary dead end in the sense that they did not give rise to a diversity of plants better adapted to terrestrial life than themselves. Bryophytes are able to absorb water from the soil by means of cell filaments, called rhizoids, which grow into soil and function as roots. There is an elementary sort of conducting system, but no xylem or phloem tissues. They are similar to thallophytes in that they lack true roots, stems, and leaves, but they also resemble tracheophytes in that they are embryophytes. Here again we see a mosaic of primitive and advanced characteristics. Bryophytes are dependent on water for fertilization, because their motile sperms must swim to the eggs. This reproductive bond with the water reminds us of a comparable

Figure 19.16. Representative of the phylum Bryophyta; *Marchantia* (a liverwort) with sexual stages. (Courtesy Carolina Biological Supply Company)

bond which exists in the amphibians. Neither bryophytes nor amphibians are completely adapted to terrestrial conditions. Among vascular plants, only the *gymnosperms* (mainly cone bearing plants like pines and firs) and *angiosperms* (flowering plants) require no water for fertilization.

The earliest fossils of tracheophytes belong to the phylum Psilophyta (order Psilophytales). These were probably the first plants to be well adapted to a terrestrial life, and it seems likely that all higher vascular plants evolved from them. Curiously, the first air-breathing animals appeared almost at the same time as the Psilophytales. Within the Psilophytales are forms that variously resemble members of all other subphyla of vascular plants, and a few forms even resemble algae in some ways. From the Psilophytales, an adaptive radiation produced, along separate lineages, the club mosses (Lycophyta), and horsetails (Sphenophyta), and the Pterophyta (ferns, gymnosperms, and angiosperms). Today, only the Pterophyta are dominant land plants. But back in the Carboniferous period of the Paleozoic era, pterophytes were absent, and members of the Lycophyta and the Sphenophyta formed great forests of large trees over extensive areas of the earth. Their remains are the chief components of coal. Today, there is only one family and two living genera of the Psilophyta. They are rare plants found only in the tropics. These relics of a bygone era have no roots and the leaves may be tiny, almost scale-like structures. The underground, horizontal stem (rhizome) has numerous rhizoids and sometimes lives symbiotically with a fungus that may aid in uptake of water from the soil. The vascular tissue is simple, and the stems are green with chlorophyll.

Leaves of living members of the Lycophyta (Figure 19.17) are usually simple, small, and spirally arranged. Unlike the psilopsids, they have distinct roots, stems, and leaves. The

Figure 19.17. Representative of the phylum Lycophyta (club mosses); *Lycopodium lucidulum.* (Courtesy Carolina Biological Supply Company)

vascular system in the stems resembles that found in the roots of angiosperms. There are only two families in modern flora.

There is only a single living family and a single genus in the Sphenophyta. Most species have silica in the walls of some of the outer cells. They were used in colonial days to scour pots and pans and hence acquired the name "scouring rushes" (Figure 19.18). Their straight stems possess multiple nodes. The tissue just above these nodes remains *meristematic* (undifferentiated tissue, capable of active cell division) and structurally weak so that stems may be easily pulled apart at these points. Whorls of small, simple, nongreen leaves surround the base of each node. Many branches may arise at each node.

There are three classes in the division Pterophyta: (1) Filicinae (ferns), (2) Gymnospermae (conifers), and (3) Angiospermae (flowering plants). In ferns, the gametophyte is small, but independent of the sporophyte and photosynthetic. Ferns, such as the Psilophyta, the Lycophyta, and the Sphenophyta, require water for their motile sperms. This requirement has largely limited the spread of these organisms into drier areas of the terrestrial environment. No one looks for ferns in the desert! Clubmosses and horsetails also generally prefer moist shady locations. Unlike the leaves of flowering plants, the apical meristem of fern leaves is long lived, producing in some cases leaves nearly 100 feet long. On the underside or the margins of some fern leaves, spores are borne in sporangia. Fossil evidence indicates that ferns

Figure 19.18. Representative of the phylum Sphenophyta (horsetails); *Equisetum telmateia.* (Courtesy Carolina Biological Supply Company)

were the earliest pterophytes. Together with the clubmosses and horsetails, they formed the dominant vegetation of the Carboniferous period. Ferns apparently evolved directly from the Psilophytales and in turn gave rise to the early gymnosperms called "seed ferns" (Figure 19.19). Gymnosperms with a distinct fern-like appearance and seeds attached to the leaves are well represented in the fossil record. Since their seeds were naked, the "seed ferns" are sometimes classified as gymnosperms. There are no "seed ferns" alive today.

Seed plants (*spermatophytes*) have been by far the most successful exploiters of the terrestrial environment. Since their male gametes are not motile, they do not require free water for fertilization. The gametophytes are nonphotosynthetic, microscopic, and dependent on the sporophytes. Their embryos are packaged together with a store of nutrients in a desiccation-resistant seed coat and can remain dormant for extended periods of time until conditions are right for germination. The first fossils of spermatophytes appear in the Carboniferous period, but they probably arose during the Devonian. Two ancient spermatophyte relics survive today. The cycads first appeared in Permian times and became very abundant during the Mesozoic era. Their large palm-like leaves have prompted the name sago palm (Figure 19.20). The palm-like plants often seen in drawings with dinosaurs were sago palms and not true flowering palms. Cycads declined after the rise of angiosperms in the Cretaceous

Figure 19.19. Reconstruction of a seed fern (*Medullosa*) from upper Carboniferous fossils.

Figure 19.20. A sago palm (*Cycas revoluta*).

period. They survive today as nine genera containing over a hundred species. They are fairly common in tropical regions and are strictly dioecious. Their polliniferous cones may superficially resemble the ovuliferous cones of a sugar pine. In the more primitive cycads, the megasporophylls (leaves bearing meiospores that give rise to female gametophytes) have the same outline as vegetative leaves, but are relatively small and covered with dense pubescence. Several large seeds are developed along each leaf margin, reminiscent of the seed ferns. The other primitive spermatophyte is the ginkgo or maiden-hair tree (*Ginkgo biloba*). It is almost unknown in the wild today, but survives mainly as a cultivated lawn tree. Its fan-shaped leaves (Figure 19.21) are divided into two lobes. The sexes are separate (dioecious). The female gametophyte is almost completely surrounded by sporophyte tissue of the ovule. Microspores are carried by air currents from male trees to the mature ovules on female trees. Male gametophytes grow parasitically and release two motile sperms that swim in a fluid produced by the female sporophyte. One sperm fuses with the egg nucleus to form the zygote. Maternal integumentary tissue surrounds the embryo to form a true seed. Most of the common modern gymnosperms are woody plants with needle-like or scale-like leaves and produce naked seeds (not surrounded by an ovary wall) born by cones. Included in this group

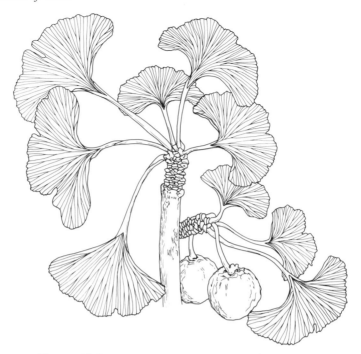

Figure 19.21. Leaves, stem, and seeds of *Ginkgo biloba.*

are the pines, hemlocks, firs, spruces, junipers, yews, redwoods, and many others. Ginkgos are more closely related to modern gymnosperms than are cycads (see Figure 19.22).

The ancestry of the angiosperms is still a mystery. Presumably they arose from early gymnosperms. They are found in fossils of Cretaceous deposits, but no older transitional forms are known. It is believed that the order Ranales (buttercups and magnolias) is most primitive. In the classification of angiosperms, great importance is placed on floral parts because their structure tends to be less subject to modifications by environmental variables than the vegetative parts. According to one popular system of classification developed by Charles E. Bessey, there were three main lines of advance from the primitive ranalian type of flower. These lines led to (1) the mints (Labiatae), (2) the asters (Compositae), and (3) the orchids (Orchidaceae). The major trends in floral evolution are summarized in Table 19.3.

Within the class Angiospermae, there are two major subclasses characterized by the number of leaves possessed by the seedling. *Monocotyledonous* plants have one seed leaf; *dicotyledonous* plants have two. There are several other distinctive differences between these two subclasses. For example, monocots usually have parallel leaf venation and flower parts in groups of three or multiples thereof; dicots tend to have netted leaf venation and flower parts in groups of four or five. Examples of familiar monocots are the grasses; lilies, irises, and orchids; familiar dicots are radishes, cotton, tomatoes, willows, mints, roses, peas, melons, parsley, and dandelions. It is now generally believed that the monocots evolved from primitive dicots. Primitive monocotyledonous water weeds such as the arrowheads and water buttercups have much in common with the primitive dicotyledenous buttercups and marsh marigolds.

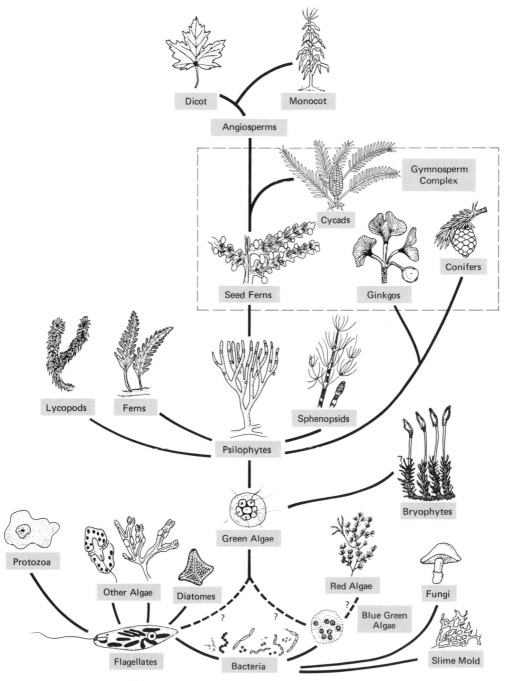

Figure 19.22. A phylogeny of the plant kingdom.

TABLE 19.3. Summary of Major Trends in Floral Evolution

Floral Character	Primitive Form	Advanced Form
1. floral axis	elongated	shortened
2. floral parts	spiral	whorled
3. stamens and carpels	numerous, separate	few, coalesced
4. sepals and petals	numerous, separate	few, coalesced
5. number of parts*	complete, perfect	incomplete, imperfect
6. corolla	regular	irregular
7. position of parts relative to the ovary†	hypogynous	epigynous

* Complete flowers have four organs (sepals, petals, stamens, pistils); incomplete flowers are missing one or more of these organs. Perfect flowers have both pistils and stamens (bisexual). Imperfect flowers possess either pistils or stamens (unisexual).
† Hypogynous flowers have floral parts attached below the ovary; epigynous flowers have floral parts attached near the summit of the ovary.

Broccoli

Brussels Sprouts

Wild Mustard

Kale

Cabbage

Cauliflower

Kohlrabi

Figure 19.23. Varieties of truck crops derived by artificial selection from the European rock cabbage (*Brassica oleracea*).

Under artificial selection by humans, the wild European rock cabbage (*Brassica oleracea*) has been transformed into several varieties of truck crops that appear so different in vegetative parts that they deceive one into thinking that they are at least different species, if not members of different higher categories (Figure 19.23). The wild ancestor looks something like a wild mustard plant. By selecting for edible, succulent branches of the inflorescence, humans have produced broccoli. Kale was developed as a leaf vegetable for greens or chicken feed. Selection for edible stems and leaf bases produced kohlrabi. Cabbage evolved by selection for a single enormous edible terminal bud. Brussels sprouts came from selection for numerous small edible lateral buds. Selection for clusters of edible flower buds and the fasciated (grown together) branches supporting them produced cauliflower. If humans can produce such astonishing varieties by selection within a single species during their short time on earth, could not natural selection, working through vast periods of time, be responsible for the biological diversity we now see around us?

Animal Evolution

Vertebrate evolution was discussed in some detail earlier in this chapter, because this group is better documented in the fossil record than almost any other (shelled mollusks excepted). Much less is known about the evolution of most soft-bodied invertebrates. Nonetheless, the major outlines of animal evolution are fairly clear from studies on comparative anatomy and embryology of living forms. Animals that possess multicellular reproductive structures and develop via distinct embryonic stages are classified as *metazoans*. Protozoans are generally considered to be the group from which the metazoans evolved. Sponges probably evolved from the protozoan collar flagellates, because the interior lining (epithelium) consists of flagellated collar cells. There is no evidence that sponges gave rise to any taxonomically higher group. They represent an evolutionary dead end, but a fairly successful one that still survives today. All other metazoans (eumetazoans) arose from protozoan ancestors (one or more) different from the sponge lineage. Therefore, metazoans are at least diphyletic (i.e., evolved from at least two different protozoan ancestral sources).

There are several theories on the origin of eumetazoans. One theory proposes that they arose from ciliated protozoans. Modern ciliates are multinucleate, and the ancestral ciliates may have become multicellular by internally partitioning nuclei into uninucleate cells. Motile, multinucleate cells with this partitioning ability are called *coccine* types. The first metazoans thus formed are thought to have been bilateral *acoel* (i.e., without a body cavity or *coelom*) flatworms. One line of evolution from the acoels produced the radiates (animals with radial symmetry; coelenterates and ctenophores), and another line produced the higher bilaterate animals such as the roundworms (with a complete digestive system). Modern free-living acoel flatworms have a ciliated epidermis with a single opening into the digestive tract. But the formation of partitions or *septa* in modern ciliates is completely unknown. Multicellularity is not known to occur by septation in any modern plant or animal. Some fungi and algae are multinucleate, but septation in them is primarily a device to disperse cells, not produce a multicellular organism. The multicellularity that does exist in these forms never reaches the tissue level of organization. On the other hand, acoel flatworms and higher eumetazoans do possess tissues. Consequently, the "ciliate-flatworm hypothesis" does not have many supporters.

A more popular theory proposes that flagellate protozoa (with amoeboid tendencies) became aggregated into a multicellular colony (either a hollow colony such as the algal *Volvox* or a solid colony such as the algal *Synura*). Most modern coelenterates develop a solid morula (stereoblastula), and some develop a hollow coeloblastula, but eventually they both become a flagellated, solid stereogastrula called the *planula larva.* According to this "flagellate-coelenterate hypothesis," the first eumetazoans were coelenterates. Later acoels and other bilateria evolved along a separate lineage from the planula-like ancestors. All modern metazoans develop either flagellate or amoeboid gametes, as well as other motile cell types. Furthermore flagellate-amoeboid protists are known to form multicellular colonies by aggregation. However, like the "ciliate-flatworm hypothesis," no flagellate-amoeboid organism is known to be capable of attaining a tissue level of organization.

Some protists are nonmotile, uninucleate forms (*sporine* condition) known to be capable of evolution beyond the colonial level. No sporine types exist today in the Protozoa or in any other colorless protists. Therefore, if sporine forms were to be ancestors of the metazoans, they would have to have been early colorless preprotozoan protists. Being saprotrophic or holotrophic, they could have been solid nonmotile morula-like aggregates typical of many modern eumetazoan embryos. Later, at least some of the exterior cells of these aggregates might have acquired flagella and thus come to resemble modern planuloid larvae. In support of this "sporine theory," some modern sporine algae are known to become flagellate. This theory is attractive, because it proposed that both metazoans and protozoans evolved in parallel from generalized primitive early colorless protists rather than from more advanced specialized ancestors. If sponges evolved from more advanced flagellated protozoans, they might not have appeared until after the eumetazoan lineage was well established. Most of the currently popular theories of early metazoan evolution favor the idea of solid multicellular aggregates, with locomotor cells on the exterior surface. This stands in sharp contrast to the Haeckelian concept of hollow aggregates like *Volvox.*

If all eumetazoans had a monophyletic origin from planula-like ancestors, then what symmetry is most primitive? According to the "coelenterate hypothesis," the first eumetazoans had radial symmetry and the bilaterally symmetrical acoel flatworms developed from them. An alternative "flatworm hypothesis" proposes that bilateral acoels were primitive and the radial symmetry of coelenterates was secondarily derived. Since the embryos of all modern coelenterates and acoels show no sign of bilateral symmetry, most zoologists accept the proposition that radial symmetry was the primitive condition and bilaterality evolved secondarily. Of course, it is also possible that eumetazoans did not have a monophyletic origin. Perhaps radial coelenterates and early bilateral flatworms arose as independent lineages from early protistan ancestors. There is still no general agreement among zoologists on the origin of the first Bilateria. However, it is generally accepted that acoel flatworms gave rise to all other Bilateria.

Gastrulation forms, among other things, the beginning of a primitive gut (*archenteron*). The opening of the archenteron is called the *blastopore.* The fate of the blastopore divides the bilaterate metazons into two major developmental groups; in the *Protostomia,* the blastopore becomes the mouth; in the *Deuterostomia,* the blastopore becomes the anus. Adult protostomes generally share additional features in common: a ventral, ladder-type nervous system, protonephridial and metanephridial excretory systems, and main blood vessels or hearts in dorsal locations.

There are three developmental subdivisions based on the type of body cavity: (1) acoelomate, (2) pseudocoelomate, and (3) eucoelomate. Animals without a body cavity are acoelomates. Those that possess a body cavity lined with mesoderm have a true coelom and are called *eucoelomates*. *Pseudocoelomates* have a false body cavity (pseudocoel) that is not lined with membranes, although patches of mesoderm may be found in the pseudocoel. Flatworms (phylum Platyhelminthes) and ribbon worms (phylum Nemertina) are classified as acoelomates. Rotifers, roundworms, and hairworms (phylum Aschelminthes) and entoprocts (phylum Entoprocta) are pseudocoelomates. All other major bilaterate metazoans are eucoelomates.

The coelom in the eucoelomates develops by two major modes: (1) schizocoely and (2) enterocoely. A *schizocoel* forms by a split in the mesoderm layer; an *enterocoel* forms by paired lateral pouches growing out from the gut. Among the schizocoelomates are mollusks (clams, snails, chitons, octopus, etc.), annelids (earthworms, leeches, clamworms, etc.), and arthropods (crustaceans, insects, centipedes, spiders, horseshoe crabs, etc.). Echinoderms (starfish, brittle stars, sea urchins, sea cucumbers, etc.), hemichordates (acorn worms), and protochordates (tunicates, amphioxus) are enterocoelomates. A group of less-common invertebrates (phoronids, ectoprocts, and brachiopods) have various other patterns of coelom formation, which have not been given any special names. Schizocoelomate protostomes typically develop from *trochophore* larvae. Deuterostomes are postulated to have evolved from protostomes through a hypothetical *dipleurula* larval ancestor. The dipleurula larva bears a strong resemblance to both the *tornaria* larvae of hemichordates and to the *bipinnaria* larvae of starfish (Figure 19.24).

Although primitive chordates and echinoderms share many similar embryological features, modern adult chordates appear quite different from their echinoderm cousins. For one thing, adult chordates have bilateral symmetry, whereas modern echinoderms show it only in the larval form (however, fossil echinoderms are known to be bilateral as adults). Furthermore, chordates are distinctively segmented, whereas echinoderms are not. For these and other reasons, an annelid ancestry had been earlier proposed for the evolution of the chordates. Annelids were bilaterally symmetrical, segmented, with a true coelom. If the annelid body is turned upside down, then the nervous system would come to lie dorsal to the major

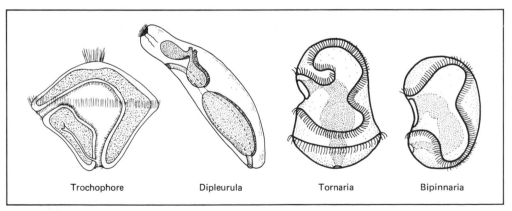

| Trochophore | Dipleurula | Tornaria | Bipinnaria |

Figure 19.24. Larval forms. The hypothetical dipleurula larva is considered to be ancestral in deuterostome evolution; its coelomic cavities are shaded.

blood vessel just as it does in the chordates. But the "annelid theory" fails to account for the fact that not all chordates are segmented (e.g., tunicates). Furthermore it ignores many developmental differences between modern annelids and vertebrates: mosaic vs. regulative eggs, spiral vs. bilateral cleavage, protostomial vs. deuterostomial developmental patterns, teloblastic vs. enterocoelous mesoderm, and trochophore vs. tadpole larvae. There are few supporters of the "annelid hypothesis" today. One thing is fairly clear; segmentation appears to have evolved at least twice independently, once in the protostomial annelid-arthropod group and once in the deuterostomial chordate group.

Among invertebrates, only the annelids and arthropods are major groups with segmentation. It is probable that arthropods evolved from early annelid ancestors. Organisms intermediate between annelids and arthropods are now known. They belong to the phylum Oncopoda (subphylum Onychophora) and are commonly called *Peripatus* (Figure 19.25). There are about seventy species of Peripatus extant. They prefer to live in moist, dark places. They have been called "the missing link" between annelids and arthropods, but probably should be thought of as a third line of descent from the ancestral stock that produced the modern annelids and arthropods. They bear a resemblance to annelids in their segmentally arranged excretory nephridia, ciliated reproductive ducts, muscular body wall, soft cuticle, short unjointed appendages each bearing a pair of claws, lack of a well-defined head, and simple eye structure. They resemble arthropods in their tubular heart with ostea (openings), tracheal respiratory system, mouth parts modified from appendages, reduced hemocoel (body cavity through which blood percolates), dorsal heart, large brain size, and paired antennae.

Mollusks are generally considered to be unsegmented. In 1952, a few specimens of a living species (*Neopilina galatheae;* Figure 19.26) were dredged from two miles deep off the west coast of Mexico. It has been placed in a new taxonomic class Monoplacophora ("bearing a single plate"), because it is covered by a single limpet-like shell. The body is bilaterally symmetrical with a broad flat foot. There are five pairs of dorsoventral muscles associated with the foot. There are five pairs of small gills, each associated with a nephridium (excretory structure), one member of each pair situated on either side of the foot. The body itself consists of five segments, three in the head. There are also five pairs of gill hearts. This is clear evidence of internal segmentation. A more recently discovered species (*N. ewingi*) has six pairs of gills. These recent discoveries suggest the possibility that ancestral mollusks were segmented and modern forms have secondarily lost the segmented condition. If this is true, then the mollusks would be closer to the annelids than had been previously suspected.

It is currently popular to view deuterostomes as having evolved from early coelomate protostomes that were ancestral to the primitive *lophophorates* (a lophophore is a tentacle-bearing arm in the anterior region of certain coelomates, that serves in food gathering). In this group, various methods of mesoderm and coelom formation are employed, and one of these

Figure 19.25. *Peripatus.* (Courtesy Carolina Biological Supply Company)

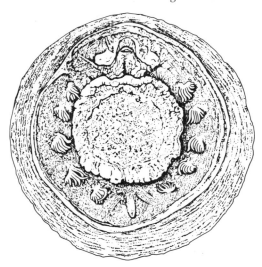

Figure 19.26. *Neopilina galatheae,* an extant mollusk with internal segmentation in several body parts; five pairs of gills are easily seen.

lines could have been enterocoelous leading to the chordates. A lophophorate ancestry would, of course, require a reversal of anteroposterior polarity to convert a protostomial into a deuterostomial embryo.

Just as there is no universal agreement among botanists concerning the evolution of the Metaphyta (higher plants), there is no universal agreement among zoologists concerning the evolution of the Metazoa. An attempt is made in Fig. 19.27 to depict the broad outline of possible and probable phylogenetic relationships among major groups of animals.

Evolutionary Trends

Perhaps now is a good time to summarize some of the more important generalizations about evolutionary tendencies that have been encountered in the textual discussions thus far. One of the most fundamental principles, obvious on *a priori* grounds, is the fact that evolution along any specific lineage is essentially irreversible. This has come to be known as *Dollo's law.* Now we know that domestic animals, for example, have been evolved under artificial selection for certain characters of economic importance to humans. It is commonly observed that when domestic animals escape back into the wild, natural selection begins to move the gene pool of the feral group back toward the adaptive peak for natural (rather than man-made) environmental conditions. Feral animals thereby tend to revert to a phenotype that resembles the ancestral forms from which they were domesticated (*atavism*). Relatively few genetic changes have occurred in domestic stock compared to the changes nature has wrought over eons of geological time. Therefore, it is theoretically possible for feral animals to become exactly like their ancestors, but in reality this is extremely unlikely to happen. In order for any evolutionary lineage to retrace its path, there would have to be exact reversals every evolutionary step of the way, including selection intensities, gene mutations, gene flow, population size, breeding systems, structural changes in chromosomes, etc. Of course, if any genetic material has been

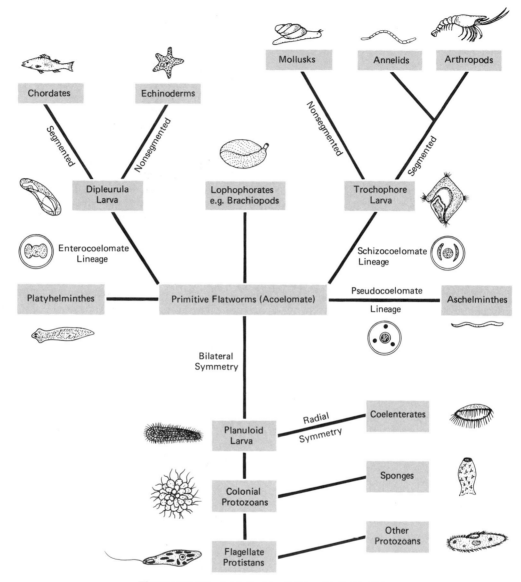

Figure 19.27. A phylogeny of the animal kingdom.

lost in the production of the domestic line, it would be essentially impossible to regain precisely that same information by chance. It is obvious that there are so many evolutionary steps to retrace that it is not possible in practice to go back along a lineage and arrive at exactly the same population base from which it began. Now if this is true for evolution within a species, it must certainly also be true for splitting of one ancestral group into different species and for the evolution of higher categories. Given any two isolated finite populations with exactly the same genetic and population structures, they would, in time, be expected to evolve distinctive differences, because evolution is often a stochastic rather than a deterministic process and chance events in any two populations are highly unlikely to be repeated exactly in both.

> Some reptiles, such as the extinct ichthyosaurs, became adapted to an aquatic way of life. The same is true for mammals such as whales and porpoises. Even some birds (e.g., penguins) have acquired adaptations for flying in water rather than in air. They all have evolved streamlined fusiform body contours that slip through the water with relatively little resistance (convergent evolution). But these external changes belie the skeletal structures typical of the group to which each of these animals belong. They obviously have not retrogressed back to a fish-like genotype. From different starting points it is possible to derive analogous features and structures that bear a superficial resemblance. However, too many genetic changes have occurred to allow such widely divergent groups as reptile, mammal, and bird to ever reverse the evolutionary process and return to the genetic condition of their fish-like ancestors. Similarly, many angiosperms have invaded water habitats and acquired superficial algae-like appearances, but their internal morphologies and life cycles remain distinctively that of tracheophytes, not thallophytes. Once major evolutionary steps are taken, they represent unique, unrepeatable events (Dollo's law).

A second major evolutionary principle is that of extinction. The overwhelming majority of species in existence only a few million years ago do not survive today. Extinction seems to have been the fate of most species since the origin of life. It is through extinction of intermediate groups that higher taxonomic categories become more easily defined. Among plants, the angiosperms are presently the most actively evolving group. Consequently many mosaic types in that division create problems for the taxonomist. The same is true of the actively evolving rodent order in the mammalian class. Extinction may arise as a consequence of phyletic evolution, where one species, in time, gradually acquires so many genetic changes that it becomes transformed into a new species. As an example, primitive mammals replaced the ictidosaurs, which in turn replaced the advanced theriodont reptiles, etc. In this case, the extinction of the old forms was probably due to the adaptive superiority of the descendants who replaced them. Much more common than phyletic replacement is extinction of forms that left no descendants. The demise of the mighty dinosaurs at the end of the Cretaceous period may be a case in point. One theory suggests that the more active "warm-blooded" mammals were better able to exploit the environment than the "cold-blooded" reptiles and thereby replaced many of them. In such cases, extinction of one group is a consequence of evolution in the replacement group(s). Although many species, genera, and even families have suffered extinction, most of the major adaptive types, represented by taxonomic classes and phyla, have survived.

> Robert T. Bakker has recently suggested that the dinosaurs were not "cold blooded" and that they have not died out but live on in a modified form called "birds." He presents three lines of evidence to support the theory that mammal-like reptiles (order Therapsida) of the late Permian had relatively high heat production (endothermy): (1) bone histology, (2) latitudinal zonation, and (3) predator-prey ratio. The compact bone of modern ectothermic reptiles is characterized by seasonal growth rings in the outer layers, analogous to growth rings in the wood of trees. Furthermore, bones of modern reptiles have a low density of blood vessels and few Haversian canals. In contrast, the bones of endothermic mammals and birds almost never have growth rings;

they are rich in blood vessels and Haversian canals. Large ectothermic reptiles today are probably confined to warm equatorial zones largely because they cannot warm up to an optimal body temperature during short winter days elsewhere. The distribution of large reptilian fossils during Permian times is therefore expected to be correlated with Permian temperature zones. Some fossil deposits contain thousands of individuals representing a single community. Estimates of the biomass of predators (carnivores) and prey species (herbivores) can be obtained therefrom. The predator-prey biomass ratio is a constant characteristic of the metabolism of the predator, irrespective of the body size of the animals in a predator-prey system. Endothermic predators require a "standing crop" of prey roughly an order of magnitude more than ectothermic predators. All early therapsids and dinosaur species investigated thus far have fully endothermic bones. Some rhinoceros-sized lizard fossils have been found in rocks within the Cretaceous Arctic Circle. Calculations of some Triassic dinosaur predator-prey ratios are 3 per cent or less (in the same range as modern mammals and birds). Bakker finds the evidence from all three indicators overwhelming support for the theory of dinosauran endothermy. Mammals did not really begin an adaptive radiation until after the disappearance of the dinosaurs (birds excepted). Bakker therefore argues against the notion that the dinosaurs disappeared because they could not compete with the more energetic endothermic mammals. The first evidence of feathered reptiles is in the Upper Jurassic fossil *Archaeopteryx*. Its shoulder joints were identical to those of carnivorous dinosaurs and adapted for grasping prey rather than for flight. Bakker suggests that feathers evolved originally as insulation to aid endothermy; much later they became useful preadaptations for flight when the shoulder joint became modified for that function during the Cretaceous period. Birds have been traditionally thought to be descendent from early thecodonts, not from dinosaurs. Endothermy (shaded area in Figure 19.28) is traditionally thought to have evolved late in the development of birds and mammals. Bakker believes it is time to reclassify the land vertebrates primarily on the basis of bioenergetic considerations (see Figure 19.28).

A third generalization concerns the dendritic (branching) schemes used to depict divergent evolutionary lines. They are sometimes called "tree-like" diagrams; but a better description would be "bush-like," because major adaptive shifts seem to have occurred, not near the tips of branches as in trees, but rather near the base of the major stems as in a shrub. In other words, quantum evolution is more likely to have its origins in relatively unspecialized (generalized) primitive members of a group rather than from highly specialized (narrowly adapted) advanced members of the group. We have seen that mammals did not evolve from advanced reptiles like the dinosaurs, but rather from generalized stem reptiles, very close to the transitional forms between amphibians and reptiles. Furthermore, we know that some degree of hybridization (introgression; allopolyploidy) has probably been extremely important in evolution of the higher plants (though of little or no consequence in the evolution of higher animals). Therefore, the evolutionary bush should, at some points, have fused branches. The cladogram thereby becomes somewhat reticulate (net-like), reflecting to some degree the extreme complexity of the situation.

A fourth principle of evolution is the fact that organisms do not always progress in complexity. Although the bulk of evidence indicates that complexity generally confers greater resourcefulness in utilization of resources and exploitation of a niche, this is not invariably necessary. Sometimes degenerate forms have carved out successful niches for themselves in competition with more complex organisms. Most examples of *retrogressive* or *degenerative*

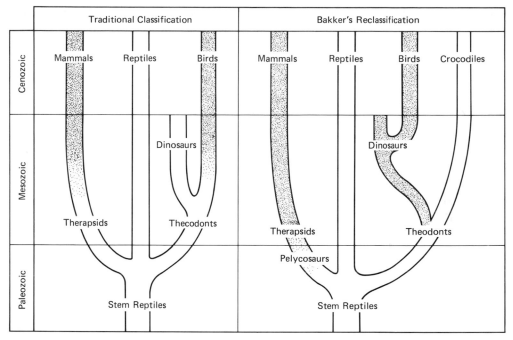

Figure 19.28. Bakker's reclassification of amniotic vertebrates (right) based on bioenergetics compared with a more conventional classification (left).

evolution involve parasites (evolved from free-living ancestors) or neoteny. Reduction in the number of parts has often been of adaptive advantage in both plants and animals. For example, the sacrum of mammals is formed by the fusion of three to five sacral vertebrae. This structure forms a more solid base for the attachment of the pelvic girdle. Many flowering plants have reduced the number of separate petals by fusing them into a corolla tube, as in the lilies.

Modern tapeworms are intestinal parasites that undoubtedly evolved from free-living flatworms by retrogressive evolution. The free-living ancestors probably had well-developed digestive, nervous, reproductive, muscular, and other systems, now lacking in tapeworms. These parasites are essentially an absorptive sac of gonads. Nonetheless, they are an extremely successful group today. Wingless fleas are believed to be derived from winged ancestors. Mistletoe is an example of an angiosperm, parasitic on certain trees, that has also evolved from free-living ancestors. The fungi may have evolved from algae through the loss of chlorophyll (and hence loss of photosynthetic ability). Grasses probably evolved from lily-like ancestors by reduction of parts, especially in the flowers. Some extremely simple marine worms (class Archiannelida) lack parapodia (segmental appendages) and setae (stiff chitinous bristles). They lack the external segmentation characteristic of the phylum Annelida to which they belong (although they have retained internal segmentation). It was formerly believed that they were the most primitive members of this phylum. Currently, however, they are considered to be specialized worms that have become symplified from a polychaete (sandworm) level of organization by neoteny. That is, they have apparently retained larval (trochophoral) traits as adults, including absence of external segmentation, presence of cilia on the ventral side of the body, persistence of the nervous system as part of the epidermis, absence of parapodia and setae, etc.

A fifth evolutionary trend is one toward increasing size of individuals, called *Cope's Law* (Figure 19.29). This trend was originally observed in vertebrates, but has also been subsequently noted in many groups of invertebrates and plants. The fossil record of almost any group reveals that the largest members were not the oldest ones, although not necessarily the youngest ones either. It seems that at present some members of the vertebrates (e.g., whales), crustaceans (king crab), echinoderms (starfish), pelecypods (giant clams), cephalopods (squid), etc. are larger than any representatives known from fossils. This trend is far from universal, however. The largest land vertebrates were the dinosaurs (some up to nearly ninety feet in length and perhaps almost fifty tons in weight). The largest reptiles today (crocodiles, turtles) are much smaller. Probably the oldest angiosperms were trees; herbs and shrubs are usually much smaller than trees and are inferred to have evolved from tree-like ancestors. An *herb* is an annual plant with a soft stem. *Shrubs* and *trees* are perennial woody plants. A tree has a single main trunk with branches only on the upper part of the plant. A shrub has a number of major branches arising from ground level (but no main trunk). Generally speaking, larger

Figure 19.29. Edward Drinker Cope (1840–1897). Zoologist, paleontologist. (From the *Dictionary of American Portraits,* Dover Publications, Inc. 1967)

organisms have fewer predators with sufficient size to prey on them as adults. Perhaps this is part of the explanation for Cope's law.

Although evolution is usually characterized by gradual changes through time, this does not necessarily mean that the rates of evolution have always been uniform. This is a sixth basic principle of evolution. From the standpoint of contemporary time, all organic evolution appears to be slow. During long periods when environments were fairly stable, evolution in most lines was probably slow; stabilizing selection would be expected to maintain populations on their adaptive peaks. These quiescent times would be characterized by relatively slow (*bradytelic*) evolutionary rates. Organisms that are broadly adapted (relatively unspecialized) for a rather generalized mode of life in variable environments would also tend to exhibit bradytelic evolution. "Living fossils," such as the brachiopod *Lingula,* cockroaches, silverfish, horseshoe crabs, and the opossum, have apparently changed very little over millions of years (Figure 19.30). Probably the most stable environment is to be found in warm shallow seas. When major geological changes (revolutions) occurred (as they did during times separating geological eras, periods, and epochs), the accompanying environmental changes caused

Figure 19.30. Some "living fossils"; examples of extremely slow rates of evolution (bradytelic). Figures are not drawn to the same scale.

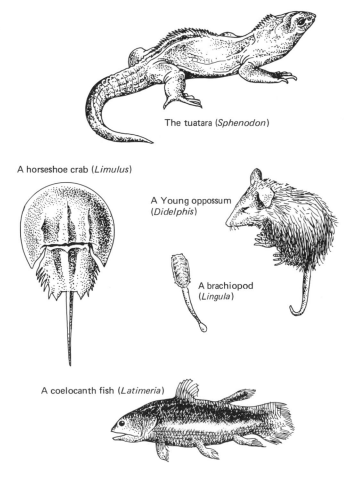

The tuatara (*Sphenodon*)

A horseshoe crab (*Limulus*)

A Young oppossum (*Didelphis*)

A brachiopod (*Lingula*)

A coelocanth fish (*Latimeria*)

pronounced alterations in many adaptive landscapes. Directional or disruptive selection would be expected to play paramount roles in altering gene pools during such times. This is one explanation for the sudden demise of certain groups and the equally abrupt appearance of new species during times of great geological activity. At present, many segments of the biotic world appear to be in rapid (*tachytelic*) evolution perhaps spurred by recent widespread glacial activity and/or by human activities. Rapid evolution seems to be characteristic of narrowly specialized organisms or those in transition from one major adaptive zone to another. Flying reptiles (pterosaurs) and birds seemed to have both made the transition from nonflying ancestors very quickly. The emergence of humans also seems to have involved tachytelic evolution. Over long periods of geological time, moderate (*horotelic*) rates of evolution are commonplace. A classic example of horotelic evolution is seen in the rise of the modern horse from *Eohippus* (Hyracotherium) over a moderate span of 55 to 60 million years. Here again we also see a trend toward increase in size (Cope's law).

Essay Questions

1. *What are the "higher taxonomic categories," how are they evolved, and how do they become well defined?*

2. *How may major adaptive shifts be accomplished without passing through "nonadaptive" zones on the fitness landscape?*

3. *Explain what is meant by an "adaptive radiation?" What conditions predispose a group to an adaptive radiation?*

4. *Outline the major steps in the evolution of (a) plants and (b) animals.*

5. *Discuss at least three general evolutionary trends that apply to both plants and animals.*

MAN

Ascent of the Hominids

Most people are naturally curious about their own origins, and their interest in the theory of evolution (if any) often tends to be kindled by a desire to know more about their own biological heritage. Whenever good human examples of genetic or evolutionary principles were available, an attempt has been made to use them in previous chapters. Much more frequently, however, the best examples are known in other organisms. This is certainly true of paleontological evidence. The fossil record of human evolution is highly fragmentary, and many gaps remain to be filled before we can retrace with confidence our own gens. Scientists are trained to be as objective as possible in the evaluation of data. However, it is usually much easier to be objective about other organisms than it is about ourselves or our own immediate ancestors. The study of man (*anthropology*) tends to be more highly emotionally charged (and therefore perhaps less critically objective) than other disciplines. With this admonition in mind, let us proceed to review some of the major anthropological fossil discoveries and the theories of how they might be related to the evolution of modern humans.

Modern man (*Homo sapiens*) is classified in the kingdom Animalia, phylum Chordata,

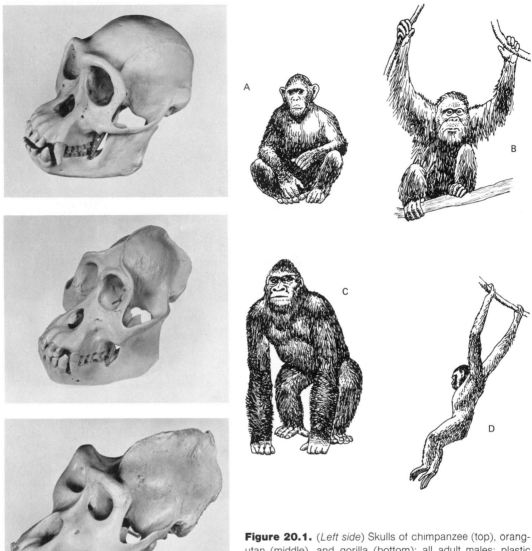

Figure 20.1. (*Left side*) Skulls of chimpanzee (top), orangutan (middle), and gorilla (bottom); all adult males; plastic reproductions. (Courtesy Carolina Biological Supply Company) (*Right side*) Sketches (not drawn to exact scale) of modern anthropoid apes. (A) chimpanzee; (B) orangutan; (C) gorilla; (D) gibbon.

class Mammalia, order Primates, and family Hominidae. Today humans are the only living species in this family. Our nearest primate relatives are in the family Pongidae, which includes the anthropoid (man-like) apes. There are only four living genera of pongids: gibbons, orangutan, gorilla, and chimpanzee. Structurally and biochemically, the chimpanzee is probably closer to humans than any other ape, although the gorilla also appears to be a near relative (Figure 20.1).

Primates

What are primates and whence did they come? Unlike most orders of vertebrates, the order Primates has no singular characteristic distinguishing all its members. For example, rodents (order Rodentia) have a pair of continuously growing gnawing incisor teeth separated by a wide gap from the molar teeth. Horses and other members of the order Perissodactyla have an odd number of hoofed toes. Bats (order Chiroptera) have forelimbs modified for flight, etc. But primates are distinctive mainly by their lack of specialization (generalized body plan). Both hands and feet are still pentadactyl (five-digits) as in primitive mammals. These digits are capped by flattened nails rather than claws. Their teeth are relatively unspecialized for an omnivorous diet. They have four groups of teeth: incisors, canines, premolars, and molars. The thumb and great toe are commonly opposable and capable of grasping objects. Primates possess collarbones or clavicles. There are rings of bone, or orbits, around the eyes. The eyes are usually toward the front of the head, rather than on the side of the head as in most mammals. This allows the field of vision of each eye to overlap to some degree giving binocular, stereoscopic vision. This visual system is adaptive for animals that spend much of their time in trees; depth perception is obviously important for brachiation or jumping from one branch to another. Development of the brain is most marked in the primates.

Four general levels of organization, or grades, are recognized in the primates: (1) prosimians, (2) platyrrhine monkeys, (3) catarrhine monkeys, and (4) hominoids. The word *simian* pertains to monkeys or apes; hence, the *prosimians* are a lower organizational level than the simian groups. Among the prosimians are the lemurs, lorises, aye-ayes, galagos, pottos, and tarsiers (Figure 20.2). Prosimians generally have long muzzles and furry faces that appear more fox-like than those of the other primates. The word *platyrrhine* refers to a wide nasal cartilage between the nostrils. Platyrrhine monkeys, or ceboids (after their principle genus, *Cebus*), are found today in South and Central America, and hence are called New World monkeys. Most of the larger ceboids have grasping (prehensile) tails. They probably descended from Eocene lemurs whose remains have been found in North America. They are considered more primitive than catarrhine monkeys, but evolved separately in parallel with them, rather than being ancestral to them. Among the platyrrhine monkeys are the howlers, woolly and spider monkeys, capuchins (organ grinder's monkeys), squirrel monkeys, and marmosets. *Catarrhine* monkeys have a narrow nasal septum and are native to Africa and Eurasia; hence, they are called Old World monkeys. They do not possess prehensile tails. There are two major ecological groups. Arboreal forms include the guenons, the colobus, and mangabeys of Africa and the langurs of Southern Asia. Macaques and baboons are more terrestrial forms. Humans and the great apes are classified as hominoids. Since most of the early hominoid fossils have been found in Africa, it is believed that humans and contemporary apes had their origins among a common ancestor of the Old World.

Shrews, hedgehogs, and moles belong to the order Insectivora and are the most primitive living mammals. It is from insectivore-like ancestors that both marsupial and placental mammals are thought to have arisen independently during the Cretaceous period. Shrews are small, sharp-snouted animals with little dental differentiation. They feed mainly on insects and spend much of their lives underground. Tree shrews are considered, by some zoologists,

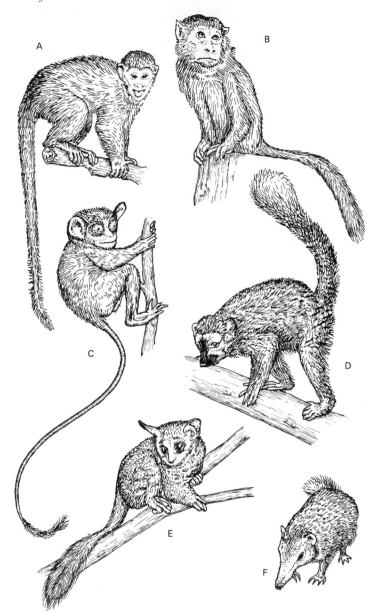

Figure 20.2. Some representatives of the lower primate groups prosimians and monkeys (not drawn to same scale). (A) squirrel monkey; (B) macaque monkey; (C) tarsier; (D) lemur; (E) galago; (F) tree shrew.

the most primitive of living Primates; others classify them among the Insectivora. However, they differ from insectivores in several respects. The digits are more flexible, and the thumb and great toe are opposable. They still retain claws on the digits rather than nails. Their eyes are larger, but their olfactory sense is diminished. Today they are found only in the Oriental Realm. Lemurs are unquestionably prosimians, but are little advanced over the tree shrew-like ancestors from which they probably arose. Lemurs have well-developed hands. Most of their digits have nails, but the second toe retains a primitive claw. Some fossil tarsiers from the Eocene epoch had 44 teeth, as did primitive insectivores; others had reduced the number to 32, characteristic of higher primates. They were the earliest known primates to develop bicuspid premolars. Some of these tarsiers also had square-shaped molars with four cusps, one in each corner of the tooth. This trait is characteristic of all higher primates. It is not known whether these tarsiers evolved directly from tree shrews or from very primitive lemurs. From these primitive primates, we gain some insight into the kinds of ancestral transitional forms that probably gave rise to the modern primates.

Early Anthropoids

Perhaps the earliest remains of anthropoid apes are those of two mandibles found in rocks of Oligocene age in Egypt. The dental formula was characteristic of anthropoids: in each half jaw were two incisors, one canine, two premolars, and three molars. The canines were the same size as the other teeth rather than enlarged as in prosimians and modern apes. Since the premolars resembled those of tarsiers, it is suggested that perhaps apes evolved from tarsoid-ancestors independent of the monkeys. These primitive Oligocene apes have been assigned to two genera, *Parapithecus* and *Propliopithecus,* the latter being larger and more specialized and generally regarded on the line of descent toward modern apes and possibly humans.

Oligocene ape fossils are extremely rare, but by the ensuing Miocene epoch apes were comparatively widespread and common. They ranged in size from that of a small gibbon to a gorilla, but were of much lighter build than modern apes. Their arms were not specialized for brachiation as are those of modern apes. Their generalized body structure suggests that they could very well have been the ancestral group from which modern apes and humans evolved. It used to be fashionable for anthropologists to give each new fossil find a different specific, if not generic, name. A tendency to overemphasize the importance of individual variations is characteristic of taxonomic "splitters." As a result, the literature on fossil hominids is replete with so many names that it has become a taxonomic mess! Recently, a trend toward recognizing similarities, rather than differences, has resulted in lumping together many specimens under common generic labels. For example, most of the Miocene pongids have been reclassified into the genus *Dryopithecus* ("oakwoods ape"). One of the most complete fossil anthropoids (including much of the skull and face) formerly called *Proconsul* is now reclassified as a dryopithecine (Figure 20.3). Five cusps are on the molar teeth of dryopithecines, arranged in a pattern called Y-5 (Figure 20.4). Although the Y-5 pattern is characteristic of later hominids, the canine teeth were large and the shape of the jaw was still rectangular in dryopithecines, just as it is in modern apes. Proconsul cannot be called man like, but was actually more so than are modern apes.

In addition to the dryopithecines, three other genera of ape-like primates are known from Miocene and Pliocene times. One of these, called *Ramapithecus,* appears to be very close to the direct ancestors of modern humans. The palates of ramapithecines were definitely hominid in shape; i.e., parabolic, rather than rectangular as in apes. Furthermore the canine teeth were much reduced in size and did not require a *diastema,* or gap, in the dental series for interlocking canines as they do in apes (Figure 20.5). One possible explanation for the

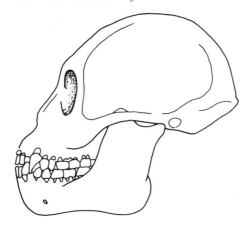

Figure 20.3. The Miocene dryopithecine *Proconsul.*

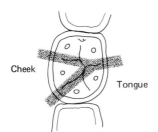

Cheek

Tongue

Figure 20.4. The Y-5 tooth pattern typical of the dryopithecines and later hominids (except those in which cusp reduction has occurred).

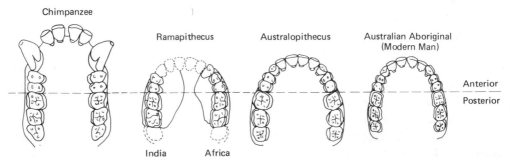

Chimpanzee

Ramapithecus

Australopithecus

Australian Aboriginal (Modern Man)

Anterior

Posterior

India   Africa

Figure 20.5. Palates (upper dental arcades); drawn to approximately the same scale. Note that in the chimpanzee (and most pongids) the anterior region (including incisors, canines, and premolars) is larger than the posterior region (molars). Also note the typical elongated, rectangular shape of the pongid palate, and the smoothly rounded (parabolic) shape of the hominid palates. In the hominids, the canine teeth are about the same height as the other teeth, and there is no space (diastema) in the dental series.

reduction in size of the canines involves the assumption of a dietary shift. In order to grind harsh food, primitive man may have found it advantageous to reduce the canines so that the lower jaw could more freely rotate its molars against those of the upper jaw. It is postulated that relinquishing canine functions (used for fighting and piercing rinds of hard fruits) could only follow the freeing of hands for tool use; this, in turn, requires a shift to some degree of bipedal posture and locomotion. It will be extremely interesting to see whether these predictions will be fulfilled when the pelvic bones of ramapithecines are found.

Australopithecines

What we know of Ramipithecus is based on one fragment of a maxilla from India and another from East Africa. A much larger number of fossils belong to a primate group containing some members difficult to distinguish from Ramapitheus. The first member of this group was found in 1924 and named *Australopithecus africanus* by Professor Raymond Dart. The name "australopithecine" refers to "southern ape," because the first find was in South Africa. Some members of the australopithecine group are dated at $3\frac{1}{2}$ million years old. Dart's original specimen was a skull of an infant (probably female, about six years old) from a limestone quarry at Taung, South Africa. From the pelvic bones of australopithecines, they most assuredly had bipedal posture (Figure 20.6). The palate was parabolic, and the canines were reduced in size. The premolars (bicuspids) were grinding teeth rather than shearing teeth as in the pongids. They were small (four to five feet tall) and estimated to weigh 70–120 pounds. The *foramen magnum* (opening at the base of the skull through which the spinal cord enters the brain) is positioned well underneath the skull, as in modern humans. In other words, the skull was fairly well balanced on the vertebral column rather than protruding from it as in modern apes. Cranial capacity of australopithecines averages about 500 cubic centimeters (cc). This is about the same size as that of an average gorilla, but we must remember that a gorilla weighs perhaps three times as much as an australopithecine. Orangutans and chimpanzees have cranial capacities averaging about 400 cc, and they too are much heavier than australopithecines. Therefore, the australopithecines were relatively larger brained (for their size) than any of the great apes. Evidence of simple tool-making behavior is available in roughly chipped stone pebbles found with some australopithecine remains.

Pithecantropines

More than thirty years before Dart identified the first australopithecine, Eugene Dubois found some skeletal fragments in a terrace of the Solo River in central Java near the town of Trinil. A higher structural grade of hominids was evident in its cranial capacity (approximately 1,000 cc), about twice that of australopithecines. Maximal breadth of the skull occurs at the level of the earholes rather than well above them as in modern humans. This is a hallmark characteristic of the pithecanthropine grade. Bipedal locomotion had been perfected, so their bodies were like ours. Their tools were somewhat more advanced than the australopithecines (including the first axes) and there is some hint of a possible use of fire. Dubois named this group *Pithecanthropus erectus* (erect ape man). Similar fossils found by Davidson Black (1927) and Franz Weidenreich in China near Peking were originally named

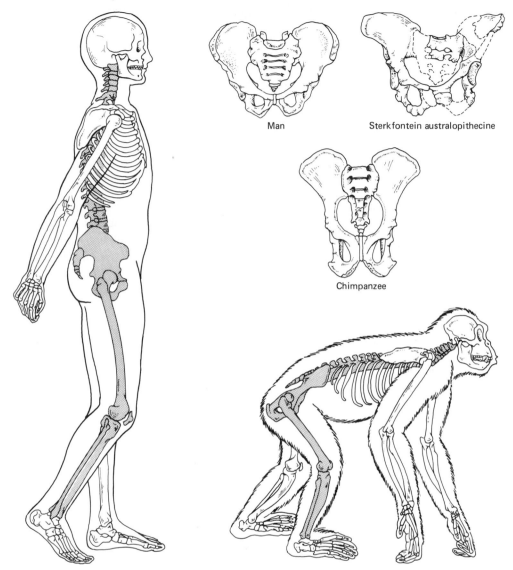

Figure 20.6. Pelvic structure as an indicator of posture and mode of locomotion. Note the similarity of the australopithecine pelvis to that of the human; hence, upright posture and bipedal locomotion.

Sinathropus pekinensis (Figure 20.7). Today, both Java man and Peking man are grouped together as subspecies of *Homo erectus* (*H. erectus javanensis* and *H. erectus pekinensis*). The term *pithecanthropine* is sometimes applied to all fossils of the structural grade corresponding to *Homo erectus*. Pithecanthropine fossils have been found in beds as old as Early Pleistocene.

Sapiens

By Middle Pleistocene times, hominids with cranial capacities in the range 1300 to 1500 cc became dominant in Europe. This structural grade of hominid is now called *Homo sapiens* ("wise man") and is the species to which modern humans belong. One of the first

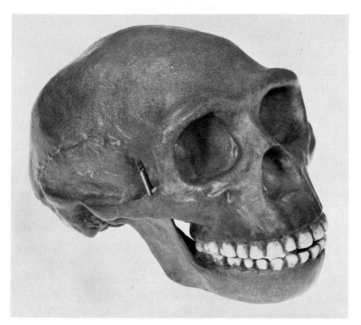

Figure 20.7. Sinanthropus skull (*Homo erectus pekinensis*); plastic reproduction. (Courtesy Carolina Biological Supply Company)

fossils to be found of *H. sapiens* was obtained in 1856 from a cave in the Neanderthal Valley of Germany. It was originally called *Homo neanderthalensis* or "Neanderthal man," but today is regarded as *H. sapiens neanderthalensis*. Neanderthal man stood about five feet tall and had a very large brain (approx. 1,450 cc) but no chin. They had prominent brow ridges and receding foreheads. Their teeth were larger than ours, but otherwise very similar. They stood erect as we do, not stooped as often seen in some artist's reconstructions. Their body build was very stocky, as judged from their robust bone structure. At the back of the skull, the occiput was bun-shaped (Figure 20.8). In addition to stone scrapers, the first pointed stone flakes (spearheads?) seem to have been part of their culture. Bone needles found with Neanderthal

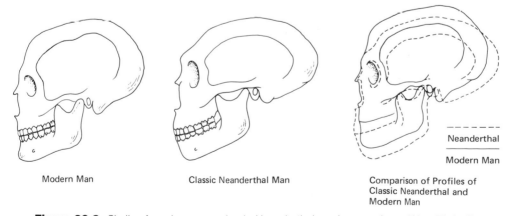

Modern Man   Classic Neanderthal Man   Comparison of Profiles of Classic Neanderthal and Modern Man

Neanderthal

Modern Man

Figure 20.8. Skulls of modern man, classic Neanderthal, and comparison. (After Birdsell)

remains indicate that they tailored fur clothing. The first indication of religion is seen in the way they intentionally buried their dead. About 40,000 years ago, the Neanderthals mysteriously disappeared from Europe.

No trace of human skeletons has been found in Europe in the interval dating 40,000 to 25,000 years old. Then a totally different kind of *H. sapiens* appears that is, in all respects, essentially like modern Europeans. These fossils are called "Cro-Magnon man" after the French cave in which they were first discovered (Figure 20.9). These are undoubtedly our ancestors that produced the famous paintings on the walls of caves. One of the distinctive features of modern humans is the development of a protruding chin on the lower jaw, a structure unknown in any other mammal. Cro-Magnon man does not have heavy brow ridges, is of light body build, has a high forehead, and a flat face. In these and many other ways, Cro-Magnon man is quite different from Neanderthal man. The main reason they are grouped together in *Homo sapiens* is that they have attained approximately the same level of cranial capacity. Very few anthropologists currently believe that classic Neanderthal man evolved into modern Europeans (the "ongoing regional evolution" model). Instead, the majority believe in the "replacement radiation" model, which proposes that a wave of immigrants replaced the Neanderthal inhabitants. The origin of these immigrants is unknown, but they are thought to have come from nearby and to the east (perhaps in the region around the Black Sea). It is probable that some gene flow occurred between Neanderthals and Cro-Magnons before the former became extinct. Perhaps the Cro-Magnons had a superior societal organization that allowed them to hunt more effectively than the Neanderthals. As their population density increased, they spread to other regions and because of their numbers and cohesive social bonds they were able to replace the Neanderthals.

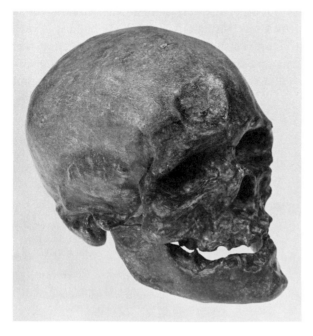

Figure 20.9. Cro-Magnon skull (*Homo sapiens sapiens*); plastic reproduction. Approximately 20,000 to 30,000 years old. A lower jaw is included, though no teeth are intact. Facial structures are well preserved, with nasal and orbital regions complete. Zygomatic processes are incomplete, but the cranial structure is undamaged. The original skull is believed to be of a twenty- to thirty-year old individual. Cro-Magnon and modern man are said to be counterparts, at least in anatomical structure. (Courtesy Carolina Biological Supply Company)

Some Complications

From the preceding cursory sketch we glimpse the major steps through which humans are thought to have evolved. The human lineage probably joins the pongid lineage in the early dryopithecines. The first structural grade that is definitely on the path to humans is that of the australopithecines. Following that, we jump to the pithecanthropines, and from there to sapient man (Figure 20.10). This evolutionary sequence seems reasonable to most anthropologists. However, many gaps remain to be filled with intermediate, transitional forms. Absolute dating of fossils from the ice ages (Pleistocene), where most human evolution occurred, has often been impeded for technical reasons. Both Pliocene and Pleistocene periods experienced extensive erosion and denudation of the earth's surface. Consequently, many geological sites that may have contained fossils of intermediate populations have been destroyed. Furthermore, it has been very difficult to correlate ice age strata in various parts of the world. This makes questionable, in many cases, whether two fossils belong to the same or different geological age. Humans have probably always been polytypic, i.e., a species consisting of differentiated geological races. But it is generally conceded that the human lineage is monophyletic, i.e., only one gene pool has been evolving since the time it separated from its common ancestors with the pongids. It is very likely that many races at various stages in human evolution died out without making any substantial contribution to higher forms. Perhaps at any one point in time, only one or a few races were responsible for the evolution of higher forms. Because of this possibility, we cannot say that any known fossils are from a population on the direct line to modern man.

If Neanderthals and Cro-Magnons are considered racial variants of *Homo sapiens,* then we must concede that much more racial variation existed in our ancestors than exists in human races today. The skulls of modern humans, regardless of race, are very similar. One of the most widely used measures of head form is called the *cephalic index* (if measured on a living person) or a *cranial index* (if measured on a skull). If the maximum breadth is x and the maximum length is y, then the index is $100 \, x/y$. Relatively short (broad) heads, called *brachycephalic,* have index values over 82; intermediate head forms, called *mesocephalic,* have indexes from 77 to 82; relatively long (narrow) heads, called *dolichocephalic,* have index values under 77. Cranial indexes run about two units lower. Some Asian Mongoloids have the broadest heads (averaging 86.8); some West African Negriforms have the narrowest heads (averaging 71.8). However, these differences in relative proportions are only minor variations compared with the striking structural differences between a Neanderthal skull and one of Cro-Magnon man.

It is far beyond the scope of this text to elaborate on all of the uncertainties of various hominid fossils. However, we should not be deceived into believing that all fossils fit neatly into the broad evolutionary scheme outlined above. Tobias and Napier (1964) found two skull bones (parietals), a lower jaw, and some foot bones at Olduvai Gorge, Tanganyika (now Tanzania), in East Africa. They named their new hominid *Homo habilis* ("handyman"). It probably dates about 1.75 million years old. L. S. B. Leakey judged *H. habilis* to be intermediate between the structural grades of australopithecines and pithecanthropines. Leakey and his wife, Mary, were the first to find fossil men at Olduvai Gorge in 1959. They named their find, *Zinjanthropus boisei;* later it was reclassified as an australopithecine. Most anthropolo-

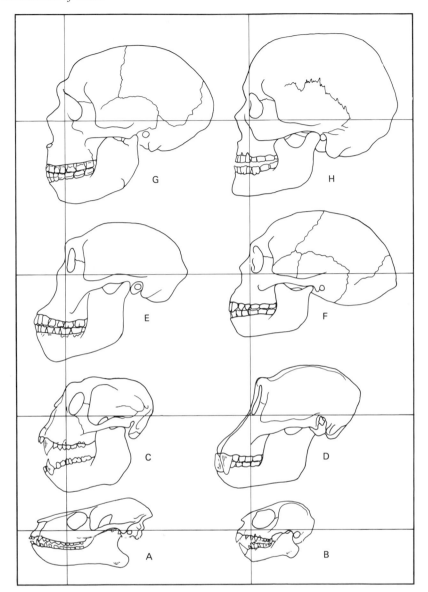

Figure 20.10. Comparison of skulls from various primates; A, B, and C are approximately one-half natural size, the others about one-fourth natural size. (A) An Eosine lemuroid *Notharctus*. (B) An Eosine tarsoid *Tetonius*. (C) A Pliocene Old World monkey *Mesopithecus*. (D) A modern chimpanzee *Pan*. (E) A Pleistocene man-ape *Australopithecus*. (F) *Homo erectus* or "Pithecanthropus." (G) Neanderthal man. (H) Cro-Magnon man. (After Colbert)

gists now believe that there were two species of australopithecines; a small (gracile) form weighing fifty to ninety pounds, and a large (robust) form weighing forty to fifty pounds more. Zinjanthropus is the largest australopithecine found to date. The large form, renamed *Australopithecus robustus* (or *A. boisei*), had heavy jaws and was probably herbivorous. The gracile form (*A. africanus*) was light boned and chiefly carnivorous (Figure 20.11). Leakey

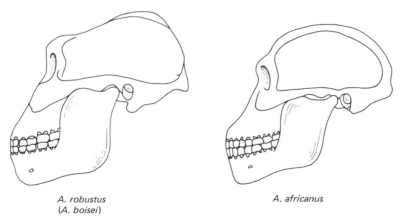

A. robustus
(A. boisei)

A. africanus

Figure 20.11. Two major kinds of australopithecines; a large (robust) form at left, and a smaller (gracile) form at right. (After Birdsell)

thought that *Homo habilis* represented a third line of human evolution in the Lower Pleistocene, in addition to the ancestors of the pithecanthropines and the dying line leading to the robust australopithecines. He maintained that *H. habilis* was a direct ancestor of humans and denied that pithecanthropines were involved. Most anthropologists disagree with Leakey's interpretation and consider the habiline fossil merely a large-brained and progressive variant of the gracile australopithecines.

Skulls with cranial capacities of about 1,250 cc have been found over great distances in Africa. This is high for the pithecanthropine grade, but low for the sapiens grade. The most complete and best known of this series is the so-called "Rhodesian man" (Figure 20.12). Its brow ridges and the height of its upper face are among the largest ever recorded. Rhodesian

Figure 20.12. Rhodesian Man skull (*Homo sapiens rhodesiensis*); plastic reproduction. This skull is approximately 120,000 to 150,000 years old. Referred to as the Broken Hill skull, it is without the lower jaw. As its primary location is in doubt, the association of this African Neanderthal is uncertain. (Courtesy Carolina Biological Supply Company)

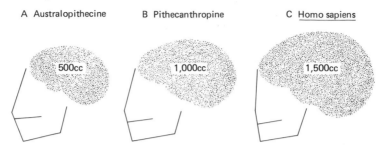

A Australopithecine   B Pithecanthropine   C Homo sapiens

500cc   1,000cc   1,500cc

Figure 20.13. Approximate brain size and generalized face shape of three structural levels of hominids.

man apparently lived as late as 30,000 to 40,000 years ago. This is quite disturbing, because perfectly modern-appearing fossil men are known from about this same time.

Many fossils are too fragmentary to permit unambiguous classification. The famous chinless jaw found near Heidelberg, Germany is thought to represent the earliest substantial relic of fossil man in Europe. But its depth and breadth do not closely resemble any pithecanthropine. Yet the teeth have moderately large pulp cavities, a trait frequently associated with Neanderthal types. A single, complete occipital bone from Vertesszöllös, Hungary has a rounded form and open angle typical of modern humans. It must have been a very large brained individual (at least 1,477 cc) and therefore classified at the *Homo sapiens* grade (Figure 20.13). It dates from 400,000 to 700,000 years old, a time when the rest of the inhabited world apparently was populated by pithecanthropines. Birdsell believes that Vertesszöllös man was probably on the direct ancestral line leading to modern humans. Parts of a skull found near London, England in 1935 are known to be at least 250,000 years old.

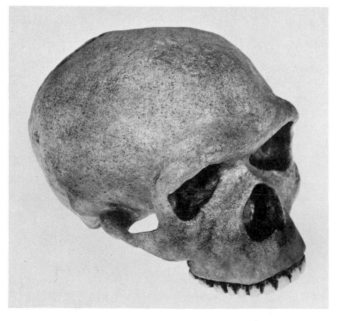

Figure 20.14. Steinheim skull (*Homo sapiens steinheimensis*); plastic reproduction. (Courtesy Carolina Biological Supply Company)

This so-called Swanscombe skull is estimated to have a cranial capacity of 1,300 cc, just a little less than the average for modern man. However, the vault is very thick boned. Another more complete skull retaining the greater portion of its face was found in 1933 near Steinheim, Western Germany (Figure 20.14) The Steinheim skull is narrower and lower than the Swanscombe skull. Its cranial capacity is estimated at about 1,150 cc, a value that is borderline between classical pithecanthropines and modern humans. However, the shape of the Steinheim skull has so little in common with the Far Eastern pithecanthropines that it must be placed in the *Homo sapiens* grade. Many other fossils in addition to those mentioned above demonstrate an impressive variability among hominids of the Middle Pleistocene. Pithecanthropines have been found throughout the Far East, East Africa, and perhaps in the Middle East and North Africa. The fact that sapient grade fossils are found in Europe with no evidence of a pithecanthropine occupancy is very strange. It is clear that human history has been extremely complex. It will require the best efforts of anthropologists, geologists, and evolutionists (in addition to a great amount of luck and technological innovation) to piece the fragmentary record into a unified theory. An attempt to display some of the complexity of hominid evolution is presented in Figure 20.15.

The Piltdown Hoax

Several large fragments of a brain case and part of a lower jaw were found in a gravel pit near Piltdown, England in 1912. The thick-boned skull cap was of modern size and appearance, but the jaw was definitely ape like, including the simian shelf not found in human mandibles. The stratum in which they were found was estimated to be Late Pliocene to Early Pleistocene in age. Primitive-looking stone tools and other animal bones were also found nearby. Some anthropologists thought that *Piltdown man* represented the most primitive human fossil known at that time and named it *Eoanthropus* (dawn man). Others could not accept the idea of such a transitional stage between ape-like and man-like forms. After many years of debate, the bones were subjected to the *fluorine test*. Fluorine, present in ground water, percolates through buried bones and forms deposits in proportion to the length of time involved. The amount of fluorine in various water systems differs widely from one locality to another. Therefore it is not a very useful method for absolute dating of fossils, but can readily be used to determine if a fossil belongs to the time zone of the stratum in which it is found. The human bones of Piltdown man contained only about one-fifth the fluorine found in the animal bones of the same deposit. This proved that the human bones were of much more recent origin. Later is was determined that the skull bone contained the amount of fluorine generally found in Late Pleistocene deposits. The mandible was actually the jaw of a young adult orangutan that had been stained to give the appearance of antiquity. A separate canine tooth associated with the find had been filed down to alter its ape-like appearance and also stained to look old. The perpetrator of this elaborate hoax remains unknown.

> An almost complete skeleton was found in a Middle Pleistocene deposit not far from London in 1888. This was another accidental find in a gravel quarry. It came as a great

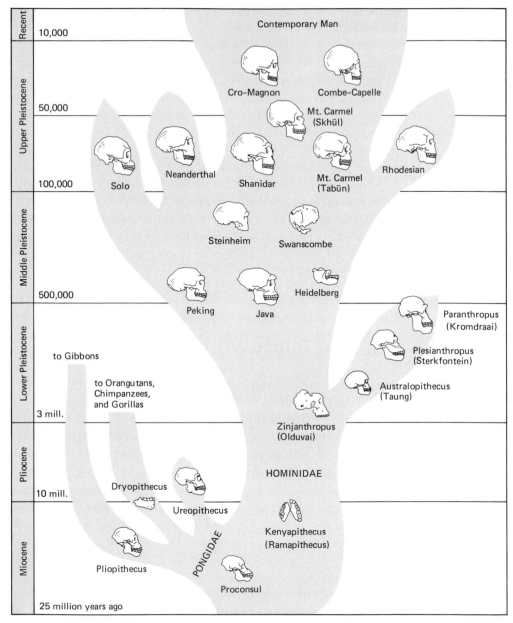

Figure 20.15. A phylogeny of the hominids. Many uncertainties concerning the relationships between the various fossils remain to be resolved. (After Volpe)

surprise that such modern forms existed that long ago (then thought to be 250,000 to 300,000 years old). Much later, the fluorine test revealed that this so-called *Galley Hill man* was much younger than the stratum in which it was found. Apparently these bones had been buried intrusively in the top level of the gravel terrace.

Only 500 yards distant in the same gravel terrace, part of a skull (parietal and occipital bones) was found in 1935. This so-called *Swanscombe* skull was later proven by fluorine test to belong to the second interglacial period (now estimated to be about 800,000 years old). It has been estimated that the cranial volume of the Swanscombe skull was about 1,300 cubic centimeters. This is only a little less than that of modern humans. It represents an authentically ancient find of a surprisingly modern type.

Races of Modern Man

Geographical variations in human populations can be grouped in many ways, depending on the nature of the anthropologist's research. For general discussions in a text such as this, it is sufficient to recognize only three or four major races primarily on the basis of hair structure.

1. *Negriforms* tend to have spiral hair and broad noses. Their skin color may be black, brown, or yellow. Most of them have relatively narrow skulls. Most peoples in Africa south of the Sahara, the Negrillo of Malaysia and Africa, and the Melanesians are usually included in this group.
2. *Europiforms* (or *Caucasoids*) tend to have wavy hair and narrow noses. They may have white or brown skins and vary greatly in breadth of skull. These people are found mainly in Europe, the Mediterranean countries, Asia Minor and India; North America is mainly populated by immigrants of this group.
3. *Mongoliforms* tend to have straight hair and noses of intermediate width. Their skin often has a yellow tint, because it contains more carotene pigment than that of other races; their skulls are commonly broad.
4. Some authorities claim that the inhabitants of the subcontinent of India deserve to be recognized as a fourth major race; others say that distinction belongs to the Australian Aborigines. East Indians tend to be as dark skinned as Negriforms, but have fine hair that is straight to wavy in form. Australiforms tend to have wavy, Europiform hair, but broad noses like Negriforms.

Obviously, races are not characterized by a singular phenotypic difference. Rather, they are recognized on the basis of many traits distributed over the earth in geographical patterns. An individual does not represent a race—that is typological thinking! Races are Mendelian populations. Some physical traits, such as body build and color of skin, are highly modifiable by environment and, consequently, are relatively poor indicators of genetic relationship. Other traits, such as hair structure, are not significantly influenced by the environment. It really is not necessary to see the group under study in order to classify them. Suppose that a random sample of one-hundred or more individuals was taken from one of the three major races. From each individual we obtain a sample of blood, urine, ear wax, and fingerprints. If a high incidence of Rh-negative blood types is found, the group is probably Caucasoid because this blood type is a low-frequency trait in Negriforms and Mongoliforms. Similarly Duffy blood group factors are virtually absent in African Blacks, but are found in approximately 43 per cent of Europeans. Populations of Asians and American Indians contain a relatively high percentage of excretors (in urine) of the amino acid BAIB (beta-amino-isobuteric acid). Many Chinese and Japanese populations have a dry, crumbly type of ear wax, whereas Negriforms and Caucasoids tend to have a more sticky and adhesive type of ear wax. The fingerprints of Mongoliforms have mostly whorled patterns; in Caucasoids and Negriforms, looped patterns predominate.

The adaptational value of the above traits are unknown. Earlier in this text, reference was made to a possible advantage of sickle-cell hemoglobin in areas where malaria is prevalent. Today, about 10 per cent of North American Blacks still carry the sickling gene derived mainly from their ancestors in which it was, and in some cases still is, adaptive in the heterozygous condition. Carriers of this gene can be detected by a simple blood test.

> About one person in thirty of the Jewish population of New York carries a potentially harmful gene for Tay-Sachs disease (amaurotic idiocy). Homozygotes for this recessive gene are deficient in an enzyme (hexosaminidase) necessary for the proper processing of fatty acids. Without this enzyme fatty deposits (gangliosides) accumulate in the myelin sheaths around nerves and interfere with neural functions. Within a few months after birth, there begins a gradual loss of motor activity and mental abilities. The word *amaurotic* refers to blindness, which is also part of the disease syndrome. Death by age three is common. This Jewish population probably acquired the gene from ancestors or immigrants from northeastern Poland (Silesia) and southern Lithuania, where the gene is in relatively high frequency. The reason for this high incidence is not known, but it has been suggested that it may have been advantageous to heterozygotes exposed to diets poor in fats. Overdominance in this environment may have led to the increase of this gene in the gene pool and establishment of a balanced polymorphism.
>
> Jews are people that share a common culture, but not a common gene pool. They are not a race in the biological sense. The Nazi purge of Jews during World War II was touted as a racial event. Too many people still fail to understand the fundamental difference between "biological race" and the term "race" as used by bigots.

Climate seems to have been a factor in selecting for body size and shape in spite of human culture. According to Bergmann's rule, slender bodies tend to loose heat more rapidly than stout bodies because of their greater surface-to-volume ratio. Allen's rule recognizes that short extremities tend to be adaptive in conserving body heat in cold climates. Natives of warm, tropical climates tend to be lean with long limbs and shallow bodies. Arctic people tend to be short and heavy. Some human populations, however, seem to violate these rules. The Mongoliforms of eastern Siberia tend to have flat faces, which reduces the chances of frostbite. Their slitty, fat-lidded eyes may serve as protection against sun glare as well as frost. Their scanty facial hair might reduce the condensation of moisture from breath around the mouth and also aid in the prevention of frostbite. Perhaps these traits evolved during the rigors of the ice ages and have now spread into warmer parts of Asia and into the Americas. A broader nose (platyrrhine) seems to be correlated with warmer and moister climates and is thought to be a primitive condition. A narrower nose (leptorrhine) seems to be correlated with colder climates and may be advantageous in warming the air before it reaches the lungs.

Skin-color differences depend on polygenic and environmental conditions. The ability to produce melanin (dark pigment) varies in different human races. Pigment granules, called melanosomes, can differ in size, number, composition of the protein portion, and degree of polymerization of melanin molecules; the ratio of melanocytes to epidermal cells may vary; melanin production itself is influenced by hormones (e.g., ACTH) and by ultraviolet light (tanning). The protein keratin in the horny outer layer of the epidermis is influential in its transparency to light. It is generally agreed that melanin particles in the outer layers of the skin serve to protect the deeper layers from damage by ultraviolet light. Dark-skinned tropical people also tend to have thicker skins than northern Europeans. Caucasoids of the United States have seven to eight times the frequency of skin cancer as Negroes living in the same cities at various latitudes. However, skin cancer has probably not been a strong selective factor for skin color, because it is a relatively mild and slow-acting form of cancer that seldom causes death. But a melanoma (cancer of malanin-producing cells) kills quickly and may also strike the young. Its incidence increases as one approaches the tropics. Dark-skinned people usually tend to have dark eyes and hair as well. This correlation is probably adaptive too. Blue-eyed people do not have as much malanin in the iris of their eyes and often must wear dark glasses to reduce glare and enhance vision in bright sunlight. It is possible that pigment in the retina of the eye is more important to Eskimos than pigment in the skin itself.

> An alternate hypothesis has been proposed to account for the evolution of skin color in various human races. Vitamin D is synthesized in the skin under the influence of ultraviolet rays. This vitamin functions to regulate the absorption of calcium and phosphorous from the gut and the deposition of these materials in bones. Too little vitamin D causes a softening and bending of bones (rickets); too much vitamin D can produce thickening of bones (causing them to become brittle), and calcification of soft tissues (especially in the aorta) and stone formation in the kidneys. In regions of intense sunlight, dark skins aid in the prevention of excessive vitamin D synthesis. People in northern latitudes would find it advantageous to have lighter skin so that the less intense rays of the sun could produce sufficient vitamin D in the few parts of the body not covered by clothing. Apparently Eskimos can be relatively dark skinned because their diet is usually rich in fish (fish liver oil is a potent source of vitamin D). Tanning would then be a physiological mechanism to prevent the overproduction of vitamin D. Until the 1930s, when milk began to be irradiated to improve its content of vitamin D, United States Black children suffered far more commonly from rickets than did White children.

No one knows which of the three major races is anatomically most like our fossil ancestors (primitive). Most of the traits currently used to characterize races involve soft parts that are unavailable as fossils. Some people think that primitive humans are those that are most ape like. They forget that modern apes are highly evolved primates like ourselves and may bear relatively little resemblance to our common ancestors. Modern apes are mostly dark colored with broad noses; therefore some people may be inclined to think that Blacks are closer to apes. However, the chimpanzee is actually lighter than some African Blacks. Both apes and Caucasians have thin lips, wavy hair, and a tendency to be hairy. These traits are least ape like in African Negroes. The notion that culturally advanced peoples are also biologically more advanced is bunk!

Cultural Evolution

Cultural evolution, unlike biological evolution, can be extremely rapid. Furthermore, once a technological innovation has been devised, all members of the population can share its rewards. The tradition can be handed down intact from one generation to the next by the processes of teaching and learning. This process would be markedly enhanced by the development of speech, language, and the arts (drawing, acting, sculpting, etc.). At first, cultural evolution was extremely slow; perhaps 90–95 per cent of human cultural history was spent in the Lower Paleolithic (old stone) age, typified by very simple pebble tools (chipped stones). Paleolithic culture is sometimes called *savagery,* though not in a disparaging sense. Savage humans are typically nomadic, gathering plant food and hunting animals. Everyone is engaged in the full-time task of securing food. They may have containers (gourds, shells, skulls, baskets), but no pottery. They may wear clothing of skins, grasses, or bark, but there are no textiles. There are no domestic animals either for food or draft. This type of culture was characteristic of Neanderthal man. Cro-Magnon man appeared in the Upper Paleolithic at a stage called "higher savagery" and continued to evolve culturally with little biological change to the present.

Barbarism is the cultural state of the Mesolithic (middle stone) age. Primitive agricultural practices began during this stage. Wheats and barley were domesticated early; rye and oats were merely weeds in the wheat and barley fields. Much later rye and oats were cultivated for their own sake, largely because they gave good yields in climates that were more northern than wheat could stand. The dog was domesticated early for use in hunting. Later, other animals (cattle, sheep, goats, pigs) were domesticated about the same time as cereal cultivation began. There were no cities; cities are large gatherings of individuals that are not engaged in food production. At first, agriculture was probably an adjunct to hunting, fishing, and gathering. Later, as their agrarian skills increased, they became more dependent on agriculture and less on other ways of obtaining food. Manual skills were advanced over Paleolithic cultures as evidenced by skilled secondary flaking of stone tools; use of bones, antlers, and teeth as tools; pointed spears and arrows; sinews for bows and harpoon lines; nets, hooks, and lines for fishing; well-constructed huts; canoes, skis, and sleds for transportation.

The middle stone age lasted only a relatively short time before it gave way to the agricultural revolution about 4,000 B.C. About this time the traction plow was developed to be pulled by oxen (later by horses). Irrigation was another innovation that allowed one person (family or tribe) to raise more food than needed to support itself. This allowed some people to engage in nonagricultural occupations and become specialists such as weavers, potters, flint polishers, and metal workers. Cities began to form; civilization has been called the culture of cities. So the agricultural revolution was responsible for the rise of civilization. The Neolithic (new stone) age began with the rise of early civilizations in Egypt and Mesopotamia and continued until about 2,000 B.C. when iron began to be used (bronze age); copper smelting had been developed about 5,000 B.C. Agriculture probably arose independently in several regions. Cultivation of maize and potatoes in Central and South America apparently developed independently of ancient Egypt, because these were crops peculiar to the New World. Perhaps the same could be said of the early millet and pig agriculture in China.

Knowledge and technology has been increasing exponentially since the agricultural revolution. Once power had been harnessed in the form of steam engines (and later internal combustion engines), the industrial revolution allowed many more people to pursue diverse occupations in business, industry, education, religion, law, politics, sociology, science, the arts, etc. As a result, our modern world has become very complex, creating many problems of global magnitude, such as pollution and the degradation of environmental quality, the population explosion and the threat of mass starvation (as predicted by Malthus), the energy crisis, the proliferation of nuclear weaponry and the threat of mass annihilation.

Our technology seems to have exceeded our ability to control its use for the sole benefit of mankind. We have the technology to limit our population growth, but we are unable socially to implement programs to accomplish that goal. Limiting family size by fiat is repugnant in today's society. If the time comes when global famine, pestilence, disease, and wars as a result of overcrowding cause deaths on a scale heretofore unknown, we may wish that we had earlier instituted laws and become subservient to them relative to limiting family size. The piper will be paid one way or another. Either we (all peoples of the world) face the fact that every benefit extracted from nature must be paid for (in some form or another) and take the lesser of "long-term evils," or we will reach the aforementioned crisis much sooner than we think.

For the vast majority of our evolutionary history, our ancestors lived in harmony with nature, as do all creatures. Humans did not overexploit resources, but gleaned what they could from nature's abundance. Population size was subject to the laws of nature. Human culture had not yet developed to the stage at which it could manipulate the environment to suit its own needs. Instead, at the earlier stages, hominids mainly adapted to changing conditions as do all other organisms—through biological evolution. Once human intelligence became sufficient to develop agriculture, civilization, and industry, they no longer were in harmony with nature. It is imperative that humans learn nature's laws and its limits, and that they devise ways to live in accord with them! This is much easier said than done. The human intellect can devise wonderful inventions, but the acquisition of social consciousness and moral obligations seems to be rudimentary by comparison. It is these latter aspects of human cultural evolution that are in great need of acceleration. Humanity now has the knowledge to direct its own biological evolution if it chooses.

Plasmids were first reported in bacteria by Joshua Lederberg in 1953 (Nobel laureate, 1958). These are circular DNA molecules that carry from 2 to 250 genes. Not all DNA segments have the ability to replicate themselves. At least one controlling element must be present in each plasmid to initiate DNA replication in a replicon. There may be one to about thirty plasmids per bacterial cell. The more plasmids per cell, the greater chance that at least one will be present in each daughter cell of a clone. Plasmids are nonessential to the life of the cell, but may provide the cell with a transient evolutionary advantage under certain environmental conditions. One of the most intensively studied class of bacterial plasmids are those that confer resistance to specific antibiotics on their host cells. Several years ago, the common human pathogen *Staphylococcus aureus* was not resistant to the antibiotic penicillin. Today, over 90 per cent of "staph" germs found in hospitals contain plasmids for resistance to penicillin. Indeed some bacteria now have plasmids containing genes for resistance to several antibiotics. These are called RTF (resistance transfer factors). Indiscriminate use of many antibiotics has resulted in strains of microorganisms acquiring multiple resistance factors by stepwise mutation and selective retention. Their existence poses a major medical problem for humanity's future.

In 1973, the first successful attempts to hybridize plasmids were reported. Recently discovered enzymes have allowed microbial geneticists to splice together plasmids containing different resistance factors from different strains of *Escherichia coli*. By plating bacteria on a medium containing two antibiotics, only the cells with hybrid RTF can grow into clones. Not only have they been able to hybridize plasmids from different strains of the same species, but also to introduce plasmid genes from one bacterial species to another (e.g., *S. aureus* into *E. coli*). Furthermore, genes from a toad (*Xenopus laevis*) have been successfully grafted into *E. coli* plasmids, and these vertebrate genes have been transcribed into RNA within the bacterial cell. We are, therefore, rapidly approaching the time when we may be able to "create" artificially new forms of life by this process of gene transfer. A new discipline is emerging under the titles of *genetic engineering*, genetic manipulation, molecular cloning, etc. Naturally, we would like to use these new methods for correcting our own genetic defects, enhancing immunological capacities, perhaps making us less dependent on dietary supplements (vitamins), etc. The potential is there, but with it comes new social problems to be faced and new threats to our well-being. During February 1975, an international group of scientists and other concerned specialists met at the famous Asilomar Conference (near Pacific Grove, California) to formulate guidelines for future research in the area of artificial DNA hybridization. Their report asked for voluntary deferring of two types of experiments until much more is known about their potential hazards to human health. Type I experiments involve the production of novel organisms containing combinations of RTF and genes responsible for toxin synthesis. Type II experiments are concerned with introducing DNA from tumor viruses or other animal viruses into bacteria.

As we have learned, organic evolution is usually an exceedingly slow process in terms of human lifetimes, and much remains to be discovered about the genetic treasures hidden in gene pools of various human races. It would be foolish therefore to embark on any eugenic programs without full knowledge of the possible consequences. On the contrary, all that we have learned from the sciences of genetics and evolution should caution us from tampering with the human genetic system, lest in the process we lose the genetic heterogeneity that represents the evolutionary potential of all gene pools. Humanity could much more profitably put its talents to use in attempting to create optimal conditions for nurturing the kind of social and moral awarenesses that will be required of future citizens to cope effectively with the problems created by an ever-increasing technology. The term *euthenics* is applied to control of the physical, biological, and social environments for the improvement of humanity. The term *euphenics* is employed for the amelioration of genotypic maladjustments brought about by efficacious treatment of genetically defective individuals. The most immediately rewarding way to improve the productivity of crops and livestock is to optimize the environment for maximal expression of their genetic qualities. After this is done, we can begin the longer process of changing their heredities. So it is with humans. We can make much immediate progress in improving human welfare through euthenics and euphenics. Perhaps it is best to ignore eugenic proposals until we have fully implemented the first principles.

Human cultural evolution has now matured to the stage at which science is part of everyday life. The scientific concept of evolution by natural selection has been ranked as one of the greatest philosophical achievements of all time. The human mind has become so complex that it can comprehend notions of its own origins through natural processes. How valid are these evolutionary theories? Are they so rigorous and so highly predictive that they virtually solve all biological problems? A partial answer to these questions is presented in the next two sections.

Mathematical Challenges to the Evolution Theory

It would be quite wrong for the student of evolution to obtain the impression from reading this or any other text on evolution that adequate answers have now been found to all the problems in this field. Nothing could be further from the truth! Many of the questions that troubled Darwin are still being raised today. Are the processes of random variation and natural selection really sufficient to account for the diversity of life that we see around us? One of the most widely publicized recent debates of this age-old question took place in April 1966 at the Wistar Institute of Anatomy and Biology in Philadelphia between a group of mathematicians and biologists. The mathematicians charged that if natural selection had to choose from the astronomically large number of available alternative systems by means of the mechanisms described in current evolution theory, the chances of producing a creature like ourselves is virtually zero. Murray Eden (Professor of Electrical Engineering at M.I.T.) was especially concerned about the element of randomness, inherent in the mutational process, which presumably provides the raw materials for the process of evolution. Attempts to simulate this process on computers have not been very successful, and the mathematicians therefore concluded that current evolutionary theory is inadequate; biologists have failed to provide sufficient information for efficient computer simulation. Eden contends that "No currently existing formal language can tolerate random changes in the symbol sequences which express its sentences. Meaning is almost invariably destroyed. Any changes must be syntactically lawful ones." He charges that an adequate theory of adaptive evolution would supply a computer programmer with a correct set of rules for "genetic grammaticality" that has a deterministic explanation, rather than owing its stability to selection pressure acting on random variations.

Some simple calculations were presented to demonstrate the high degree of improbability of generating useful proteins from selection of random variations. If we consider an average protein ("word") to consist of 250 amino acid residues ("letters"), each letter chosen from an alphabet of 20 amino acids, about 20^{250} (or 10^{325}) possible protein words could be uttered by the genetic system. By making some further assumptions, an upper limit to the number of different proteins that have ever existed was estimated to be about 10^{52}. If these estimates are realistic, then only a tiny fraction of all possible protein molecules has ever been formed. Furthermore, it is argued that of the 10^{52} proteins that possibly were formed, again only a very tiny fraction have ever had any utility. Therefore, Eden contends, instead of hunting randomly throughout the universe of all possible proteins in search of viable systems, there must be some determinate feature, yet undiscovered, by which evolution has focused rather quickly on this infinitesimally small proportion of all conceivable proteins.

As a specific example, the similarity of the α and β chains of normal human hemoglobin was cited. When these two chains are arranged to optimize their similarities, it is found that they agree in 61 places, they disagree in 76 places, and there are 9 gaps. From our knowledge of the genetic code, a minimum of 42 single mutations, 33 double changes, and 1 triple alternation (plus 9 gaps) would be required to convert one chain into the other or derive both from a common ancestral molecule. In comparing the estimate of required mutations with the observed statistical distribution of amino acids along both chains, there is a mean difference of only about $1\frac{1}{2}$ per amino acid type. Eden contends that a random process could not produce so good an agreement after only 120 attempts. Of all the possible paths that evolution might have taken in establishing the useful proteins, it seems to have selectively moved along the shortest one. He suggests, therefore, that the principle task of the evolutionist is to discover the mechanisms that constrain variation of phenotypes to a very small class and to relegate the notion of randomness to a minor and noncrucial role.

Important chemical laws are derivable from the assumption of random collisions between reacting molecules, but the explanation of how unique products result from these collisions cannot be sought in the random rearrangement of the reactants. It must be found by studying the relation between specific molecular structure and reactivity. The genetic system is not simply a blueprint for producing an organism, but rather an *algorithm* (i.e., a minimal generative procedure that makes use of organizational tendencies having a high probability of occurrence). To the computer programmer, an algorithm is the least set of instructions that, like a formula, is used to generate a table of numbers (i.e., for every *x* value, a corresponding *y* value can be computed). A blueprint-type of genetic system, on the other hand, would specify *a priori* the whole table. If the genetic system is an algorithm, then biologists have yet to discover the abstract language in which it is written. The genetic code represents the alphabet of this language. However, knowledge of an alphabet probably furnishes little information about the language in which it is used. According to Eden, until the natural laws governing this language have been found, neo-Darwinian evolution is only a restatement in modern terminology of Darwin's germinal insight that the origin of species can have a naturalistic explanation.

Eden also claims that the evolutionary concept of fitness is *tautologous,* i.e., it simple restates the fact that only the properties of organisms that survive to produce offspring *do survive.* To the question, "Which organisms leave more offspring?," evolutionary theory can only answer that it is those that leave more offspring. He believes that this kind of conceptual system is vacuous and incapable of disproof. In contrast with the theories of physics, no crucial experiment can be devised to determine if the theory is valid or not.

> J. C. Fentress provides us with an example of how the flexibility of the evolutionary theory can be used to explain antithetical data. While he was in Cambridge, England, studying behavior of British voles (a kind of field mouse), he noticed that a species (which normally inhabits woodlands) would freeze (immobilize), and another species (which normally lives in open fields) would run when a test object moved overhead. Fentress took his data to some zoologist friends, but reversed the observations. He asked them why, when the converse was actually true, that the field species froze and the woodland species ran. These zoologists were able to give an elaborate and quite satisfying explanation using conventional ideas about evolutionary theory.

Several eminent geneticists responded to the challenges raised at the Wistar symposium. Peter Medawar agrees that the simplistic phrase "survival of the fittest" can appear to be tautologous when used in a very elementary level of discussion. When the Darwinian theory is fully presented, however, the principle of natural selection is neither tautologous nor vacuous. According to Medawar, evolution by natural selection means two things: (1) all people living today will not contribute equally to human populations of the future; (2) the unequal contributions people make to future populations is correlated with their genotypes. Neither statement is tautologous. The contributions that a given genotype makes to future populations might be strictly proportional to its present frequency, but in fact it is not.

Another of England's well-known geneticists, C. H. Waddington, maintains that the calculations that produced an estimate of 120 amino acid subsitutions to homologize the α and β chains of hemoglobin were done under the assumption that it was necessary for each mutation to spread through the entire population before the next one could be processed. "We don't need 120 changes one after the other. We know perfectly well of 12 changes which exist in the human population at the present time." He also submits that the predictive value of the evolutionary theory is beyond question. For example, if a new insecticide was applied to a large population of flies, it is highly probable that they would evolve a resistance within a few generations. Evolutionary theory cannot predict how they will adapt—perhaps by changing the permeability of their cuticles or by synthesizing an enzyme to detoxify the chemical—but, in Waddington's view, this is a perfectly good evolutionary prediction. Biological systems are much more complex than physical systems. To expect the same degree of predictive value from a biological theory (where so many variables exist in the experimental materials) as a physical theory (where all but one or a few variables can be controlled) is not realistic.

Eden had proposed, in an analogy, that "the chance of emergence of man is like the probability of typing at random a meaningful library of one-thousand volumes using the following procedure. Begin with a meaningful phrase, retype it with a few mistakes, make it longer by adding letters, and rearrange subsequences in the string of letters; then examine the result to see if the new phrase is meaningful. Repeat the process until the library is complete." In response to this challenge, S. Wright proposes a "numbers game" of his own. "On the principle of the children's game of twenty questions in which it is possible to arrive at the correct one of about a million objects by a succession of twenty yes-or-no answers, it would require less than 1250 questions to arrive at a specified one of these proteins. While this is not a perfect analogy to natural selection, it is enormously more like natural selection than the typing at random of a library of one thousand volumes with its infinitesimal chance of arriving at any sensible result."

The foregoing charges and countercharges are merely a small sample of the dialogue at the Wistar symposium. Most biologists are satisfied with theories that are consistent with observable facts and have some degree of predictive value. To challenge a theory for lack of an effective working model may not be warranted because it may be the model rather than the theory that is defective. It does not necessarily follow that the prosperity of any research field implies that its theories are adequate or even meaningful. We can imagine that a thriving experimental nuclear physics could exist with, or without, its abstract models. It is questionable whether this symposium has substantially influenced modern evolutionary theories, but it may have prompted improvement of future models.

More Critique

It is not within the scope of this or any other textbook to attempt to answer, one by one, all of the criticisms that have been leveled at the theory of evolution. This would indeed be a herculean task. Darwin did an admirable job, for his time, in anticipating the jabs of his critics. More recent publications by other evolutionists have usually been lax by comparison in this regard. Several current publications are written for the express purpose of criticizing evolutionary theory, rather than explaining its basic principles. This is a healthy situation, for these authors bring to light many of the unresolved problems that are a source of embarrassment to evolutionists. Scientists, perhaps more than other professionals, should try to remain unbiased; among other things this demands that they display for public consumption not just the strengths and successes of their theories, but also the weaknesses and failures of same. As stated earlier, any profession that does not supply its own criticism and iconoclasm will find that someone else will do the job, and usually in a way it does not like. So the student of evolution should read what some of these critics have to say and become acutely aware of the shortcomings of modern evolutionary theories.

In the remainder of this section, a sampling of quotes, remarks, and criticisms will be presented from a book by Norman Macbeth entitled *Darwin Retried* (1971). No attempt will be made either to justify these excerpts or to argue in rebuttal. They are simply presented to give the reader some feeling for the kinds of challenges being directed at neo-Darwinism.

1. A critic or skeptic may peck at any aspect of the theory, testing it for flaws. He is not obligated to set up any theory of his own.
2. One and the same ancient stock split into a group with astonishing plasticity and another group with almost total rigidity. This is very hard to swallow. Therefore there was a tendency, while phylogenetic trees were still being used, to move known forms out on the branches and reserve the trunk for malleable forms not yet discovered.
3. Professor Ernst Mayer of Harvard rules out admittedly valid objections on the ground that the objectors have not advanced a better suggestion. Macbeth cannot take this view seriously. If a theory conflicts with the facts or with reason, it is entitled to no respect.
4. Macbeth objects to the double standard of simplifying evolutionary concepts for lay consumption at the expense of presenting a misleading picture (as horse orthogenesis), but providing the detailed qualifications which clarify the picture only for professionals.
5. Macbeth points to the disagreement on the definition of a species as a major problem for evolutionists. Realists regard species as something real, as Linnaeus did. Nominalists regard the species name as only a convenient label. Louis Agassiz pointed out long ago the dangers of nominalism when he asked "If species do not exist at all, as the supporters of the transmutation theory maintain, how can they vary?"
6. Mayr said that Darwin lacked understanding of the nature of species. Darwin failed to solve the problem indicated by the title to his work. G. G. Simpson said Darwin's book *On the Origin of Species* is not really on that subject.
7. Breeders can make only small changes. They never tried to make big changes like adding wings to a horse.
8. Darwin apparently believed that organisms could vary in all directions and to an unlimited degree. He extrapolated from micro changes observed by breeders to macro changes to account for gaps in the fossil record. If you observe the growth of a baby during its first months, extrapolation into the future will show that the child will be eight feet tall when it is six years old.

9. Bernhard Rensch, in his book *Evolution Above The Species Level,* has no actual examples, but he asserts that transspecific evolution should not be regarded as impossible. This hardly constitutes proof that it does occur.

10. Richard B. Goldschmidt (1878–1958), after observing mutations in fruit flies for many years, said the changes were so hopelessly micro that if a thousand mutations were combined in one specimen, there would still be no new species. This led him to propose the hypothesis of the "hopeful monster," whereby a huge change might have occurred all at once and been preserved by a favoring environment.

11. Mayr points out that animal populations have the property of resisting sudden or drastic change (genetic homeostasis); there is a tendency of animals and plants to balk at being bred too far in any one direction.

12. Edward S. Deevey says that some remarkable things have been done by crossbreeding and selection inside the species barrier or within a larger circle of closely related species, such as the wheats. But wheat is still wheat, and not, for instance, grapefruit.

13. Simpson is forced to postulate a short period of rapid change in formation of the bat wing. It has changed little since the Eocene. But it must not be too rapid or it would be considered a saltation, to which he is unalterably opposed.

14. Macbeth's thesis seems to be summarized in the following statement. "I cannot assert that the biologists have expressly abandoned Darwin's position. Indeed, it seems likely that most of them would say that he simply *must* be correct. But on the other hand, they would all recognize the limits of variability, the curse of sterility, the dangers of extrapolation, the hopelessness of trying to convert bears into whales or of breeding winged horses, and the strong inertia of genetic homeostasis. I do not see how these points can be reconciled with Darwin's position, and I suggest that the time has come for a retreat."

15. Macbeth says that natural selection for Darwin was differential mortality (some die early, some late). Therefore differential reproduction among organisms with exactly the same life span is neo-Darwinian selection.

16. G. L. Stebbins says that while the demonstration that selection has occurred is not excessively difficult, the nature of its action and the causes of this selective process are much harder to discover or prove.

17. Trying to ascribe adaptive characteristics to every feature of an organism right down to the number of spots or bristles, has resulted in wild speculations that have brought the whole idea of selection into contempt.

18. Weak selection might be highly effective over long periods of time, but quite beyond our powers of observation. Sir Julian Huxley rebukes Lamarckians who plead a similar case, saying "To plead the impossibility of detection is a counsel of dispair. It is also unscientific."

19. Macbeth admires a willingness to view things as marvels and refrain from utilitarian explanations.

20. Macbeth suggests that the works of mathematicians (Fisher, Ford, Haldane, Wright) should be investigated to see if they have really added anything to our understanding of nature. Are their assumptions rooted in a firm basis of fact, or will it turn out to be garbage as was the case with Galton and Pearson's efforts?

21. Macbeth states that neither Darwin nor his immediate followers had much feeling for

the internal stability and harmony of the organism. For example, Roux saw the organs struggling with each other for nourishment. Simpson says that Darwin's concepts had ethical, ideological, and political repercussions which were and continue to be in some cases unfortunate.

22. Nature is not consistent enough (witness examples of both strife and cooperation) to yield a firm basis for a theory.

23. Macbeth asserts that no one can decide, with the naked eye or with instruments or with mathematics, whether a given trait is adaptive, inadaptive, or neutral. Since having selective value is the same thing as being adaptive, this comes down to saying that we do not know what is and what is not adaptive. Can a solid theory be built on such a base?

24. Macbeth sees three different degrees of adaptation: (1) clowns—seemingly poorly adapted (e.g., gorilla's arms seem adapted to swinging but spends most of the time on the ground), (2) craftsmen—exemplified by camouflage and mimicry, and (3) wizards—animals which behave in incredibly complicated and mysterious ways. All of these types seem to get along in much the same way. The clowns do not die out and the craftsmen do not take over the earth. It is possible to infer from this that crafty adaptation is not really a matter of life or death, but little attention is paid to this line of thought. LaFontaine's fable of the ant and the grasshopper is very impressive for children, but the grasshoppers do not die out despite their improvidence. No one pays much attention to the clowns. The Darwinians, who have always had a strong utilitarian bias, stress the work of the craftsmen. Anti-Darwinians stress the work of the wizards, rejoicing in the lack of utilitarian value and in the difficulty of explaining the magic.

25. The problem of the human eye defies anyone to explain the step by step changes which occurred to produce such a complex and highly integrated structure. Macbeth says that explanations for all such "adaptations" must employ the "wave the wand" method.

26. Evolutionists are soothed and sustained by the "best-in-field fallacy." The most famous rivals were vitalism, fundamentalism, Lamarckism, and hopeful monsters. Darwinism appears better than the others (special creation being excluded as a scientific explanation). Is there any glory in outrunning a cripple in a foot race? Simpson says ". . . scientists rarely are psychologically capable of accepting a phenomenon as a fact and also accepting it as inexplicable."

27. A watch indicates it was created by a watchmaker. How can man, much more complicated than a watch, be created by anything less than a supreme watchmaker?

28. Fisher has pointed out that natural selection is a mechanism for generating improbabilities. But if you take Fisher's pithy phrase seriously, nothing is unlikely. The greater the improbability, the greater the glory of natural selection. The effect of the pithy phrase is to seal off discussion and tests, since theories that are not testable are not scientific. A theory which is not refutable by any conceivable event is not scientific. Irrefutability is not a virtue of a theory (as people often think), but a vice.

29. Natural selection tends to make each organic being as perfect as, or slightly more

perfect than, the other inhabitants of the same country with which it has to struggle for existence. It therefore should have endowed the savage with a brain just a little better than the ape, but in fact he has one very little inferior to that of the average member of our learned societies. Wallace (1869) challenged the whole Darwinian position by insisting that artistic, mathematical, and music abilities could not be explained by natural selection. Wallace remained a skeptical Darwinist until his death in 1913. Darwinism has no predictive power.

30. Evolutionists cannot explain the actual cause of a single case of natural extinction. Mayr says "ultimately their extinction is due to an inability of their genotype to respond to new selection pressures." Macbeth says this is meaningless because the same could be said of every extinct species or person including Julius Caesar and Abraham Lincoln. Darwinism cannot explain extinction or survival, although these phenomena are the essence of evolution.

31. Evolutionists harm their cause by wild swings from arrogance to humility, from boasts of wisdom to confessions of ignorance. A constant position should be maintained; they should cultivate sobriety.

32. Simpson says "Life and its environment are in such ceaseless flux that it is simply inconceivable that a permanent equilibrium will ever be reached." How then can the persistent types (horseshoe crab, etc.) be explained?

33. Mayr states that "the basic theory is in many instances hardly more than a postulate." Macbeth adds that a postulate is a position assumed without proof.

34. Is there a moral duty to offer an explanation for every biological phenomenon or are evolutionists badgered by Simpson's admonition that it is not enough to look at things with the unreasoning wonder of a child? This passion to explain has injurious consequences: (1) it often makes the profession look foolish, and (2) all of this explaining takes the charm out of biology. How many people find lectures on evolution lively or profitable?

35. The public may rightly feel that is has been paltered with if the complexity and insecurity of the theory are not laid before it. Scientists are not expected to be infallible; confession is good for the soul, and candor is always highly valued.

36. Goldschmidt states that "The development of the evolutionary theory is a graphic illustration of the importance of the *Zeitgeist*. A particular constellation of available facts and prevailing concepts dominates the thinking of a given period to such an extent that it is very difficult for a heterodox viewpoint to get a fair hearing. . . . The fact that the synthetic theory is now so universally accepted is not in itself proof of its correctness."

Religion and Evolution

Macbeth also lists the hallmarks of a religious attitude among evolutionists:

1. No middle ground—the attitude that all who are not with me are against me. Simpson rejects the suggestion that "one could gather more facts and suspend judgment as to what meaning they might eventually have."

2. Reproof of the fainthearted—Simpson says of those who fail to go along with his views that they have succumbed to dispair and faintness in the search and have embraced defeatism or escapism.
3. Missionary zeal—advocating that evolution be taught in every high school because (as Julian Huxley says) it is evolution, in the broad sense, that links inorganic nature with life, and the stars with earth, and matter with mind, and animals with man.
4. Perfect faith—Simpson believes, in his more euphoric states, that the synthetic theory is the ultimate solution to all biological riddles.
5. Millenarianism—the idea of a heaven on earth, seems to be embodied in some rash statements concerning man's ability to direct his own future evolution.

These indeed are danger signs, for scientists cannot allow themselves the privilege of irrevocable commitment to any one idea, theory, or model. Their minds must ever remain open to the possibilities in alternative conceptual schemes. Bigotry in other areas of human endeavor can sometimes be excused on the basis that no methodological framework is available to prohibit same. But scientists cannot use this excuse; bigotry has no place in science.

This text began with an example of bigotry in the form of religious intolerance responsible for the Scope's "monkey trial." Most of the history of science is dominated by the gradual retreat of vitalism in the face of hard-won victories of the mechanistic school. Today there is still a sizable group preaching that one cannot embrace the theory of evolution and at the same time believe in a Diety. In the author's opinion, this attitude does not help people to cope with the real world and harmonize the findings of modern science with religion. Even in this enlightened generation, too few are aware of the distinctive roles science and religion play in philosophy. Darwin wrote in the *Origin of Species,* "I see no good reason why the views given in this volume should shock the religious feelings of any one." He also quotes an anonymous source that seems to convey his own feelings: ". . . it is just as noble a conception of the Deity to believe that He created a few original forms capable of self-development into other and needful forms, as to believe that He required a fresh act of creation to supply the voids caused by the action of His laws." If there be any conflict between religion and science, it would seem to reside in the minds of those whose concept of their Deity is too limited to allow Him the power to create gradually through an evolutionary process. If God is God, then He can do as He wishes with His universe; He could have brought all living things into being instantaneously, or He could have unfolded His creativity through the geological ages according to "natural laws." And what is the source of these natural laws? To those who believe in a Deity, they are part of the way that He sustains His creation. Darwin saw the natural laws of growth, reproduction, inheritance, variability, natural selection, and evolution operative in the formation of "the most exalted object which we are capable of conceiving, namely, the production of the higher animals." Darwin concludes his most famous book with these words, "There is grandeur in this view of life, with its several powers, having been originally breathed by the Creator into a few forms or into one: and that, whilst this planet has gone cycling on according to the fixed law of gravity, from so simple a beginning endless forms most beautiful and most wonderful have been, and are being evolved."

Essay Questions

1. *Describe three structural grades of hominids and their temporal relationships, one to another.*

2. *Which modern human race is most primitive? Explain fully.*

3. *In what ways can cultural evolution affect the course of human biological evolution?*

4. *Is the evolutionary concept of fitness a tautology? Explain fully.*

5. *Explain the principle of the fluorine test and its utility in anthropological research.*

REFERENCES

Adams, F. T. 1968. *The Way to Modern Man.* Teachers College Press, New York.

Alfvén, H. 1966. *Worlds-Antiworlds.* W. H. Freeman, San Francisco.

Allard, R. W. 1960. *Principles of Plant Breeding.* John Wiley & Sons, Inc. New York.

Allison, A. C. 1956. Sickle cells and evolution, *Sci. Amer.* **195** (August) 87.

Alsberg, P. 1970. *In Quest of Man.* Pergamon Press, New York.

American Scientific Affiliation. 1950. *Modern Science and Christian Faith.* Van Kampen Press, Wheaton, Illinois.

Anderson, E. 1949. *Introgressive Hybridization.* John Wiley & Sons, Inc. New York.

Anfinsen, C. B. 1959. *The Molecular Basis of Evolution.* John Wiley & Sons, Inc., New York.

Appleman, P. 1970. *Darwin.* W. W. Norton & Company, Inc., New York.

Ardrey, R. 1961. *African Genesis.* Atheneum, New York.

Avers, C. J. 1974. *Evolution.* Harper & Row, Publishers, New York.

Ayala, F. J. 1972. Competition between species. *Amer. Scientist* **60**(3):348–57.

Azêvedo, E., H. Kreiger, M. P. Mi, and N. E. Morton. 1965. PTC taste sensitivity and endemic goiter in Brazil. *Am. J. Human Genetics* **17**(1):87–90.

Bakker, R. T. 1975. Dinosaur renaissance. *Sci. Amer.* **232**(4):58–78.

Barnett, S. A. 1971. *The Human Species.* Harper & Row, Publishers, New York.

Beer de, G. R. 1951. *Embryos and Ancestors.* Oxford University Press, London.

Benson, L. 1943. The goal and methods of systematic botany. *Cactus and Succulent Journal* **15**(7):99–111.

———. 1957. *Plant Classification.* D. C. Heath and Company, Boston.

Bernal, J. D. 1967. *The Origin of Life.* The World Publishing Company, New York.

Birdsell, J. B. 1972. *Human Evolution.* Rand McNally & Company, Chicago.

Bishop, J. A., and M. Laurence. 1975. Moths, melanism and clean air. *Sci. Amer.* **232** (January) 90.

Bleibtreu, H. K. 1969. *Evolutionary Anthropology.* Allyn and Bacon, Inc., Boston.

Boyd, W. C. 1950. *Genetics and the Races of Man.* Little, Brown, and Company, Boston.

Brace, C. L. 1967. *The Stages of Human Evolution.* Prentice-Hall, Inc. Englewood Cliffs, New Jersey.

Bresler, S. E. 1971. *Introduction to Molecular Biology.* Academic Press, New York.

Brewbaker, J. L. 1964. *Agricultural Genetics.* Prentice-Hall, Inc., Englewood Cliffs, New Jersey.

Broom, R. 1933. *The Coming of Man.* H. F. & G. Witherby, London.

Brooks, J., and G. Shaw. 1973. *Origin and Development of Living Systems.* Academic Press, New York.

Brown, W. L., Jr., and E. O. Wilson. 1956. Character displacement. *Systematic Zool.* **5**:49–64.

Bryson, V., and H. J. Vogel. 1965. *Evolving Genes and Proteins.* Academic Press, New York.

Buvet, R. and C. Ponnamperuma. 1971. *Molecular Evolution I. Chemical Evolution and The Origin of Life.* American Elsevier Publishing Co., Inc., New York.

Calvin, M. 1969. *Chemical Evolution.* Oxford University Press, New York.

Cameron, T. W. M. (editor). 1960. *Evolution: Its Science and Doctrine.* University of Toronto Press, Canada.

Camp de, L. S. 1969. The end of the monkey war. *Sci. Amer.* **220** (February) 16–21.

Campbell, B. G. 1966. *Human Evolution.* Aldine Publishing Company, Chicago.

———. (editor). 1972. *Sexual Selection and the Descent of Man.* 1871–1971. Aldine Publishing Company, Chicago.

Cavalli-Sforza, L. L. 1969. "Genetic drift" in an Italian population. *Sci. Amer.* **221** (August) 30.

———, and W. F. Bodmer. 1971. *The Genetics of Human Populations.* W. H. Freeman and Company, San Francisco.

Chambers, R. 1969. *Vestiges of the Natural History of Creation.* Leicester University Press, New York.

Changeux, Jean-Pierre. 1965. The control of biochemical reactions. *Sci. Amer.* **212** (April) 36.

Chardin de, Pierre Teilhard. 1966. *Man's Place in Nature.* Harper & Row, Publishers, New York.

Clark, R. E. D. 1950. *Darwin: Before and After.* The Paternoster Press, London.

Clark, W. E. L. 1955. *The Fossil Evidence for Human Evolution.* The University of Chicago Press, Chicago.

Clausen, J., D. D. Keck, and W. M. Heisey. 1948. Experimental studies on the nature of species. III. Environmental responses of climatic races of *Achillea. Carnegie Inst. Wash. Publ.* **581**:1–129.

Cleland, R. E. 1972. *Oenothera; Cytogenetics and Evolution.* Academic Press, New York.

Cohen, S. N. 1975. The manipulation of genes. *Sci. Amer.* **233** (July) 24.

Colbert, E. H. 1955. *Evolution of the Vertebrates.* John Wiley & Sons, Inc. New York.

Cox, C. B., I. N. Healey, and P. D. Moore. 1973. *Biogeography—an Ecological and Evolutionary Approach.* John Wiley & Sons, New York.

Crow, J. F. 1959. Ionizing radiation and evolution. *Sci. Amer.* **201** (September) 138.

———, and C. Denniston, 1974. *Genetic Distance.* Plenum Press, New York.

———, and M. Kimura. 1965. Evolution in sexual and asexual populations. *Am. Naturalist* **99**:439–50.

———, and M. Kimura. 1970. *An Introduction To Population Genetics Theory.* Harper & Row, Publishers, New York.

Darlington, C. D. 1958. *The Evolution of Genetic Systems.* Basic Books, Inc., Publishers, New York.

———. 1961. *Darwin's Place in History.* Macmillan Publishing Co. Inc., New York.

Dart, R. A. 1959. *Adventures with the Missing Link.* Harper & Brothers, New York.

Darwin, C. 1874. *The Descent of Man and Selection in Relation to Sex.* The Werner Company, Akron, Ohio.

———. 1900. *The Variation of Animals and Plants Under Domestication.* D. Appleton and Company, New York.

———. 1902. *The Origin of Species by Means of Natural Selection or the Preservation of Favored Races in the Struggle for Life.* D. Appleton and Company, New York.

———. 1909. *The Foundations of the Origin of Species.* Cambridge University Press, England.

———, and A. R. Wallace. 1958. *Evolution by Natural Selection,* Cambridge University Press, London.

Dauvillier, A. 1965. *The Photochemical Origin of Life.* Academic Press, New York.

Davenport, C. B. 1972. *Heredity in Relation to Eugenics.* Arno Press & The New York Times, New York.

Day, M. H. 1972. *The Fossil History of Man.* Oxford University Press, London.

Dayhoff, M. O. 1969. Computer analysis of protein evolution. *Sci. Amer.* **221** (July) 86.

Dickerson, R. E. 1972. The structure and history of an ancient protein. *Sci. Amer.* **226** (April) 58.

Dietz, R. S., and J. C. Holden. 1970. The breakup of Pangaea. *Sci. Amer.* **223** (October) 30.

Dillon, L. S. 1973. *Evolution, Concepts and Consequences.* C. V. Mosby Company, Saint Louis.

Dobzhansky, T. 1948. Genetics of natural populations. XVIII. Experiments on chromosomes of *Drosophila pseudoobscura* from different geographic regions. *Genetics* **33**:588–602.

———. 1950. The genetic basis of evolution. *Sci. Amer.* **182** (January) 32.

———. 1951. *Genetics and the Origin of Species* (3rd edition, revised) Columbia University Press, New York.

———. 1952. Nature and origin of heterosis. In *Heterosis* (editor, J. W. Gowen) Iowa State College Press, Ames.

———. 1955. *Evolution, Genetics, and Man.* John Wiley & Sons, Inc., New York.

———. 1970. *Genetics of the Evolutionary Process.* Columbia University Press, New York.

———, and O. Pavlovsky. 1957. An experimental study of interaction between genetic drift and natural selection. *Evolution* **11**:311–19.

———, and M. L. Queal. 1938. Genetics of natural populations. II. Genic variation in populations of *Drosophila pseudoobscura* inhabiting isolated mountain ranges. *Genetics* **23**:463–84.

———, and B. Spassky. 1947. Evolutionary changes in laboratory cultures of *Drosophila pseudoobscura. Evolution* **1**:191–216.

Dodson, E. O. 1952. *A Textbook of Evolution.* W. B. Saunders Company, Philadelphia.

Dose, K., S. W. Fox, G. A. Deborin, and T. E. Pavlovskaya (editors). 1974. *The Origin of Life and Evolutionary Biochemistry.* Plenum Press, New York.

Dowdeswell, W. H. 1958. *The Mechanism of Evolution* (2nd edition). Harper and Row, Publishers, New York.

Dun, R. B., and A. S. Fraser. 1958. Selection for an invariant character—"vibrissa number"—in the house mouse. *Nature* **181**:1018–1019.

Dunn, L. C. 1965. *A Short History of Genetics.* McGraw-Hill Book Company, New York.

East, E. M., and D. F. Jones. 1919. *Inbreeding and Outbreeding.* J. B. Lippincott Company, Philadelphia.

Eaton, T. H. Jr. 1970. *Evolution.* W. W. Norton & Company, Inc., New York.

Ehrlich, P. R., and R. W. Holm. 1963. *The Process of Evolution.* McGraw-Hill Book Co., Inc., New York.

———, R. W. Holm, and P. H. Raven (editors). 1960. *Papers on Evolution.* Little, Brown and Company, Boston.

Eiseley, L. C. 1956. Charles Darwin. *Sci. Amer.* **194** (February) 62.

———. 1959. Charles Lyell. *Sci. Amer.* **201** (August) 98.

Endler, J. A. 1973. Gene flow and population differentiation. *Science* 179(4070):243–250.

Eglinton, G., and M. Calvin. 1967. Chemical fossils. *Sci. Amer.* **216** (January) 32.

Eyde, R. H. 1975. The foliar theory of the flower. *Amer. Scientist* 63(4):430–37.

Falconer, D. S. 1960. *Introduction To Quantitative Genetics.* The Ronald Press Company, New York.

Fincham, J. R. S. 1966. *Genetic Complementation.* W. A. Benjamin, Inc., New York.

Fisher, R. A. 1930. *Genetical Theory of Natural Selection.* Clarendon Press, Oxford.

Fisher, R. A. 1949. *The Theory of Inbreeding.* Oliver and Boyd, Edinburgh.

Florkin, M. 1966. *A Molecular Approach to Phylogeny.* Elsevier Publishing Company, New York.

Ford, E. B. 1940. Polymorphism and taxonomy. In *The New Systematics* (editor, J. Huxley). Clarendon Press, Oxford.

———. 1964. *Ecological Genetics.* Methuen, London.

———. 1965. *Genetic Polymorphism.* The Massachusetts Institute of Technology Press, Cambridge, Mass.

Fox, S. W. 1965. *The Origins of Prebiological Systems* and *Their Molecular Matrices.* Academic Press, New York.

———, and K. Dose. 1972. *Molecular Evolution and the Origin of Life.* W. H. Freeman and Company, San Francisco.

Gardner, E. J. 1960. Organic Evolution and the Bible. Utah State University Press, Logan.

———. 1962. *Mechanics of Organic Evolution.* Utah State University Press, Logan, Utah.

———. 1965. *History of Biology* (2nd edition). Burgess Publishing Company, Minneapolis.

———. 1968. *Principles of Genetics* (3rd edition). John Wiley & Sons, Inc., New York.

Gause, G. F. 1964. *The Struggle for Existence.* Hafner Publishing Company, New York.

Giesel, J. T. 1974. *The Biology and Adaptability of Natural Populations.* C. V. Mosby Company, Saint Louis.

Glass, B. 1953. The genetics of the Dunkers. *Sci. Amer.* **189** (August) 76.

———, and C. C. Li. 1953. The dynamics of racial intermixture—an analysis based on the American Negro. *Am. J. Human Genet.* **5**:1–20.

———, O. Temkin, and W. L. Straus, Jr. (editors) 1959. *Forerunners of Darwin: 1745–1859.* The Johns Hopkins Press, Baltimore.

Goldsby, R. A. 1971. *Race and Races.* Macmillan Publishing Co. Inc., New York.

Goldschmidt, R. 1960. *The Material Basis of Evolution.* Pageant Books, Inc. New Jersey.

Goodenough, U., and R. P. Levine. 1974. *Genetics.* Holt, Rinehart and Winston, Inc., New York.

Grant, V. 1963. *The Origin of Adaptations.* Columbia University Press, New York.

Gray, A. 1963. *Darwiniana.* Harvard University Press, Cambridge, Massachusetts.

Greene, J. C. 1959. *The Death of Adam.* The New American Library of World Literature, Inc., New York.

Gurdon, J. B. 1968. Transplanted nuclei and cell differentiation. *Sci. Amer.* **219** (December) 36.

Haldane, J. B. S. 1932. *The Causes of Evolution.* Harper & Row, New York.

———. 1948. The theory of a cline. *J. Genet.* **48**:277–84.

Hallam, A. 1972. Continental drift and the fossil record. *Sci. Amer.* **227** (November) 56.

Hamilton, T. H. 1967. *Process and Pattern in Evolution.* Macmillan Publishing Co. Inc., New York.

Hardy, G. H. 1908. Mendelian proportions in a mixed population. *Science* **28**:49–50.

Harpstead, D. D. 1971. High-lysine corn. *Sci. Amer.* **225** (August) 34.

Hecht, M. K., and W. C. Steere (editors). 1970. *Essays in Evolution and Genetics.* Appleton-Century-Crofts, New York.

Heirtzler, J. R. 1968. Sea-floor spreading. *Sci. Amer.* **219** (December) 60.

Herskowitz, I. H. 1960. *Study Guide and Workbook for Genetics.* McGraw-Hill Book Co., Inc., New York.

———. 1973. *Principles of Genetics.* Macmillan Publishing Co. Inc., New York.

———. 1965. *Genetics.* (2nd edition). Little, Brown and Company, Boston.

Hickman, C. P. 1970. *Integrated Principles of Zoology* (4th edition). C. V. Mosby Company, Saint Louis.

Hildemann, W. H. 1970. *Immunogenetics.* Holden-Day, San Francisco.

Hodge, M. J. S. 1972. The universal gestation of nature: Chamber's *Vestiges and Explanations. J. Hist. Biol.* **5**(1):127–52.

Hooton, E. A. 1946. *Up From The Ape.* Macmillan Publishing Co. Inc., New York.

Holton, N., III. 1968. *The Evidence of Evolution.* American Heritage Publishing Co., Inc. New York.

Hull, D. L. 1973. *Darwin and His Critics.* Harvard University Press, Cambridge.

Hurley, P. M. 1968. The confirmation of continental drift. *Sci. Amer.* **218** (April) 52.

Huxley, J. S. 1943. *The New Systematics.* Harper & Brothers, New York.

———. 1953. *Evolution in Action.* Harper & Brothers., New York.

———. 1963. *Evolution: The Modern Synthesis* (2nd edition) Allen & Unwin, London.

———, and H. B. D. Kettlewell. 1965. *Charles Darwin and His World.* The Viking Press, New York.

Huxley, T. H. 1968. *On The Origin of Species* or, *The Causes of the Phenomena of Organic Nature.* The University of Michigan Press, Ann Arbor.

Jacquard, A. 1974. *The Genetic Structure of Populations.* Springer-Verlag, New York.

Jägersten, G. 1972. *Evolution of the Metazoan Life Cycle.* Academic Press, New York.

Jepsen, G. L., E. Mayr, and G. G. Simpson (editors). 1949. *Genetics, Paleontology and Evolution.* Princeton University Press, New Jersey.

Jinks, J. L. 1964. *Extrachromosomal Inheritance.* Prentice-Hall, Inc., Englewood Cliffs, New Jersey.

Jukes, T. H. 1966. *Molecules and Evolution.* Columbia Univ. Press, New York.

Keeton, W. T. 1967. *Biological Science.* W. W. Norton & Company, Inc. New York.

Keosian, J. 1964. *The Origin of Life.* Reinhold Publishing Corporation, New York.

Kettlewell, H. B. D. 1958. A survey of the frequencies of *Biston betularia* (L) (Lep) and its melanic forms in Great Britain. *Heredity* **12**:51–72.

———. 1959. Darwin's missing evidence. *Sci. Amer.* **200** (March) 48.

Kimura, M., and T. Ohta. 1971. *Theoretical Aspects of Population Genetics.* Princeton University Press, Princeton, New Jersey.

King, J. C. 1971. *The Biology of Race.* Harcourt Brace Jovanovich, Inc., New York.

King, R. C. 1972. *A Dictionary of Genetics* (2nd edition) Oxford University Press, New York.

Knight, G. R., A. Robertson, and C. H. Waddington. 1956. Selection for sexual isolation within a species. *Evolution* **10**:14–22.

Koopman, K. F. 1950. Natural selection for reproductive isolation between *Drosophila pseudoobscura* and *Drosophila persimilis. Evolution* **4**:135–48.

Kraus, B. S. 1964. *The Basis of Human Evolution.* Harper & Row, Publishers, New York.

Kurtén, B. 1969. Continental drift and evolution. *Sci. Amer.* **220** (March) 54.

Lack, D. 1953. Darwin's Finches. *Sci. Amer.* **188** (April) 66.

———. 1961. *Darwin's Finches.* Harper & Row, Publishers, New York.

Lasker, G. W. 1961. *The Evolution of Man.* Holt, Rinehart and Winston, New York.

Lawless, J. G. 1973. Thermal synthesis of amino acids from simulated primitive atmosphere. *Nature* **243**:405.

Leakey, L. S. B. 1960. *Adam's Ancestors* (4th edition). Harper & Row, Publishers, New York.

Leigh, E. G. 1971. *Adaptation and Diversity.* Freeman, Cooper & Company, San Francisco.

Leone, C. A. (editor). 1964. *Taxonomic Biochemistry and Serology.* The Ronald Press Co., New York.

Lerner, I. M. 1954. *Genetic Homeostasis.* John Wiley & Sons, Inc., New York.

———. 1958. *The Genetic Basis of Selection.* John Wiley & Sons, Inc., New York.

Lewis, A. E. 1966. *Biostatistics.* Reinhold Publishing Corporation, New York.

Lewontin, R. C. 1964. The role of linkage in natural selection. *Proc. 11th Intern. Congr. Genet.* **3:**517–525.

Li, C. C. 1955. *Population Genetics.* The University of Chicago Press, Chicago.

Lindsey, A. W. 1952. *Principles of Organic Evolution.* C. V. Mosby Company, St. Louis.

Loewenberg, J. 1959. *Darwin, Wallace and the Theory of Natural Selection.* Arlington Books, Cambridge.

Lorenz, K. Z. 1958. The evolution of behavior. *Sci. Am.* **199** (December) 67.

Lull, R. S. 1947. *Organic Evolution.* Macmillan Publishing Co. Inc., New York.

Lush, J. L. 1945. *Animal Breeding Plans.* Iowa State College Press, Ames.

Macbeth, N. 1971. *Darwin Retried—An Appeal To Reason.* Gambit Incorporated, Boston.

Manwell, C. 1970. *Molecular Biology and the Origin of Species.* University of Washington Press, Seattle.

Margulis, L. 1971. Symbiosis and evolution. *Sci. Amer.* **225** (August) 48.

Markert, C. L., and H. Ursprung. 1971. *Developmental Genetics.* Prentice-Hall, Inc., Englewood Cliffs, New Jersey.

Mason, F. 1928. *Creation by Evolution.* Macmillan Publishing Co. Inc., New York.

Mather, K. 1973. *Genetical Structure of Populations.* Chapman and Hall, London.

Mather, W. B. 1964. *Principles of Quantitative Genetics.* Burgess, Minneapolis.

Matsunaga, E., Y. Hiraizumi, T. Furusho, and H. Izumiyama. 1962. Studies on selection in ABO blood groups. *Ann. Rept. Natl. Inst. Genet. Japan* **13:**103–106.

Mathews, S. W. 1973. This changing earth. *Nat. Geographic* **143**(1):1–37.

Mayr, E. 1942. *Systematics and the Origin of Species.* Columbia Univ. Press, New York.

———. (editor) 1957. *The Species Problem.* Amer. Assn. for the Advancement of Science, Washington.

———. 1963. *Animal Species and Evolution.* Harvard Univ. Press, Cambridge.

———. 1964. *Systematics and the Origin of Species from the Viewpoint of a Zoologist.* Dover Publications, New York.

Merrell, D. J. 1962. *Evolution and Genetics.* Holt, Rinehart and Winston, New York.

Mettler, L. E., and T. G. Gregg. 1969. *Population Genetics and Evolution.* Prentice-Hall, Inc., Englewood Cliffs, New Jersey.

Miller, H. 1971. *Footprints of the Creator: or, The Asterolepis of Stromness.* Gregg International Publishers Limited, Westmead, England.

Miller, S. L. 1953. A production of amino acids under possible primitive earth conditions. *Science* **117:**528–529.

Montagu, A. 1952. *Darwin, Competition and Cooperation.* Henry Schuman, New York.

Moody, P. A. 1970. *Introduction To Evolution* (3rd edition) Harper & Row, Publishers, New York.

Moore, J. A. 1957. An embryologists view of the species concept. *AAAS Publ.* **50:**325–338.

Moorhead, P. S., and M. M. Kaplan (editors). 1967. *Mathematical Challenges To The Neo-Darwinian Interpretation of Evolution.* The Wistar Institute Symposium Monograph No. 5; The Wistar Institute Press, Philadelphia.

Morris, H. M., W. W. Boardman, Jr., and R. F. Koontz. 1971. *Science and Creation.* Creation-Science Research Center, San Diego, California.

Müller, F. 1869. *Facts and Arguments for Darwin.* John Murray, London.

Muller, H. J. 1927. Artificial transmutation of the gene. *Science* **66:**84–87.

———. 1939. Reversibility in evolution considered from the standpoint of genetics. *Biol. Rev.* **14:**261–80.

———. 1950. Our load of mutations. *Am. J. Human Genet.* **2:**111–76.

————. 1958. The mutation theory re-examined. *Proc. 10th Intern. Congr. Genet.* **1**:306–317.

————. 1958. How much is evolution accelerated by sexual reproduction? AAAS General Program, p. 205.

Murray, J. 1972. *Genetic Diversity and Natural Selection.* Oliver & Boyd, Edinburgh.

Newman, H. H. 1932. *Evolution Genetics and Eugenics.* Greenwood Press, Publishers, New York.

————. 1939. *The Phylum Chordata.* Macmillan Publishing Co. Inc., New York.

Nicholson, A. J. 1955. Density governed reaction, the counterpart of selection in evolution. *Cold Spring Harbor Symp. Quant. Biol.* **20**:288–93.

Odum, E. P. 1963. *Ecology.* Holt, Rinehart and Winston, New York.

Ohno, S. 1970. *Evolution By Gene Duplication.* Springer-Verlag, New York.

Olson, E. C. 1965. *The Evolution of Life.* The New American Library, New York.

Oparin, A. I. 1938. *The Origin of Life* (*2nd edition*). Dover Publications, Inc. New York.

————. 1968. *Genesis and Evolutionary Development of Life.* Academic Press, New York.

Orgel, L. E. 1973. *The Origins of Life: Molecules and Natural Selection.* John Wiley & Sons, Inc. New York.

Papageorgis, C. 1975. Mimicry in Neotropical butterflies. *Amer. Scientist* **63**(5):522–32.

Parsons, P. A., and W. F. Bodmer. 1961. The evolution of overdominance: Natural selection and heterozygote advantage. *Nature* **190**:7–12.

Perutz, M. F. The hemoglobin molecule. *Sci. Amer.* **211** (November) 64.

Pfeiffer, J. E. 1969. *The Emergence of Man.* Harper & Row, Publishers, New York.

Pianka, E. R. 1974. *Evolutionary Ecology.* Harper & Row, Publishers, New York.

Pirchner, F. 1969. *Population Genetics in Animal Breeding.* W. H. Freeman and Company, San Francisco.

Ponnamperuma, C. 1972. *The Origins of Life.* E. P. Dutton, New York.

Provine, W. B. 1971. *The Origins of Theoretical Population Genetics.* The University of Chicago Press, Chicago.

Rasmuson, M. 1961. *Genetics On The Population Level.* Svenska Bokförlaget, Stockholm.

Reed, S. C., and E. W. Reed. 1950. Natural selection in laboratory populations of *Drosophila.* II. Competition between a white-eye gene and its wild type allele. *Evolution* **4**:34–42.

Rensch, B. 1960. *Evolution Above The Species Level.* Columbia University Press, New York.

Robbins, W. W., T. E. Wier, and C. R. Stocking. 1957. *Botany* (2nd edition) John Wiley & Sons, Inc., New York.

Romer, A. S. 1972. *The Procession of Life.* Doubleday & Company, Inc., Garden City, New York.

Ross, H. H. 1962. *A Synthesis of Evolutionary Theory.* Prentice-Hall, Inc., Englewood Cliffs, New Jersey.

————. 1966. *Understanding Evolution.* Prentice-Hall, Inc., Englewood Cliffs, New Jersey.

Sahakian, W. S. 1968. *History of Philosophy.* Barnes & Noble, Inc., New York.

Sanghvi, L. D. 1963. The concept of genetic load: A critique. *Am. J. Human Genet.* **15**:298–309.

Savage, J. M. 1963. *Evolution.* Holt, Rinehart and Winston, New York.

Scharloo, W. 1964. The effect of disruptive and stabilizing selection on the expression of a *cubitus interruptus* mutant in *Drosophila. Genetics* **50**:553–62.

Schmalhausen, I. I. 1949. *Factors of Evolution.* The Blakiston Company, Philadelphia.

Schuchert, C. and C. O. Dunbar. 1941. *A Textbook of Geology; Part II-Historical Geology* (4th edition). John Wiley & Sons, Inc., New York.

Schwartz, D., and W. J. Laughner. 1969. A molecular basis for heterosis. *Science* **166**:626–27.

Sheppard, P. M. 1958. *Natural Selection and Heredity.* Hutchinson University Library, London.

Sibley, C. G. 1954. Hybridization in the red-eyed towhees of Mexico. *Evolution* **8**:252–90.

Simpson, G. G. 1949. *The meaning of evolution.* Yale Univ. Press, New Haven.

————. 1951. *Horses.* Oxford University Press, New York.

———. 1953. *The Major Features of Evolution.* Columbia Univ. Press, New York.

———. 1953. *Life of the Past.* Yale Univ. Press, New Haven.

———. 1962. *Evolution and Geography.* Oregon State System of Higher Education, Eugene.

———. 1965. *Tempo and Mode in Evolution.* Hafner Publishing Company, New York.

Singleton, W. R. 1967. *Elementary Genetics* (2nd edition) D. Van Nostrand Company, Inc., Princeton, New Jersey.

Smith, H. H. (editor). 1972. *Evolution of Genetic Systems.* Gordon and Breach, New York.

Smith, J. M. 1966. *The Theory of Evolution* (2nd edition). Penguin Books, Baltimore.

Sokal, R. R., and F. J. Rohlf. 1969. *Biometry.* W. H. Freeman and Company, San Francisco.

Solbrig, O. T. 1966. *Evolution and Systematics.* Macmillan Publishing Co. Inc., New York.

———. 1970. *Principles and Methods of Plant Biosystematics.* Macmillan Publishing Co. Inc., New York.

Srb, A., R. D. Owen, and R. S. Edgar. 1952. *General Genetics* (2nd edition). W. H. Freeman and Company, San Francisco.

Stahl, B. J. 1974. *Vertebrate History: Problems in Evolution.* McGraw-Hill Book Company, New York.

Stebbins, G. L., Jr., 1950. *Variation and Evolution in Plants.* Columbia University Press, New York.

———. 1966. *Processes of Organic Evolution.* (2nd edition) Prentice-Hall, Inc., Englewood Cliffs, New Jersey.

———. 1969. *The Basis of Progressive Evolution.* The University of North Carolina Press, Chapel Hill.

Stent, G. S. 1971. *Molecular Genetics.* W. H. Freeman and Company, San Francisco.

Streams, F. A., and D. Pimentel. 1961. Effects of immigration on the evolution of populations. *Am. Naturalist* **95**:201–210.

Strickberger, M. W. 1968. *Genetics.* Macmillan Publishing Co. Inc., New York.

Sturtevant, A. H. 1965. *A History of Genetics.* Harper & Row, Publishers, New York.

———, and K. Mather. 1938. The inter-relations of inversions, heterosis, and recombination. *Am. Naturalist* **72**:447–52.

Sved, J. A., T. E. Reed, and W. F. Bodmer. 1967. The number of balanced polymorphisms that can be maintained in a natural population. *Genetics* **55**:469–81.

Swanson, C. P. 1957. *Cytology and Cytogenetics.* Prentice-Hall, Inc., Englewood Cliffs, New Jersey.

Taylor, G. R. 1963. *The Science of Life.* McGraw-Hill Book Company, Inc., New York.

Thoday, J., and T. B. Boam. 1959. Effects of disruptive selection. II. Polymorphism and divergence without isolation. *Heredity* **13**:205–18.

———, and J. B. Gibson. 1962. Isolation by disruptive selection. *Nature* **193**:1164–66.

Tobias, P. 1971. *The Brain in Hominid Evolution.* Columbia University Press, New York.

Tompkins, J. R. (editor). 1965. *D-Days at Dayton.* Louisiana State University Press, Baton Rouges.

Trujillo, J. M., S. Ohno, J. H. Jardine, and N. B. Atkins. 1969. Spermatogenesis in a male hinny: histological and cytological studies. *J. Heredity* **60**(2):79–85.

Turesson, G. 1922. *Genotypical Response of the Plant Species to the Habitat.* Berlingska Boktryckeriet, Lund.

Urey, H. C. 1952. The origin of the earth. *Sci. Amer.* **187** (October) 53.

Van Valen, L. 1963. Introgression in laboratory populations of *Drosophila persimilis* and *D. pseudo-obscura. Heredity* **18**:205–214.

Volpe, E. P. 1967. *Understanding Evolution.* Wm. C. Brown Company Publishers, Dubuque, Iowa.

Vorzimmer, P. J. 1970. *Charles Darwin: The Years of Controversy.* Temple University Press, Philadelphia.

Vries De, H. 1909. *The Mutation Theory.* The Open Court Publishing Company, Chicago. (New York, Kraus Reprint Co., 1969)

———. 1912. *Species and Varieties. Their Origin by Mutation.* The Open Court Publishing Company, Chicago.

Waddington, C. H. 1953. Genetic assimilation of an acquired character. *Evolution* 7:118–26.

———, and E. Robertson. 1966. Selection for developmental canalization. *Genet. Res.* 7:303–12.

Wald, G. 1954. The origin of life. *Sci. Amer.* **191** (August) 44.

Wallace, A. R. 1870. *Contributions to the Theory of Natural Selection.* Macmillan Publishing Co. Inc., New York.

———. 1910. *The World of Life.* Chapman and Hall, Limited, London.

Wallace, B. 1952. The estimation of adaptive values of experimental populations. *Evolution* **6**:333–341.

———. 1968. *Topics in Population Genetics.* W. W. Norton & Company, Inc., New York.

———. 1970. *Genetic Load.* Prentice-Hall, Inc., Englewood Cliffs, New Jersey.

———, and Th. Dobzhansky. 1962. Experimental proof of balanced genetic loads in *Drosophila. Genetics* **47**:1027–1042.

———, and A. M. Srb. 1964. *Adaptation,* (2nd edition). Prentice-Hall, Englewood Cliffs, N.J.

Waterman, T. H. and H. J. Morowitz (editors). 1965. *Theoretical and Mathematical Biology.* Blaisdell Publishing Company, New York.

Watson, J. D. 1970. *Molecular Biology of the Gene* (2nd edition). W. A. Benjamin, Inc., New York.

Weismann, A. 1891. *Essays Upon Heredity.* Clarendon Press, Oxford.

———. 1904. *The Evolution Theory.* Edward Arnold, London.

Weisz, P. B. 1959. *The Science of Biology* (4th edition). McGraw-Hill Book Company, New York.

———. 1966. *The Science of Zoology.* McGraw-Hill Book Company, New York.

Weller, J. M. 1969. *The Course of Evolution.* McGraw-Hill Book Company, New York.

Wells, K. D. 1973. The historical context of natural selection: the case of Patrick Matthew. *J. Hist. Bio.* **6**(2):225–58.

White, M. J. D. 1948. *Animal Cytology and Evolution.* Cambridge University Press, London.

———. 1968. Models of speciation. *Science* **159**: 1065–70.

Wickler, W. 1968. *Mimicry in Plants and Animals.* McGraw-Hill Book Company, New York.

Williams, G. C. (editor). 1971. *Group Selection.* Aldine/Atherton, Chicago/New York.

Wills, C. 1970. Genetic load. *Sci. Amer.* **222** (March) 98.

Wilson, E. O. 1975. *Sociobiology—The New Synthesis.* Harvard University Press, Cambridge, Mass.

Wilson, J. T. 1963. Continental drift. *Sci. Amer.* **208** (April) 86.

Wright, S. 1931. Evolution in Mendelian populations. *Genetics* **16**:97–159.

———. 1932. The roles of mutation, inbreeding, crossbreeding, and selection in evolution. *Proc. 6th Intern. Congr. Genet.* **1**:356–66.

———. 1934. Physiological and evolutionary theories of dominance. *Am. Naturalist* **68**:24–53.

———. 1958. *Systems of Mating and Other Papers.* The Iowa State College Press, Ames.

———. 1964. The distribution of self-incompatibility alleles in populations. *Evolution* **18**:609–619.

———. 1968. *Evolution and The Genetics of Populations.* The University of Chicago Press, Chicago.

Wynne-Edwards, V. C. 1964. Population control in animals. *Sci. Amer.* **211** (August) 68.

Yanofsky, C. 1967. Gene structure and protein structure. *Sci. Amer.* **216** (May) 80.

Zimmerman, P. A., J. W. Klotz, W. H. Rusch, and R. F. Surburg. 1959. *Darwin, Evolution, and Creation.* Concordia Publishing House, Saint Louis, Missouri.

Zuckerkandl, E. 1965. The evolution of hemoglobin. *Sci. Amer.* **212** (May) 110.

INDEX